Animal Sonar Systems

NATO ADVANCED STUDY INSTITUTES SERIES

A series of edited volumes comprising multifaceted studies of contemporary scientific issues by some of the best scientific minds in the world, assembled in cooperation with NATO Scientific Affairs Division.

Series A: Life Sciences

Recent Volumes in this Series

The series is published by an international board of publishers in conjunction with NATO Scientific Affairs Division

A Life Sciences	Plenum Publishing Corporation
B Physics	London and New York
C Mathematical and Physical Sciences	D. Reidel Publishing Company Dordrecht, Boston and London
D Behavioral and Social Sciences	Sijthoff & Noordhoff International Publishers
E Applied Sciences	Alphen aan den Rijn, The Netherlands, and Germantown U.S.A.

Animal Sonar Systems

Edited by
René-Guy Busnel
Ecole Pratique des Hautes Etudes
Jouy-en-Josas, France

and

James F. Fish
Sonatech, Inc.
Goleta, California

SPRINGER SCIENCE+BUSINESS MEDIA, LLC

Library of Congress Cataloging in Publication Data

International Interdisciplinary Symposium on Animal Sonar Systems, 2d, Jersey, 1979.
  Animal sonar systems.

  (NATO advanced study institutes series: Series A, Life sciences; v. 28)
  Symposium sponsored by the North Atlantic Treaty Organization and others.
  Bibliography: p.
  Includes indexes.
  1. Echolocation (Physiology) – Congresses. I. Busnel, René Guy. II. Fish, James F.
III. North Atlantic Treaty Organization. IV. Title. V. Series.
QP469.I57 1979   599'.01'88   79-23074
ISBN 978-1-4684-7256-1   ISBN 978-1-4684-7254-7 (eBook)
DOI 10.1007/978-1-4684-7254-7

This work relates to Department of the Navy Grant N00014-79-G-0006
issued by the Office of Naval Research. The United States Government
has a royalty-free license throughout the world in all copyrightable
material contained herein.

Proceedings of the Second International Interdisciplinary Symposium on
Animal Sonar Systems, held in Jersey, Channel Islands, April 1–8, 1979.

ORGANIZING COMMITTEE
-=-=-=-=-=-=-=-=-=-=-=-=-=-=-=-=-=-=-

- R.G. Busnel
  Laboratoire d'Acoustique Animale
  E.P.H.E. - I.N.R.A. - C.N.R.Z.
  78350 Jouy-en-Josas, France

- J.F. Fish
  U.S. Navy
  Naval Ocean Systems Center
  Kailua, Hawaii 96734, U.S.A.

- G. Neuweiler
  Department of Zoology
  University of Frankfurt
  D-6000 Frankfurt, F.R.Germany

- J.A. Simmons
  Department of Psychology
  Washington University
  St. Louis, Mo. 63130, U.S.A.

- H.E. Von Gierke
  Aerospace Medical Research Lab.
  Wright Patterson Air Force Base
  Dayton, Ohio 45433, U.S.A.

A C K N O W L E D G M E N T S

-=-=-=-=-=-=-=-=-=-=-=-=-=-=-=-=-

This symposium was sponsored by various Organizations, National and International. The organizing Committee would like to thank them and their representatives :

- North Atlantic Treaty Organization, N.A.T.O.

- Advanced Study Institutes Programme

- Office of Naval Research of the U.S.A.

- United States Air Force

- Volkswagen-Stiftung from F.R. of Germany

- Ecole Pratique des Hautes-Etudes, Laboratoire d'Acoustique Animale, I.N.R.A. - C.N.R.Z. - 78350 JOUY-en-JOSAS, France

Preface

Thirteen years have gone by since the first international meeting on Animal Sonar Systems was held in Frascati, Italy, in 1966. Since that time, almost 900 papers have been published on its theme. The first symposium was vital as it was the starting point for new research lines whose goal was to design and develop technological systems with properties approaching optimal biological systems.

There have been highly significant developments since then in all domains related to biological sonar systems and in their applications to the engineering field. The time had therefore come for a multidisciplinary integration of the information gathered, not only on the evolution of systems used in animal echolocation, but on systems theory, behavior and neurobiology, signal-to-noise ratio, masking, signal processing, and measures observed in certain species against animal sonar systems.

Modern electronics technology and systems theory which have been developed only since 1974 now allow designing sophisticated sonar and radar systems applying principles derived from biological systems. At the time of the Frascati meeting, integrated circuits and technologies exploiting computer science were not well enough developed to yield advantages now possible through use of real-time analysis, leading to, among other things, a definition of target temporal characteristics, as biological sonar systems are able to do.

All of these new technical developments necessitate close cooperation between engineers and biologists within the framework of new experiments which have been designed, particularly in the past five years.

The scientists who have been working on these problems in various fields (electronics experts, signal processors, biologists, physiologists, psychologists) have produced new and original results, and this second symposium furnished the opportunity of cross-disciplinary contacts permitting an evaluation of the state of present research.

The Jersey meeting in April, 1979, brought together more than 70 participants from 8 different countries. This meeting was particularly necessary considering the number of new research groups that have appeared in various fields. I mention in particular: the Federal Republic of Germany, where two young scientists who participated at the Frascati meeting, Gerhard Neuweiler and Hans-Ulrich Schnitzler, have since become professors and have founded two schools of highly productive research; the United States, where funded by the U.S. Navy, studies on dolphins have spread to San Diego and Hawaii, given impetus by Bill Powell, Forrest Wood, Sam Ridgway, C. Scott Johnson, Bill Evans, and Ron Schusterman, and where certain of Donald Griffin's most gifted students, such as Alan Grinnell and Jim Simmons, are continuing his work on bats at various universities; Canada, where Brock Fenton is performing outstanding research.

Although much research has been carried out in the Soviet Union since 1969-1970, it is most unfortunate that, for reasons independent of their wishes, our colleagues from this country, who have moreover published several excellent reviews of their work, were not able to participate in our discussions.

The tendancy which appeared at Frascati towards a certain zoological isolation corresponding to a form of animal specialization dominated by either dolphins or bats, has partially regressed, thanks to several physicists who use both bat and dolphin signals in their theoretical approaches. While this segregation by field still remains a dominant behavior made obviously necessary, up to a certain point, by the different biological natures of these two groups of animals, the phenomenon is aggravated by the use of semantics specific to each group. Nevertheless, common interests and attempts at mutual understanding which appeared are encouraging and should be congratulated.

The proceedings of the Jersey meeting demonstrate the extent to which technology has advanced in the past decade, in performances of transducers, various microphones, hydrophones, as well as in recording apparata, analytical methods, particularly in the use of signal processing techniques, and in the application of new ideas such as time-domain, auditory processing, frequency-domain, Doppler compensation, target-acoustic imaging, and so on.

It is also interesting to note to what extent experimental strategies have been refined, and one can only admire the elegance of certain demonstrations carried out on dolphins as well as on bats. Many aspects of the performances of diverse species continue, however, to intrigue us, as they reveal sensory abilities whose fine analysis still eludes us, particularly the central mechanisms which control and regulate them. As an example of this, I would particularly like to mention how enriching was the experience that we were able to have using Leslie Kay's apparatus for the blind, which gave spatiodirectional sensations analogous to those of airborne animal sonar systems. I do not doubt in the least that those several minutes will lead to a totally new concept of biosonar problems.

As Henning von Gierke pointed out during the last session and in a personal communication, the trends appearing at the Jersey meeting indicate that pattern recognition theory is becoming more and more important to biosonar research, replacing range finding and echo theories as the promising research areas of the future. Pattern recognition specialists should be included in future meetings as well as experts on the spatial frequency analysis of acoustic and visual perception. Although animal sonar might be used to a large extent for acoustic imaging of space, we know from ultrasonography that acoustic images are different from optical images. Acoustic space perception therefore differs from visual space perception. Since the acoustic space is scanned sequentially, the total "acoustic image" depends primarily on memory capability, which is the major difficulty encountered by Leslie Kay in his device for the blind. The findings of this Symposium may have a major impact on general auditory physiology regarding, on one hand, the example of sharp filter (acoustic fovea) and on the other, peripheral processing.

A reading of the present set of volumes presenting the current state of research will bring out at the same time the unknowns of the problem, the uncertainties, the hypotheses, and will allow veterans of Frascati to measure the progress made since then.

I am most happy to thank here my American colleagues who agreed to respond to my call in 1977, particularly Henning von Gierke, who was our referee and support for N.A.T.O. as well as the U.S. Air Force, and Bill Powell and Forrest Wood for the U.S. Navy, who later recommended to me Jim Fish, a young associate, dynamic and efficient, who held a preponderant place in the Organizing Committee.

I also would particularly like to express my gratitude to Gerhard Neuweiler for his constructive participation in our planning group. It is thanks to his outstanding reputation that our Symposium was able to obtain funding from Volkswagen for the active and highly productive German delegation. He assumed as well the heavy responsibility of financial management of our funds.

I also wish to thank, on behalf of our Committee and myself, the diverse individuals from my laboratory who, with devotedness, took on countless tasks, often thankless and lowly: Michèle Bihouée, Annick Brézault, Marie-Claire Busnel, Sophie Duclos, Diana Reiss, and Sylvie Venla. Leslie Wheeler, who assisted me in editing these volumes, deserves special mention, as without her help I would not have been able to publish them so rapidly.

During the last plenary session in Jersey, the Organizing Committee and the Co-chairmen of the different sessions decided on the publication format of the proceedings, and suggested to the Symposium participants to dedicate this book to Donald R. Griffin. Our colleagues unanimously rendered homage to the spiritual father, the inventor, of echolocation.

The scientific wealth brought out during the three half-day poster sessions, bears witness to the interest and importance of the work of numerous young scientists and makes me optimistic for the future of the field of animal sonar systems. For this reason, and thanks to the two experiences which the majority of you have considered successful, I wish good luck to the future organizer of the Symposium of the next millenium.

René-Guy Busnel
Jouy-en-Josas
France

Contents

INDEXES

On behalf of the members of the organizing Committee and
all the Symposium participants, we dedicate this book to

Donald R. GRIFFIN

in   homage to his pioneering work in the
field of echolocation.

Donald R. Griffin

DEDICATION

by Alan D. Grinnell

Just over 40 years ago, a Harvard undergraduate persuaded a
physics professor to train his crystal receiver and parabolic horn
at active bats. They detected ultrasonic pulses, and the contemporary field of echolocation research was born. The undergraduate was
Donald R. Griffin, the professor, G. W. Pierce, and their first report appeared in 1938. In recognition of his founding role and his
countless important contributions to echolocation research this volume is dedicated to Don Griffin. Fortuitously, this coincides with
the approach of his 65th birthday, a time when particularly popular
and influential figures in a field are often honored with a "Fest-
schrift" volume.

We were fortunate, at this meeting, that Don Griffin was persuaded to add a few recollections of his experiences during the in-
fancy of the field. These are included in this volume. Additional
perspective on some of the early years has been volunteered by his
partner in the first experiments demonstrating echolocation, Dr.
Robert Galambos, then a Harvard graduate student, now a well-known
auditory neurophysiologist and Professor of Neuroscience at UCSD.

"In early 1939 Don found out that Hallowell Davis at the
Harvard Medical School was teaching me how to record electric
responses from guinea pig cochleas and asked if I could slip
in a few bats on the side. He wanted to know whether their
ears responded to the "supersonic notes" he and Professor G. W.
Pierce had just discovered. So I asked Dr. Davis, who said
"go ahead" and thus Don found himself a collaborator in some
unforgettable adventures. (This was not the only time Don
enlisted me in one of his enterprises--I remember spending
3 days helping him build a bird blind on Penekeese Island

in weather so foul the Coast Guard had to come rescue us.)

During that spring I worked out the high frequency respon-
sivity of the bat cochlea, and in the fall we assessed the
obstacle avoidance capabilities of the 4 common New England
species (using an array of wires hanging from the ceiling to
divide a laboratory room into halves). We recorded the inau-
dible cries bats make in flight and demonstrated (by inserting
earplugs or tying the mouth shut) that they must both produce
and perceive them if their obstacle avoidance is to be successful. (We also made a sound movie of all this which nobody can
find.) That research yielded my Ph.D. thesis and launched Don
on the career this book honors.

All the crucial new measurements we made used the unique
instruments devised by Professor Pierce. A physics professor
who, like Don, seemed to me really a naturalist at heart, Pierce
had designed his "supersonic" receivers in order to listen to
the insects singing around his summer place in Vermont (or was
it New Hampshire?). How Don found Professor Pierce and then
talked him into letting us use his apparatus I do not know.

In the spring of 1940, when we felt we understood how bats
avoided obstacles in the laboratory, it occurred to us to test
our ideas in the field. So we made an expedition to a cave
Don knew in New York State taking along the portable version of
Pierce's supersonic receiver shown in the accompanying picture
(Don was a skillful photographer even in those days).

The cave in question opened just beyond the bank of a
mountain brook. Once inside you first climbed down a ways and
then up. We had no trouble getting in and, after the small
upward climb, found ourselves looking along a straight tunnel
as far as our flashlights could penetrate. We set up the
equipment and Don went further on to the gallery where the bats
roosted. He sent several of them flying, one by one, down the
tunnel in my direction and as they approached Professor Pierce's
machine emitted the chattering clamor we hoped for. Then Don
and I changed places so he could hear the noise too. Since
the readout of Pierce's portable device came only via earphones,
the sole record of the first bat ultrasonic cries ever heard
outside a laboratory is the one engraved in our memories.

Throughout all of this Don kept urging me not to waste
time, and I thought he closed off the exercise and moved us
toward the entrance after an almost indecently short interval.
He explained his behavior once we had scrambled down and then
up and then out to cross the brook: the spring sun was rapidly
melting the snow, the brook was rising fast, and he had been
worried from the moment we arrived that the water might fill

the entrance to the cave and trap us inside. My captain, I
thought to myself, always looks out for the safety of his crew."

.During the nearly four decades since that first demonstration
of echolocation, the field has flourished. This is evident from
the quality and diversity of work described in this volume. It can
also be seen, quantitatively, in the accompanying graph, which shows
the number of publications on echolocation in microchiropteran bats
during each 3-year period since 1938. This graph, kindly prepared
by Uli Schnitzler, shows several surges in productivity - one about
1961, when the second generation of U.S. scientists began publishing,
and a very large one in 1967, when, following Frascati, the second
generation of German scientists and the Russian group entered the
field. Now, with the Jersey Symposium, the third generation is beginning to contribute importantly. Throughout this entire period,
while moving professionally from Harvard to Cornell, back to Harvard,
and then to the Rockefeller University, and while making major contributions to a number of other fields, Don Griffin has continued to
guide development of the field of echolocation with imaginative, incisive experiments: demonstrating the usefulness of echoes for orientation in the lab and in the field, showing that echolocation was
used for insect capture, documenting the sensitivity of the system
and its resistance to jamming, developing techniques to show how
accurately targets can be discriminated, and analyzing the laryngeal

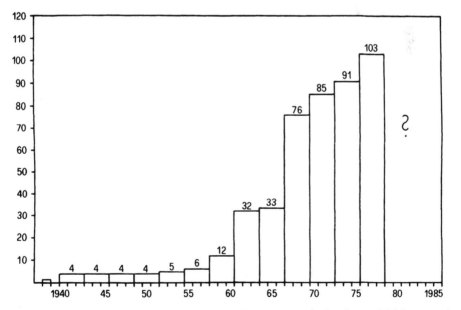

Histogram of papers published per 3-year period since 1938 on echolocation in microchiropteran bats.

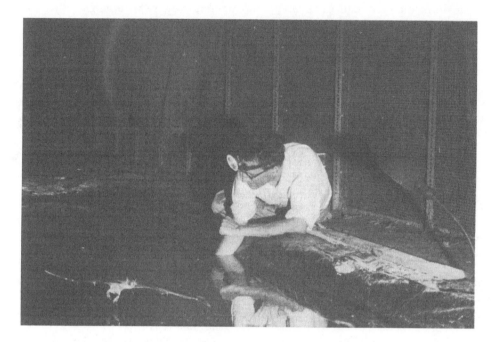

D. R. Griffin at Simla, Trinidad, studying capture of food by
<u>Noctilio</u> (1960).

mechanisms involved in ultrasonic orientation. It was in his lab,
with his encouragement, that the first major advances were made in
understanding the neural adaptations for echolocation, an aspect of
the field that has grown to enormous and impressive proportions. In
association with Al Novick, he established the great value of comparative approaches to echolocation, traveling the world not only to
document the variety of sounds and skills shown by bats, but to do
the first careful studies of echolocating birds, as well. Out of
early interests in the feeding habits, home ranges, and seasonal mi-
grations of bats (not to mention a long-standing fascination with
homing and migration in birds) came pilot studies on the role of
echolocation, passive hearing and vision in homing behavior. Indeed,
in countless instances, he has broken new ground and set standards
of rigor and experimental elegance that have served the field well.

A major milestone in the field of animal behavior was the publication, in 1958, of <u>Listening in the Dark</u>, Griffin's monograph on
his experiments and thoughts on echolocation, and winner of The Daniel
Giraud Elliot Medal of the National Academy of Sciences. With the
perspective of 20 years subsequent work in the field, it is astonishing to anyone re-reading this classic to realize how fully Griffin
already understood the phenomenon of echolocation, how many critical
discoveries he had already made, and how profound were his insights.
This was one of five books Don Griffin has written, the most recent

of which is "The Question of Animal Awareness: Evolutionary Continuity of Mental Experience" (1976).

It was not only for his incisive early work that Don Griffin deserves recognition as "father" of the field (or Godfather as he was also described during the meeting); he has also been the academic father, or colleague, of a high percentage of those who have become active in the field. Most of the contributors to this volume who have worked on bats have felt the imprint of Griffin's personality and experimental approach directly, as graduate students or postdoctoral research associates working in his laboratory, or as their students, and it has been a powerful influence on their lives and careers.

Don is a great storyteller, as his chapter in this volume attests. More than that, however, his approach to science is guaranteed to never leave a dull moment. Whether it is fighting to stay aloft after releasing birds from a small plane piloted by Alex Forbes, or trying to convince suspicious authorities that he had valid reason for clamboring around the fire escapes of a mental hospital, or barely escaping a flooded cave, or covering for a student who had fallen through the ceiling of the Mashpee Church, or searching the tombs of an ancient Italian cemetery for Rhinolophus, his approach to experimentation is as direct and audacious as his ideas. Many of the breakthroughs in echolocation research were the result of his introduction of new or unfamiliar technology, from home-made ultrasonic microphones to high speed tape recorders, sonagraphs, and information theory. This applies in all of his other fields, as well. In recent years, for example, much of his energy has been devoted to radar identification and tracking of migrating birds. One of his collaborators in this research, Charles Walcott, tells of just one of the complications that this has gotten them into, and of Don's characteristic aplomb in solving the problem:

"That season Don arrived in Stony Brook with a trailer carrying what looked like a giant coffin. Actually the box contained a helium filled kitoon - a sort of hybrid balloon - kite combination that could be flown up to several thousand feet in the air. Its purpose was to place instruments in the same air mass that the radar showed birds to be flying in. Indeed on its trials in Stony Brook the kitoon worked splendidly - we were able to get it up at least a thousand feet or more - amply high enough to be where birds flew. Unfortunately they aren't the only things that fly there - a helicopter soon appeared flying significantly lower than the kitoon. As it came close and inspected the kitoon we noticed that the helicopter bore the inscription "Police" in large letters. We rapidly hauled down the kitoon and anxiously awaited the arrival of sirens and patrol cars; fortunately they never came. But on other occasions, the law has become

involved and Don had been read the Federal Air Regulations
which prohibit the flying of balloons, kites or other such
devices more than a few feet above ground. The State Police
in Millbrook, New York next to The Rockefeller field station
had alerted the FAA about Don's activity. Don promised the
FAA representative that it wouldn't happen again. It was the
following Easter Sunday when Don and his colleagues were
following a kitoon with radar that the kitoon's string broke.
Following the kitoons progress with the radar, they saw that
it suddenly stopped drifting in the wind and became stationary.
It was hanging only a few hundred feet in the air directly over
the State Police headquarters. The string had caught in the
upper branches of a small sapling. With a typical display of
ingenuity, Don rigged a second kitoon with a grappling iron
and managed to snag the string of the escaped kitoon retrieving
both just before dawn on Easter Sunday!"

To those of us who have had the privilege of working with Don
Griffin, perhaps the greatest lesson has been his emphasis on rigor
in experimentation. No one is more acutely aware of the ambiguities
of an experiment, the difficulties involved in demonstrating something convincingly. Nor does he ever jump to conclusions. His intellectual vigilance has been characterized by the claim that if he
were in a car passing a flock of sheep in a field, and a travelling
companion commented on the fact that among the sheep were two that
were black, he would reply, "They're black on the side facing us,
anyway." (Attributed to Don Kennedy.) On the other hand, he is not
afraid to consider some of the most complex formsof animal behavior,
and to take on established dogma, decrying the excessive use of "sim-
plicity filters" in interpreting behavior, a position he argues elo-
quently in his recent writing on animal awareness.

This mixture of insistence on careful unbiased observation of
what is really there and rigorous proof of any conclusions, combined
with a brilliant imagination and willingness to adapt new technology
to biological problems, has done much to develop the field of animal
behavior research. It has served as particular inspiration for all
of us.

We are honored to dedicate this volume to Donald R. Griffin.

THE EARLY HISTORY OF RESEARCH ON ECHOLOCATION

Donald R. Griffin

The Rockefeller University
New York, N.Y. 10021
U.S.A.

It has been very gratifying to hear and read about the exciting new discoveries concerning echolocation by both bats and cetaceans from so many active laboratories on both sides of the Atlantic. I am particularly impressed because from about 1940 to 1960 very few scientists were actively investigating these fascinating problems, despite what seemed to me their obvious significance. The Frascati and Jersey conferences in 1966 and 1979 have not only confirmed but exceeded my expectations.

I was asked to describe the early history of echolocation, but unfortunately I cannot bring you firsthand recollections of Lazzaro Spallanzani or his contemporaries. As you know he was the true pioneer who first discovered that there <u>was</u> a problem, and that bats could navigate dexterously without the use of their eyes. This history has been reviewed, both in my book <u>Listening in the Dark</u> (1958) and by Dijkgraaf (1960). I can only recount experiences during the Twentieth Century, but it is sobering to realize that forty-five years has elapsed since I first began to study the orientation and navigation of bats. Another forty-five years before that time would take us back well into the "Dark Ages" when the brilliant early work of Spallanzani and Jurine had been virtually forgotten, a time when Cuvier's tactile theory was generally accepted.

As an undergraduate in the 1930s my interest in bat migration and homing led to banding bats in substantial numbers. Several friends at Harvard kept telling me that I really should visit a physics professor, George Washington Pierce, who had developed apparatus to analyze what were then called supersonic sounds. His was probably the only apparatus in the entire world which could detect and analyze sounds over a broad frequency range extending from the upper limit of human hearing to about 100 kHz. He was already deeply engaged in studying the high frequency sounds of insects (Pierce, 1948), and once I visited his laboratory he enthusiastically invited me to bring some bats to his apparatus. A cage full of little brown bats (Myotis lucifugus) set the apparatus buzzing and clicking. But when bats were allowed to fly their high frequency sounds were picked up only occasionally. Therefore our first paper (Pierce and Griffin, 1938) now seems absurdly cautious, for it then seemed quite possible that these newly discovered high frequency sounds had nothing to do with orientation in the dark.

Discussions with a fellow student, Robert Galambos, led to much more thorough investigations. Turning somewhat reluctantly from my program of banding bats, I plunged into what was completely joint work with Galambos during the years from 1938-1940. He was already an expert in auditory physiology, and I was familiar with the natural history of bats, knew where to catch them, how to handle them, and what questions in their natural behavior were of primary interest. One of our first steps was to repeat and extend the type of quantitative experiments described by Hahn (1908) to measure obstacle avoidance ability under various conditions. We began by repeating with minor variations the experiments pioneered by Jurine and extensively replicated by Hahn, and confirmed that covering the ears of bats produced almost total disorientation. It was one of Galambos' major contributions to suggest that we should also interfere with the production of high frequency sounds; and indeed this had almost the same effect as plugging the ears. We soon found that the high frequency sounds could be picked up from flying bats provided we located Pierce's microphone, which was equipped with a parabolic horn, in front of the flying bat. When we took care to do this the brief pulses of high frequency sound proved to be invariable accompaniments of flight and to increase in repetition rate whenever the bat faced a difficult problem such as dodging small obstacles.

This is now so rudimentary that it is difficult to realize how unexpected was the discovery that bats actually orient their flight by hearing echoes of sounds generated for this purpose. For example, the distinguished physiologist Selig Hecht was so incredulous when he heard our report of these experiments at a meeting of the American Zoological Society around Christmas, 1940 that he seized Galambos by the shoulders and shook him while complaining that we could not possibly mean such an outrageous suggestion. Radar and sonar were still highly classified developments in military technology, and the notion that bats might do anything even remotely analogous to the latest triumphs of electronic engineering struck most people as not only implausible but emotionally repugnant. It was partly to overcome this sort of resistance that I suggested the general term echolocation to cover the wide variety of orientation mechanisms, natural and artificial, based upon the emission of probing signals and the location of distant objects by means of echoes (Griffin, 1944).

One amusing incident occurred when Galambos borrowed Pierce's apparatus for the production of high frequency sounds in order to stimulate bats from which he was recording cochlear microphonics. Such recordings were still relatively new and exciting aspects of auditory physiology and the only available apparatus was in the laboratory of Hallowell Davis at the Harvard Medical School. It is also necessary to realize that Pierce's apparatus was absolutely unique, as far as we knew in the entire world, both for detecting high frequency sounds and also for generating them under anything approaching controlled conditions. Thus the only way in which Galambos could look for cochlear microphonics in bats at high frequencies was to take one of Pierce's devices across Boston to the Harvard Medical School. We were somewhat hesitant to ask for this privilege, but after the importance of the enterprise was explained to Professor Pierce he readily agreed.

Unfortunately Galambos did not realize that some of the innocent-looking power outlets in Davis' laboratory supplied DC rather than AC power, and when Pierce's sound producing apparatus was first plugged in the result was a puff of white smoke and a serious diplomatic problem. I remember Galambos saying "Don, I think I can fix the thing, but won't you please be the one who tells Pierce what happened." This was an obviously expedient division of labor, and to our great relief Pierce was quite sympathetic and reasonable, worrying only that the repair work be done properly.

I worked during World War II in the psychoacoustic laboratory at Harvard directed by S. S. Stevens and learned a great deal about acoustic apparatus and measurements which I had never heard of during my graduate studies in the biology department. Francis Wiener was especially helpful and arranged to let me borrow a Western Electric 640AA microphone, preamplifier, and oscilloscope for improved measurements of the acoustic properties of the high frequency sounds we had been able to detect but only crudely to characterize with Pierce's apparatus. It was during one of these unofficial sessions that a Myotis lucifugus escaped into Leo Beranek's enormous anechoic chamber. The bat had no difficulty detecting the supposedly anechoic walls, and indeed it was quite a problem to retrieve it since the chamber was so large that direct approach to the walls was impossible. Later when Beranek asked to borrow a bat in order to demonstrate to the press the total lack of sound reflection from the walls of this chamber, I was able to spare him the embarrassment of a totally unconvincing demonstration.

As I think you all know, Sven Dijkgraaf in Utrecht independently discovered bat echolocation without the benefit of any electronic apparatus (Dijkgraaf, 1943, 1946). By listening carefully as bats flew about in quiet surroundings he was able to hear the faint audible component which accompanies the ultrasonic orientation sounds. The Ticklaut as he named this audible sound, and what Galambos and I called the audible click accompanying the physically much more intense ultrasonic signals were clearly one and the same. After the war we straightened out these matters by cordial correspondence, and at one point I even sent Dijkgraaf a live Myotis lucifugus to compare with the very closely related European species he had studied.

Something I should like to emphasize, because it may have implications for the future, is that after these basic facts had been generally accepted there was what now seems in retrospect an incredible lack of interest in further studies of echolocation. A typical reaction was "Well, Griffin you've convinced us that bats really do detect stalactites in caves by hearing echoes of these high frequency sounds, and that's all very well, but why do you want to spend any more time on this sort of thing? Who cares whether they can detect slightly smaller wires or whether one species uses a slightly different kind of sound from another? Why do you want to go to the tropics just because some of the bats there have funny looking excrescences on their faces? Isn't it time you turned your attention to something really important and forgot about all those silly bats."

This attitude was so pervasive that I found it difficult to maintain my own motivation. It seemed silly to worry about how bats might catch their insect prey, and almost unthinkable that they might use echolocation. It was really only because of the failure of some other, indoor experiments during the summer of 1950 that I persuaded myself to drag all of the necessary apparatus out-of-doors in order to study the high frequency sounds of bats when they were catching insects under natural conditions. I cannot overemphasize the intellectual inertia and difficulty in justifying the necessary effort to continue studies of echolocation and inquire what further ramifications might develop after a wider variety of species had been studied under a wider range of natural conditions. We should be alert in the future for similar mental blocks that may restrict imagination and thus retard progress.

One exception was Harold Trapido who had been working for several years at the Gorgas Laboratories in Panama and who persuaded me after rather extensive correspondence to come to Panama with the necessary apparatus to inquire whether echolocation might be different in any of the neotropical bats conveniently available there for further study. My apparatus filled 15 or 20 large

packing cases, and the first of my two months in Panama was consumed
in tracking the air freight shipment which had gone astray. Nevertheless during that trip, and a follow-up trip the next year by
Alvin Novick, it was possible to sketch out something of the
enormous range of orientation sounds used by various families and
genera of neotropical bats (Griffin and Novick, 1955). Yet highly
distinguished experts had attempted to discourage me from wasting
the time and funds necessary for this trip to Panama. "Why" they
said "a bat is a bat and what makes you think there will be any
significant differences just because they live in the tropics and
have funny looking faces?" This sort of resistance extended to
arguing that alleged differences in frequency patterns were of no
significance, and that all these high frequency sounds were simply
noise bursts. This argument was advanced even after Mohres had
clearly demonstrated that the European horseshoe bats used sounds
of much longer duration and of almost constant frequency in contrast
to all bats previously studied which had been members of the family
Vespertilionidae or Molossidae.

One episode from the early 1950s is especially appropriate to
relate at this second International Conference on Animal Sonar
Systems. Among the very first European visitors who came to see
something of our experiments on echolocation in bats were Rene-Guy
and Marie-Claire Busnel who spent several hours with me and Novick
at the Harvard Biological Laboratories. Wishing also to show them
how we study bats under natural conditions I drove to Kenneth
Roeder's home in Concord, Massachusetts where he was studying the
hearing of moths. A colony of Eptesicus fuscus emerged every
evening from a neighboring house and many of them flew over Roeder's
yard where he often set up moth preparations and recorded action
potentials in response to the passing bats. In order to elicit
insect pursuit maneuvers I had for some time been using a slingshot
to throw pebbles into the air 2 or 3 meters in front of an approaching
bat. This very often elicited a buzz and often a spectacular
pursuit maneuver.

As so often happens in such cases we arrived a little late and
the bats were already emerging for the evening. Hastily assembling
in Roeder's yard the apparatus which then consisted of a 640AA
microphone, battery operated amplifiers, and oscilloscope I took
careful aim with my slingshot for a point three meters in front of
an approaching bat. To the surprise of all concerned (except the
bat itself) there was a sharp "plunk" as the pebble collided with
one of the bat's flight membranes. The Busnels obviously thought
that I was so proficient with my slingshot that I could hit a
flying bat at a range of 10 or 15 meters. I was happy to make
excuses about not using the slingshot again so as to leave the
Busnels with the glowing impression that a typical American field
biologist was such a crack shot that he could hit bats with pebbles

from a slingshot. What none of us realized at the time was that
bats actively reach out with their wing membranes to catch insects.
Undoubtedly this Eptesicus had done so, but it was only some years
later thanks to the photographic talents of Frederic Webster that
we learned about one more unsuspected aspect of bat echolocation.

A long time elapsed after Galambos' original studies of cochlear
microphonics before anyone else took up the neurophysiology of
audition in bats. I can only explain this in terms of the same sort
of conservatism and lack of interest in unexpected new comparative
studies which I mentioned earlier. The man who opened up the
neurophysiology of bat hearing came to me first as a freshman
student asking to be excused from taking the elementary course I was
then teaching. He also expressed interest in whatever I might be
doing, and soon became actively involved in studies of obstacle
avoidance and sensitivity of echolocation. During this interval an
amusing episode occurred which, while it does not relate directly to
echolocation, shows how throughly involved in studies of bat behavior
this undergraduate student became. Approximately twenty years
earlier I had banded a large number of bats in certain caves in
Vermont. John Hall of the University of Massachusetts recaptured
one of these which must have had a minimum age of about 19 years.
I therefore returned to these caves with several colleagues equipped
with records of all band numbers so that suspected longevity records
could be examined with maximum care before the bat was released.
We found three or four individual Myotis lucifugus that were 19 or
20 years old. When one of them was positively identified the man
who was holding it was heard to remark "Why it's older than I am."

But bat banding and obstacle avoidance tests were not
sufficiently challenging to hold the full attention of this student,
and while still an undergraduate he wanted to record action
potentials from the brains of bats. I discouraged this for some
time, explaining patiently that recording from the central nervous
system was a complex and difficult business which no one in the
Harvard Biological Laboratories was equipped to undertake. But
finally he wore down my resistance and we anesthetized a Myotis
lucifugus with Nembutal and exposed the dorsal surface of the brain.
Using a piece of copper wire (or perhaps a paper clip) as an
electrode we displayed on an oscilloscope whatever potentials might
result when the anesthetized bat was stimulated by short duration
high frequency sounds or by the orientation sounds of another bat.
To my amazement beautiful action potentials could easily be
recorded from anywhere on the bat's brain -- from cerebellum to
olfactory lobes. This was the beginning of Alan Grinnell's studies
of N_4 and later other auditory evoked potentials. This first crude
experiment certainly served to hook us both, and Grinnell only
escaped with difficulty several years later for postdoctoral work
with Bernard Katz in London while I sought to compensate for his

loss by writing Professor Katsuki in Tokyo to ask whether he had
a young postdoc who would like to extend Grinnell's studies of
auditory responses in bat brains. Fortunately he did and the
result was the long and brilliant series of studies by Nobuo Suga
and later his several students and colleagues. The latest of their
many extremely significant results have been presented at the Jersey
meetings.

Dr. Suga had occasion to learn that studying bat echolocation
leads one into a variety of challenging situations and opportunities.
Shortly after he had arrived from Japan we were approaching a
dilapidated old building in rural Massachusetts to collect bats
when he saw what he thought was a cat with an unusual pattern of
black and white markings. He was rescued barely in time from too
close an approach to a skunk (Mephitis). Later, with his charming
wife, he was introduced to the tropics at the William Beebe
Tropical Research Station in Trinidad. This former estate had well
equipped laboratory buildings and very pleasant gardens where one
could investigate bats and other animals under ideal conditions.
For instance, the director, my wife Jocelyn Crane, saw to it that
the food did not make any of the northern visitors sick. The high
point of the day usually was the evening cocktail hour on the
terrace over which flew more species of bats than we could ever
identify. But in listening to them with a bat detector, usually
with a rum punch known as a "Simla special" in hand, we often heard
the high frequency sounds of many kinds of insects. Most were also
audible to the unaided ear, but one in particular was not.
Frequently this insect-like sound came from small potted bushes
which could be approached closely from all sides. By quietly
approaching with the bat detector it was easy to ascertain that
the sound came from the bush in question, but close examination
of every part of the bush failed to disclose any signs of the animal
which had been making the sound before it was disturbed by our
close approach.

At times, especially after a second Simla special, we even
entertained the possibility that the sound might be generated by
the plant! The Sugas were especially intrigued by this unidentified
sound source called the "Snarley Buzz." I suspect that Nobuo
secretly hoped to discover the first echolocating plant. But for
whatever reason he and his wife worked long and hard to discover
the actual source, often climbing into tall trees on rickety ladders,
being attacked by biting ants and stinging bees, but in the end
locating the insect responsible -- a small long horned grasshopper
of the genus Pflugus (Suga, 1966).

In short, echolocation has been a most rewarding subject to
investigate, both in the laboratory and in the field. It is now
flourishing and ramifying in many significant directions. I am

confident that this happy state of affairs will continue well into
the future, and I hope all students of echolocation will have as
satisfying experiences as I have been privileged to enjoy.

REFERENCES

Dijkgraaf, S., 1943, Over een merkwaardige functie van den gehoorsin
    bij vleermuizen, Verslagen Nederlandische Akademie van
    Wetenschappen Afd. Naturkunde, 52:622-627.
Dijkgraaf, S., 1946, Die Sinneswelt der Fledermäuse, Experientia,
    2:438-448.
Dijkgraaf, S., 1960, On Spallanzani's unpublished experiments on
    the sensory basis of object perception in bats, Isis, 51:9-20.
Griffin, D. R., 1944, Echolocation by blind men, bats, and radar,
    Science, 100:589-590.
Griffin, D. R., 1958, "Listening in the Dark," Yale University
    Press, New Haven, Conn. (reprinted 1974 by Dover Publications,
    New York).
Griffin, D. R., and Novick, A., 1955, Acoustic orientation of
    neotropical bats, J. Exptl. Zool., 130:251-300.
Hahn, W. L., 1908, Some habits and sensory adaptations of cave-
    inhabiting bats, Biol. Bull., 15:135-193.
Pierce, G. W., 1948, "The Songs of Insects," Harvard University
    Press, Cambridge, Mass.
Pierce, G. W., and Griffin, D. R., 1938, Experimental determination
    of supersonic notes emitted by bats, J. Mammalogy, 19:454-455.
Suga, N., 1966, Ultrasonic production and its reception in some
    neotropical Tettigoniidae, J. Insect Physiol., 12:1039-1050.

Chapter I
Performances of Animal Sonar Systems

Chairman : F.G. Wood

co-chairmen

R.J. Schusterman and H.U. Schnitzler

- <u>Underwater</u>
   Behavioral methodology in echolocation by marine
   mammals.
   R.J. Schusterman

   Detection range and range resolution of echolocating
   bottlenose porpoise.
   A.E. Murchison

   Odontocete echolocation performance on object size,
   shape and material.
   P.E. Nachtigall

   Cetacean obstacle avoidance.
   P.W.B. Moore

- <u>Airborne</u>
   Performance of airborne animal sonar systems
   I. MICROCHIROPTERA.
   H.U. Schnitzler and O.W. Henson, Jr.

   Performance of airborne animal sonar systems
   II. VERTEBRATES other than MICROCHIROPTERA
   O.W. Henson, Jr. and H.U. Schnitzler

BEHAVIORAL METHODOLOGY IN ECHOLOCATION BY MARINE MAMMALS

Ronald J. Schusterman
Department of Psychology and Biology
California State University
Hayward, California 94542

INTRODUCTION

Despite the fact that a preponderance of the investigations dealing with echolocation by marine mammals has been done with intact animals performing some characteristic behavior, only a single paper written some fifteen years ago (Turner, 1964) has ever dealt in any profound way with problems of behavioral methodology. The 1964 paper by R. N. Turner addressed itself primarily to conditioned responses and included response measures, techniques of target or signal presentation, threshold measurement and reinforcement contingencies. When Turner's paper was published, there had only been a few pioneering studies done on porpoise echolocation and hearing, and by today's standards, these studies were, for the most part, relatively unsophisticated. A recent Russian publication (Lekomtsev and Titov, 1974) attempted to update Turner's work, but in reality the paper served primarily to highlight Russian behavioral procedures to the study of porpoise sonar abilites.

Even though numerous advances in the behavioral analysis of sensory and perceptual capabilities of a wide variety of animal species have been made since the mid 1960's, there has been no attempt to review and evaluate procedural variables in the behavioral study of animal sonar systems. In particular, modern operant conditioning techniques (Skinner, 1961) have become very efficient, and methodological advances have been applied to many problems and many animal species. With these advances has come a marked interest in a relatively new field of animal psychology which has been termed "animal psychophysics" (Blough and Blough, 1978; Stebbins, 1970).

The purpose of this paper is to review and evaluate behavioral techniques used to study echolocation by marine mammals, particularly dolphins, within the context of this new field of animal psychophysics. The goal of these techniques is to establish and maintain stimulus control of an animal's behavior. Stimulus control is essentially a convenient expression for saying that a stimulus change brings about a change in some measurable aspect of behavior. The degree to which a stimulus exerts control over an animal's behavior may be measured in a variety of ways including probability, latency, amplitude, or rate of some specified response and it is usually assumed that a controlling stimulus either signals the animal that a particular class of responses will be reinforced or it signals the animal that a particular class of responses will not be reinforced. Once stimulus control has been established, the experimenter may want to identify the manner in which the animal classifies or dimensionalizes the stimulus. This has been particularly true in recent echolocation experiments in which the controlling aspects of the stimulus targets were interpreted in one way by the experimenter and in quite a different way by the porpoise (Schusterman and Kersting, 1978). In animal psychophysics, the answer to the question about what the porpoise "sees" when it is stimulated with suprathreshold echoes depends on a series of perceptual investigations in which tests of equivalence, transfer or generalization are given. Thus, the porpoise is placed in experimental situations in which different stimulus configurations, as defined by the experimenter, are responded to in the same fashion.

The emphasis in this paper will be on problems associated with the types of indicator responses used by echolocating animals and the ways in which stimulus targets have been presented. In this regard, it should be pointed out that the major difference between a passive hearing task and an active sonar task is that in the former task it is not necessary to preclude the animal's dependence on vision unless visual stimuli are correlated with acoustic stimuli in signalling reinforcement. However, in echolocation tasks the targets invariably have a visual analogue, and if the animal can use visual cues, it may not depend on its acoustic auto-communication cues, i.e., on echoes which return following its own sound emissions. When using marine mammals in echolocation tasks, enucleation has rarely, if ever, been used and the traditional way of occluding or eliminating vision has been to either have the animal work in turbid water or in water where the light level was very low, place a blindfold or eye cups on the animal, use targets which are quite different acoustically but appear visually to be the same, or place the targets behind a visually opaque but acoustically transparent screen. All of these methods have costs and benefits and the technique used will frequently depend on the species under investigation and the type of problem being studied.

This paper will not discuss some procedural problems of echolocation experiments which have been reviewed elsewhere (Turner, 1964; Lekomtsev and Titov, 1974). These include randomization of stimulus presentation and motivation or food deprivation variables.

UNCONDITIONED RESPONSES

In procedures with unconditioned responses, the indicator response appears to require no special training, i.e., it is reflex-like in its action, being directly elicited by a variety of submerged targets and sounds.

Perhaps the first systematic use of reflexes as indicator responses were the experiments by Kellogg (1961) with two Atlantic bottlenose dolphins (Tursiops truncatus) which were conducted in a pool with very limited visibility. Kellogg's targets included: (a) polelike sheet-metal devices, triangular in cross section; (b) BB shot; (c) wooden streamlined shapes which could be silently submerged; (d) food fish, and (e) human swimmers. Kellogg recorded two classes of behavior: (a) bursts of sonar clicks, and (b) changes in movement or swimming patterns. The latter behavior was subdivided into either approach or avoidance responses. These experiments suggested that almost any target "noisely" submerged would elicit echolocation pulses by Tursiops sometimes followed by approach (e.g. fish) or avoidance (human swimmer or long metal pole). Oscillating head movements were frequently noted as the porpoise, emitting clicks, swam toward a hand-held fish. If the water was splashed without submerging a target, then the animals did not sustain their clicks. However, when reflecting targets were submerged, sustained clicking occurred, and it was assumed by the experimenter that the porpoises detected targets and differentiated between them on the basis of the sound signals emitted and the type of swimming movement (approach or avoidance) observed.

In retrospect, it is likely that one of the most vexing problems in Kellogg's experiments was a lack of an objective and quantitative evaluation of the porpoises's behavior under these extremely variable stimulus conditions. On occasion, the experimenter must have been forced to make a subjective judgment as to whether there was indeed an avoidance response or an approach response—particularly if the porpoise did not take the hand-held fish. The fact that Kellogg did not present quantitative data in reporting these unconditioned responses to submerged targets and water splashing supports the notion that the use of unconditioned responses is of limited value in the study of echolocation by marine mammals. Another confounding factor in these experiments

is the possibility that these reflex-like responses habituated
with repeated stimulation by similar sounds or similar targets.
Habituation of sudden changes in the swimming movements detected
visually by the experimenter when relatively loud pure tones were
presented may have led to an initial underestimate of the upper
frequency limit of hearing in Tursiops (Kellogg and Kohler, 1952;
Kellogg, 1953).

CONDITIONED RESPONSES

Unconditioned responses depend on a relatively simple and
direct stimulus-response relationship in which echoes from the
target or other auditory stimuli elicit the indicator response,
which may not be easy to define and quantify. When conditioned
responses are used, the experimenter usually chooses some carefully
defined motor response which can be readily measured. In classical
or Pavlovian procedures, the unconditioned stimulus (US) or
reinforcer is contingent on the conditioned stimulus (CS) i.e.,
the CS and US are paired. In operant or Skinnerian procedures,
reinforcement is made contingent on some class of responses. If
a stimulus sets the occasion for reinforcing a class of responses,
then the stimulus becomes a discriminative stimulus (S^D) and
serves as a signal very much as the CS serves as a signal in
classical conditioning procedures. Following repeated pairings
on the CS and US or the S^D and reinforcement, a highly stereotyped
indicator response is elicited or emitted whenever the CS or S^D
is presented.

Operant procedures may involve the use of positive or negative
reinforcers. When positive reinforcement is used, if the animal
makes a response in the presence of an S^D, it receives a reward.
When negative reinforcement is used, if the animal makes a response
in the presence of an S^D, it avoids an aversive stimulus event.

Classical conditioning in which the US is shock and the CS
is an acoustic signal has been used extensively by the Russians
to study the sound reception by echolocating porpoises. For
example, sound detection thresholds were obtained in the common
porpoise (Phocaena phocaena) by pairing auditory stimuli (CS)
with shock (US) and measuring a restrained animal's galvanic skin
response (Supin and Sukhorunchenko, 1970). In another experiment
on restrained bottlenose dolphins the characteristics of directional
sound signals in the horizontal plane were investigated by measuring
the cardiac component of what the Russians call "motor-defensive
conditioned reflexes" (Ayrapet'yants, et al., 1973). All of
these experiments have involved passive hearing in porpoises and
I do not know of a single experiment that has used classical
conditioning procedures to determine the sonar capabilities of

marine mammals. It seems that only operant conditioning procedures have been used to investigate the performance of echolocating marine mammals.

RESPONSE MEASURES

Examination of a variety of response measures and their related reinforcement contingencies reveals remarkable similarity among different experimenters in different laboratories around the world in their use of operant procedures to study echolocation in both marine mammals and in bats. Investigators from the Soviet Union and United States have used very different terminology in describing essentially the same procedure. For example, when a Russian researcher says he or she used the "method of motor-food conditioned reflexes in unrestrained dolphins" this is nearly the equivalent of an American researcher saying he or she used "a two-alternative forced-choice procedure",

Retrieval

Typically, this procedure involves the arrangement of reinforcement contingencies such that if the porpoise picks up a single submerged target with its mouth or rostrum and returns it to the experimenter or trainer, then the animal is positively reinforced with a food fish. In theory, at least, the retrieval technique could be used as a way of determining the echo dimensions along which auditory quality is represented in the echolocating porpoise. This could be done by repeatedly presenting a blindfolded Tursiops with a variety of targets differing in some aspect of fine target structure and differentially reinforcing the animal for retrieving targets differing in this single aspect. In fact, however, the retrieval procedure has been used infrequently to study the sonar abilities of marine mammals, and its use has generally been limited to simple detection tasks.

An experiment by C. Scott Johnson (1967) used the retrieval procedure in an attempt to determine the relationship between pulse rate and target range. The targets used in the experiment included a weighted air-filled foam rubber ring and single vitamin pill. Scronce and Ridgway (this volume) failed in their attempt to demonstrate echolocation in a blindfolded gray seal (Halichaerus gyrpus) by requiring the seal to retrieve an air-filled plastic ring 20 cm in diameter situated randomly in one large section of a 10 m diameter redwood tank. Most recently Dziedzic and Alcuri (this volume) studied changes in the click burst of T. truncatus when the task was to retrieve different shaped objects.

Obstacle Avoidance

The most definitive early demonstrations of echolocation in bats were those reported by Griffin and Galambos (1941) in which enucleated bats were required to navigate through a maze of wires. Similar experiments have been done with free-swimming T. truncatus and P. phocaena under a variety of conditions (Busnel et al., 1965; Busnel and Dziedzic, 1967; Kellogg, 1961). The complexity of these obstacle avoidance tasks makes it extremely difficult to determine what aspects of the targets were important for the occurrence of the complex swimming patterns of porpoises which included short and difficult turns in order to avoid colliding wth poles, wires and nets. The obstacle avoidance technique is extremely limited from this standpoint and is rarely used today in echolocation experiments with marine mammals where the emphasis is the determination of a well-defined response and a carefully controlled target or signal dimension.

Go/No-Go

Surprisingly, the "go/no-go" procedure has rarely been chosen to measure the echolocation performance of marine mammals despite the fact that it is, perhaps, the most simple and straight forward procedure to study discriminative echolocation. There are several variations of the go/no-go paradigm. In the study of marine mammal bioacoustics, the paradigm operates within the context of a discrete trial procedure in which target or signal presentation is relatively brief and is contingent on the animal's maintaining a reasonable fixed position between target presentations. During a trial, indicated by a sound or a light, the animal orients from its starting position towards the target area. Then it either performs a carefully defined action, e.g. pressing a paddle (a "yes" response) in the presence of a standard target, or it inhibits or withholds the action (a "no" response) in the presence of a comparison target or in the absence of a target. Following the indicator response the animal stays or returns to its starting position. Response measures include the likelihood of a response occurring on each trial and the latencies of "yes" responses. Reinforcement may be programmed so that "yes" responses in the presence of a target are positively reinforced and are extinguished, i.e., are not reinforced in the absence of a target, or reinforcement may be programmed symmetrically so that it follows "yes" responses in the presence of a target ("hits") and "no" responses in the absence of a target ("correct rejections"). Errors (which may be punished with "time-outs" or other aversive events) are of two types: a "yes" response in the absence of a target ("false alarms") or a "no" response in the presence of a target ("misses").

The go/no go procedure is usually employed when targets or

stimuli are presented sequentially and is thus used most extensively
in the earlier stages of training when only a single standard
target is presented on repeated trials (Lekomstev and Titov, 1974).

One type of go/no-go procedure is exemplified by an experiment
on discriminative echolocation (Ayrapet'yants, et al., 1969) in
three Black Sea dolphins (Tursiops truncatus), under high and
very low illumination. The dolphins were required to tug at a
ring in the presence of a standard target (a cylinder 25 mm in
length and 110 mm in diameter) and to withhold this response when
presented with comparison targets (cylinders 28, 30, 35, 45, 55
and 75 mm in length and 110 mm in diameter). A 5 kHz tone
signalled the dolphin to swim "from the starting position at a
distance of 20 meters to the experimenter who immersed different
targets into the water...." (Ayrapet'yants, et al., 1969).
Tugging the ring when the standard target was presented resulted
in a fish reward. The manner in which this paper was written
makes it difficult to be sure about the reinforcement contingencies,
but in my interpretation the contingencies were not symmetrical
which means that withholding the "ring-tug response" in the
presence of the comparison targets did not result in a fish
reward. Following each response, the dolphin was required to
swim back to the start position. Eventually, all dolphins accurately
differentiated all but the 28 and 30 mm comparison cylinders from
the standard 25 mm standard cylinder. The authors state that "it
was more difficult to differentiate the 30 mm cylinder which was
only 5 mm taller than the positive one. Such differentiation was
not absolute (correct answers constituted 70%). Presentation of
the 28 mm cylinder, which was 3 mm taller than the positive one,
elicited no recognition. Thus, the threshold of differentiation
of targets according to height was 5-10 mm" (Ayrapet'yants, et
al., 1969).

The Russian experiment highlights some of the least desirable
aspects of the go/no-go procedure. With only a single well-
defined response, it is sometimes difficult to separate "no"
reports from a failure to respond, and there is no measurable
latency for the "no" response. Moreover, it would seem that
unless the reinforcement contingencies or what signal detection
theorists call the payoff matrix was symmetrical there might be a
preference for the "yes" response, thereby driving the threshold
estimate down. Furthermore, the statistical definition of a
"threshold" is sorely missing in this particular paper.

Multiple-Response Forced-Choice Procedures

The multi-response technique which has probably been used
most extensively for doing experiments on marine mammal biosonar

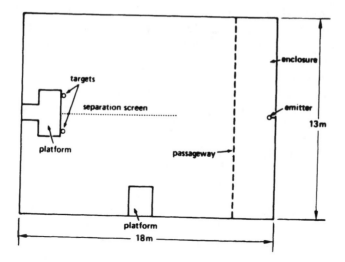

Figure 1. Schematic of a testing tank illustrating the use of the
two-response forced-choice procedure with simultaneous
target presentation. Prior to the onset of an 8 kHz
tone, the dolphin positioned itself in front of the
emitter. The targets were lowered silently into the
water at two points 2 m apart. In response to the
audio signal the dolphin emerged from the enclosure
through the passageway and headed for the platform
where it detected the targets from a certain distance
(adapted from Lekomtsev and Titov, 1974).

is the two-response forced-choice procedure. The major variation
on this procedure depends on the simultaneous or successive
presentation of targets.

     Simultaneous Target Presentation. Following the stationing
of an animal each of two spatially separated targets are presented
near their associated response manipulanda and the animal is
reinforced for responding to the "correct" stimulus. Frequently,
a vertical partition or separation screen has been placed between
the targets so the animal is forced to make its choice at some
minimum specified distance (see Figure 1).

     Variables affecting the echolocation performance of several
species of dolphins, including T. truncatus and D. delphis, have
frequently been investigated by the two-response forced-choice
procedure using a simultaneous presentation technique. These
factors include object detection with noise disturbance (Titov,
1975), angular and range resolution (Bel'kovich et al, 1970;
Lekomtsev and Titov, 1974; Murchison, this volume), size discrimination of planometric and stereometric figures (Kellogg, 1961;

Turner and Norris, 1966; Barta, unpublished ms.; Ayrapet'yants et
al, 1969), shape discrimination of planometric and stereometric
figures made of metal or plastic (Gurevich, 1969; Barta, unpublished
ms.; Nachtigall and Murchison, this volume; Bagdonas et al.,
1970) and material compostiion of plates and solid elastic spheres
(Evans and Powell, 1967; Dubroskiy et al, 1971).

Perhaps the only disadvantage of the two-alternative forced-
choice procedure with a simultaneous presentation of targets is
that even with a partition between the targets it is extremely
difficult to force a dolphin or seal without eye cups to make its
choice at some fixed distance from the targets when the experimenter
has difficulty visually tracking the animal in turbid water or at
very low levels of light (see Schusterman, 1967).

Successive Target Presentation. A group of American investigators at the Naval Ocean Systems Center (NOSC) in Hawaii have
done a wide variety of echolocation experiments by presenting
targets on successive trials and requiring the dolphin to respond
to one of two spatially separated response manipulanda. In
"presence-absence" experiments, a response to one manipulandum
(e.g. to the animal's right) is reinforced if a target is present
("yes" response) and a response to the other manipulandum (e.g.
to the animal's left) is reinforced if the target is absent ("no"
response). In so-called discrimination or recognition experiments,
targets usually differing multi-dimensionally control differential
responding and the porpoise is reinfoced for responding to the
manipulanda associated with a specific target. Despite spatially
disparate response manipulanda, targets are presented in the same
location. Thus, unlike the two-alternative forced-choice method
with simultaneous presentation, where stimulus targets and associated
response manipulanda have approximately the same relative spatial
location (so that different reflected sounds come from different
places), the two-alternative forced-choice technique with successive
stimulus presentation requires the animal to make a bilateral or
spatially different response to different sounding reflected
echoes emanating from the same locus in space.

An illustration of this procedure within the context of a
detection experiment has been reported by Penner and Murchison
(1970), who used it to determine the smallest diameter copper
wire that could be detected by an Amazon River dolphin, Inia
geoffrensis. The Inia stationed between the response manipulanda
(mounted on the opposite side of the pool from the stimulus
display) until a trial began with the onset of a 1-sec burst of a
30 kHz tone. Then the Inia swam across the 7.3 diameter tank
toward the target area emitting echolocation pulses, and returned
to the response manipulanda area. Activation of the "yes" manipulandum in the presence of a target was a hit and the dolphin was
reinforced. In the absence of a target activation of the "no"

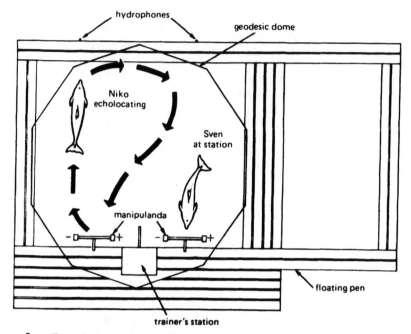

Figure 2.   Testing situation showing the use of the two-response
forced-choice procedure with successive presentation
of targets. Presence of a target was reported on the
"yes" manipulandum and the absence of a target was
reported on the "no" manipulandum. For further
details of this procedure see Murchison (this
volume).

manipulandum was a correct rejection and the dolphin was reinforced.
Errors (false alarms as well as misses) were never reinforced .

An interesting variation of this procedure was used to
calculate the furthest distance at which solid and water filled
spheres could be detected by Tursiops (Murchison, this volume).
The study used two Tursiops working alternately on the same task
(see Figure 2). In an attempt to determine whether cooperative
or cueing behavior occurred in this type of paired-porpoise range
detection task, Penner (1977) conducted a similar experiment
comparing performance when animals worked on alternate trials,
simultaneously or alone, and he found that echolocation performance
appeared unaffected by these conditions.

Finally, Hammer (1978) and Au and Hammer (this volume)
applied the two-alternative forced-choice procedure with successive
target presentation to the study of salient target characteristics
of different-size cylinders. One set of cylinders (A) controlled
A reports and the other set of cylinders (B) controlled B reports
(see Figure 3).

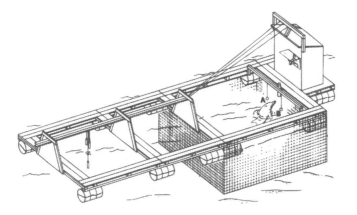

Figure 3.   Experimental setting and apparatus showing porpoise
             stationing, A and B manipulanda, experimenter's
             shack and the lowering of hollow cylindrical targets
             attached to monofilament lines mounted on I-beam
             support.

Collateral Behavior

     As previously indicated, a response-contingent reinforcement
procedure within the context of a discrete trial procedure is the
method by which echolocation is studied in marine mammals.   For
example, following the intake of a fish reward, the beginning of
the next trial depends upon the dolphin resuming its stationing
or starting position.   At the start of the trial, the dolphin
makes an "observing response" by emitting echolocation pulses at
the target, and on the basis of its echo perception, the dolphin
makes a choice (e.g. hitting one of two rubber balls mounted on
flexible stalks as shown in Figure 3) which results in either a
fish reward, no reward, or a time-out period (negative reinforcement).
In this type of response-contingent procedure, the form or topography
as well as the temporal patterning of each response--stationing →
observing → choice → feeding or time-out--are all likely to
influence the relationship between the "choice" response, the
target variables and the reinforcement in either unknown or at
least in rather imprecisely measured ways.   Thus, at the start of
a trial the dolphin leaves its starting position, swims to its
right, begins to emit pulses while swinging its head from side to
side as well as up and down, turns its entire body around, swims
back to the manipulandum, breaks the water surface by spinning on
its vertical axis and terminates the behavioral sequence by
striking one of the rubber balls with its rostrum while simultaneously emitting squeaking sounds.   Then the dolphin immediately
opens its mouth, swallows the fish reward, begins swimming along

the right wall of the tank before returning to station where the
animal subsequently begins bobbing its melon and rostrum up and
down prior to the start of the next trial. This bewildering
array of responses, their sequential dependencies and their
duration are, to some extent, important regarding an interpretation of the dolphin's auditory perception of echo complexes.

In a discriminative echolocation experiment with different
sized spheres presented simultaneously to an Atlantic bottlenose
dolphin, Turner and Norris (1966) called some of these behaviors
"collateral behavior" and found a relationship between "disruption
of a stable pattern of collateral behavior" and a substantial
increase in errors. Moreover, these experimenters could accurately
predict incorrect choices or errors on the basis of the dolphin's
swimming pattern. Evans and Powell (1967) also used the term
"collateral behavior" to refer to all behaviors except target
pressing and showed that there was an inverse correlation between
"performance scores" (correct target presses) and "scanning
rate". They also found that choices of a target (metal plates)
were more rapid for easy discriminations as compared to difficult
ones, but that this was largely due to the greater head scanning
at distances of less than one meter from the plates for the
difficult comparisons. Here again we see that responses occurring
before the originally conditioned operant were predictive of the
animal's performance.

Although Evans and Powell (1967) emphasized various aspects
of the observing response in an echolocating dolphin and showed
correlations between this behavior and the "terminal" response,
Penner and Murchison (1970) noted that an echolocating Inia
geoffrensis sometimes swam upside down, jaw gapped, jaw snapped
and yawned. In terms of predicitng whether their fresh water
dolphin would hit the "yes" or "no" manipulandum, Penner and
Murchison found that Inia terminated its observing response
sooner or later depending on the absence or presence of a target.
Similar observations have been reported in echolocation experiments
dealing with the effects of distance and reverberation on object
detection in Tursiops truncatus (Murchison, 1979).

Collateral behavior in echolocating porpoises may be viewed
in the larger context of environmentally induced sequential
patterns of behavior. Falk (1961; 1969; 1971) coined the term
"adjunctive behavior" in reference to excessive drinking in a
hungry rat when the animal was lever pressing on a schedule of
intermittent food reinforcement. The distinction between adjunctive
or interim activities and terminal responses was first made by
Staddon and Simmelhag (1971) who showed that during interfood or
intertrial intervals predictable sequential behavior patterns
increased and decreased with terminal responses (e.g. a conditioned
operant or Pavlovian response) occurring just before the delivery

of food. In a recent review Staddon (1978) has shown that temporal
sequences of behavior depend on their interactions with other
behaviors that are induced in the situation. Interim activities
for an echolocation dolphin would include all aspects of the
observing response (pulse emission, head scanning, etc.) as well
as head bobbing, jaw clapping, jaw gapping, side swimming, etc.
Various measures of the terminal or indicator response might
include its form or topography, its duration, and its strength.
Any and all of these response measures may be correlated with the
"choice" response in an echolocation task and are likely to help
in our understanding of the behavior of a porpoise following its
analysis of complex echo inputs in relation to its previously
stored information.

To illustrate the interaction between interim activities and
terminal responses, I will describe one of a series of discriminative
echolocation experiments which I conducted at the Naval Ocean
Systems Center in Kanoehe Bay, Hawaii. Hollow steel and hollow
bronze targets were presented successively to an experienced
adult male porpoise (Sven) in a two-alternative forced-choice
procedure. Figure 3 shows a schematic diagram of the experimental
setting. The dolphin's task was to differentiate echoes from
hollow steel and hollow bronze targets--each having two different-
sized outer diameters (OD) and wall thicknesses. The targets
were described in detail by Hammer (1978). All targets were 17.8
cm long. The smaller OD's were 3.81 cm with a wall thickness of
0.32 cm and the larger OD's were 7.62 cm with a wall thickness of
0.40 cm. Both the large and small steel targets were to be
reported on the A or left manipulandum and the large and small
bronze targets were to be reported on the B or right manipulandum.
In earlier experiments Sven had been trained to report the presence
of a target by hitting the A manipulandum and the absence of a
target by hitting the B manipulandum. To take advantage of those
previously established response contingencies, Sven was first
trained to make an A report when presented with either a large or
a small hollow steel target submerged about 1.14 m from the
surface of the water and to make a B report in the absence of a
target. Once Sven perfected this differentiation, both the large
and small bronze targets were submerged gradually, i.e., the
vertical positions of the bronze targets were initially always
closer to the surface of the water than the steel targets. The
vertical position of the bronze targets gradually approached that
of the steel targets until eventually all targets were presented
in the same position. In general, this type of stimulus fading
procedure minimizes errors (see below) during the acquisition
phase of a learned discrimintation.

Figure 4 shows the acquisition of the discrimination between
steel and bronze targets regardless of size. Because a fading
procedure was used, the number of errors was minimized even when

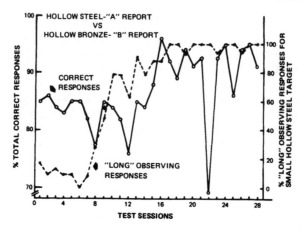

Figure 4. Acquisition of a hollow bronze and hollow steel
differentiation regardless of the size of the outer
diameter of the targets. "Long" observing responses
occurred only in the presence of the small hollow
steel target ("A" reports) and were a precursor
of a significant increase in total correct responses.

the porpoise was first confronted with the discrimination.
Nevertheless, the percentage of total correct responses, i.e.,
for all four targets, rose significantly from the first 14 test
sessions to the last 14 sessions (p < .01; sign test). In addition
to measuring the terminal choice responses as a function of the
type of target presented, we also recorded response latencies and
classified observing responses into two categories. The most
frequently occurring response was one in which the turn from
station toward the target area at the beginning of a trial was
quickly followed by another abrupt turn to the manipulanda.
During the course of these rapid observing responses, Sven never
swam more than two meters from station and the response latencies
were invariably less than four seconds. This observing response
occurred most frequently in the presence of both bronze targets
and the large steel target. However, about one-quarter of the
way through the experiment, Sven began to make a different observing
response. When he made a complete turn from station toward the
target area at the start of a trial, he rolled onto his left side
and with his head tucked slightly he faced the target area.
Following a pause of between one to two seconds, after swimming
at least three meters from station, Sven swung his entire body
around in a very deliberate fashion before returning to the
manipulanda. The response latencies for these observing responses
were usually more than four seconds. These "long" observing
responses occurred primarily in the presence of the small steel
target. As Figure 4 shows, "long" observing responses were very

infrequent during the first seven test sessions and then increased dramatically. The increased probability of "long" observing responses in the presence of the small steel target was a precursor to the significant increase in total correct responses. Thus, in this experiment, there was a strong interaction between interim activities and terminal responses as a function of specific target stimuli. Moreover, a measure of the interim responses showed that the porpoise was actually differentiating between the large and small hollow steel targets even though the terminal response (A-report) was the same for both echo complexes.

METHOD OF TARGET PRESENTATION

The development and maintenance of stimulus control in echolocation experiments makes it absolutely necessary to ensure that the animal is appropriately oriented to the target area and that targets are chosen and presented in such a way that target parameters cover a wide range of values including threshold values. Target detectability must be assured if reinforcement contingencies are to produce the desired results of establishing and maintaining stable responding. The spacing of target parameter values around threshold and the sequence of presentation are also crucial considerations in training and maintaining stimulus control. Once established, the degree of stimulus control by an echo complex can be tested or probed by presenting novel targets and finding out the degree to which targets contain similar echo configurations (Hammer, 1978). In general, experimenters using porpoises in studies of echolocation have used traditional psychophysical methods to arrive at threshold estimates.

Observing Response

In an early experiment on discriminative echolocation with different sized spheres (Turner and Norris, 1966), a blindfolded dolphin had to deflect a lever corresponding to the standard sphere in order to collect a fish which was thrown to a specific "zone" or area of the pool. Once the dolphin collected the fish, it oriented to the apparatus to make another response. Obviously, this type of control of the animal's positioning while it echolocates is crude and makes it difficult to specify the properties of the echoes returned to the dolphin which controlled the animal's final choice response. Without controlling the position from which a discrimination is made, it is not even possible to be certain that the intensity of the echo return, depending as it does on the distance between the porpoise and the targets and the manner in which the porpoise faced the targets, is the critical cue in differentiating objects of different target strength (e.g. see Evans and Powell, 1967). Much of the early Russian research

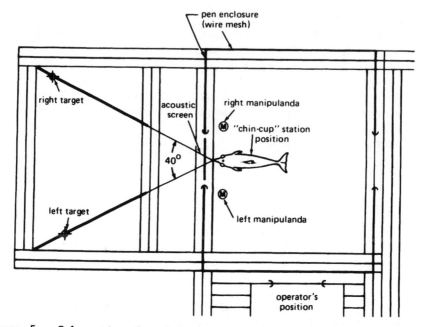

Figure 5.   Schematic of a dolphin assuming a fixed stationing posi-
            tion in a chin cup behind an acoustic screen during a
            range resolution task. The animal's task was to report
            whether the left or right target was closer and the
            report was made by striking the appropriate manipulan-
            dum. On this trial the correct report is on the left.
            The two-alternative forced-choice procedure with
            simultaneous target presentation was used.

suffers from the same problem, i.e., the ambiguous nature of the
echo return in relation to the position of the dolphin and the
targets at the time the discrimination is made.

     During the past few years, much more refined and unambiguous
control of an animal's orientation with respect to the targets
has been accomplished by making the observing response contingent
upon target presentation. In a range resolution task, Murchison
(1979) made target presentation (consisting of two identical 7.6
cm diameter polyurethane spheres) contingent on "chin-cup" stationing
(see Figure 5). Once the porpoise was stationed in the fiberglass
chin cup and the targets were in position, an acoustic screen was
dropped and the target display was presented to the dolphin.
Thus, the acoustic screen prevented the animal from prematurely
scanning the target display and has the potential of allowing the
experimenter to make rather precise measurements of pulse emission
and the echo return at the relatively precise position of pulse
emission (Au, Floyd and Haun, 1978).

HOOP STATION

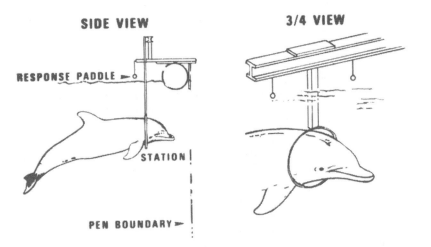

Figure 6.   Diagram of an Atlantic bottlenose dolphin (T. truncatus)
            maintaining a fixed position in a water-filled plastic
            hoop. The dolphin remained in the hoop up to its
            flippers and when given an audio signal it emitted
            a burst of echolocating clicks at a target 6 m
            directly in front of its melon. Following the termi-
            nation of the last pulse within a burst of pulses,
            the dolphin backed out of the hoop and reported the
            presence of a sphere or cylinder by hitting the "A"
            or "B" response paddle or manipulandum with its rostrum.

Another technique of controlling not only the postural
aspect but also the pulse emission aspect of the observing response
in an echolocation task has been developed by Schusterman, Kersting
and Au (this volume). In a sphere-cylinder discrimination task
the porpoise had to position itself in a 41 cm water-filled
plastic hoop with its rostrum directly in line with the center of
the target located 6 m away and inhibit the emission of echolocation
pulses until an audio cue was given (see Figure 6). The preciseness
of the observing response just described permitted a determination
of the type of signal return the dolphin used to make its differentiation (Au, Schusterman, and Kersting, this volume).

Most recently, Penner and Kadane, in an evaluation of biosonar
detection in noise (this volume), were able to record pulse train
latencies and response latencies following the last pulse of a
train from two bottlenose dolphins who were trained to make a
precise postural observing response. Thus, these investigators
were and are capable of making relatively exact measurements of

the moment in time in which the dolphins make their decision
regarding the presence or absence of a target.

Fading Procedures

Perhaps the most powerful technique currently available
which minimizes the number of incorrect responses during the
acquisition phase of a detection or discrimination task via ·
echolocation is the technique of fading. In general terms the
technique depends on initiating training or stimulus control
procedures with an easy detection or discrimination (high signal/
noise ratio) and gradually changing some feature of the targets
on successive trials so that stimulus control of responding is
transferred from one feature or characteristic to another. For
example, in the visual mode Schusterman (1965; 1966) showed that
sea lions could repeatedly reverse their responses to different
visual shapes (circles and triangles) errorlessly if the animals
were transferred from a previously well-established size discrimination. These and other experiments suggest that, in a wide
variety of species using a wide variety of stimuli, interdimensional
training frequently results in less errors than intradimensional
training probably because of the greater differences between
stimulus configurations in the interdimensional condition (see
Castillo and Pinto-Hamuy, 1978, for a recent review of the use of
fading procedures).

In, perhaps, the best controlled experiment on discriminative
echolocation of planometric shapes, Simmons and Vernon (1971)
found that enucleated big brown bats (Eptesicus fuscus) initially
could not be readily trained to distinguish between different
shaped triangles (made of polystyrene) having the same surface
area. However, when the experimenters presented the bats with
two triangular targets differing both in size and shape, the
discrimination was acquired rapidly. Thereafter, the size dimension
was gradually reduced so that eventually only differences in
triangular shape controlled responding by the bats.

In a dolphin counterpart to the bat echolocation experiment
on fading from a size discrimination to a shape discrimination,
Barta (unpublished ms.) and Evans (personal communication) used
the type of fading technique first described by Schusterman
(1965; 1966). After the successful discrimination of planometric
circular targets (made of neoprene cemented to aluminum) differing
in size, a blindfolded Tursiops truncatus readily distinguished
between equal-size triangles and circles following the fading or
gradual elimination of the size dimension.

Several studies on echo ranging in bottlenose dolphins have
shown that thresholds were considerably lowered by presenting

progressively more difficult detections or discriminations, i.e.,
by incorporating fading procedures. Lekomtsev and Titov (1974)
report that the detection of a steel sphere 1.1 cm in diameter
occurred at a significantly greater range if the separation
screen was advanced by successively smaller steps (0.5 m) than if
it was advanced by successively large steps (1.0 m). Murchison
(1979) showed that in a range resolution study daily sessions of
small incremental changes of distance in the near and far targets
(ΔR) had to be preceded by large ΔR trials and extensive fading
of the ΔR in order to maintain stable responding and the obtaining
of reliable estimates of ΔR thresholds.

Probe Trials

In virtually all previously described studies, the behavioral
methodologies allowed for interpretation of the dolphin's echolocation capabilities primarily in terms of the systematic covariation
between values of target parameters (i.e., size, material, shape,
etc.) and performance. These methods are unsuitable for identifying
the manner in which dolphins classify target or echo parameters.
In animal psychophysics, many investigators depend on generalization
tests or test of response equivalence to determine the dimensions
along which some sensory quality is represented in a given species
or population. One variation on the generalization test which
has been used in echolocation experiments is to establish an
extremely stable discrimination between two different targets or
sets of targets and then, interspersed with trials in which
targets from the original or baseline discrimination are presented,
the investigator presents novel targets on the "probe" or test
trials.

Recent research by Hammer (1978) and Au (Au and Hammer, this
volume) serves as a fine illustration of the probe-trial technique.
These investigators applied the two-alternative forced-choice procedure with successive target presentation to the study of discriminative echolocation in which the object of the study was to have the
porpoise classify echo complexes in terms of response equivalence.
An adult male Tursiops, who had been well-trained on a range detection task, was presented with two sets of different-sized cylindrical targets. One set of cylinders (A) controlled responses to the
left response manipulandum (A) and the other set of cylinders (B)
controlled responses to the right response manipulandum (B) (see
Figure 3). In one very interesting experiment, the porpoise was
first trained to differentiate between hollow aluminum cylinders and
solid coral rock. The animal was then presented with probe trials
of different-sized hollow cylinders of glass, bronze and steel and
the probability of trials on which the porpoise gave the aluminum
response was calculated. On the basis of these ranked probabilities,
Hammer and Au found that the porpoise classified glass most like
aluminum and steel least like aluminum, regardless of target

size. The results of synthesized broadband dolphin-like signals
bounced off the same targets and subjected to a matched-filter
analysis indicated a relatively close resemblance to the dolphin's
performance on the task.

Psychophysical Procedures

    Once maximal stimulus control is attained, then an estimate
of the animal's absolute or differential sensitivity is sought.
Regardless of the signal or noise parameters involved, the threshold
is usually taken as some interpolated stimulus value associated
with some arbitrary criterion of detectability or discriminability.
Most threshold estimates are arrived at by one of several graphic
methods in which a broad range of target or noise values yield a
broad range of correct choices extending from chance to perfect
performance.

    As previously noted, fading procedures are usually incorporated
in the application of psychophysical methods to the study of
echolocation parameters. Frequently, the dolphin is initially
confronted with detections of discriminations with high signal/noise
ratios followed by decreasingly lower signal/noise ratios. Thus,
in general, the most widely used traditional psychophysical
procedure for studying the echolocation abilities of dolphins has
been some variation of the descending "method of limits". In
order to maintain reliable performance, investigators have frequently
used blocks of several trials prior to decreasing the signal/noise
ratio.

    Titration methods have been used on occasion in order to
maintain tight stimulus control in an efficient manner. In
titrating (also known as "stair-case" or "tracking"), the signal/
noise ratio is decreased (either in terms of single trials or in
blocks of trials) until the animal makes an error or reaches some
arbitrary error criterion; then the signal/noise ratio is stepped
up. Thus, in titration methods, the signal/noise ratio is either
decreased or increased, depending on the animal's performance.
Murchison (1979) incorporated a titration procedure as a warm-up
technique in his study of range resolution in Tursiops truncatus
and Barta (unpublished ms.) used a titration procedure to obtain
a size discrimination threshold in the same species.

    In the psychophysical method of constant stimuli, a set of
stimulus values occurs in random order, either in terms of single
trials or in blocks or trials. Interestingly, Barta (unpublished
ms.), in an attempt to obtain a size discrimination threshold,
began with the method of constant stimuli by using comparison
target values considerably above threshold and when an asymptotic
level of performance had not been attained, he shifted to a

titration method. A recent exception to this trend of using the descending method of limits in psychophysical studies of dolphin echolocation performance is illustrated by an experiment by Penner and Kadane (this volume) on the effects of noise on the detection of a 7.65 cm steel water-filled sphere at a distance of 16.5 m. These researchers, in using a modified method of constant stimuli, randomly presented five levels of white noise (ranging from 67 to 87 dB re 1 μPa) in 10-trial blocks and generated a very predictable relationship between noise levels and detectability in two experienced Atlantic bottlenose dolphins.

PROCEDURAL PROBLEMS

I have already focused in on some fairly important problems related to the behavioral methods used in the study of echolocation in marine mammals. These have included the desirability of explicitly controlling in precise measurable ways various aspects of observing responses, indicator or terminal responses and ways of measuring what the animal "sees" when it is stimulated with suprathreshold echoes. There are other related problems that I would like to deal with in my concluding remarks, and these include learning, response bias, and the subtle difference between the recognition or discrimination of targets and the detection of targets.

Baseline Behavior

There have been several measurements of thresholds related to such target parameters as distance, size, resolution, wall thickness, etc. which have resulted in considerably detailed data on the echolocation capabilities of several species of porpoises (see the papers by Moore, Murchison and Nachtigall in this chapter). In all the psychophysical experiments reviewed, it usually takes months before stable asymptote or baseline is achieved. Several psychophysical studies have reported learning effects over a relatively long period of time (e.g. see Turner and Norris, 1966, and Lekmotsev and Titov, 1974).

The continued learning effect on the echolocation capability of a bottlenose dolphin is clearly illustrated in an experiment reported by Lekmotsev and Titov (1974). Following several days of training, the dolphin could reliably detect a lead sphere 5.0 cm in diameter at a distance of 11 m. After two additional months of training, the dolphin reliably detected an 0.8 cm sphere of the same material at the same distance, a greater than fourfold increase in detection acuity.

In perceptual experiments of echolocation in which generaliza-

tion techniques are used, it is extremely important, as Hammer
(1978) has pointed out, that probe trials be "superimposed" on
unchanging baseline behavior. It is meaningless to attempt to
interpret responses to novel targets either in terms of sound
emission parameters or echo return parameters if response to the
original training targets are no longer reliable.

Detection, Recognition and Discrimination

In the literature of animal sonar systems, sometimes the
terms "detection", "discrimination", and "recognition" are used
interchangeably. However, traditional, as well as operational,
use of these concepts in the psychophysical literature suggests a
distinction between these concepts (Galanter, 1962). Classically
speaking, detection (in echolocation tasks) should refer to the
problem of how much echo strength is necessary in order for an
animal to reliably differentiate the target or signal from the
ambient noise. Recognition and discrimination, on the other
hand, assume detectability and refer to the question of how
distinctive or how similar echoes emanating from two or more
targets are from each other. Even a cursory review of the literature
leads me to believe that relatively few investigators working on
animal sonar systems have realized that behavioral methods,
particularly in terms of a whole host of variables dealing with
the animal's previous training experience, critically determine
the way in which an animal perceives the common as well as the
distinctive or salient features of different target-reflected
echoes.

In an experiment on shape discrimination in bats (Myotis
oxygnathus), the animals were first food reinforced for approaching
squares and not triangles (presented simultaneously) of equal
surface area (Konstantinov and Akhmarova, 1968). After learning
this discrimination to a high degree of accuracy, when the nonreinforced triangle was replaced with a circle, the probability of
the bats still approaching the square remained unchanged. Later,
when the square was presented simultaneously with both the circle
and the triangle, performance still remained relatively unchanged.
A similar experiment was conducted under water with T. truncatus
and the results were essentially the same, i.e, the introduction
of a new target presented simultaneously with the previously
reinforced target did not effect performance deleteriously (Barta,
unpublished ms.). These results might be expected if during or
following original learning it was primarily the presence of
distinctive echo characteristics from the positive or reinforced
target which controlled responding and not the distinctive echo
characteristics of the negative or nonreinforced targets or the
echo characteristics which both positive and negative targets had
in common. However, the results are surprising if it is understood

that during and following original training both positive and
negative targets were equally or nearly equally effective in
controlling discriminative performance (Mackintosh, 1974). The
question of what would have happened in these and other similar
experiments if the positive target were replaced and a novel
target was paired with the negative target seems never to really
have been considered. My guess is that performance would have
deteriorated. It would appear, therefore, that the animals in
these experiments were listening for distinctive echoes from the
positive target and that echoes from other targets acquired the
status of background or ambient noise similar to the situation
that prevails in detection experiments.

Perhaps the clearest example of a dolphin being trained
inadvertently to listen for distinctive echoes reflected from one
set of targets and not another comes from the work of Hammer
(1978). As previously mentioned, Hammer's Atlantic bottlenose
dolphin (Sven) had previously been trained in a series of detection
tasks to report the presence of a target to its left and the
absence of a target (ambient noise) to its right, and was then
trained by Hammer in a two-alternative forced-choice procedure to
report the presence of small and large hollow aluminum cylinders
(A targets) to its left (A response manipulandum) and to report
the presence of small and large solid coral rock (B targets) to
its right (B response manipulandum). After the porpoise had
perfected this baseline differentiation, Hammer conducted a
series of experiments in which he used probe trials to determine
what characteristics the probe targets had in common with the
baseline targets of rock and aluminum. In almost all instances,
probe targets—even hollow aluminum targets with a wall thickness
of only 0.16 cm difference from the standard hollow aluminum
targets—were classified as the standard solid rock targets,
i.e., as B targets or ambient noise. These results suggest that
most targets were classified as ambient noise because the dolphin,
Sven, was listening for a salient portion of the echo complex to
make its differentiation and that in the absence of the salient
or distinctive feature, Sven reported on the B manipulandum.
Since Sven's previous training probably played a crucial role in
the way the dolphin reported on the A and B manipulandum, I
suspect that if Hammer had trained Sven to report solid rock
cylinders on A and hollow aluminum cylinder on B, most of the
probe targets would have been classified as aluminum, or, in
other words, as ambient noise. Thus, Hammer's results and his
interpretations may have strongly depended simply on the way he
constructed the original or baseline reinforcement contingencies.

I tested the hypothesis that Sven made his classification on
the basis of a distinguishing feature or features emanating from
the A targets and on the basis of irrelevant features or ambient
noise emanating from the B targets in the following way. First,

I trained the dolphin to differentiate between successively
presented small and large hollow steel cylinders (A targets) and
small and large hollow bronze cylinders (B targets). The dolphin
perfected and continued this baseline behavior during each test
session for 56 of the 64 trials, but during the other eight
trials, "compound probe" trials were presented. On these compound
probe trials, instead of presenting a single A or B target, both
the large and small A targets were presented side by side approximately 4-8 cm apart. If Sven was listening for a specific echo
feature from a single A target (large or small) in order to
report A, this feature, according to my hypothesis, should be
blocked or in some way masked or negated by the presence of both
the large and small A targets and Sven should classify such
reflected echoes as irrelevant noise and report on the B manipulandum. If, on the other hand, Sven was indeed distinguishing
the A target features from the B target features, then reflected
echoes from two A targets should be classified as A and reported
as such. During eight compound probe trials with both A targets,
Sven classified all such presentations as B (p < .01, binomial
test); thus confirming the hypothesis. A control procedure in
which compound probes consisted of large and small B targets
(hollow bronze) resulted in 7 of 8 B reports (p < .05, binomial
test).

These experiments clearly demonstrate the importance of
considering both the previous training and the current behavioral
techniques in use in interpreting results from so-called "discriminative" echolocation tasks. Apparently the way echolocating
dolphins perceived and classify echo complexes from specified
targets depends, not only on previous experience, but also on the
manner in which these targets are presented in relation to the
specific responses required by the experimenter, i.e., the response-
contingent reinforcement procedures. It might very well be the
case that bilateral responding to different echo complexes emanating
from the same location in space (as is the case in the two-
alternative forced-choice procedure with successive presentations)
has a different effect on the processing of echo information by
dolphins than does bilateral responding to different echo complexes
emanating from different locations.

Response Bias

When sonar targets are presented successively in order to
determine a threshold, e.g., the maximum range at which a given
sized object is detected by a dolphin on 50 percent of the trials
(see Murchison, this volume), response bias may affect the threshold estimate. The bias can be manifested either as a position
preference in a forced choice procedure or as a preference for
responding or not responding in a go/no-go procedure. A bias in

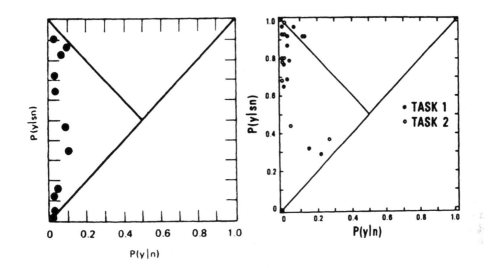

Figure 7.   Hits or P(y/sn) plotted against false alarms or P(y/n)
            for echolocating Amazon River porpoise in a size
            detection task (left panel) and two echolocating
            Atlantic bottlenose dolphins (T. truncatus) in a
            maximum detection range task (right panel).

favor of a "no" response would result in an increased threshold
while a bias in favor of a "yes" response would result in a
decreased estimated threshold. The theory of signal detectability
is a relatively recent approach which permits one to evaluate
independently the contributions to discriminative behavior of an
animal's sensitivity and its response bias (Swets, 1973). Schusterman (1974) has pointed out that in some echolocation experiments
biases introduced by training conditions and reinforcement contingencies have led to underestimates of sensitivity.

     Figure 7 presents conventional signal detection plots with
"hits" (y/sn) on the ordinate and "false alarms" (y/n) on the
abscissa. The data were originally plotted as ogives and involved
an Amazon River dolphin in a size detection task (Penner and
Murchison, 1970) and two Atlantic bottlenose dolphins performing
on two range detection tasks using different sonar targets (Murchison, this volume). The data points lie between the upper left-
hand corner (perfect detection) and the major diagonal (chance
detection). Points lying along the minor diagonal (the line

drawn from the upper left-hand corner to the major diagonal)
would represent "zero" bias since the two possible types of error
(false alarms and misses) would be equally probable. Both graphs
show that all three echolocating dolphins (regardless of the
task) have a strong and relatively consistent bias against making
false alarms. This is particularly true of the Amazon River
dolphin. The sensitivity index, called $\underline{d}'$, is a measure of
sensory effects uncontaminated by bias and represents the distance
of a data point from the major diagonal. The signal detection
arrangement of plotting hits in relation to false alarms is a
convenient way to display response bias and to estimate its
interaction with apparent sensitivity to stimulus events. A
careful inspection of the right panel of Figure 7 shows that the
smallest $\underline{d}'$ values also yielded the weakest bias against making
false alarms. This analysis suggests that for the two echo
ranging detection tasks by Tursiops there is an interaction
between response bias and apparent sensitivity to the strength of
the returning echo.

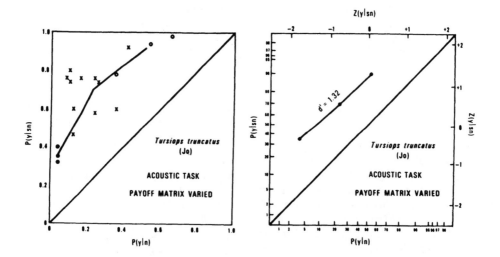

Figure 8. ROC functions for an Atlantic bottlenose dolphin
           (T. truncatus). The animal attempted to detect
           an 8-kHz signal varying between -9 to -11 dB re 1 μbar
           with ambient noise varying by as much as 15 dB. Each
           data point in the left plot is based on 100 trials
           The x's represent ratios of number of fish rewards for
           hits; correct rejections of 1:1. The open circles
           represent ratios of 4:1 and the solid circles
           represent ratios of 1:4. The right plot shows the
           same data plotted on Z-score coordinates.

Several recent studies with marine mammals have shown that in underwater passive hearing tasks with acoustical signals having a relatively low signal/noise ratio, variation in such nonsensory variables as signal probability, amount of reinforcement and probability of reinforcement, resulted in highly predictable changes in response bias without appreciably altering their ability to detect the acoustic signal (Schusterman, 1976; Schusterman, 1974; Schusterman and Johnson, 1975; Schusterman, Barrett and Moore, 1975). If signals of similar strength are repeatedly presented while the amount of reinforcement for "yes" and "no" responses is varied, then a series of data points along a "receiver operating characteristic" (ROC) or "isosensitivity" curve are produced. Figure 8 shows ROC curves for a dolphin. The curves were generated by holding signal strength constant (the ambient noise levels were relatively unstable) and varying the number of fish for hits and correct rejections. According to the theory of signal detectability, when the averaged data points in the left panel of Figure 8 are transformed into standard deviation units or Z-scores, they should yield an ROC curve parallel to the major diagonal as shown in the right panel of Figure 8 (Green and Swets, 1966).

In order to obtain a threshold estimate uncontaminated by response bias, one could extrapolate a $\underline{d}'$ value from a family of ROC curves where the curves cross the minor diagonal, i.e., at the point of zero bias. Unfortunately, ROC curves have not, as yet, been generated from dolphin performance on echolocation tasks. It seems that a signal detection approach would certainly further the understanding of signal processing in echolocating marine mammals by disentangling the effects of bias (e.g. motivation and attention) from psychophysical data dealing with their sonar receiver, i.e., the peripheral and central mechanisms of their auditory system.

ACKNOWLEDGEMENTS

There is not enough space to thank all those people who made my sabbatical (1977-78) at the Naval Ocean Systems Center, Kailua, Hawaii, so successful from almost every standpoint. However, I am especially indebted to Whitlow Au, Bob Floyd, Jim Fish, Earl Murchison, Paul Nachtigall, and Ralph Penner for their stimulating ideas and helpful criticisms which ultimately led to the writing of this paper. My productivity would have been nil without the help of Debbie Kersting in Hawaii and Patrick Moore at the California State University, Hayward. I acknowledge the financial support of ONR Contract N00014-77-C-0185, and I am forever grateful to those two purposeful porpoises - Sven and Ekahi.

REFERENCES

Aronsohn, S, Castillo, O., and Pinto-Hamuy, T., 1978, Fading pro-
    cedure effects on a visual pattern discrimination reversal
    in the albino rat, Anim. Learn. and Behav., 6:72.
Au, W. W. L., Floyd, R. W., and Haun, J. E., 1978, Propagation of
    Atlantic bottlenose dolphin echolocation signals, J. Acoust.
    Soc. Am., 64:411.
Au, W. W. L., and Hammer, C., 1979, Target recognition via echo-
    location by Tursiops truncatus, this volume.
Au, W. W. L., Schusterman, R. J., and Kersting, D. A., 1979,
    Sphere-cylinder discrimination via echolocation by Tursiops
    truncatus, this volume.
Ayrapet'yants, E. Sh., Golubkov, A. G., Yershova, I. V., Zhezherin,
    A. R., Zvorkin, V. N., and Korollev, V. I., 1969, Echo-
    location differentiation and characteristics of radiated
    pulses in dolphins, Report of the Academy of Science of the
    USSR, 188:1197 (English translation JPRS 49479).
Ayrapet'yants, E. Sh., Voronov, V. A., Ivanenko, Y. V., Ivanov,
    M. P., Ordovskii, D. L., Sergeev, B. F., and Chilingiris,
    V. I., 1973, The physiology of the sonar system of Black
    Sea dolphins, J. Evol. Biochem. Physiol., 9:364. (English
    translation JPRS 60298).
Bagdonas, A., Bel'kovich, V. M., and Krushinskaya, N. L., 1970,
    Interaction of analyzers in dolphins during discrimination
    of geometrical figures under water, J. Higher Neural Act.,
    20:1070 (English translation in: "A Collection of Trans-
    lations of Foreign Language Papers on the Subject of Bio-
    logical Sonar Systems", K. J. Diercks, 1974, ed., Applied
    Research Laboratories, U. of Texas, Austin, Tech. Rept.
    74-9).
Barta, R. E., 1969, Acoustical pattern discrimination by an At-
    lantic bottlenosed dolphin, unpublished manuscript, Naval
    Undersea Center, San Diego, Ca.
Bel'kovich, V. M., Borisov, V. I., Gurevich, V. S., 1970, Angular
    resolution by echolocation in Delphinus delphis, in: "Proc-
    eedings of Scientific Technical Conference, Ministry of
    Higher and Secondary Specialized Education RSFSR, Leningrad"
    (English translation in: "A Collection of Translations of
    Foreign Language Papers on the Subject of Biological Sonar
    Systems", K. J. Diercks, 1974, ed., Applied Research Lab-
    oratories, U. of Texas, Austin, Tech. Rept. 74-9).
Blough, D. and Blough, P., 1978, Animal psychophysics, in: "Hand-
    book of Operant Behavior", W. K. Honig and J. E. R. Staddon,
    eds., Prentice-Hall, Englewood Cliffs, N. J.
Busnel, R.-G., and Dziedzic, A., 1967, Résultats métrologiques
    expérimentaux de l'écholocation chez le Phocoena phocoena
    et leur comparaison avec ceux de certaines chauves-souris,
    in: "Animal Sonar Systems, Biology and Bionics", R.-G. Bus-
    nel, ed., Laboratoire de Physiologie Acoustique,

Jouy-en-Josas, France.

Busnel, R.-G., Dziedzic, A., and Andersen, S., 1965, Seuils de perception du système sonar du Marsouin Phocoena phocoena, en fonction du diamètre d'un obstacle filiforme, C. R. Acad. Sci., 260:295.

Dubrovskiy, N. A. and Krasnov, P. S., 1971, Discrimination of elastic spheres according to material and size by the bottlenose dolphin, Trudy Akusticheskogo Institute, 17, as cited in: Bel'kovich, V. M., and Dubrovskiy, N. A., 1976, "Sensory Bases of Cetacean Orientation", Nauka, Leningrad (English translation JPRS L/7157.).

Dubrovskiy, N. A., Krasnov, P. S., and Titov, A. A., 1971, On the question of the emission of ultrasonic ranging signals by the common porpoise, Akusticheskiy Zhurnal, 16:521 (English translation JPRS 52291).

Dziedzic, A., and Alcuri, G., 1979, Variations in characteristics of the pulsed emissions of a T. truncatus during the approach process and acoustic identification of different polygonal shapes, this volume.

Evans, W. E., and Powell, B. A., 1967, Discrimination of different metallic plates by an echolocating delphinid, in: "Animal Sonar Systems, Biology and Bionics", R.-G. Busnel, ed., Laboratoire de Physiologie Acoustique, Jouy-en-Josas, France.

Falk, J. L., 1961, Production of polydipsia in normal rats by an intermittent food schedule, Science, 133:195.

Falk, J. L., 1969, Conditions producing psychogenic polydipsia in animals, Ann. N. Y. Acad. Sci., 157:569.

Falk, J. L., 1971, The nature and determinants of adjunctive behavior, Physiol. and Behavior, 6:577.

Galanter, E., 1962, Contemporary psychophysics, in: "New Directions in Psychology", Holt, Rinehard, and Winston, New York.

Green, D. M., and Swets, J. A., 1966, "Signal Detection Theory and Psychophysics", John Wiley and Sons, New York.

Griffin, D. R., and Galambos, R., 1941, The sensory basis of obstacle avoidance by flying bats, J. Exp. Zool., 86:481.

Gurevich, V. S., 1969, Echolocation discrimination of geometric figures in the dolphin, Delphinus delphis, Moscow, Vestnik Moskovskoga Universiteta, Biologiya, Pochovedeniye, 3:109. (Translation JPRS 49281).

Hammer, C. E., 1978, Echo-recognition in the porpoise (Tursiops truncatus): An experimental analysis of salient target characteristics, Naval Ocean Systems Center, San Diego, Tech. Rep. 192.

Johnson, C. S., 1967, Discussion to paper by Evans and Powell, in: "Animal Sonar Systems, Biology and Bionics", R.-G. Busnel, ed., Laboratoire de Physiologie Acoustique, Jouy-en-Josas, France.

Johnson, C. S., 1967, Sound detection thresholds in marine mammals, in: "Marine Bio-Acoustics, Proc. of the Second Symposium on Marine Bio-Acoustics, New York", W. N. Tavolga, ed.,

Pergamon Press, New York.

Kellogg, W. N., 1953, Ultrasonic hearing in the porpoise, J. Comp. Physiol. Psychol., 46:446.

Kellogg, W. N., 1960, Auditory scanning in the dolphin, Psych. Rec., 10:25.

Kellogg, W. N., and Kohler, R., Responses of the porpoise to ultrasonic frequencies, Science, 116:250.

Konstantinov, A. I., and Akhmarova, N. I., 1968, Object discrimination with the aid of echolocation by bats (Myotis oxygnathus), Nauchnye Doklady Vysshei Shkoly, Biol. Naak., 11:22.

Lekomtsev, V. M., and Titov, A. A., 1974, Procedures for studying the echolocation apparatus of dolphins, Bionika, 8:83, (Translation JPRS 63492).

Mackintosh, N. J., 1974, "The Psychology of Learning", Academic Press, New York.

Murchison, A. E., 1979, "Maximum Detection Range and Range Resolution in Echolocating Tursiops truncatus (Montague), Ph. D. Dissertation, U. of Calif. at Santa Cruz, Ca.

Nachtigall, P. E., Murchison, A. E., and Au W. W. L., Cylinder and cube shape discrimination by an echolocating blind-folded bottlenose dolphin, this volume.

Penner, R. H., 1977, Paired simultaneous echo ranging by Tursiops truncatus, Proc. (Abstracts) 2nd Conf. on the Bio. of Mar. Mammals, 2:38.

Penner, R. H., and Kadane, J., 1979, Tursiops biosonar detection in noise, this volume.

Penner, R. H., and Murshison, A. E., 1970, Experimentally demonstrated echolocation in the Amazon river porpoise Inia geoffrensis (Blainville), Proc. 7th Ann. Conf. Bio. Sonar and Diving Mammals, 7:17.

Schusterman, R. J., 1965, Errorless reversal learning in a California sea lion, Proc. A. Conf. Am. Psychol. Ass., 73(1): 139.

Schusterman, R. J., 1967, Perception and determinants of underwater vocalization in the California sea lion, in: "Animal Sonar Systems, Biology and Bionics", R.-G. Busnel, ed., Laboratoire de Physiologie Acoustique, Jouy-en-Josas, France.

Schusterman, R. J., 1974, Low false-alarm rates in signal detection by marine mammals, J. Acoust. Soc. Am., 55:845.

Schustermnan R. J., 1976, California sea lion auditory detection and variation of reinforcement schedules, J. Acoust Soc. Am., 59:997.

Schusterman, R. J., Barrett, B., and Moore, P. W. B., 1975, Detection of underwater signals by a California sea lion and a bottlenose porpoise: variation in the payoff matrix, J. Acoust. Soc. Am., 57:1526.

Schusterman, R. J., Johnson, B. W., 1975, Signal probability and response bias in California sea lions, Psych. Rec., 25:39.

Schusterman, R. J., and Kersting, D., 1978, Selective attention in discriminative echolocation by the porpoise (Tursiops trun-

catus), Paper read at the Animal Behavior Society Annual
  Meeting, June 19-23, U. of Wash., Seattle.

Schusterman, R. J., Kersting, D. A., and Au, W. W. L., 1979,
  Stimulus control of echolocation pulses in Tursiops truncatus, this volume.

Scronce, B. L., and Ridgway, S. H., 1979, Grey seal, Halichoerus
  grypus: echolocation not demonstrated, this volume.

Simmons, J. A., and Vernon, J. A., 1971, Echolocation: discrimination of targets by the bat, Eptesicus fuscus, J. Exp. Zool.,
  176:315.

Skinner, B. F., 1961, "Cumulative Record", Appleton-Century-Crofts,
  New York.

Staddon, J. E. R., 1978, Schedule-induced behavior, in: "Handbook
  of Operant Behaviour", W. K. Honig and J. E. R. Staddon, eds.,
  Prentice-Hall, Englewood Cliffs, N. J.

Staddon, J. F. R., and Simmelhag, V. L., 1971, The "superstition"
  experiment: A re-examination of its implications for the
  principles of adaptive behavior, Psychol. Rev., 78:3.

Stebbins, W. C. (ed.), 1970, "Animal Psychophysics", Prentice-Hall,
  Englewood, N. J.

Supin, A. Ya., Sukhoruchenko, M. N., 1970, The determination of
  auditory threshold on Phocoena phocoena by the method of
  skin-galvanic reaction (Russian), Trudy Akousticheskogi
  Inst., 12:194.

Suthers, R. A., 1965, Acoustic orientation by fish-catching bats,
  J. Exp. Zool., 158:319.

Swets, J. A., 1973, The relative operating characteristic in
  psychology, Science, 182:990.

Titov, A. A., 1975, Recognition of spherical targets by the
  bottlenose dolphin in the presence of sonic interference,
  in: "Marine Mammals, Proceedings of the Sixth All-Union
  Conference on the Study of Marine Mammals", G. B. Agarkov,
  ed., Naukova Dumka, Kiev (English translation JPRS L/6049).

Turner, R. N., 1964, Methodological problems in the study of
  behavior, in: "Marine Bio-Acoustics", W. N. Tavolga, ed.,
  Pergamon Press, Oxford.

Turner, R. N., and Norris, K. S., 1966, Discriminative echolocation
  in a porpoise, J. of the Exp. Anal. of Beh., 9:535.

DETECTION RANGE AND RANGE RESOLUTION OF ECHOLOCATING BOTTLENOSE PORPOISE (TURSIOPS TRUNCATUS)

A. Earl Murchison

Naval Ocean Systems Center

Kailua, Hawaii   96734

MAXIMUM DETECTION RANGE

Most of the quantifiable information available on object detection by echolocating odontocetes has resulted from studies conducted in tanks, and only a few of them have had maximum range of detection as their primary goal (Evans, 1973). The results of Zaslavskiy et al. (1969) as reported by Ayrapet'yants and Konstantinov (1974), indicated detection ranges (90% correct response) of 11 m for a metal cylinder (75 mm in height and diameter), 7.3 m for a plastic cylinder (115 mm high, diameter not given) and 6.8 m for a wooden cylinder (115 mm high, diameter not given). The subject of this experiment was a Tursiops truncatus. No other information was given.

Morozov et al. (1972), in their study of the pulse repetition rate of Tursiops truncatus as a function of food object distance, reported a maximum "starting position" of 30 m for approach and consumption of horse mackeral (species not given) of "about 20 cm in size." Although their assumption of detection by echolocation of the food fish for 100% of the only 33 runs at 30 m by the animal may have been justified, their vaguely defined target strength, and small number of trials makes their results only prefigurative as they relate to maximum echolocation range of Tursiops truncatus.

Titov (1972 - in Bel'kovich and Dubrovskiy, 1976) reported 50-mm spherical target detection by a bottlenose porpoise. The maximum detection ranges given were 5.9 m for rubber targets, 7.8 m for wax targets, 11.1 m for lead targets, and 12.4 m for steel targets.

Bel'kovich and Dubrovskiy (1976) also presented the bottlenose porpoise maximum detection range results reported by Babkin
and Dubrovskiy (1971). This experiment determined the maximum
detection ranges of several sizes (diameters from 4.0 mm, to 12.3
mm) of steel (presumably solid) spheres and lead spheres (diameters from 3.0 mm, to 12.7 mm) by two Tursiops truncatus. The
maximum detection range for the smallest steel sphere was 3.5 m,
and for the largest it was 7.5 m. The maximum detection range
for the smallest lead sphere was 3.0 m, and 9.0 m for the largest.
Bel'kovich and Dubrovskiy (1976) reported that the 75% correct
detection level of all of the steel spheres was near "5' [minutes
of arc] angular dimension." This refers to the expression:

$$\theta = \frac{d}{R}$$

where θ is the angular dimension (in radians), d = sphere diameter,
and R = maximum detection range. They reported 2.5' for lead
spheres.

The Babkin and Dubrovskiy (1971) study also reported (according to Bel'kovich and Dubrovskiy, 1976) a 9.8 m maximum detection
range by a bottlenose porpoise for a mackerel (species not given)
13.5 +1 cm long, 3.8 +0.5 cm wide and 1.8 +0.3 cm thick with the
swim bladder removed. They reported that the porpoise-mackerel
aspect relationship was very important with the maximum detection
range being attained when the fish was presented "lateral" to the
porpoise. For multiple fish targets, Bel'kovich and Dubrovskiy
applied an echo magnitude increase of $\sqrt{n}$ (n = number of fish) to
predict a 100 m maximum detection range for a school of 4000
mackerel, 19 cm long, by Tursiops truncatus. Assuming a live
mackerel could be detected from 12 m, and utilizing the same
school echo magnitude ($\sqrt{n}$) estimate, Ayrapet'yants and Konstantinov (1974) predicted a range of 350 meters for a porpoise's
detection of a 4000-fish school.

Titov (1972 - as reported in Bel'kovich and Dubrovskiy,
1976) conducted experiments to determine the effect of reverberation resulting from other objects near a target on maximum
detection range by a Tursiops truncatus. Smooth round rocks
between 5 and 30 mm in diameter were strewn on the bottom of the
test tank in circles 40 cm in diameter. The targets to be
detected were presented at the center of these rock circles, on
the tank bottom, amid the rocks, as well as various distances
above the rock circles. The 75% correct response detection range
for a lead sphere (50 mm) and a steel sphere (33 mm) was reported
to be "over 11 m" with the targets "30 cm beneath the surface"
(depth of tank not given), but the porpoise could detect both
targets "5 m away only if they extended at least 2.0 cm above the
largest stones." More complete results were not given. However,

the wording suggests that the porpoise could <u>not</u> detect the target on the bottom, among the stones.

The only other report in the literature giving any information on maximum echolocation detection range of <u>Tursiops</u> is a field observation. Eberhardt and Evans (1962) and Evans and Dreher (1962) observed five Pacific bottlenose porpoises behaving in a manner that strongly suggested their detection, from approximately 366 m, of 15 air-filled aluminum tubes, 5.1 cm in diameter and 4.6 m long placed vertically 3.7 m deep at 15.2 m intervals.

Detection Range Experiments in Open Waters

Two male <u>Tursiops truncatus</u> (Sven and Niko) were the subjects of three maximum detection range experiments conducted in a floating porpoise pen (Steele, 1971) at the Naval Ocean Systems Center's (NOSC) facility, at Kaneohe Bay, Hawaii. Two of these psychophysical experiments were designed to determine the maximum detection range of two different spherical targets, and the third was designed to determine the effects of target depth (or nearness to bottom) on maximum detection range. The rectangular, wire-mesh, floating pen (Figure 1) had an aperture (2 m x 1.2 m)

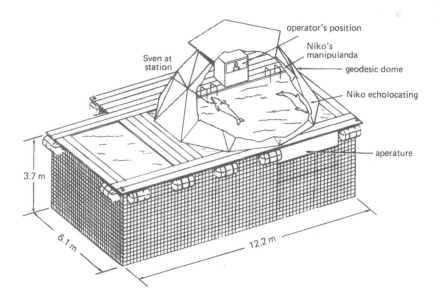

Figure 1. Floating pen used in the maximum detection range experiments with cutaway view through the geodesic dome.

that was opened during echolocation trials to afford the echolocating porpoises an unobstructed "view" of the target range. The
targets were presented at various distances from the pen aperture
by means of a "skyhook" cable and pully arrangement between two
metal poles 158 m apart (Figure 2). The mean water depth along
the range was 6.3 m. Mean water visibility (Secchi disk) was 3.2
m. The bottom was covered with soft sandsilt up to 2 m deep along
the range. The reef slope at the end of the range consists of
coral rubble, live coral and benthic algae. The measured mean
sound velocity was 1542 m/sec and the transmission loss was 0.044
dB/m (Au et al., 1974).

Two detection versus distance ogives were determined utilizing
a 2.54-cm diameter solid steel sphere (target strength (TS) = -41.6
dB) and a 7.62-cm diameter water-filled stainless steel sphere
(TS = -28.3 dB). A third target, used to determine the effects of
depth or nearness to the bottom on range of detection, was a 6.35-
cm diameter solid steel sphere (TS = -34.8 dB). Target strengths
were determined by the method of Au and Hammer (1978).

Each animal was conditioned to hold position at a station
area near the side of the pen opposite the pen's aperture; each
station had two response manipulanda. Sixty-trial sessions (30
trials/animal) were conducted, and target-present and target-
absent trials were randomly presented. Trials were alternated
between animals during sessions. While one animal's trial was
being conducted, the other animal remained at its station. During
the first two experiments targets were presented at 1.22 ± 0.15 m
depth at all distances. One or two sessions were conducted per
day, preceded by warm-up trials. Details of the operant conditioning
techniques used are presented by Murchison (1979).

In the third experiment (6.35-cm solid steel sphere), the
same procedure was repeated. Sessions were conducted with this

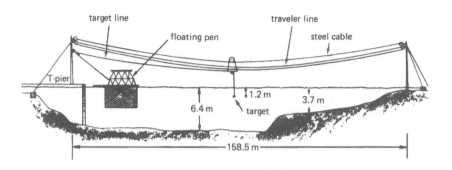

Figure 2.   Skyhook detection range (not to scale).

target suspended at depths of 1.22 ± 0.15 m, 2.43 ± 0.15 m, 3.66
± 0.15 m, 4.88 ± 0.15 m, and on the bottom 6.3 ± 0.76 m. Separate
detection distance ogives were determined for each of the 5
target depth increments.

    The mean hit rates for each distance increment and target
size, computed by the method of Woodworth and Schlosberg (1954),
is presented in Figure 3 as a composite curve for both animals.
Utilizing the same method, mean hit rates were plotted as a
function of distance and target depth for the third experiment.
The distances at which the 0.50 probabilities of detection occurred
for each depth is shown in Figure 4, as a function of depth,
distance and bottom contour.

    <u>Noise Limited or Reverberation Limited?</u>   Although the
echolocation systems of both bats and odontocetes have evolved to
reduce the effects of noise (Grinnell, 1967; Evans, 1967; Bel'ko-

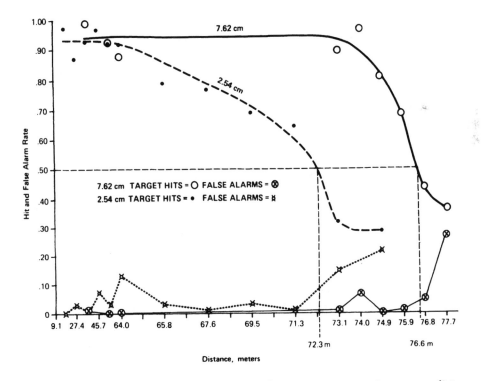

Figure 3.   Composite (both <u>Tursiops</u>) detection performance (hit
            rate) as a function of distance for the 2.54-cm
            and 7.62-cm targets.

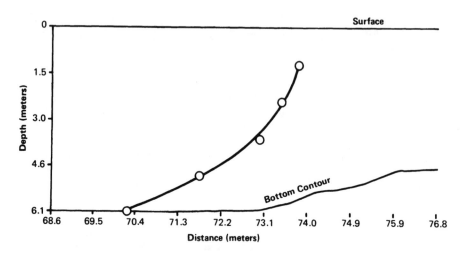

Figure 4.   Detection thresholds (0.50 hit rate) for the 6.35-cm
            target for each depth increment tested as a function
            of distance and bottom contour.

vich and Dubrovskiy, 1976) on their detection and discrimination
of objects, they have not completely eliminated it.

   The two noise dimensions important to odontocetes are the
ambient noise and reverberation. The echo level from a sonar
signal decreases with target range but the ambient noise is
independent of target range. On the other hand, reverberation
level varies as a function of complex interactions of target
range, oceanographic conditions, and objects and boundaries
(surface, bottom, reefs) (Urick, 1975). The _Tursiops_ used in
these experiments produced echolocation pulses with maximum
energy between 120 and 130 kHz. This is much higher in frequency
than previously reported for _Tursiops_ and may have been an
adjustment by the animals that optimized the signal-to-noise
ratio in the ambient noise environment of Kaneohe Bay (Au et al.,
1974).

   Reverberation from the surface, bottom, or water volume can
mask, distort or otherwise interfere with target echoes. Additionally, echoes from objects other than the target may be distinctly
separable by an echolocating porpoise, yet affect target detection
by increasing the difficulty of recognition. That is, detection
becomes confounded by discrimination (see Schusterman, this
volume).

   The detection threshold (DT) of an active sonar occurs when
the returning echo level equals the masking level of the noise _or_
reverberation in the environment. For the noise-limited case,
the appropriate sonar equation is (Urick 1975):

$$SL - 2TL + TS = NL - DI + DT$$
$$\text{(Echo Level)} = \text{(Noise Masking Level)}$$

Assuming that during the skyhook experiments the porpoises maintained a constant DT and that the source levels (SL) of their pulses, their receiver directivity index (DI) and the mean ambient noise level (NL) were the same during the sessions with the 2.54 cm and 7.62 cm targets, the only parameters of the equation different for the two targets were transmission loss (TL) and target strength (TS), if the porpoises were noise limited. Transmission loss involves range-dependent losses due to spherical spreading of acoustic waves and to absorption (TL = 20 log R + αR; R = range, α = absorption coefficient) (Urick, 1975). Au et al. (1974) measured the TL from absorption for simulated porpoise pulses with peak energy at 120 kHz in the Kaneohe Bay test area as 0.044 dB/meter. Therefore, for the 2.54-cm target, two-way transmission loss was 80.7 dB when the target was at the 0.50 hit rate range. The target strength of the 7.62-cm sphere was 13.3 dB greater than that of the 2.54-cm sphere and, therefore, if the detection level is the same for both targets, the 7.62-cm target should have a 121 m detection range. That is 44.4 m farther than the mean detection range actually attained during the 7.62-cm target sessions (Figure 3). Because the noise-limited form of the sonar equation does <u>not</u> predict the porpoises' maximum detection range for the 7.62-cm target based on their performance with the 2.54-cm target, it is likely that the porpoises' detection of the larger target was limited by reverberation rather than ambient noise.

It is not surprising that the porpoises' biosonar systems were probably reverberation limited at their maximum, 7.62-cm sphere, detection range. Porpoise hearing is highly directional (Bullock et al., 1968; Ayrapet'yants et al., 1969; Ayrapet'yants and Konstantinov, 1974) and resistant to masking by noise from directions outside of their cone of best reception (Morozov et al., 1975). Their biosonar pulses are also very directional (Norris et al., 1961; Norris and Evans, 1967; Evans et al., 1964; Romanenko et al., 1965; Zaslavskiy, 1974 – according to Bel'kovich and Dubrovskiy, 1976; Au et al., 1978; Au, this volume). Thus, the greatest amount of potential masking noise will be created by backscatter from objects and surfaces that are in the animal's cone of inspection.

The effects of reverberation from the reef at the end of the skyhook range is further indicated by calculating the predicted detection range of the 6.35-cm target. Based on results from the 1.2 m (depth) sessions only, the detection range of the 2.54-cm target predicts a noise limited 0.50 hit rate for 6.45-cm target at 95 m. However, the two animals' mean detection range

was 21 m less than that (Figure 4). Thus, based on the animals'
detection range for the 2.54-cm target, the TL + TS model predicts
maximum detection ranges that are 22% greater for the 6.35-cm
target and 40% greater for the 7.62-cm target than were attained.
The degradation of detection ability with range suggests that the
closer the targets were to the reef, the greater were the possible
effects of reverberation. However, it is not possible to determine to what extent reverberation from the reef was affecting the
animals' maximum detection ranges for the 2.54-cm target or even
whether its detection range was reverberation limited at all. It
may be that reverberation limiting occurred only at greater
ranges, when the targets were closer to the reef.

Odontocetes may have "two functionally (and possibly to some
extent morphologically) distinct subsystems" of hearing, one
functional for the perception of low frequencies and the other
for the perception of higher frequencies (Vel'min and Dubrovskiy,
1975; 1976; Bel'lovich and Dubrovskiy, 1976). Vel'min and
Dubrovskiy propose that the lower frequency "passive" subsystem
is important for emotional and/or communicative sound perception,
and the other, "active hearing", subsystem is especially adapted
for echolocation. An important function of this "active hearing"
subsystem may be the reduction of the effects of reverberation
(Vel'min, 1975; Vel'min and Dubrovskiy, 1975; 1976; Bel'kovich
and Dubrovskiy, 1976). Giro and Dubrovskiy (1972 - according to
Vel'min, 1975) proposed an odontocete reverberation-limiting
mechanism of temporary gating or strobing of hearing sensitivity
whereby the auditory system is blocked during periods of potential
reverberation and operates with increased sensitivity during echo
reception. This mechanism implies a timing capability for the
interval between pulse emission and the "appearance of the
strobe" (Vel'min, 1975). Webster (1967) proposed similar predictive time gates for bats, describing them as "the anticipated
interval(s) of echo reception". Norris (Norris and Evans, 1967)
also mentioned the possibility of gating in a Tursiops.

Vel'min and Dubrovskiy (1975; 1976) reported the results of
several experiments and concluded there is an interval within
which odontocetes fuse "the indiviudal pulses comprising an echo
into the complete acoustic shape of an object." They conclude
that 300 microsec echo interpulse interval is the time that
determines the maximum size of an object perceived by porpoises
as an "integral echoranging object". These experiments are also
discussed by Vel'min and Titov (1975), Vel'min (1975) and Bel'kovich and Dubrovskiy (1976).

Variable Target Depth Experiment. The results of the
variable target depth experiment, as summarized in Figure 4, also
show the effects of reverberation. The closer the target was to

the bottom, the shorter the animals' maximum detection range was. This strongly suggests that the reef and bottom were a source of detection-limiting reverberation. It should be noted that although the maximum range of target detection was reduced when the target was on the bottom, the porpoises still detected it at a mean distance of 70.2 m. This is considerably greater than would be supposed from the results of Titov's (1972) bottom-clutter experiment. However, in his experiment the bottom composition was quite different from the skyhook range bottom. Reverberation level in any location is influenced by complex physical, environmental and acoustical factors and predicting it for any particular area or time is very difficult (Urick, 1975).

There was considerable temporal variability in the detection ranges of Sven and Niko. It was suspected that this was due, in part, to factors that resulted in fluctuating reverberation levels (e.g., fish, entrained gas bubbles, volume reverberation) in any area, particularly shallow, inshore areas. Future psycho-physical studies conducted in a field setting such as the one described here, should attempt to quantify possible environmental effects on echolocation and search for convarience between them and the porpoises' detection and/or discrimination performance.

If these environmentally influenced differences and fluctuations in biosonar capabilities also occur in the porpoises' natural habitats, the accuracy of the predictions made here (and those made by Babkin and Dubrovskiy, 1971, and Ayrapet'yants and Konstantinov, 1974) of odontocete detection ranges in those habitats will vary considerably with time and place. Ecologically, this means that the availability of prey species might not be the only factor affecting the hunting success of porpoises and that bioacoustic parameters are also important. Just as bats avoid flying in fog (Pye, 1971; Sales and Pye, 1974), porpoises may avoid areas during high bubble entrainment or that have other conditions adverse to their biosonar capabilities.

With these reservations concerning differences and fluctuations in detection capabilities and assuming that the detection range for the 2.54-cm target was not reverberation limited, rough predictions of detection ranges of fish by Tursiops can be made from these experiments. Using Love's (1971) formula for the side aspect target strength of a 19-cm fish, at the pulse frequencies produced by Sven and Niko, a detection range of 98 meters is predicted for a single fish and 292 m for a school of 4000 fish. This is almost three times the maximum detection range predicted by Bel'kovich and Dubrovskiy (1976) for the same school and 17% less than that predicted by Ayrapet'yants and Konstantinov (1974), for a 25-30 kHz biosonar pulse. The net-detection capabilities of Sven and Niko predicted from their 2.54-cm target detection

performance and the net-target-strength measurements given by
Leatherwood et al. (1977), produce considerable variability in
the detection ranges predicted for different nets. The extreme
examples are the ranges of 130 m predicted for a 11.4-cm, twisted
net (TS = -26dB at 120 kHz) and 310 m for a 5.1-cm knotless net
joined in the middle to a 2-inch (5.1-cm) braided net (TS = 4 dB
at 120 kHz). Little wonder that Norris and Brocato had such a
difficult time catching Tursiops (Norris, 1974). Although it may
be safe to assume that Niko and Sven might detect such nets at at
least these ranges in quieter, less reverberant, open ocean
waters, the detection ranges of the species (Stenella attenuata,
S. longirostris, Delphinus delphis) of concern to Leatherwood et
al. may have different detection ranges. However, if one net
were predicted as more than twice as detectable as another for
the two Tursiops used in the present experiments, similar ratios
might be expected for other odontocetes.

Performance predictions based on the sonar equations should
be considered as only "best-guess time average(s)." (Urick,
1975). The detection ranges predicted above should be considered
as rough estimates of the potential detection ranges of the two
Tursiops under conditions similar to Kaneohe Bay, but possibly
with less bottom and reef reverberation. They do, in general,
support Ayrapet'yants' and Konstantinov's (1974) prediction and
statement that the delphinid biosonar system has detection ranges
not exceeding several hundred meters.

Bel'kovich and Dubrovskiy (1976) reported that the 75%
correct response detection level for Tursiops truncatus was 5' of
arc for steel spheres. This relationship predicts a 17.5 m
detection range for a 2.54-cm sphere such as used in the skyhook
experiments. Although the experimental design and methodology
underlying the relationship was not given by Bel'kovich and
Dubrovskiy, environmental (particularly reverberation) or other
test differences could account for my measured difference with
their prediction.

Target Range Inspection Behavior. Body movements and head-
tucking behaviors were observed during Sven and Niko's detection
trials and were probably related to their acoustic inspection of
the target range. These head and body movements may facilitate
internal scanning (Bel'kovich and Nestereno, 1972; Evans, 1973)
or may optimize echo reception (Norris, 1968). These behaviors
occurred more often when targets were at greater distances (and
closer to the reef), and may have helped reduce the effects of
reverberation, or aided detection at longer ranges.

The utility of head scanning movements observed in Tursiops
by a number of authors (Schevill and Lawrence, 1956; Kellogg,

1960; Norris, 1968, Norris et al., 1961) toward reducing the
effects of reverberation is indirectly indicated by an observation
by Evans (1973). He points out that Tursiops and Inia geoffrensis
(the other species reported to have head scanning movements like
Tursiops; Penner and Murchison, 1970) have much less fusion of
their cervical vertebrae than most pelagic delphinids. Both of
these species live in relatively shallow, inshore or riverine
environments where reverberation levels are high and multisourced.
Marine mammals' echolocation behaviors in open deep water may be
quite different than in "tidal flats and muddy bays" (Norris,
1968). It would be interesting to learn if the other odontocete
species that live in reverberant environments exhibit more distinctive posturings or elaborate scanning movements than pelagic
species. The experience of Norris (1968) with Delphinus delphis,
Lagenorhynchus obliquidens, and Tursiops truncatus gilli that were
unable to echolocate with sufficient efficiency to keep from
hitting tank walls or find objects when they were first placed in
reverberant tank environments, suggested that these pelagic
animals were ill equipped and/or had not learned to cope with high
reverberation levels. It is tempting to speculate that because of
their inshore ecology that Tursiops truncatus may have a biosonar
system that is among the better adapted relative to reverberation,
particularly bottom and volume reverberation. The role of learning
in scanning behaviors was illustrated by a gradual changes in the
scanning strategies used by Niko and Sven during the initial
training of their target detection and reporting behaviors.

Pulse Repetition Rate. A number of studies have reported
odontocete echolocation pulse repetition rates that increase with
decreasing animal-to-target range (Busnel and Dziedzic, 1966;
Evans and Powell, 1967; Norris et al., 1967; Johnson, 1967;
Morozov et al., 1972). However, the pulse repetition rate does
not always increase in the systematic fashion that calculations of
animal-to-target sound travel time would predict (Norris, 1964;
Johnson, 1967). Thus, Norris (1964) suggested that repetition
rate may be, at least in part, related to the degree of discrimination desired by the porpoise at any particular time.

In their study of pulse repetition rate as a function of
target distance with five Tursiops truncatus, Morozov et al.
(1972) found that the interpulse interval varied between 20
milliseconds and 3 milliseconds longer than the time necessary for
the pulse to travel to the target and the echo to return. They
expressed their results by the formula:

$$T_o = (2L/c) + 3\text{-}20 \text{ milliseconds}$$

(T_o = interpulse interval; L = the distance from the target to the
animal; c = speed of sound in seawater). Au et al. (1974),

recording animals echolocating at greater ranges than the Morozov
et al. animals, also found that the interpulse interval was
always just greater than the calculated two-way transit time (30–
50 milliseconds). Morozov et al. concluded that the time difference between echo reception and production of the following pulse
represented processing time, and "that a dolphin during echo
ranging emits each successive pulse only _after_ (italics theirs)
reception of the echo signal from the preceeding pulse." Diercks
(1972) in a discussion of the Morozov et al. conclusions suggests
that pulse generation by porpoises may be triggered by feedback
from the animal's receiver system; however, he qualifies this
suggestion with the observation that test animals seem to have
some control of their repetition rate.

If, as Morozov et al. (1972) suggests, odontocete pulses are
produced only upon the reception of the echo from the previous
pulses, the interpulse interval would always be reflective of the
nearest object in the porpoise projecting-receiving field, not
necessarily the salient target object being tested. Their model
could be modified to include some sort of "neural template" (Simmons,
1969), or "spectral copy" (Mermoz, 1967), thus allowing optimal
reception of echoes from a particular target object while ignoring
echoes from other objects. Even so modified, the target-echo
pulse triggering model of Morozov et al. does not account for the
interpulse intervals recorded from the _Tursiops_ during the maximum
detection range experiments reported here. During the skyhook
experiments, the animals continued to produce pulses during trials
when no target was present in the test range and interpulse intervals during target-present trials and target-absent trials were
essentially the same (Au, this volume, Figure 9). The important
meaning of those target-absent trials' repetition rates and intertrial intervals is that the animals maintained intertrial intervals that allowed sufficient time for echoes to return from an
expected target, at an expected approximate range. If an echo was
not received 30–50 milliseconds beyond that interval, another
pulse was produced without waiting for an echo. The animals'
target-absent range expectancies in those experiments were set
(learned) by the perceived target range during target-present
trials of the same and recent detection sessions.

The important point here is that the interpulse interval for
an echolocating porpoise is not automatically controlled by the
distance of a target, but it is a measurable function of range of
attention as indicated by measurable interpulse intervals. The
relationships between echolocation signal parameters and target
dimensions may offer an ideal sensory vehicle with which to study
selective attention and expectancy in echolocating animals. The
Schusterman et al. (this volume) study illustrates that potential.

In a discussion of the "search image" concept of von Uexküll

(1934) and L. Tinbergen (1960), Hinde (1970) points out that such
a selective attention concept implies the existence of a mechanism
in the absence of an activating stimulus, yet the evidence for
such mechanisms has all been based on results obtained during the
presence of activating stimuli. The interpulse intervals maintained by Sven and Niko in the maximum range study, (Figure 9, Au,
this volume) during target-stimulus-absent trials, might be
considered to support the "search image" concept. The porpoises
had an expected or "searched-for" echo "image," at least along the
pulse-to-echo delay-time dimension they possibly perceived as
target-stimulus distance. This all assumes that interpulse intervals
are indeed a function of the distance of the target or area of
attention.

The adoption of the word "target" into biosonar terminology
may have brought a subtle bit of inappropriate conceptual simpli-
city with it because it implies that the experimenter knows the
one stimulus, among all the stimuli available, that the echolocating animal is attending to. This implies that the stimulus
defined as target by the experimenter is, at all times during the
experiment, defined as target by the echolocating animal. Analysis of repetition rate/target-to-animal distance relationships,
might be affected by these assumptions and implications. As
Schusterman (1968) has pointed out, echolocation cannot be understood as a phenomenon isolated from the developmental processes,
learning and memory characteristics, attention mechanisms, and
other sensory modalities of the echolocating animal. The only
thing that defines a biosonar target as such is the selective
attention of the echolocating animal.

If the interpulse intervals of odontocete echolocation
pulses are behaviorally controlled by the animal and indicative of
its area of attention or search (distance and/or "depth of field")
the variability reported for Tursiops truncatus pulse repetition
rates by Johnson (1967) may simply reflect momentary attention
shifts by that animal as it swam toward the target. If viewed in
this light, Johnson's statement that pulse repetition rate is not
used for range determination may need re-examination.

The variable relationships between pulse repetition rates and
target distances found by Johnson may have been the results of
momentary attention shifts as reflected in interpulse intervals.
This variability would have been confounded if there were variability in the time lags between echo reception and production of the
following pulse as reported by Morozov et al. (1972) and Au et
al. (1974 this volume).

Experiments designed to compare the pulse repetition rate of
relatively static porpoises in a search mode (i.e., the animals
know the target, its direction and depth, but not its distance)

with those of porpoises detecting targets at known distances might
clarify the relationship of repetition rate and detected object
distance.

RANGE RESOLUTION

Although Ayrapet'yants and Konstantinov (1974) present the
results of a number of Soviet experiments designed to determine
porpoises' abilities to discriminate target differences in azimuth
(horizontal separation), elevation (vertical separation), and
range, there is insufficient information available in the English
translation of their work to make a meaningful evaluation of these
results as they relate to the range resolution experiment reported
here. They report range resolution for spheres of various sizes
and materials that were presumably 5 to 7 cm apart, but the absolute
porpoise-target distances were not given. Range resolution results,
without absolute target range, are of little value and the discrimination of target separation in studies with targets so close
together could be based on differences in echo spectra of the
multiple target complex rather than the determination of range
differences between individually perceived targets.

There are two sets of competing hypotheses of echo-informa-
tion processing by bats in the determination of target distance.
The hypothesis that bats process their FM signal echoes in a
manner that allows a direct measure of echo travel time (Hartridge, 1920), led Strother (1961), McCue (1966), and others to
postulate a receiving system incorporating a neural equivalent of
a matched filter or template of the outgoing signal. The competing models of animal echolocation all propose some indirect
determination of distance by the perception of neural or acoustic
interactions between outgoing signals and echoes. Trains of brief
acoustic pulses produce the sensation of an apparent tone. The
frequency of this tone is determined by the temporal distance
between the pulses of the train (Thurlow and Small, 1955). Most
of the indirect biosonar models of distance determination are
based on some variation of this time-separation-pitch (TSP)
phenomenon, or frequency-difference beat tones. In general, these
models propose that the echolocating animal determines target
distance by perceiving these time-separation-pitch or frequency
beat tones resulting from the interaction of the echoes with the
outgoing signal, or its replica (Nordmark, 1961; Pye, 1961; Kay,
1962; Stewart, 1968; Johnson and Titlebaum, 1976). The Johnson
and Titlebaum model emphasizes spectrum analysis of the echolocation
pulse plus its echo as the proposed processing mechanism and is
related to other TSP models by their conclusion that TSP can be
explained by frequency domain, correlation-equivalent, spectrum
analysis; however, their model does not necessitate the assumption
that the animal perceive a pitch. Simmons (1969) proposed that

the autocorrelation function of a bat's echolocation signal approxi-
mates the crosscorrelation function of emitted and received signals
and can be used to predict the animal's abilities to discriminate
distances between objects (i.e., range resolution). The results
of his experiments with Eptesicus fuscus supported his matched-
filter model, and later experiments (Simmons, 1971a, 1971b, 1973)
included other species and distances.

    Caine (1976) reported the results of pulsepair interpulse
interval discrimination by a Tursiops truncatus. She interpreted
her results according to a TSP processing model of distance
determination. The interpulse-interval difference thresholds she
reported were 100 microsec (standard = 0.65 millisec), 170 microsec (standard = 1.35 millisec), and 130 microsec (standard = 4.0
millisec). These temporal difference thresholds predict echolocation range resolutions of 7 cm (absolute range = 0.49 m), 13 cm
(absolute range = 1.01 m), and 10 cm (absolute range = 3.00 m).

Range Resolution Experiments

    A Tursiops truncatus (Heptuna) was the subject of three
experiments designed to determine the range resolution capabilities
of an echolocating porpoise at three different distances.

    The experiments were conducted in the same area of Kaneohe
Bay in a pen (Figure 5) similar to the one used for the maximum
range experiments. Two identical spherical, polyurethane foam
targets, 7.62-cm in diameter (TS = -33.8 dB), were used as

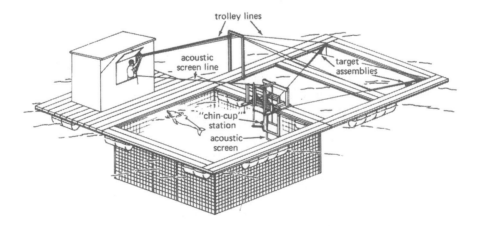

Figure 5. Floating pen used in the range resolution experiments.

discriminative stimuli. Target stability was enhanced by suspending
each of them on 4 monofilament lines 40 cm below the water (Figure
6). The targets were attached to aluminum travelers, and the
relative target distances could be adjusted by line and pulley
arrangements from the operator's position in an instrument shack
(Figure 5).

The porpoise was conditioned to place the anterior part of
his lower jaw snugly into a molded fiberglass "chin-cup" (Figure
6). An acoustically opaque panel obscured the targets until their
relative distances were adjusted. It was then dropped, presenting
the target display to the porpoise. The fiberglass chin-cup was
mounted 50 cm beneath the surface in a plexiglass swivel and could
rotate horizontally to allow the porpoise to remain positioned in
the chin-cup while scanning back and forth between the two target
spheres. The targets were always 40° from one another relative to
the center of the chin-cup. The porpoise was conditioned to
indicate his choice of which of the two targets was the closer by
striking one of two manipulanda located at the right and left of
the chin-cup position. His choice was indicated by striking the
manipulandum on the same side as the closer (right or left) of the
two spheres. Details of the operant conditioning techniques used
are given by Murchison (1979).

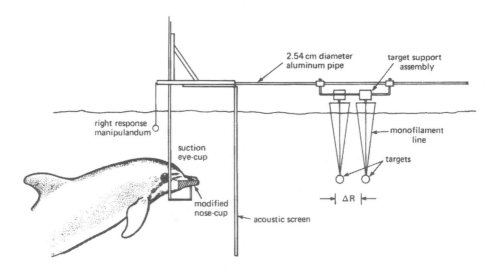

Figure 6.   Diagramatic view of the porpoise's position during the
            range resolution trials. The acoustic screen was
            removed (dropped) and targets presented to the porpoise
            after ΔR was adjusted.

Sessions were conducted in blocks of 10 trials (limited random right-left, 5 trials each) utilizing a modified, descending method of limits. One-meter absolute range means that the target closest to the porpoise during any particular trial (and therefore the "correct" response on that trial) was always one meter from the tip of the chin-cup. The same is true for the 3-meter and 7-meter sessions.

Heptuna's target range discrimination performance results are summarized by Figures 7, 8 and 9. The three absolute range experiments were plotted separately, and the 75% correct response level ΔR's (range resolution) estimated from them (Woodworth and Schlosberg, 1954). The theoretical curve in Figure 7 was calculated from the envelope of the autocorrelation function from one of Heptuna's echolocation pulses in the manner proposed by Simmons (1969, 1971a, 1971b, 1973); the pulse and its autocorrelation function is shown by Figure 10.

The 75% correct ΔR's were plotted in Figure 9 with the absolute ranges and ΔR's converted to two-way pulse-echo travel

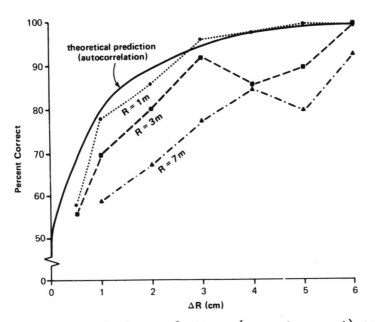

Figure 7. Range resolution performance (percent correct) as a function of ΔR (range differences) for absolute ranges (R) of 1, 3 and 7 meters. The theoretical prediction curve was taken from the envelope of the autocorrelation function of the echolocation pulse shown in Figure 10 in the manner given by Simmons (1969).

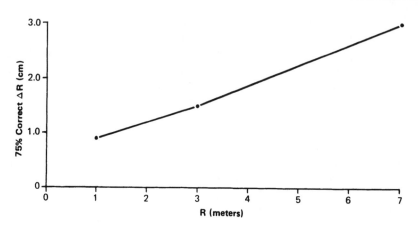

Figure 8. Range difference (ΔR) thresholds (75 percent correct) as a function of absolute ranges (R) of 1, 3 and 7 meters.

time. This figure also shows the results of R. Johnson's (this volume) study of time-separation-pitch (TSP) discrimination in humans. The abscissa is the standard pulse-pair time separations (T) and the ordinate is discrimination performance (75% correct level) of subjects, converted to the pulse-pair time-separation differences (ΔT). Figure 9 also includes the reported results of the range resolution experiments of Simmons (1969, 1971b, 1973) with bats (Eptesicus fuscus). Simmons' range resolution results were also converted to two-way pulse-echo travel time using a two-way travel time of 5.8 microsec/1 mm of range (Simmons, 1973). The dashed line on Figure 9 is Heptuna's performance predicted by Simmons' autocorrelation technique.

The most accurate prediction of porpoise range resolution was made by Kenshalo (1967) who predicted that the "resulting function would resemble the standard Weber-Fechner function of ΔD/D=K or conform to the Stevens Power Law." Figures 8 and 9 confirm his prediction.

The comparisons of the porpoises' range resolution performance with that predicted by Simmons' model (1969, 1971a, 1971b, 1973), (Figures 8 and 9) argue strongly against the same type of signal processing by porpoises and bats. The range resolution performance of Eptesicus fuscus reported by Simmons (1969, 1971b, 1973) and shown in Figure 9, seems puzzling in light of Kenshalo's prediction and the results of the porpoise experiments reported here. However, porpoises and bats are very different creatures. About the only two things they have in common are mammalian ancestry and the ability to echolocate. As Pye (1967) pointed out, it may

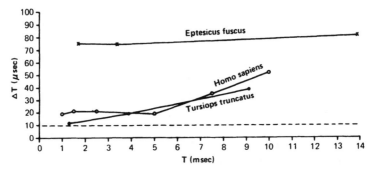

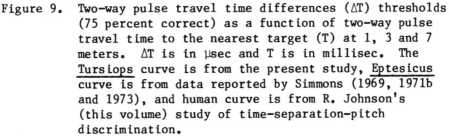

Figure 9.   Two-way pulse travel time differences (ΔT) thresholds
(75 percent correct) as a function of two-way pulse
travel time to the nearest target (T) at 1, 3 and 7
meters. ΔT is in μsec and T is in millisec. The
Tursiops curve is from the present study, Eptesicus
curve is from data reported by Simmons (1969, 1971b
and 1973), and human curve is from R. Johnson's
(this volume) study of time-separation-pitch
discrimination.

not be valid to assume that there is only one signal processing
mechanism used by all species of echolocating animals. On the
other hand, Johnson and Titlebaum (1976) argue that it would be
"convenient" to have one model for all echolocators. The function
generated by human TSP experiments and the one generated by
Heptuna's range resolution performance (Figure 9) show greater
similarity than any other proposed predictive models or evidence.
Although Simmons et al. (1976) and Johnson and Titlebaum (1976)
each caution that their particular model may not uniquely explain
the results of range resolution experiments like the one reported
here, comparison of their results indicates that neither of their
models predicts or explains the other's results (Figure 9).

The ideal detector (matched filter) model of echolocation
range determination proposed by Simmons (1969, 1971a, 1971b, 1973)
relies heavily on the envelope of the calculated autocorrelation
function of the produced pulse to predict the range resolution
capabilities of an echolocator. However, as Diercks (1972) has
emphasized, no proposed sonar signal processing technique can
yield a meaningful analog in the absence of considerations of the
biological, ear-brain processor. The matched filter, model
proposed by Simmons (1969, 1971a, 1971b, 1973) and others, seems
to have left out an important consideration in this regard. The
abcissas of calculated, predictive, autocorrelations are based on the
physical continua, i.e. measured time intervals. However, the
output of any type of time-domain processing by echolocating

animals is going to be based on a <u>psychological</u> <u>continua</u>, i.e.
sensed time intervals. (See Stevens, 1975 for a discussion of the
non-linearity of the relationship between the two). As Green and
Swets (1966) have pointed out, the unrealistic assumption of ideal
detector models is that they "measure with infinite accuracy with
no computational errors."

If Heptuna's sensation of duration is assumed to be non-
linear; that is, if he does not have a perfect internal clock his
judgments of relative pulse-to-echo delay times would be different
at different absolute ranges. If the differences between his
sensation of duration and the calculated abcissa of the autocorrelation function predicted by Simmons model are considered, the
difference in resolution at the different absolute ranges might
still be accounted for by a time (perceived) domain correlation
similar to that proposed by Simmon's model. That is, Heptuna
could have been utilizing pulse-to-echo correlation-equivalent
processing with an imperfect timing mechanisms, and the ΔT as a
function of T plot could be a measure of that imperfection.

These may have been the kinds of considerations that prompted
Kenshalo (1967) to predict Weber-Fechner or Stevens Power Law

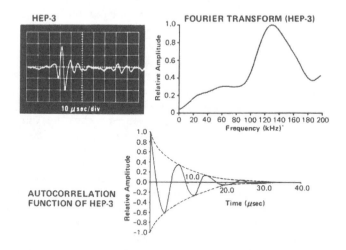

Figure 10.   Echolocation pulse produced by Heptuna during range
              resolution experiment and its Fourier transform
              and autocorrelation function. (Pulse recorded and
              Fourier transform and autocorrelation function
              calculated by Whitlow W. L. Au, Naval Ocean
              Systems Center, Hawaii Laboratory).

results from such a range resolution experiment as the one reported
here. This still leaves the puzzle of Simmons' (1969, 1971a,
1971b, 1973) results, as shown in Figure 9. Do bats have perfect
clocks?

If they don't, perhaps the pulse-to-echo delay times that the
bats in Simmons' experiments were discriminating were in a region
of optimal and/or constant time judgment. Just as the frequency
difference thresholds (Δf) plots of many vertebrates show frequency
regions of relatively constant difference thresholds (Fay, 1974),
there may be regions of relatively constant duration (pulse-to-
echo) difference thresholds (ΔT). It could be that the plot of E.
fuscus' time resolution, shown in Figure 9, represents such a
region. If this were the case, and the porpoise has such a region,
it might be at shorter (or longer) pulse-to-echo delay times than
represented by the ranges tested.

The only similarity between Caine's (1976) predicted Tursiops
truncatus range resolution as a function of absolute distance and
the actual performance of Heptuna was that she predicted an
increase in ΔR between the first two absolute ranges (0.49m to
1.01m). However, the similarity ended there. All of the predicted
ΔR's were much larger (less acuity) than Heptuna's actual performance.
Caine predicted range resolutions of 7 cm at R = 0.49m, 13 cm at
R = 1.35m, and 10 cm at R = 3m. As she points out, the reverbera-
tions of her experimental tank may have affected the porpoise's
resolution of the pulse time separations presented to him.

Vel'min and Titov (1975) concluded in their report of their
pulse-pair interval difference threshold experiment that Tursiops
truncatus differential threshold "was approximately 10%." This
predicts T. truncatus' range resolution of 10 cm at R = 1m, 30 cm
at R = 3m, and 70 cm at R = 7m. As Figure 8 shows, Heptuna's
range resolution was more acute than that. However, Vel'min and
Titov worked with pulse-pair intervals much smaller than the
pulse-to-echo delay times that Heptuna was discriminating. They
stated that they could not train their porpoise to discriminate
pulse pair intervals longer than 500 microsec. It was for this
reason that they concluded that their results supported the
critical-interval hearing hypothesis of Vel'min and Dubrovskiy
(1975, 1976).

The fact that Heptuna was obviously able to discriminate
pulse-to-echo delay intervals longer than 500 microsec does not
necessarily invalidate Vel'min's and Dubrovskiy's critical interval
hypothesis. It could indicate that the Tursiops in the Soviet
experiments had learned to discriminate pulse-pair intervals
with delay times less than 500 microsec (Vel'min and Titov, 1975)
or 300 microsec (Vel'min and Dubrovskiy, 1976) utilizing signal
processing that was optimal at those delay times (but not longer

delay times) and, therefore, was unable to discriminate longer
pulse delay-times because they necessitated a different signal
analysis that the animal may have been capable of but had not
learned (or been trained) to utilize to accomplish the discrimination task. This is consistent with that part of the critical-
interval hypothesis that proposes different signal processing of
pulse-to-echo delay times less than 300 microsec than for intervals
greater than that. The ability of Heptuna to obviously discriminate
pulse-to-echo delay times considerably greater than 300 microsec,
and the reported inability of the Soviet porpoises to discriminate
pulse intervals greater than 300-500 microsec might also indicate
that the open hearing-sensitivity window ("temporarily gating")
hypothesis of Giro and Dubrovskiy (1972), proposed as one method
that odontocetes use to combat reverberation, is variable in
duration according to the acoustic environment, especially reverberation. That is, the 300-microsec interval that Vel'min and
Dubrovskiy (1975, 1976) proposed might have been the duration of
the strobed sensitivity interval their animals learned in order to
optimize their performance in the reverberant tank environment.
The duration of the proposed range-gating, hearing sensitivity
window may also vary as a function of target distance. In the
Vel'min and Dubrovskiy (1975) 40mm sphere detection experiment
with variable artificial noise-pulse delay times, the animals-to-
target distance was the same for all noise-to-pulse delay times;
therefore, this possibility was not eliminated.

SPECULATIONS

Since echolocation is a part of the behavior of the animals
that have that ability, perhaps behavioral as well as mammalian
hearing and sonar engineering models should be included for
consideration. An open acoustic sensitivity window, as proposed
by Giro and Dubrovskiy, (1972) could be opened for any expected
target distance. Echolocation range of attention, regulated by
interpulse intervals, and a time gating of echo sensitivity
regulated by expectancy (or search depth), would reduce reverberation resulting from echoes from objects or surfaces behind or in
front of the target range of interest.

The critical interval hypothesis (Vel'min and Dubrovskiy,
1975) can be combined with the strobed-hearing hypothesis (Giro
and Dubrovskiy, 1972) in the following way. Heptuna's range
resolution performance showed that a porpoise can discriminate
pulse-to-echo delay times greater than 300 microsec. Whatever
mechanism he utilized to do that could be considered the timing
mechanism necessary for the Giro and Dubrovskiy (1972) strobed
sensitivity window. If it is assumed that the delay time, from
pulse production to the opening of the hearing sensitivity window,
is a function of the animal's attention or expectancy, it could be

opened at any time after the production of the pulse and the
proposed critical-interval processing could then apply to all
echoes received during the time that the sensitivity window is
open. The Vel'min and Dubrovskiy (1975) conclusion that pulses
closer together than 300 microsec result in an "integral echoranging object" is complementary to the spectrum analysis models
of construction, size, and materials proposed by Dubrovskiy et al.
(1971) and Au and Hammer (1978), because the secondary echoes that
resulted from the material and structure of the targets in those
experiments were all less than 300 microsec apart and would be,
therefore, coherently added to the processed spectrum.

A visual analogy emerges from this speculation and that on
the relationship between interpulse intervals and target range.
Funcationally, interpulse interval control may be roughly analogous
to visual focus and the strobed sensitivity window may function
somewhat like visual depth of field with all echoes received
during the period of open sensitivity (possibly adjustable) that
are closer in time than the 300 microsec critical interval perceived
as an integrated, coherent "image".

Regardless of whether this particular analogy is meaningful
or not, it does seem that echolocating Tursiops do have the
ability to attent ("focus") to one particular area or object at a
chosen distance.

REFERENCES

Au, W. W. L, 1979, Echolocation signals of the Atlantic bottlenose
    dolphin Tursiops truncatus in open waters, this volume.
Au, W. W. L., Floyd, R. W., and Haun, J. E., 1978, Propagation
    of Atlantic bottlenose dolphin echolocation signals, J.
    Acoust. Soc. Am., 64:411.
Au, W. W. L., and Hammer, C., 1978, Analysis of target recognition
    via echolocation by an Atlantic bottlenose porpoise
    (Tursiops truncatus), J. Acoust. Soc. Am., 60:587.
Au, W. W. L., Floyd, R. W., Penner, R. H., and Murchison, A. E.,
    1974, Measurement of echolocation signals of the Atlantic
    bottlenose dolphin, Tursiops truncatus (Montagu), in open
    waters, J. Acoust. Soc. Am., 56:1280.
Aryapet'yants, E. Sh., Golubkov, A. G., Yershova, I. V., Zhezherin,
    A. R., Zvorkin, V. N., and Korolev, V. I., 1969, Echolocation
    differentiation and characteristics of radiated pulses in
    dolphins, Report of the Academy of Science of the USSR,
    188:1197 (English translation JPRS 49479).
Ayrapet'yants, E. Sh., and Konstantinov, 1974, "Echolocation in
    Nature", Nauka, Leningrad (English translation JPRS 63328).

Babkin, V. P., and Dubrovskiy, N. A., 1971, Range of action and
    noise stability of the echolocation system of the bottle-
    nose dolphin in detection of various targets, Tr. Akust.
    Inst., Moscow, 17:29, as cited in: "Sensory Bases of Ceta-
    cean Orientation", Nauka, Leningrad (English translation
    JPRS L/7157).
Bel'kovich, V. M., and Dubrovskiy, N. A., 1976, "Sensory Bases of
    Cetacean Orientation", Nauka, Leningrad (English translation
    JPRS L/7157).
Bel'kovich, V. M., and Nesterenko, I., 1971, How the dolphin's
    locator operates, Priroda, 7:71 (German translation by W.
    Petri, 1974, Naturwissenschaftliche Rundschau, 0004:0143).
Bullock, T. H., Grinnell, A. D., Ikezono, E., Kameda, K., Katsuki,
    Y., Nomoto, M., Sato, O., Suga, N., Yanagisawa, K., 1968,
    Electrophysiological studies of central auditory mechanisms
    in cetaceans, Zeitschrift für vergleichende Physiol., 59:117.
Busnel, R.-G., Dziedzic, A., 1966, Acoustic signals of the pilot
    Whale Globicephala melaena and the porpoises Delphinus
    delphis and Phocoena phocoena, in: "Whales, Dolphins, and
    Porpoises", K. S. Norris, ed., U. of California Press,
    Berkeley.
Caine, N. G., 1976, "Time Separation Pitch and the Dolphin's Sonar
    Discrimination of Distance", Master's Dissertation, San
    Diego State University.
Diercks, K. J., 1972, Biological Sonar Systems: A Bionics Survey,
    Applied Research Laboratories, ARL-TR-72-34, Austin, Texas.
Dubrovskiy, N. A., Krasnov, P. S., and Titov, A. A., 1971, Dis-
    crimination of solid elastic spheres by an echolocating
    porpoise, Tursiops truncatus, in: "Proc. 7th International
    Acoust. Conf.", Budapest.
Eberhardt, R. L., and Evans, W. E., 1962, Sound activity of the
    California gray whale, Eschrichtius glaucus, J. Aud. Eng.
    Soc., 10:324.
Evans, W. E., 1967, Discussion, in: "Animal Sonar Systems,
    Biology and Bionics", R.-G.Busnel, ed., Laboratoire de
    Physiologie Acoustique, Jouy-en-Josas, France.
Evans, W. E., 1973, Echolocation by marine delphinids and one
    species of fresh-water dolphin, J. Acoust. Soc. Am., 54:191.
Evans, W. E., and Dreher, J. J., 1962, Observations on scouting
    behavior and associated sound production by the Pacific
    bottlenosed porpoise (Tursiops gilli Dall), Bull. Soc.
    Calif. Acad. Sci., 61:217.
Evans, W. E., and Powell, B. A., 1967, Discrimination of different
    metallic plates by an echolocating delphinid, in: "Animal
    Sonar Systems, Biology and Bionics", R.-G. Busnel, ed.,
    Laboratoire de Physiologie Acoustique, Jouy-en-Josas, France.
Evans, W. E., Sutherland, W. W., and Beil, R. G., 1964, The dir-
    ectional characteristics of delphinid sounds, in: "Marine
    Bio-Acoustics, Proceedings of the First Symposium on Marine
    Bio-Acoustics, Bimini", W. N. Tavolga, ed., Pergamon Press,

New York.

Fay, R. R., 1974, Auditory frequency discrimination in vertebrates, J. Acoust. Soc. Am., 56:206.

Giro, L. R., and Dubrovskiy, N. A., 1972, Correlation between recurrence frequency of delphinid echolocation signals and difficulty of echoranging problem, in: Marine Instrument Building Series, 2:84, as cited in: "Marine Mammals, Proceedings of the Sixth All-Union Conference on the Study of Marine Mammals," G. B. Agarkov, ed., Nauka Dumka, Kiev (English translation JPRS L/6049).

Green, D. M., and Swets, J. A., 1966, "Signal Detection Theory and Psychophysics", John Wiley and Sons, New York.

Grinnell, A. D., 1967, Mechanisms of overcoming interference in echolocating animals, in: "Animal Sonar Systems, Biology and Bionics", R.-G. Busnel, ed., Laboratoire de Physiologie Acoustique, Jouy-en-Josas, France.

Hartridge, H., 1920, The avoidance of objects by bats in flight, J. Physiol., 54:54.

Hinde, R. A., 1970, "Animal Behavior, A Synthesis of Ethology and Comparative Psychology", McGraw-Hill, Palo Alto.

Johnson, C. S., 1967, Discussion to paper by Evans and Powell, in: "Animal Sonar Systems, Biology and Bionics", R.-G. Busnel, ed., Laboratoire de Physiologie Acoustique, Jouy-en-Josas, France.

Johnson, C. S., 1967, Sound detection thresholds in marine mammals, in: "Marine Bio-Acoustics, Proc. of the Second Symposium on Marine Bio-Acoustics, New York", W. N. Tavolga, ed., Pergamon Press, New York.

Johnson, R. A., 1979, Energy spectrum analysis in echolocation, this volume.

Johnson, R. A., and Titlebaum, E. L., 1976, Energy spectrum analysis: a model of echolocation processing, J. Acoust. Soc. Am., 60:484.

Kay, L., 1962, A plausible explanation of the bat's echolocation acuity, Animal Behaviour, 10:34.

Kellogg, W. N., 1960, Auditory scanning in the dolphin, Psych. Rec., 10:25.

Kenshalo, D. R., 1967, Discussion, in: "Animal Sonar Systems, Biology and Bionics", R.-G. Busnel, ed., Laboratoire de Physiologie Acoustique, Jouy-en-Josas, France.

Leatherwood, J. S., Johnson, R. A., Ljungblad, D. K., and Evans, W. E., 1977, "Broadband measurements of underwater acoustic target strengths of panels of tuna nets", NOSC TR 126, San Diego.

Love, R. H., 1971, Measurements of fish target strength: a review, Fish. Bull., 69:703.

McCue, J. J. G., 1966, Aural pulse compression by bats and humans, J. Acoust. Soc. Amer., 40:545.

Mermoz, H., 1967, Discussion, in: "Animal Sonar Systems, Biology
    and Bionics", R.-G., Busnel, ed., Laboratoire de Physiologie
    Acoustique, Jouy-en-Josas, France.
Morozov, V. P., Akopian, A. I., Burdin, V. I., Donskov, A. A.,
    Zaytseva, K. A., and Sokovykh, Tu. A., 1971, Delphinid
    audiogram, Sechenov Physiol. Jour., of the USSR, 57(6):843.
Morozov, V. P., Akopian, A. I., Burdin, V. I., Zaytseva, K. A.,
    and Sokovykh, Yu. A., 1972, Tracking frequency of the lo-
    cation signals of dolphins as a function of distance to the
    target, Biofizika, 17:139 (English translation JPRS 55729).
Morozov, V. P., Akopian, A. I., Zaytseva, K. A., and Titov, A. A.,
    1975, On the characteristics of spatial directivity of the
    dolphin acoustic system in signal perception against the
    background of noise, in: "Marine Mammals, Proceedings of the
    Sixth All-Union Conference on the Study of Marine Mammals,"
    G. B. Agarkov, ed., Naukova Dumka, Kiev (English translation
    JPRS L/6049).
Murchison, A. E., 1979, "Maximum Detection Range and Range Reso-
    lution in Echolocating Tursiops truncatus (Montague), Ph. D.
    Dissertation, U. of California at Santa Cruz, California.
    (Manuscript).
Nordmark, J., 1961, Perception of distance in animal echolocation,
    Nature, 190:363.
Norris, K. S., 1964, Some problems of echolocation in cetaceans,
    in: "Marine Bio-Acoustics, Proceedings of the First Sym-
    posium on Marine Bio-Acoustics, Bimini", W. N. Tavolga, ed.,
    Pergamon Press, New York.
Norris, K. S., 1968, The evolution of acoustic mechanisms in odont-
    ocete cetaceans, in: "Evolution and Environment", E. T.
    Drake, ed., Yale Univ. Press, New Haven.
Norris, K. S., 1974, "The Porpoise Watcher", George J. McLeod,
    Limited, Toronto.
Norris, K. S., and Evans, W. E., 1967, Directionality of echo-
    location clicks in the rough-tooth porpoise Steno bredan-
    ensis (Lesson), in: "Marine Bio-Acoustics, Proceedings
    Second Symposium on Marine Bio-Acoustics, New York", W. N.
    Tavolga, ed., Pergamon Press, New York.
Norris, K. S., Evans, W. E., and Turner, R. N., 1967, Echolocation
    of an Atlantic bottlenose porpoise during discrimination,
    in: "Animal Sonar Systems, Biology and Bionics", R.-G. Bus-
    nel, ed., Laboratoire de Physiologie Acoustique, Jouy-en-
    Josas, France.
Norris, K. S., Prescott, J. H., Asa-Dorian, P. V. and Perkins, P.,
    1961, An experimental demonstration of echolocation behav-
    ior in the porpoise, Tursiops truncatus (Montagu), Biol.
    Bull., 120:163.
Penner, R. and Murchison, A. E., 1970, Experimentally demonstrated
    echolocation in the Amazon river porpoise Inia geoffrensis
    (Blainville), Proc. 7th Ann. Conf. Biol. Sonar and Diving
    Mammals, 7:17.

Pye, J. D., 1961, Perception of distance in animal echolocation, Nature, 190:362.

Pye, J. D., 1967, Discussion, in: "Animal Sonar Systems, Biology and Bionics", R.-G. Busnel, ed., Laboratoire de Physiologie Acoustique, Jouy-en-Josas, France.

Pye, J. D., 1971, Bats and fog, Nature, 229:572.

Romanenko, Ye. V., Tomilin, A. G., and Artemenko, B. A., 1965, in: "Bionica", M. G. Gaaze-Rapoport and V. E. Yakobi, eds., Nauka, Moscow (English translation JPRS 35125).

Sales, G., and Pye, D., 1974, "Ultrasonic Communication by Animals", Chapman and Hall, London.

Schevill, W. E., and Lawrence, B., 1956, Food-finding by a captive porpoise (Tursiops truncatus), Breviora Museum Comp. Zool., 53:1.

Schusterman, R. J., 1968, Experimental laboratory studies of pinniped behaviour, in: "The Behaviour and Physiology of Pinnipeds", R. J. Harrison, R. C. Hubbard, R. S. Peterson, C. E. Rice, and R. J. Schusterman, eds., Appleton-Century-Crofts, New York.

Schusterman, R. J., 1979, Behavioral methodology in echolocation by marine mammals, this volume.

Simmons, J. A., 1969, "Depth Perception by Sonar in the Bat Eptesicus fuscus", Doctoral dissertation, Princeton University, University Microfilms, Inc., Ann Arbor.

Simmons, J. A., 1971a, The sonar receiver of the bat, Ann. N. Y. Acad. Sci., 188:167.

Simmons, J. A., 1971b, Echolocation in bats: Signal processing of echoes for target range, Science, 171:925.

Simmons, J. A., 1973, The resolution of target range by echolocating bats, J. Acoust. Soc. Am., 54:157.

Simmons, J. A., Altes, R. A., Beuter, K. J., Bullock, T. H., Capranica, R. R., Goldstein, J. L., Griffin, D. R., Konishi, M., Neff, W. D., Neuweiler, G., Schnitzler, H. U., Schuller, G., Sovijarvi, A. R. A., and Suga, N., 1977. Localization and identification of acoustic signals, with reference to echolocation, group report, in: "Dahlem Workshop on Recognition of Complex Acoustic Signals, Berlin, 1976", T. H. Bullock, ed., Abakon Verlagsgesellschaft, Berlin.

Steele, J. W., 1971, Marine-environment cetacean holding and training enclosures, NUC TP, 227, San Diego.

Stevens, S. S., 1975, "Psychophysics: Introduction to its Perceptual, Neural, and Social Prospects," G. Stevens, ed., John Wiley and Sons, New York.

Stewart, J. L., 1968, Analog simulation studies in echoranging, T. D. R. N°. AMRL-TR-68-40, Aerospace Medical Div., U. S. A. F. Systems Command.

Strother, G. K., 1961, Note on the possible use of ultrasonic pulse compression by bats, J. Acoust. Soc. Am., 33:696.

Thurlow, W. R., and Small, A. H., 1955, Pitch perception for
    certain periodic auditory stimuli, J. Acoust. Soc. Am.,
    27:132.

Tinbergen, L., 1960, The natural control of insects in pinewoods,
    I. Factors influencing the intensity of predation by song
    birds, Arch. neerl. Zool., 13:265.

Titov, A. A., 1972, Investigation of sonic activity and phenomeno-
    logical characteristics of the echolocation analyzer of
    Black Sea delphinids. Canditorial dissertation, Karadag, as
    cited in: Bel'kovich, V. M., and Dubrovskiy, N. A., 1976,
    "Sensory Bases of Cetacean Orientation", Nauka, Leningrad
    (English translation JPRS L/7157).

von Uexküll, J., 1934, Streifzuge durch die Umwelten von Tieren
    und Menschen, Translated in: "Instinctive Behaviour", 1957,
    C. H. Schiller, ed., Methuen, London.

Urick, R. J., 1975, "Principles of Underwater Sound", McGraw-Hill
    Co., New York.

Vel'min, V. A., 1975, Target detection by the bottlenose dolphin
    under artificial reverberation conditions, in: "Marine
    Mammals, Proceedings of the Sixth All-Union Conference on
    the Study of Marine Mammals", G. B. Agarkov, ed., Naukova,
    Dumka, Kiev, (English translation JPRS L/6049).

Vel'min, V. A., and Dubrovskiy, N. A., 1975, On the auditory anal-
    ysis of pulsed sounds by dolphins, Doklady Akad. Nauk.,
    SSSR, 225(2):470.

Vel'min, V. A., and Dubrovskiy, N. A., 1976, The critical interval
    of active hearing in dolphins, Sov. Phys. Acoust., 22(4):
    351.

Vel'min, V. A., and Titov, A. A., 1975, Auditory discrimination of
    interpulse intervals by bottlenose dolphin, in: "Marine
    Mammals, Proceedings of the Sixth All-Union Conference on
    the Study of Marine Mammals", G. B. Agarkov, ed., Naukova
    Dumka, Kiev (English translation JPRS L/6049).

Webster, F. A., 1967, Interception performance of echolocating
    bats in the presence of interference, in: "Animal Sonar
    Systems, Biology and Bionics", R.-G. Busnel, ed., Labora-
    toire de Physiologie Acoustique, Jouy-en-Josas, France.

Woodworth, R. S., and Scholsberg, H., 1954, "Experimental Psychol-
    ogy", revised edition, Holt, Rinehart, and Winston, Inc.
    New York.

Zaslavskiy, G. L., 1974, Experimental study of time and space
    structure of the dolphin's echolocating signals, candidat-
    orial dissertation, Karadag, as cited in: Bel'kovich V. M.,
    Dubrovskiy, N. A., 1976, "Sensory Bases of Cetacean Orient-
    ation", Nauka, Leningrad (English translation JPRS L/7157).

Zaslavskiy, G. L., Titov, A. A., and Lekomtsev, V. M., 1969, Re-
    search on the sonar abilities of the bottlenose dolphin,
    Trudy Akusticheskogo Instituta, 8:134, as cited in: Ayra-
    pet'yants, E. Sh., and Konstantinov, A. I., 1974, "Echoloc-
    ation in Nature", Nauka, Leningrad (English trans. JPRS 63328)

ODONTOCETE ECHOLOCATION PERFORMANCE ON OBJECT SIZE, SHAPE AND

MATERIAL

Paul E. Nachtigall

Naval Ocean Systems Center

Kailua, Hawaii 96734

A cetacean swimming in its natural environment must frequently
encounter a variety of sources for returning echos. An echolocation system which only allowed for the detection of objects would
be insufficient for gathering information necessary to meet
needs like prey location and predator avoidance. The ability of
cetaceans to differentiate and recognize characteristics of
objects via echolocation has an obvious biological benefit. The
purpose of this paper is to review the behavioral experiments
that have examined delphinid abilities to differentiate between
objects differing in size, shape, or material.

Since the last Animal System Sonar Conference was held in
1966, a high percentage of the experiments examining delphinid
discrimination have been conducted within the Soviet Union. Two
fine reviews of the Soviet work, one by Ayrapet'yants and Konstantinov (1974) and the other by Bel'kovich and Dubrovskiy (1976),
have been written. Because of the thoroughness of those reviews,
the Soviet work reviewed within this paper will be confined to
those studies that are applicable or comparable to recent work
done outside of the USSR. The tables presented within this
review were prepared to be comparable extensions of the thorough
tables presented in both Soviet reviews.

The emphasis within this review is placed on animal performance.
The procedures used in animal discrimination experiments generally
require the animal to choose between at least two stimuli presented
either simultaneously or successively. The stimuli normally
differ along only one dimension and the limits of the animal's
discrimination are frequently tested by making the differences
progressively smaller. The point at which the animal can no

longer effectively make the discrimination is termed the threshold.
For comparative purposes, threshold levels for this review are
arbitrarily set at 75% correct. Although direct comparison
between performance levels and thresholds will be made within
this review, the reader must bear in mind that comparisons between
the results of differing experiments must be considered as approximate. Different animals, training procedures, environments, and
experimental procedures frequently lead to variability in performance levels independent of the actual sensitivities of the
cetacean subjects.

SIZE DISCRIMINATION

Kellogg's (1958) initial work on echolocation demonstrated
that a bottlenosed dolphin (Tursiops truncatus) could choose
between food fish of different sizes. Since that early work, a
number of experiments have demonstrated that a variety of delphinids
are quite capable of discriminating targets differing in size.
The first tests of the limits of delphinid size discrimination
ability were both presented at the previous animal sonar symposium
(Busnel, 1967). Busnel and Dziedzic (1967), using an unconditioned
obstacle avoidance technique, demonstrated that a blindfolded
harbor porpoise (Phocaena phocaena) could easily avoid wires made
of copper, iron, and steel .5 mm in diameter but frequently ran
into wires .2 mm in diameter. While Norris, Evans and Turner
(1967) demonstrated that a blindfolded Tursiops truncatus could
choose the larger of two simultaneously presented nickel-steel
ball bearings. The results of the experiment (Turner and Norris,
1966) indicated that the animal could discriminate between a 5.71
cm in diameter standard sphere and spheres larger than 6.35 cm in
diameter.

Shortly after the publication of the last animal sonar
symposium, scientists within the Soviet Union began publishing
work on the size discrimination ability of cetaceans. Ayrapet'yants,
Golobkov, Yershova, Zhezherin, Zvorykin and Korolev (1969) investigated the ability of bottlenosed dolphins to discriminate steel
disc shaped cylinders of varying heights. The animals were
trained to swim a distance of 20 meters and pull a ring to receive
a fish when a standard 25 mm high 110 mm in diameter disc-shaped
cylinder was presented and to withhold the response when a cylinder
of the same diameter but of a greater height was presented.
While Schusterman (this volume) is rightfully unsure about the
balancing of the reinforcement for "yes" or "no" responses, in
that the animals may only have received reinforcement in the
presence of the standard cylinder, the data indicate that the
animal did discriminate a 30 mm high cylinder from the 25 mm high
standard at the 70% level. That threshold estimate may be considered
as an underestimate if the animal was, in fact, biased by the

unbalanced reinforcement contingencies.

Zaslavskiy, Titov, and Lekomstev (1969) conducted an experiment using the harbor porpoise (Phocaena phocaena) discriminating the heights of a similar series of targets. Rather than using the ring-pull technique used by Ayrapet'yants et al., targets were apparently presented simultaneously in a two alternative forced choice technique. Echolocating from a distance of 2.5 meters, the porpoise was able to differentiate 75 mm in diameter cylinders 75 mm high from cylinders 95 mm high at the 80% level. Ayrapet'yants and Konstantinov (1974) indicate that the results of these two experiments (even with the disparity of species) and other experiments on cylinder height discrimination conducted within the Soviet Union, generally agree and show that the threshold height ratios are within 1.1 to 1 and 1.3 to 1.

While Soviet scientists were comparing species on cylinder height discrimination, American scientists were examining two very different species on cylinder diameter discrimination. Performance of an Atlantic bottlenosed dolphin (Tursiops truncatus), trained by Evans and Hall in San Diego, and an Amazon River dolphin (Inia geoffrensis), trained by Penner and Murchison in Hawaii, was compared using solid chloroprene cylinders varying in target strength in 1 dB increments. Both animals were trained to wear blindfolding devices and to choose the standard cylinder (see Table 1 for target size details). Once this was achieved, targets were paired with the others in the series in a two alternative forced choice simultaneous presentation procedure with position of targets varied in a modified random order. The results of the comparison with targets having a greater target strength than standard indicated that both Inia and Tursiops could discriminate targets varying by one dB with performance levels above 70% (Evans, 1973).

Sphere Diameter Discrimination

According to Ayrapet'yants and Konstantinov (1974) Soviet Scientists were unsatisfied with the spherical size discrimination work presented by Norris, Evans, and Turner (1967):

It should be kept in mind that the experiment was not set up entirely properly: The balls were connected together by a metal rod, and the distance between them exceeded the diameter of the large ball by only two times. Thus acoustic interaction through the water was not excluded. Therefore, it would be more accurate to refer not to discrimination between balls but to recognition of asymmetry in the acoustic field scattered by a system of two balls and a connecting rod.

The reviews by both Ayrapet'yants and Konstantinov (1974) and Bel'kovich and Dubrovskiy (1976) indicate that Soviet scientists conducted a series of experiments to further examine delphinid sphere size discrimination using both steel and lead balls. Unfortunately neither review paper provides sufficient methodological detail to know how the targets were presented to overcome the reviewer's objection to Norris et al.'s procedure. In the first of these experiments (Dubrovskiy, 1972), bottlenosed dolphins were trained to discriminate a 5 cm in diameter standard lead ball from balls 10, 6 and 5.5 cm in diameter. Ayrapet'yants and Konstantinov (1974) indicate that discrimination precision "dropped regularly in almost all cases as the distance to the target was increased". The animal, at a range of 8 meters from the target, discriminated a 5.5 cm in diameter sphere from the 5.0 cm in diameter standard at the 78% level. Bel'kovich and Dubrovskiy (1976) indicate that additional work with lead spheres by Dubrovskiy and Krasnov (1971) and Fadeyeva (1973) showed that bottlenosed dolphins could discriminate a 1.4 cm in diameter sphere from 1.6 cm sphere at the 70% level at a distance of 4.8 meters from the targets; and a 1.02 cm in diameter lead sphere from a 1.17 in diameter sphere at the 75% level at a distance of 3 meters. Additional tabular data presented in Ayrapet'yants and Konstantinov (1974) indicate that at a distance between 2 and 6 m to the targets, lead spheres varying in diameter from 10.2 to 14.3 cm were compared. The results indicate an orderly relationship between target size and overall percentage correct with a threshold of 75% correct being achieved when comparing 10.2 and 11.7 cm lead spheres.

A considerable amount of Soviet work was also done by Dubrovskiy et al. (1973) on the discrimination of the size of steel spheres. The size of steel spheres was varied and presented in a two alternative forced choice procedure at distances varying between 2 and 6 meters. These data were collected within the same experiments and with the same animals previously used in the lead ball diameter discrimination presented above. Tabular data presented within Ayrapet'yants and Konstantinov (1974) indicate that with steel spheres in a graded series between 10.0 and 20.7 cm the bottlenosed dolphin could discriminate between a steel sphere 10.4 cm in diameter and a steel sphere 14.3 cm in diameter at the 75% level. Further data with larger targets indicate that, with sphere radii varying between 50.8 and 63.4 cm, the animal's 77% threshold level was reached with a comparison of spheres with radii of 57.1 and 63.5 cm.

A comparison of lead vs steel extrapolated from the above data indicates a 1.5 cm threshold difference (10.2 vs 11.7) in radii for lead balls and a 3.9 cm (10.4 vs 14.3) threshold difference in radii for steel balls of similar sizes. Ayrapet'yants and Konstantinov (1974) indicate that this interaction between size

and materials was the most important discovery in the series of
experiments.

Planometric Targets

Barta (unpublished manuscript) conducted the first Western
size discrimination experiment using flat targets. Targets were
constructed of 3.18 mm neoprene (celltite) foam cemented to a .79
mm aluminum sheet metal backing. The circular targets ranged
from 15.2 cm to 25 cm in diameter. The dolphin (Tursiops truncatus)
was trained to choose the smaller of two simultaneously presented
targets. A divider placed between the targets assured that the
animal was not closer than .7 m from the targets during discrimination. The results indicated that the animal discriminated a 16.1
cm in diameter target from the standard 15.2 cm target at a 75%
correct level. Subsequent reflectivity measures indicated that
this size difference threshold corresponded to a 1 dB difference
in target strength.

Bel'kovich, Borisov, Gurevich and Krushinskaya (1969) determined the difference threshold for the discrimination of flat
plastic foam squares of different sizes. The animal was trained
to choose the larger of two simultaneously presented targets.
The standard was 100 cm^2 while the comparison targets measured
50, 81, and 90.25 cm^2. The common porpoise (Delphinus delphis)
showed an orderly decline in performance with decreasing size
differences, with performance failing to the 77% level on comparisons between the 100 cm^2 and 90.25 cm^2 targets. The size difference threshold was therefore reported to be around 10%.

Target or Wall Thickness

The first evidence that cetaceans could discriminate between
objects varying in thickness of the material was presented by
Evans and Powell (1967) at the initial Animal Sonar Systems
Meeting. Circular shaped 20 cm in diameter metal discs varying
in both type of metal and thickness were mounted within brass
backed sponge neoprene 30 x 30 cm square target holders. The
Tursiops truncatus was trained to wear eyecups and swim 7 m to
choose a .22 cm thick copper standard target. The standard and
one comparison target were presented on each trial using a two
alternative forced choice technique and the psychophysical method
of constant stimuli. The animal was unable to discriminate .16
and .27 cm thick targets from the .22 cm standard but the .32 and
.64 cm copper discs were discriminated at the 75% and 90% levels.

Although few details are available, both Ayrapet'yants and
Konstantinov (1974) and Bel'kovich and Dubrovskiy (1976) indicate

that Titov (1972) examined the bottlenosed dolphin's ability to
discriminate differences in the wall thickness of hollow cylinders.
The steel cylinders measured 50 mm high with an outer diameter of
50 mm and were presented 5 m away from the animal (presumably
using the two alternative forced choice procedure). The animal
was trained to choose the thinner of two cylinders with target
wall thickness varying from .1 to 2 mm (according to Ayrapet'yants
and Konstantinov, 1974) and was able to discriminate a wall thickness
difference of .2 mm at the 75% level.

Hammer (1978) conducted an interesting series of experiments
using hollow and solid cylinders. In the first study, he examined
a discrimination of hollow aluminum cylinders and solid cylinders
made of coral rock chunks embedded in epoxy resin (see materials
section for further details). The results of that first experiment
led him to believe that the salient cue for target "recognition"
was the thickness of the wall of the hollow aluminum cylinders.
In order to determine the limits of wall thickness, a series of
cylinders varying in wall thickness was constructed. The animal
continued the same basic task as that of the first experiment.
Four targets were presented to maintain the animal's baseline
performance. Only one target was presented on each trial. If
either of two standard aluminum hollow cylinders (see Table 1 for
description of target dimensions) was presented the animal was
required to hit a manipulandum on the left. If either of two
solid coral rock molded in epoxy targets (with the same outside
measurements as the aluminum targets) was presented the animal
was required to hit a manipulandum on the right. During each
session the animal continued to perform this task with performance
levels near 100%. Occasionally a hollow aluminum target with a
different wall thickness was inserted. Responses to either
manipulandum were not reinforced. The wall thickness of these
"probe" targets varied around the two hollow aluminum standard
targets. Those around the smaller standard varied in .16 cm
increments while those around the large standard varied in .32 cm
increments. Each session was made up of 76 trials. The first 12
were "warm-up" trials during which only standard aluminum and
coral rock targets were presented. In the following 64 trials,
56 were continued as standard baseline trials while 8 were randomly
chosen as trials. The aluminum targets varying in wall thickness
were presented within these probe trials.

Data were collected in three phases. During the first and
third phases, targets were presented 6 m from the animal. During
the second phase targets were presented 16 m from the animal.
The data from the first phase indicated that the targets most
similar in wall thickness to the hollow aluminum standards were
infrequently "classified" as those standard targets, but the data
from the second and third phases indicated that all of the hollow
aluminum targets with wall thicknesses differing from the standards

were classified along with the solid coral rock in epoxy targets.
These results would appear to be unexpected but see Schusterman's
(this volume) chapter on the importance of behavioral methodology
in the interpretation of psychophysical data.

The data indicate that the animal could classify a hollow
aluminum target with a .16 cm wall thickness difference as being
different from the standard on 100% of the trials. A difference
threshold was not obtained but that is not surprising considering
Titov's data indicating wall thickness difference thresholds as
small as .2 mm with steel hollow cylinders.

SHAPE DISCRIMINATION

Planometric Targets

Two experiments were conducted within the late 1960's examining
cetacean ability to discriminate differences between the shapes
of flat targets. Barta (unpublished manuscript) examined the
ability of a blindfolded Tursiops to choose between circles,
squares and triangles, while Bagdonas, Bel'kovich and Krushinskaya
(1970) tested Delphinus' ability to discriminate a large square
from a smaller triangle. Barta constructed triangles and squares
from the same materials previously reported in the planometric
size discrimination work. The same circular target standards
were used in comparison to triangles and squares with varying
overall surface areas (see Table 1 for target dimensions). Using
the same two alternative forced choice paradigm, (used in the
earlier reported size discrimination work) the animal reliably
chose the circular target.

Bagdonas et al. (1970) presented targets made from ebonite
10 mm thick. The animal (Delphinus delphis) was trained to swim
to a 100 cm² square as compared to a 50 cm² triangle. This
compounding of size and shape differences makes it impossible to
precisly know which stimulus dimension controlled the animal's
behavior. The animal was not blindfolded and initial presentations
were made in clear water with sessions conducted both day and
night. The fact that the animal was correct 82% of the time
during the day and 57% during the night sessions indicated to the
authors that the animal was relying on visual cues. After some
15 sessions, however, the experimenters began hearing echolocation
signals and the dolphin suddenly began to echolocate intensely.
On the 17th session the animal reached the 100% performance level
independent of visibility conditions.

An additional planometric shape discrimination experiment
was conducted in 1971. Bel'kovich and Borisov tested the common

dolphin's (<u>Delphinus delphis</u>) ability to differentiate flat squares from similar sized squares with circular holes cut in the center. The animal was initially trained to choose a foam rubber covered plexiglass 10 x 10 cm square when it was simultaneously presented with a 7 x 7 cm square. When the 7 x 7 cm square was paired with a 7 x 7 cm square with a hole cut out of the center, in unreinforced probe trials, the animal chose the (previously negative) solid square 89.5% of the time.

A subsequent series of comparisons of the hole-cut and solid squares indicated that the animal could distinguish squares with 6.5% of area cut out. A previous study, mentioned earlier in this review, (Bel'kovich et al., 1969) with discrimination based solely on the size of the squares, had indicated an area difference threshold of 10%. The authors concluded, therefore, that the animal did not make the discrimination solely on the basis of overall reflectivity but must have been provided with other cues, such as spectrum differences of the return echo, on targets with the cut out hole.

Planometric and Three Dimensional Targets

The first experiment designed to determine the delphinid's ability to discriminate three-dimensional from flat targets was conducted by Bel'kovich, Borisov, Gurevich and Krushinskaya (1969). Two <u>Delphinus delphis</u> were trained to choose a three-stepped pyramid plastic foam target when compared to a simultaneously presented flat plastic foam triangle in a two alternative forced choice procedure. Once this discrimination was established, flat squares with size equal to that of the base of the three dimensional pyramid were also easily differentiated.

Bel'kovich et al. further examined the dolphin's ability to discriminate stepped pyramids. The standard pyramid was constructed of foam plastic squares layered in three 12 mm thick steps. The base step measured 100 cm^2, the middle step measured 49 cm^2, while the top step was only 9 cm^2. This standard was compared to pyramids in which: (1) the top step was omitted (a two-step pyramid), (2) The size of the top step was made smaller, and (3) the thickness of the top two steps was changed.

When the pyramid with the top step omitted was compared to the standard 3 step pyramid, the animals made the discrimination easily at the 95% level. When area differences of the top step were made smaller (surface areas of 6.25 and 8.4 cm^2 as compared to the standard 9 cm^2) performance fell to near 70% for both target comparisons.

The thickness of the top two steps was also varied in order
to "ascertain the ability of the dolphin to detect any shift of
the echo signal in time". One comparison pyramid was varied by
reducing the thickness of each of the top two steps from 12 mm to
6 mm. The animal discriminated this target from the standard
easily (at the 100% level). A second comparison pyramid varied
only by changing the thickness of the middle step to 6 mm. This
target was discriminated at the 86.7% level. The authors concluded
that "it is reasonable to assume that the dolphins distinguished
between the figures (step thickness differences) by using the
change in spectral composition of the reflected echo signals,
including change in the relationship between the time of their
return and the elements constituting the truncated stepped pyramid".

A bottlenosed dolphin's ability to discriminate the difference
between three dimensional solid cylinders and cubes made of
syntatic foam was examined by Nachtigall, Murchison and Au (this
volume). The blindfolded animal was required to examine two
simultaneously presented targets located 38 cm apart, and two
meters from the tip of its rostrum, and to press one of two
adjacent manipulanda indicating the randomly predetermined left
or right position of a cylinder as compared to a cube. Sizes of
both the cylinders and cubes were varied to control for possible
differences in overall reflectivity (see Table 2). Each of three
different sized cylinders was repeatedly paired with each of
three different sized cubes. Performance data, presented in
Table 2, indicate an interesting interaction of shape discriminability and target size. Comparisons of cubes and cylinders of
the same sizes show increased discriminability of shape as target
sizes increase.

Once the ability of the animal to differentiate cubes from
cylinders was well established, the aspect of the targets presented
to the animal was systematically varied using a probe technique.
Baseline performance with targets presented upright was maintained
within 56 of the 63 trials per session. On the other 7 probe
trials one of the targets was either rotated or layed down horizontally. During the first two probe sessions the cube was rotated
so that the edge, rather than a flat surface, faced toward the
animal. This change did not disrupt the animal's performance.
During the next two sessions, both the cubes and the cylinders
were changed during probe trials. The cubes were once again
presented with the edge forward but the cylinders were layed down
horizontally. Overall baseline performance remained high but
performance during probe trials declined to 71%. During the
final two sessions, both the cubes and the cylinders were presented
with the flat surface (i.e., top of the cylinder) facing the
animal. Performance on these probes declined to near chance.

Following the completion of the experiment the targets were

examined acoustically with a monostatic sonar measurement system
that projected simulated dolphin echolocation signals. The echo
returns for the various targets were processed, and resulting
frequency spectra were obtained, by taking a 1024 point fast
fourier transform. Comparisons of the target spectra failed to
reveal consistent and obvious cues for shape discrimination.

Fifteen measures of target strength for each target, however,
revealed an interesting possible cue for shape discrimination.
Standard deviations around the mean of the measures indicated
high variability in measures for both the cube and the cylinder
with the flat face forward and low variability with the cylinder
straight up. These differences in variability paralleled the
performance given by the animal. The animal most likely received
repeated pulse echos varying in amplitude when scanning across
cubes and relatively uniform pulse echos when scanning across the
curved portion of cylinders.

An animal discriminating the shapes of objects has more than
one acoustic cue (dimension) available to it. Barta's work
indicated that Tursiops could readily discriminate flat squares
from flat circles, yet in Nachtigall et al.'s experiment the
dolphin did not discriminate the difference between the flat
square surface of a cube and the flat circular cylinder top.   It
seems very likely that the primary discriminative cue for differentiating flat objects of differing shapes is not the same as the
primary cue for differentiating three dimensional objects. One
may conclude that the dolphin in Nachtigall et al.'s experiment
was quite capable of making the discrimination of flat circles
and flat squares, but had not been trained to make that particular
discrimination. This is an important point. The interpretation
of the results of a behavioral experiment depend heavily upon
both the type of training that the animal received and the type
of experimental procedure used to gather the data. Had the three
dimensional targets been rotated during training, prior to presenting
them within probe trials, the animal would have most likely
learned both discriminative cues and easily made the discrimination,
but the hypothesis of differing discriminative cues for the two
tasks would not have been suggested.

MATERIALS DISCRIMINATION

Targets made of differing types of material will reflect
differing amounts of acoustic energy. A target composed of air
filled styrofoam, for instance, will return a higher amplitude
echo under water than a similarly shaped target composed of
plexiglas. Thus an animal might easily differentiate targets
made up of these differing materials solely on the amplitude
differences of the echos. Evans and Powell (1967) predetermined

the theoretical differences in the amplitude of the return of
metal plates of varying thickness and composition. They found
that when copper plate targets of varying thickness were presented
to a Tursiops truncatus, the animal could not discriminate between
copper discs differing in calculated reflectivity of +.04 and -.06.
However, when an aluminum disc, with a calculated reflectivity
difference of only .02, was compared to the standard copper disc,
the animal easily made the discrimination. This was the first
solid evidence indicating that a delphinid could discriminate
differences in target materials independent of their overall
reflectivity. According to Evans (1973) this experiment was
replicated with a second Tursiops and a Lagenorhynchus obliquidens
with comparable results for both species. Johnson (1967) stated
that the animal may have been making this discrimination based on
phase differences.

Following the initial Evans and Powell work, Soviet scientists
began a lengthy series of experiments examining delphinid ability
to discriminate spheres and cylinders made up of different materials.
These experiments are more than adequately reviewed by both
Ayrapet'yants and Konstantinov (1974) and Bel'kovich and Dubrovskiy
(1976). Zaslavskiy, Titov and Lekomtsev (1969) investigated the
common porpoise's (Phocaena phocaena) ability to discriminate
11.5 cm high by 7.5 cm in diameter cylinders composed of steel as
compared to other materials. Using an interesting technique of
determining the range at which 75% correct thresholds were obtained,
they found wood and plastic to be more discriminable from steel
than was glass. Three experiments (Babkin, Dubrovskiy, Krasnov,
and Titov, 1971; Abramov, Ayrapet'yants, Burdin, Golubkov,
Yershova, Zhezherin, Koroleve, Malyshev, Ulyanov and Fradkin,
1971; and Titov, 1972) examined the bottlenosed dolphin's ability
to discriminate 5 cm in diameter solid spheres made of various
materials (steel, aluminum, brass, ebonite, textolite, flouroplastic,
rubber, wax, paraffin, plexiglas, and lead) using the two alternative
forced choice technique. Performance on most comparisons was
well above 90% but the discrimination of steel and aluminum was
shown to be particularly difficult in all three studies.
Dubrovskiy, Krasnov and Titov (1971) noted that this difficulty
existed in spite of the fact that the specific impedance of steel
is about 2.6 times more than that for aluminum, meaning that the
"specific impedance (as well as density) of (the) material is not
an important factor in the discrimination of solid spheres".
They found, however, that a parameter Δ, which is monotonically
related to the speed of the shear waves within the target material,
was very similar for steel and aluminum targets. Dubrovskiy et
al. hypothesized that target materials were discernable as long
as differences in Δ were discriminable. According to the hypothesis
"the discrimination is performed by means of an auditory analysis
of the intensity and the spectrum of a so-called secondary echo
which contains information about free vibrations of the target

forced to vibrate by a porpoise signal". Dubrovskiy, Krasnov and
Titov (1971) indicated, however, that they were "far from the
opinion" that this difference in the secondary echo spectrum was
the only cue for discrimination. They assumed discrimination
could also be made based on: (1) the fine structure of the power
spectrum, (2) differences in average intensity of echos, and (3)
phase shifts.

Although no work on the discrimination of spheres made up of
differing materials has been done outside the Soviet Union,
Hammer (1978) conducted two experiments on the classification of
cylinders made up of various materials. Using the procedure
outlined in an earlier section of this paper, Hammer first trained
the dolphin (Tursiops truncatus) to press a paddle on its left
when either of two standard hollow aluminum targets was presented
and to press a paddle on its right if either of two solid cylinders
composed of coral rock chunks embedded in epoxy resin was presented.
The animal continued on this basic task for 56 out of 64 trials
per session. On the other 8 randomly dispersed probe trials, any
one of 8 different cylinders was presented. Four of these cylinders
were composed of aluminum and the other four were composed of
various nonmetallic materials (see Table 3, Experiment 1).

The data, once again, were collected in three phases.
During the first and third phases the targets were presented 6 m
from the animal. During the second phase the distance was
increased to 16 m. Data collected during the first phase indicated
that the animal classified a small and a large aluminum hollow
cylinder and a small solid solid aluminum cylinder along with the
standard hollow aluminum cylinders. But, in the next two phases,
all cylinders not exactly like the standard aluminum cylinders
were classified along with the coral rock epoxy cylinders 100% of
the time. These data indicate that with continued training, an
animal may narrow its classification criterion down to a particular
target type, and that, without specific training, dolphins do not
necessarily categorize stimuli along a dimension of target material.

In order to more precisely examine the animal's ability to
classify targets of differing materials, Hammer (1978) conducted
another experiment using the same basic procedure but different
targets. During this experiment the comparison probe targets,
made of bronze, glass, and stainless steel, matched the standard
aluminum targets in all dimensions other than target materials.
The data on the stainless steel targets indicated clearly and
consistently that the animal did not classify them as the same as
aluminum standards. Responses on the bronze targets were mixed
while the glass targets were consistenly classified the same as
the aluminum.

Using the same animal, experimental set up, and glass and

aluminum targets (but not the same experimental procedure) that
Hammer had previously used, Schusterman, Kersting and Au (this
volume) continued training the animal on the aluminum-glass
discrimination. They found that the animal could eventually
discriminate the smaller glass target from the two aluminum
standards but that the animal could not discriminate the large
glass cylinder from the aluminum standards. In fact, sessions,
during which the animal was presented the "insolvable" problem,
biased the animal and made the solvable problem more difficult.

INTERACTIONS

Up to this point the experiments on cetacean discrimination
have been classified into sections based on differences in target
size, shape, or material. Although we may assume these categories
to be broadly defined, each experiment is usually limited to a
particular target type. An experimenter usually varies only one
of the three dimensions. An experiment on size discrimination,
for instance, frequently contains targets of only one shape and
only one material. This procedure tends to limit the experimental
findings because the interactions between the three dimensions
are frequently not explored. Experiments which have examined
more than one dimension have frequently produced fascinating
results. The fact that Evans and Powell (1967), for example,
used targets of different materials in their experiment on the
bottlenose dolphin's ability to discriminate flat plates of
different thicknesses, led to the discovery that the animal was
capable of making the discrimination on some basis other than
reflectivity.

It is also possible to examine interactions by comparing
data across experiments. A comparison of the Soviet work on the
discrimination of solid aluminum and steel spheres and the Hammer
(Experiment 3) work on the classification of aluminum and steel
hollow cylinders, provides an interesting example of the interaction
of target materials and target shape. In the three Soviet experiments bottlenosed dolphins were shown to have great difficulty
discriminating solid aluminum and solid steel spheres, yet in
Hammer's work aluminum and steel hollow cylinders were distinguished
with comparative ease. It seems apparent that the cues that an
echolocating dolphin uses to differentiate target materials may
be particular to the target's shape or structure. Aluminum
hollow cylinders are apparently easily discriminated from steel
hollow cylinders but solid steel spheres are not easily discriminated
from solid aluminum spheres. The dolphin's ability to discriminate
steel from aluminum may depend on the shape of the targets or it
may depend on whether the targets are solid or hollow.

A similar interaction may be observed when comparing Evans

and Powell's (1967) thickness discrimination work with that
conducted by Titov (1972). Where Evans and Powell found that
copper discs were 75% discriminable at thickness differences of
.1 cm using solid flat discs, Titov found that the hollow steel
cylinder wall thickness difference threshold for a bottlenosed
dolphin was .02 cm. Once again there appears to be some sort of
interaction occurring. The better performance reported by Titov
may have been due to the difference in materials or the difference
in shape. Steel may yield more discriminable thickness difference
return echoes than does copper, or the echo returns from solid
flat plates may not be as discriminable as the echo returns from
hollow cylinders.

   The fact that these interactions occur makes it difficult to
speak strictly about a single target dimension. All three dimensions
which form the division of this paper - size, shape, and material -
must be considered as relevant to any one particular discriminative
echolocation task. Further understanding of the process of
discriminative echolocation may well be enhanced by examining the
causes of the interactions.

Table 1. Size Discrimination

SHAPE	MATERIAL	VARYING TARGET SIZE (cm) POS.	NEG.	SIZE DIMEN.	METHOD	THRES-HOLD	% CORR.	MIN. DIST.	SPECIES	LIT. SOURCE
Solid Spheres	Nickel-Steel	6.35	5.72	Dia.	2 Alt. Force Choice	.63 cm	77	Ø	Tursiops Truncatus	Turner & Norris (1966)
			5.40				90			
			5.08				93			
			4.76				98			Norris, Evans & Turner (1967)
			4.45				.100			
			4.13				100			
			3.97				100			
			3.81				100			
			3.18				100			
Wires	Iron Copper and Steel	—	.35	Dia.	Obstacle Avoidance	.035 cm	98.9	Ø	Phocaena Phocaena	Busnel, Dziedzic & Anderson (1964)
			.15				98.9			
			.075				94.7			
			.05				90.9			
			.035				78.9			Busnel & Dziedzic (1967)
			.02				46			
	Nylon		.15				72.5			

Table 1 (Cont'd). Size Discrimination

TARGET SHAPE	TARGET MATERIAL	VARYING TARGET SIZE (cm) POS.	VARYING TARGET SIZE (cm) NEG.	SIZE DIMEN.	METHOD	THRES-HOLD	% CORR.	MIN. DIST.	SPECIES	LIT. SOURCE
Solid Cylinders H=17.78 cm	Chloroprene (DC-100)	1.64	2.07 2.62 3.28 4.16 5.20	Dia. & dB -18dB -17dB -16dB -15dB -14dB	2 Alt. Forced Choice	.43 cm	85 94 89 96 90	.77 m	Tursiops Truncatus	Evans (1973)
		1.64	2.07 2.62 3.28 4.16 5.20	-18dB -17dB -16dB -15dB -14dB		.98 cm	70 77 89 90 90	.77 m	Inia Geof-frensis	
Circular Disks	Neoprene (Cell-tite) Back With Aluminum	15.2	25.0 20.0 18.0 17.2 16.1 15.7 15.2	Dia.	2 Alt. Forced Choice	.9 cm	98.3 98.5 97 94.1 74.8 57.5 50.7	.7 m	Tursiops Truncatus	Barta & Evans (Unpub-lished Manu-script)

Table 1 (Cont'd). Size Discrimination

| TARGET | | VARYING TARGET SIZE (cm) | | SIZE DIMEN. | METHOD | THRES-HOLD | CORR. | MIN. DIST. | SPECIES | LIT. SOURCE |
SHAPE	MATERIAL	POS.	NEG.							
Circular Disks D=20 cm	Copper	.22	.16 .27 .32 .64	Thick-ness	2 Alt Forced Choice	.10 cm	50 60 75 90	.7 m	Tursiops Truncatus	Evans & Powell (1976)
Hollow Cylinders H=17.78 cm	Aluminum	OD= 7.62 .95	.31 .63 1.27 1.59	Wall Thick-ness	2 Alt Succes-ive Probe	–	100 100 100 100	6 m 16 m	Tursiops Truncatus	Hammer (1978)
		OD= 3.18 .64	.32 .48 .80 .96				100 100 100 100			

Table 2.   Shape Discrimination

TARGET SHAPE & MATERIAL		TARGET SIZE CM		METHOD	DISTANCE TO TARGET	PERCENT CORRECT	SPECIES	SOURCE
POSITIVE	NEGATIVE	POSITIVE	NEGATIVE					
Circular Neoprene Disks	Flat Squares	182 cm²	428 cm² 269 cm² 182 cm² 87 cm²	2 Alt Forced Choice	.7 m	100 98.6 93.1 100	Tursiops Truncatus	Barta & Evans (Unpublished Manuscript)
Circular Neoprene Disks	Flat Triangles	182 cm²	346 cm² 187 cm² 48 cm²			97.9 92.3 100		
	Circular Disks	182 cm²	182 cm²			49.6		
Solid Styrofoam Cylinders (Straight Up)	Solid Styrofoam Cubes (Flat Face Forward)	H 4 cm D 4 cm H 5 cm D 5 cm H 6 cm	H 4 cm D 4 cm H 5 cm D 5 cm H 6 cm D 6 cm H 4 cm D 4 cm H 5 cm D 5 cm H 6 cm D 6 cm H 4 cm	2 Alt Forced Choice	2 m	75 94 91 87 87 98	Tursiops Truncatus	Nachtigall, Murchison, And Au (This Volume)

Table 2 (Cont'd). Shape Discrimination

| TARGET SHAPE & MATERIAL | | TARGET SIZE CM | | METHOD | DISTANCE TO TARGET | PERCENT CORRECT | SPECIES | SOURCE |
POSITIVE	NEGATIVE	POSITIVE	NEGATIVE					
		D 6 cm	D 4 cm H 5 cm D 5 cm H 6 cm D 6 cm		2 m	91 94 96		
Cylinders Straight Up	Cubes Edge Forward	All Sizes	All Sizes	2 Alt Forced Choice Probe		93		
Cylinders Layed Horizontal	Cubes Edge Forward	All Sizes	All Sizes	2 Alt Forced Choice Probe		71		
Cylinders Flat Top Forward	Cubes Flat Face Forward	All Sizes	All Sizes	2 Alt Forced Choice Probe		57		

Table 3. Materials Discrimination

SHAPE	TARGET MATERIAL & SIZE (CM) POSITIVE	TARGET MATERIAL & SIZE (CM) NEGATIVE	METHOD	PERCENT CORRECT	MIN. DIST. TO TARGETS	SPECIES	SOURCE
Circular Disks	Copper D-20 T.22	Aluminum D-20 T.32 D-20 T.64 D-20 T.79 Brass D-20 T.32 D-20 T.64	2 Alt Forced Choice	100 98 93 55 100	Ø	Tursiops Truncatus	Evans & Powell (1967)
Cylinders H-17.78 cm	Hollow Aluminum D-3.81 WT-.64 D-7.62 WT-.95	Solid Coral Rock/ Epoxy Resin D-3.81 D-7.62 D-3.81 D-7.62 Solid Aluminum D-3.81 D-7.62 Hollow Aluminum D-6.35 WT-.48 D-11.43 WT-.64	2 Alt Successive Presentations (Baseline) 2 Alt. Successive Presentations Probe	100 100 100 100 100 100 100 100	6 m 16 m	Tursiops Truncatus	Hammer (1978) Experiment I

Table 3 (Cont'd). Materials Discrimination

SHAPE	TARGET MATERIAL & SIZE (CM) POSITIVE	NEGATIVE	METHOD	PERCENT CORRECT	MIN. DIST. TO TARGETS	SPECIES	SOURCE
		Solid Coral Rock/Epoxy Resin D-11.43		100			
		Solid Chloroprene D-6.35		100			
		D-4.06		100			
		Polyvinyl-Chloride D-7.62 WT-.79		100			
Cylinders H-17.78 cm	Hollow Aluminum D-3.81 WT-.32 D-7.62 WT-.40	Solid Coral Rock/Epoxy Resin D-3.81 D-7.62 D-3.81 D-7.62	2 Alt Successive Presentations (Baseline)	100 100 100 100		Tursiops Truncatus	Hammer (1978) Experiment III
		Hollow Bronze D-3.81 WT-.32 D-3.81 WT-.32 D-7.62 WT-.40 D-7.62 WT-.40	2 Alt Successive Presentations Probe	75 37.5 100 62.5	6 m* 16 m 6 m* 16 m		

Table 3 (Cont'd). Materials Discrimination

SHAPE	TARGET		METHOD	PERCENT CORRECT	MIN. DIST. TO TARGETS	SPECIES	SOURCE
	MATERIAL & SIZE (CM)						
	POSITIVE	NEGATIVE					
		Hollow Glass					
		D-3.81 WT-.32		∅	6 m*		
		D-3.81 WT-.32		∅	16 m		
		D-7.62 WT-.40		12.5	6 m*		
		D-7.62 WT-.40		12.5	16 m		
		Hollow Steel					
		D-3.81 WT-.32		87.5	6 m*		
		D-3.81 WT-.32		100	16 m		
		D-7.62 WT-.40		100	6 m*		
		D-7.62 WT-.40		100	16 m		

KEY:
  D = Diameter
  T = Thickness
  WT = Wall Thickness
  H = Height
  * = 6 m data from the second series
      following the 16 meter data collection.

REFERENCES

Abramov, A. P., Ayrapet'yants, E. Sh., Burdin, V. I., Golubkov,
A. G., Yershova, I. V., Zhezherin, A. R., Korolev, V. I.,
Malyshev, Yu. A., Ul'yanov, G. K., and Fradkin, V. G., 1971,
Investigations of delphinid capacity to differentiate between three-dimensional objects according to linear size and
material, "Report from the 7th All-Union Acoustical Conference", Leningrad, as cited in: Bel'kovich, V. M., and Dubrovskiy, N. A., 1976, "Sensory Bases of Cetacean Orientation", Nauka, Leningrad (English translation JPRS L/7157).

Ayrapet'yants, E. Sh., Golubkov, A. G., Yershova, I. V., Zhezherin,
A. R., Zvorkin, V. N., and Korolev, V. I., 1969, Echolocation differentiation and characteristics of radiated echolocation pulses in dolphins, Report of the Academy of Science of the USSR, 188:1197 (English translation JPRS 49479).

Ayrapet'yants, E. Sh., and Konstantinov, 1974, "Echolocation in
Nature", Nauka, Leningrad (English translation JPRS 63328).

Babkin, V. P., Dubrovskiy, N. A., Krasnov, P. S., and Titov, A. A.,
1971, Discrimination of material of spherical targets by
the bottlenose dolphin, in: "Report from the 7th All-Union
Acoustical Conference, Leningrad".

Bagdonas, A., Bel'kovich, V. M., and Krushinskaya, N. L., 1970,
Interaction of analyzers in dolphins during discrimination
of geometrical figures under water, J. Higher Neural Act.,
20:1070 (English translation in: "A Collection of Translations of Foreign Language Papers on the Subject of Biological Sonar Systems", K. J. Diercks, 1974, ed., Applied
Research Lab, U. of Texas, Austin, Tech. Rept. 74-9).

Barta, R. E., 1969, Acoustical pattern discrimination by an Atlantic bottlenosed dolphin, unpublished manuscript, Naval
Undersea Center, San Diego, Ca.

Bel'kovich, V. M., and Borisov, V. I., 1971, Locational discrimination of figures of complex configuration by dolphins,
Trudy Akusticheskogo Institute, 17:19.

Bel'kovich, V. M., Borisov, I. V., Gurevich, V. S., and Krushinskaya, N. L., 1969, Echolocating capabilities of the common
dolphin (Delphinus delphis), Zoologicheskiy Zhurnal, 48:876.

Bel'kovich, V. M., and Dubrovskiy, N. A., 1976, "Sensory Bases of
Cetacean Orientation", Nauka, Leningrad, (English translation JPRS L/7157).

Busnel, R.-G., and Dziedzic, A., 1967, Résultats métrologiques
expérimentaux de l'écholocation chez le Phocoena phocoena
et leur comparaison avec ceux de certaines chauves-souris.
in: "Animal Sonar Systems, Biology and Bionics", R.-G. Busnel, ed., Laboratoire de Physiologie Acoustique, Jouy-en-
Josas, France.

Busnel, R.-G., Dziedzic, A., and Andersen, S., 1965, Seuils de
perception du système sonar du Marsouin Phocoena phonoeca
en fonction du diamètre d'un obstacle filiforme, C. R. Acad.

Sci., 260:295.

Dubrovskiy, N. A., 1972, Discrimination of objects by dolphins using echolocation, "Report of the 5th All-Union Conference on Studies of Marine Mammals, Part 2, Makhachkala", as cited in: Ayrapet'yants, E. Sh. and Konstantinov, A. I., 1974, "Echolocation in Nature", Nauka, Leningrad, (English Translation JPRS 63328-2).

Dubrovskiy, N. A., and Krasnov, P. S., 1971, Discrimination of elastic spheres according to material and size by the bottle-nose dolphin, Trudy Akusticheskogo Institute, 17, as cited in: Bel'kovich, V. M., and Dubrovskiy, N. A., 1976, "Sensory Bases of Cetacean Orientation", Nauka, Leningrad (English translation JPRS L/7157).

Dubrovskiy, N. A., Krasnov, P. S., and Titov, A. A., 1971, On the question of the emission of ultrasonic ranging signals by the common porpoise, Akusticheskiy Zhurnal, 16:521 (English translation JPRS 52291).

Dubrovskiy, N. A., Krasnov, P. S., and Titov, A. A., 1971, Discrimination of solid elastic spheres by an echolocating porpoise, Tursiops truncatus, in: "Proc. 7th International Conf. Acoust.", Budapest.

Evans, W. E., 1973, Echolocation by marine delphinids and one species of fresh-water dolphin, J. Acoust. Soc. Am., 54:191.

Evans, W. E., and Powell, B. A., 1967, Discrimination of different metallic plates by an echolocating delphinid, in:"Animal Sonar Systems, Biology and Bionics", R.-G. Busnel, ed., Laboratoire de Physiologie Acoustique, Jouy-en-Josas, France.

Fadeyeva, L. M., 1973, Discrimination of spherical targets with different echo signal structure by the dolphin, in: "Report from the 8th All-Union Acoustical Conference, Moscow, 1" As cited in: Bel'kovich, V. M., and Dubrovskiy, N. A., 1976, "Sensory Bases of Cetacean Orientation", Nauka, Leningrad. (English translation JPRS L/7157).

Hammer, C. E., 1978, Echo-recognition in the porpoise (Tursiops truncatus): an experimental analysis of salient target characteristics, Naval Ocean Systems Center, San Diego, Tech. Rep. 192.

Johnson, C. S., 1967, Discussion to paper by Evans and Powell, in: "Animal Sonar Systems, Biology and Bionics", R.-G. Busnel, ed., Laboratoire de Physiologie Acoustique, Jouy-en-Josas, France.

Johnson, C. S., 1967, Sound detection thresholds in marine mammals, in: "Marine Bio-Acoustics, Proc. of the Second Symposium on Marine Bio-Acoustics, New York", W. N. Tavolga, ed., Pergamon Press, New York.

Kellogg, W. N., 1958, Echo ranging in the porpoise, Science, 128:982.

Norris, K. S., Evans, W. E., and Turner, R. N., 1967, Echolocation of an Atlantic bottlenose porpoise during discrimination, in: "Animal Sonar Systems, Biology and Bionics", R.-G. Busnel, ed., Laboratoire de Physiologie Acoustique, Jouy-en-

Josas, France.

Schusterman, R. J., 1979, Behavioral methodology in echolocation
    by marine mammals, this volume.

Schusterman, R. J., Kersting, D. A., and Au, W. W. L., 1979, Stim-
    ulus control of echolocation pulses in Tursiops truncatus,
    this volume.

Titov, A. A., 1972, Investigation of sonic activity and phenomeno-
    logical characteristics of the echolocation analyzer of
    Black Sea delphinids. Canditorial dissertation, Karadag,
    as cited in: "Sensory Bases of Cetacean Orientation", Nauka,
    Leningrad (English translation JPRS L/7157).

Turner, R. N. and Norris, K. S., 1966, Discriminative echolocation
    in a porpoise, J. Exp. Anal. Behav., 9:535.

Zaslavskiy, G. L., Titov, A. A., and Lekomtsev, V. M., 1969,
    Research on the sonar abilities of the bottlenose dolphin,
    Trudy Akusticheskogo Instituta, 8:134, as cited in:
    Ayrapet'yants, E. Sh., and Konstantinov, A. I., 1974,
    "Echolocation in Nature", Nauka, Leningrad (English trans-
    lation JPRS 63328).

CETACEAN OBSTACLE AVOIDANCE

Patrick W. B. Moore

SEACO, Inc.
146 Hekili St.
Kailua, Hawaii 96734

As Schevill (introduction to McBride, 1956) has pointed out, the original ideas about cetacean echolocation were based on observations of the porpoises' ability to avoid obstacles in turbid waters. McBride (1956) noted that although porpoises (Tursiops truncatus) avoided fine mesh nets, by either rolling over the corkline or swimming directly through any openings in the net, they would charge larger mesh nets (24 cm) and were easily caught.

W. N. Kellogg in 1958 performed the first experimental obstacle avoidance task involving cetaceans. Two bottle-nosed dolphins (T. truncatus) were placed in a specially dredged pool, 16 x 21 m and 1.6 to 2.1 m deep (depending on tide level). The water was turbid (visibility less than 50 cm). Using suspended barriers of triangular metal poles or posts 1.32 m long, 5.08 cm wide and spaced 2.44 m apart, the number of animal collisions were recorded in twenty minute sessions. Only four such collisions occurred in the first session by the two animals. In those instances the animal had passed the obstacle but brushed it with its tail flukes. Sessions were also conducted during dark, moonless nights with no resulting collisions.

Kellogg (1958) also conducted experiments involving glass and plexiglass barriers. The experimenter required the porpoise to select the one of two openings in a suspended net which did not contain a barrier in order to receive a fish. Additional incentive to pass through the selected opening was generated by slowly crowding the animal with a larger net. A combined 98% correct performance level was demonstrated in the plexiglass experiment and 100% correct level using plate glass.

Interestingly enough, it is now known that even though porpoises can detect the presence of a plexiglass (or acrylic) barrier, a thin sheet of this material is sufficiently transparent to the echolocation pulses and return echoes of the animal that it can detect objects through it (Penner and Murchison, 1970). The visual analogy would be looking through a glass window; one knows it is there, but it is still transparent.

Norris, Prescott, Asa-Dorian and Perkins (1961) demonstrated echolocation unequivocally by blindfolding a female porpoise (T. truncatus) with soft rubber eye cups and requiring her to swim through a series of mazes formed by two rows of iron pipes. The rows were 3.05 m apart and pipes were suspended at 1.22 to 2.13 m intervals. The animal swam through various maze patterns several times a day over a 60-day period. Norris et al. (1961) reported that only one collision occurred during this time, and they considered that collision to be deliberate.

Busnel and Dziedzic(1967) were the first to attempt to make a defined and quantifiable evaluation of the performance of an underwater biosonar system other than T. truncatus, using a harbor porpoise (Phocoena phocoena) in an obstacle avoidance experiment with various diameter iron, steel, or copper wires (diameters of 92 mm to 0.2 mm). These obstacles were vertically suspended in a tank 1 m apart in alternate rows. The animal was either blindfolded or sighted. When blindfolds were used the animal showed "practically 100%" avoidance for wire diameters between 4 and 2.8 mm, 90% for diameters of 2.8 and 0.5 mm, 79% for diameters of 0.35 mm and finally reached 46% for diameters of 0.2 mm.

When the animal was sighted the results indicated "practically 100%" performance for wire diameters down to 0.75 mm; 90% for diameters of 0.6 mm and 0.5 mm; and 79% for diameters of 0.35 and 0.2 mm. The authors reported that "there appeared to be no correlation between the nature of the metal thread and ability of the animal to avoid it". Measurement of the acoustic signals emitted by this animal was reported to have a maximum intensity at 2 kHz.

Two problems seem to exist with Busnel and Dziedzic's original experiments. First, the reported peak frequencies around 2 kHz would indicate the detection of wire diameters of extremely small fractional sizes compared to the wavelength of the echolocating pulse. Schevill, Watkins, and Ray (1969) have pointed out that the large number of harmonic frequencies found by Busnel and Dziedzic were probably a result of their use of recordings made with systems of limited response. Schevill et al. (1969) reported occasional frequencies as high as 10 kHz in the clicks of P. phocoena with the "favored click" generally

narrow band and less than 2 kHz. Anderson's (1970) audiogram of
this species indicates their best hearing range is between 4-40
kHz, with upper frequency limits near 150 kHz. Supin and Sukhoruchenko (1974) found Phocoena hearing as high as 190 kHz, with the
best frequency range between 64 and 128 kHz; and Sukhoruchenko
(1973) says they hear from 3-190 kHz. Comparison of these Phocoena
hearing studies with the echolocation pulse frequencies reported
by Busnel and Dziedzic, and Schevill et al. indicates some discrepancy, as previous studies have indicated that other echolocating
animals produce signals that are matched to their hearing abilities
(Johnson, 1966; Diercks et al., 1973).

The second problem in the Busnel and Dziedzic obstacle
avoidance study was related to the 0.35 mm detection threshold.
Penner and Murchison (1970) during a study of Inia geoffrensis
wire detection thresholds found that data fluctuated widely on
wire diameters below 1.0 mm. The confounding variable turned out
to be minute bubble formation on the wires which occurred within
60 sec. following submersion. An elaborate procedure involving
special polishing of target wires, wetting agent treatment between
trials, and greatly restricted immersion times (less than 15 sec)
was developed to eliminate this confounding variable. However,
Busnel and Dziedzic apparently did not eliminate bubble formation
on the wires used in the Phocoena experiments (see Schusterman,
this volume).

FIELD OBSERVATIONS

Field recordings of cetacean sound and associated behavior
have been conducted in an attempt to discover the extent of
echolocation throughout this group of animals. Recordings of
blue whale (Balaenoptera musculus) clicks and Minke whale
(B. acutorostrata) clicks by Beamish and Mitchell (1971, 1973)
and gray whale (Eschrichtius robustus) "clicks" by Wenz (1964),
Gales (1966), Poulter (1968) and Fish, Sumich and Lingle (1974)
suggested that mysticete cetaceans may also be possible echolocators.

However, Eberhardt and Evans (1962) observed gray whales in
their calving ground in Baja, California, to collide with an
acoustically reflective barrier. Recordings were also made of
the associated sound activity of the gray whales. Eberhardt and
Evans erected an acoustic barrier consisting of a string of 5.08
cm diameter, 4.57 m aluminum tubes, 6.09 m apart. This barrier
was placed part way across the Piedra channel. Gray whale sounds
were recorded as were the "sounds caused by spar buoys hitting
together as whales crashed through the barrier". They noted that
"no biological sounds from the whales were evident" (Eberhardt
and Evans, 1962).

During this same research cruise, five <u>Tursiops</u> <u>gilli</u> were
observed moving toward the channel barrier and echolocation
clicks were recorded. As the animals approached the barrier they
moved over to shoal water and schooled up. After five minutes
one animal made a "sonar run" on the barrier and returned to the
group. Several more single animal runs were observed, followed
by the entire group slowly proceeding up to and through the
barrier with no collisions reported (Evans and Dreher, 1962).

The reported collisions of the gray whales do not completely
eliminate the possibility of echolocation in an animal producing
sounds of long wavelengths. It is very possible that these
animals could use the various sounds they produce to obtain echo
information about water depth, gross bottom contours and large
obstacles in much the same fashion that blind persons detect
walls and estimate room size based on the timbre and other echo
characteristics produced by their footsteps, fingersnaps or their
own voices (Rice, C. E., Feinstein, S. H., Schusterman, R. J.,
1965). Although it may not be the primary function of their
"singing", water depth information is certainly available in the
echoes of humpback whale (<u>Megaptera</u> <u>novaeangliae</u>) "songs".

Evidence for echolocation among the mysticeti may remain in
question for some time, as their size would seem to eliminate the
possibility of the kinds of captive experiments necessary to
prove echolocation.

THE TUNA-PORPOISE PROBLEM

In order to reduce the annual porpoise mortality rate due to
purse seine fishing of yellowfin tuna, several lines of investigation have been pursued that are directly related to obstacle
avoidance by the various species of porpoise (<u>Stenella</u> <u>attenuata</u>,
<u>S. longirostris</u>, <u>Delphinus</u> <u>delphis</u>) which are associated with the
tuna catch. The basic question is how the porpoise group associated
with the tuna school can be released from the net confines
without significant loss of the tuna catch.

According to Norris et al. (1978), the tuna catch procedure
is as follows: After the porpoise group has been spotted, the
seiner "tends to turn continuously counterclockwise in an ever
diminishing spiral". At the onset this spiral is quite large and
speedboats "work the area where the wake is weakest and reinforce
the wake by turning in arcs and circles at high speed". The
porpoise, approaching the bubble wake, turn and avoid the wake.
The seiner drops the net skiff "when the diameter of the spiral
reaches the appropriate size and when the wind is in the proper
quadrant", and starts to deploy the net in a circle-like tract,

using the wake (an acoustic obstacle) and the net (an acoustic
obstacle and a physical barrier) to surround the animals.
Entrapment is complete when the seiner and net skiff rejoin.

Norris et al. (1978) has clearly described the general
behavior of the porpoise group, saying that during the capture
sequence the animals stopped in the spiral-shaped bow wake of the
seiner, about 137 m from the wake. This suggests biosonar detection
of the bubble wake. They do not penetrate the wake or approach
it. Some do, however, get past the speed boats attempting to
herd the animals, passing behind the smaller boats. The majority
of the animals form an aggregate inside the net. Within this
aggregate a compression of natural behavior occurs with associated
increases of aggressivity and acoustic behavior. A few animals
dive below a rafting group (animals hanging quietly in the water)
down to approximately 60 feet.

W. F. Perrin and J. R. Hunter (1972) performed a number of
experiments to assess the escape behavior of the spinner porpoise
(S. longirostris). Perrin and Hunter tried to discover what
configuration, if any, of a rescue gate in a purse-seine net
could be best recognized and utilized by the entrapped porpoise.
The subjects were five spinner porpoises, three of which were
trained animals and two naive animals. The animals were tested
in a 24.7 m circular crowding tank, 4 meters deep at the center.
Nets were placed around the edge to create a 20 m diameter
circular crowding chamber. Two radial crowding nets, one fixed
and one pivoting, extended from a central aluminum mast to the
outer edge of the chamber. The moving net wall was used to
simulate a slowly closing net, crowding the animals closer to the
fixed net which was modified for various types of escape openings.
The net was tuna purse-seine webbing 10.8 cm, #42 thread, knotted
nylon. Flotation was by purse-seine corkline with 15.0 cm
diameter x 9 cm sponge plastic floats. The basic 5.5 m wide x
1.8 m deep escape opening was modified via a purse-seine net flap
to decrease the width and/or depth. A cork line was strung across
the surface of the opening to test responses to an overhead
surface barrier, and various low acoustic reflective panel
materials were attached for tests of responses to an occluded
opening. The panels consisted of: 8.6 cm #12 mono filament
webbing, a .38 mm or 1.04 mm thick polyvinyl sheet. Results of
the various experiments were expressed as a function of area of
chamber, at escape, in m vs. trials.

The effect of width of opening was tested and, generally,
the results suggested that the opening width was not a significant
factor effecting escape if it exceeded 1.5 m. Interesting
behavior occurred with failure to escape in these experiments;
occasionally, the test animal would expel air and sink passively

to the bottom of the chamber and would not move even when the net
was completely collapsed. Similar behavior was seen in the tuna
net at sea (Norris K., Personal Comm.).

The effect of depth was tested by presenting various hole
depths of 0.15 m to 1.8 m and holding hole width at 5.5 m. The
authors' tentative conclusion was that a depth of opening of 1 m
or greater is "optimal" for escape. When a corkline was placed
across the opening, initial escape performance always decreased.
However, with greater depth of opening, the effect of the corkline
on escape performance was lessened.

When a monofilament web panel was inserted into the opening
all animals swam into the webbing "as if it did not exist"
(Perrin & Hunter, 1972), and the animals had to be unentangled.
With a clear polyvinyl sheet placed in the opening (only for two
trials with the same animal) the porpoise hit the panel and slid
over the top as the panel buckled. Noteworthy was the fact that
with the polyvinyl sheet the animal passed back and forth over
the panel several times while the enclosure was constantly being
reduced in size.

These series of experiments were repeated with two naive
porpoises beginning the day after capture. Both these animals
showed erratic behavior in the experimental area due to what was
believed to be shock or extreme fright and in the depth experiment one of the animals when confronted with a 0.61 m deep
opening dove and became entangled in the net. Width of opening
and corkline experiments with the non-naive animals (captive for
three years) showed the same general trends as with the naive
animals.

The authors generally suggest that an escape gate in a purse
seine net should be no less than 1.5 m wide and 1.0 m deep.
There should be no corkline; self-actuating release ports, if
used, should be made of acoustically transparent materials.

This study suggests that an entrapped spinner porpoise, when
faced with a net barrier obstacle and stressed by crowding, can
locate, either visually or acoustically, a near surface void and
manage an escape, given that the void meets some general size and
configurational criteria. Also, the results of the acoustically
transparent panels and net material suggest that echolocation was
used to access the escape route.

Leatherwood, Johnson, Ljungblad and Evans (1977) measured
the target strength of various ($1\frac{1}{4}$, 2 and $4\frac{1}{2}$ inches) tuna net
panels at 0° and 35° aspect to access their "acoustic" visibility
to echolocating porpoises. Using a broadband sound source
representing a porpoise echolocation signal, Leatherwood et al.

reported that, "porpoises utilizing their echolocation system, should be able to "see" any of the mesh sizes measured regardless of the angles at which they are oriented and regardless of the degree of stretch in the nets". However, the degree of acoustic visibility varied 17dB between net types. They note that this 17 dB difference in target strength may have significant advantage in the complex acoustic environment of the purse seine net (see Murchison, this volume). Of course, to what extent the animals can or do echolocate inside the purse seine net is unknown.

Awbrey (in press) investigated the general acoustic environ of the purse seining operation during October 1976 to discover to what extent, if any, the purse seining operation would mask porpoise echolocation or communication signals. Recordings were made during one "water set" (no animals present) and nine actual porpoise-tuna sets.

Measurements of noise produced by net boat, the speed boats, and the sound of the net itself were analyzed. (Measurements reported were in dBre 1 µpa.) When the net boat approaches the tuna-porpoise aggregate, 100-200 m from the animals, 91% of the total noise energy was below 6 kHz peaking at 360 Hz. Also, low frequency pulsations were caused by propeller beats with sound pressure levels (SPL) of 120-125 dB at 200 m from the ship, 85% of the speed boat noise (Peak = 125 dB for short periods) during the set was below 6 kHz peaking at 4 kHz (SPL of 117 dB). Once the net is out and pursing begins high frequency clashing sounds of the pursing rings at the bottom of the net are produced. These transient sounds peaked at 5-6 kHz and contained energy at frequencies above 30 kHz (no SPL given). After the net is out and the pursing operation is completed the general net ambient noise is considerably reduced and noise is primarily generated from the hauling in of the net (SPL of 105-115 dB with major energy below 3 kHz). Measurements of animal whistles during the hauling operation suggest that vessel noise had little effect on the general characteristics of the vocalization. Lastly, during the "backing down" operation some low frequency noise bursts are produced by the net boat (70% energy below 6 kHz) at an SPL of 125 dB.

Awbrey (in press) summarizes his report by stating that "reducing vessel noise would probably contribute very little to reduction of porpoise mortality", in that the majority of the purse seining noise lies in a frequency range below the animal's vocalizations and echolocation clicks and probably does not prevent the animals from hearing their own whistle or echolocating on the net. However, the noise can be considered an aversive stimulus to the animals because loud transients do elicit fright responses.

Perryman and Au (1977) observed porpoises in the Commission Yellowfin Regulatory Area beginning to avoid the seiners at distances of 5-6 km. Awbrey (in press) reports that both spinner and spotters now began avoidance behavior (running or lying quietly) at greater distances, suggesting that with repeated exposure to purse seining the animals gain sophistication with respect to avoiding the tuna fishing operation. This notion is also supported by observations made by Pryor and Norris (1978). During the initial chase the animals are herded by the speed boats and are eventually stopped by the vast bubble curtain created by the bow wake of the large seiner, the net skif is set free and while the speed boats herd the animals, the net is set around the tuna-porpoise pod; however, some animals have "learned....to go under the bubble curtain of the ship" (Pryor and Norris, 1978).

Ingrid Kang, (Personal Comm.) an observer on the vessel Queen Mary, reported spinners occasionally jumping and flipping inside   the net. Spotters, on the other hand, tended to raft. When a small dingy was inside the net with the animals, both spotters and spinners would avoid the dingy, but if divers entered the water, the animals did not seem to try to avoid them, and the rafting animals could easily be approached. Additionally, the tighter the net confines, the less difficult it was to approach the animals. Also, at no time was an animal observed to charge the net. During some net sets on combined spinners and spotters, large amounts of small bubbles were present throughout the animal aggregate, generated by the spinners whistling vocalizations; obviously these bubbles would have reduced the acoustic transmission quality of the surrounding water and hamper the animal's echo-location signals when it would seem to have been needed the most. Results from the Perrin and Hunter (1972) study and the Leatherwood et al. (1977) study coupled with Awbrey's (in press) noise measurements suggests that the porpoises could have used echolocation to find and exit the tuna net. The extent to which these animals utilize information, gathered via the various sensory modalities of passive hearing, echolocation, and vision, to evaluate and react to the tuna net operation is unknown. Also, the relative dominance of these sensory information channels, if any, is a purely speculative matter at this point.

In general, the entire tuna fishing operation is a complex of avoidance behaviors for the porpoise, the exact meaning of which has yet to be completely explained, but certainly relates to the acoustic abilities and obstacle avoidance strategies used by these animals. Also, to what degree the visual, echolocation and passive hearing systems share prominance in information acquisition is a question of current interest (see Schusterman, this volume).

Presently, Japan is involved in deep sea salmon fishing, using 16 km x 6 m monofilament gill nets. This fishing operation accounts for a seasonal mortality of Dall's porpoise of 10-20 thousand when animals became entangled and drowned over night in the nets. Various other small cetaceans are also subject to gill net fishing operations from Greenland to Northern Australia (Evans, W. E. and Norris, J., Personnal Comm.). The Perrin and Hunter (1972) study, using a monofilament panel, shows similar non-avoidance to these presumably acoustically transparent nets.

CETACEAN STRANDING

W. H. Dudock Van Heel (1962, 1966) put forth the hypothesis that mass strandings of Cetaceans were due to the inability of the animal to correctly navigate or identify a gently shelving shoreline due to either the physical properties of the returned echo (i.e. reduced strength, reverberation) or the animals not "attending" to the low frequency components of the return pulse which would, presumably, give an indication of a gently shelving shoreline. Other explanations, such as group sickness or parasitic infestations coupled with ideas of a disoriented dominant or leader animal stranding a generally healthy following, also abound (Wood, 1978).

F. G. Wood (1978) suggested a non-acoustic stress hypothesis to explain mass strandings. The stress hypothesis suggests that highly stressful situations would cause basic and primitive behavioral responses, and the stranding act is seen as a stress or fear response, which is a primitive and evolutionarily regressive response, to seek safety on land. Stress may also act as the catalyst to produce other behaviors. The observed behaviors of rafting inside the tuna net or the passive sinking of the spinner porpoise noted in the Perrin and Hunter (1972) study may also be stress induced.

CONCLUSION

The use of the obstacle avoidance paradigm in cetacean echolocation research has declined in recent years, since the technique does not yield the rich and precise results of more rigorous experimental designs (see Schusterman, this volume). It does, however, offer the advantage of less animal training.

It should be noted, however, that to date the only behavioral evidence relating to echolocation capabilities in mysticetes has come from field-conducted obstacle avoidance studies (Eberhardt and Evans, 1962) suggesting that obstacle avoidance has not yet

outlived its usefulness as a field observational technique
intended to glean echolocation behavior in less well documented
species.

ACKNOWLEDGEMENTS

     I would like to thank Ronald Schusterman, Paul Nachtigall
and Earl Murchison for the helpful advice and support in the
preparation of this manuscript, Phyllis Johnson for her cooperation and understanding in typing this subchapter, Jack Harmon of
SEACO, Inc., for his financial support and the Naval Ocean Systems Center for providing this opportunity.

REFERENCES

Andersen, S., 1970, Directional hearing in the harbor porpoise,
     Phocoena phocoena, in: "Investigation on Cetacea, Vol. II",
     G. Pilleri, ed., Benteli Ag., Berne.
Awbrey, F. T., (in press), Background study of acoustical and
     bioacoustical factors in tuna fishing productivity and
     associated porpoise mortality, Fish. Bull.
Beamish, P., and Mitchell, E., 1971, Ultrasonic sounds recorded
     in the presence of a blue whale Balaenoptera musculus, Deep-
     Sea Res., 18:803.
Beamish, P., and Mitchell, E., 1973, Short pulse length audio
     frequency sounds recorded in the presence of a Minke whale
     Balaenoptera acutorostrate, Deep-Sea Res., Abstra., 20:375.
Busnel, R.-G., and Dziedzic, A., 1967, Résultats métrologiques
     expérimentaux de l'écholocation chez le Phocoena phocoena
     et leur comparaison avec ceux de certaines chauves-souris,
     in: "Animal Sonar Systems, Biology and Bionics", R.-G. Bus-
     nel, ed., Laboratoire de Physiologie Acoustique, Jouy-en-
     Josas, France.
Diercks, K. J., Trochta, R. T., and Evans, W. E., 1973, Delphinid
     sonar: measurement and analysis, J. Acoust. Soc. Amer.,
     54:200.
Dudok van Heel, W. H., 1962, Sound and Cetacea, Neth. Jour. Sea
     Res., 1:407.
Dudok van Heel, W. H., 1966, Navigation in Cetacea, in: "Whales,
     Dolphins, and Porpoises", K. S. Norris, ed., U. of Calif.
     Press, Berkeley.
Eberhardt, R. L., and Evans, W. E., 1962, Sound activity of the
     California gray whale, Eschrichtius glaucus, J. Aud. Eng.
     Soc., 10:324.
Evans, W. E., and Dreher, J. J., 1962, Observations on scouting
     behavior and associated sound production by the Pacific
     bottlenosed porpoise (Tursiops gilli Dall), Bull. Soc.
     Calif. Acad. Sci., 61:217.

Fish, J. F., Sumich, J. L., and Lingle, G. L., 1974, Sounds pro-
    duced by the gray whale, Eschrichtius robustus, Mar. Fish
    Rev., 36:38.
Gales, R. S., 1966, Pickup analysis and interpretation of under-
    water acoustic data, in "Whales, Dolphins, and Porpoises",
    K. S. Norris, ed., U. of Calif. Press, Berkeley.
Johnson, C. S., 1966, Auditory thresholds of the bottlenosed
    porpoise, (Tursiops truncatus Montagu), N. O. T. S. TP 4178.
Kellogg, W. N., 1958, Echo ranging in the porpoise, Science,
    128:982.
Leatherwood, J. S., Johnson, R. A., Ljungblad, D. K., and Evans,
    W. E., 1977, "Broadband measurements of underwater acoustic
    target strengths of panels of tuna nets", NOSC TR 126, San
    Diego.
McBride, A. F., 1956, Evidence for echolocation by cetaceans,
    Deep-Sea Res., 3:153.
Murchison, A. E., 1979, Detection range and range resolution of
    echolocating bottlenose porpoise (Tursiops truncatus),
    this volume.
Norris, K. S., Prescott, J. H., Asa-Dorian, P. V., and Perkins, P.,
    1961, An experimental demonstration of echolocation behav-
    ior in the porpoise, Tursiops truncatus (Montagu), Biol.
    Bull., 120:163.
Norris, K. S., Stuntz, W. E., and Rogers, W., 1978, The behavior
    of porpoises in the eastern tropical Pacific yellowfin tuna
    fishery - Preliminary Report, U.S. Marine Mammal Comm.,
    Report PB-283 970, Washington, D.C.
Penner, R., and Murchison, A. E., 1970, Experimentally demonstrated
    echolocation in the Amazon river porpoise Inia geoffrensis
    (Blainville), Proc. 7th Ann. Conf. Biol. Sonar and Diving
    Mammals, 7:17.
Perrin, W. F., and Hunter, J. R., 1972, Escape behavior of the
    Hawaiian spinner porpoise (Stenella cf. S. longirostris),
    U.S. Nat. Mar. Fish. Bull., 70:49.
Perryman, W., and Au, D., 1977, Aerial observations of evasive
    behavior of dolphin schools, Proc. 2nd Conf. Biol. Mar.
    Mammals, Abstr.
Poulter, T. C., 1968, Vocalization of the gray whales in Laguna
    Oja de Liebre (Scammon's Lagoon), Baja, California, Mexico,
    Norsk Hvalfangsttid, 58:53.
Pryor, K. and Norris, K. S., 1978, The tuna porpoise problem:
    Behavioral aspects, Oceanus, 21:31.
Rice, C. E., Feinstein, S. H., and Schusterman, R. J., 1965, Echo-
    detection ability of the blind: size and distance factors,
    J. Exp. Psych., 70:246.
Schevill, W. E., 1956, Evidence for echolocation by cetaceans,
    Deep-Sea Res., 3:153.
Schevill, W. E., Watkins, W. A., and Ray, C., 1969, Click structure
    in the porpoise, Phocoena phocoena, J. Mamm., 50:721.

Schusterman, R. J., 1979, Behavioral methodology in echolocation
        by marine mammals, this volume.
Sukhoruchenko, M. N., 1973, Frequency discrimination in dolphins
        (Phocoena phocoena), Sechenov Physiological Jour. of the
        U.S.S.R., 59(8):1205.
Supin, A. Ya., and Sukhoruchenko, M. N., 1974, Characteristics of
        acoustic analyzer of the Phocoena phocoena L. dolphin, in:
        "Morphology, Physiology, and Acoustics of Marine Mammals",
        V. Ye. Sokolov, ed., Nauka, Moscow (English translation
        JPRS 65139).
Wenz, G. M., 1964, Curious noises and the sonic environment in
        the ocean, in: "Marine Bio-Acoustics, Proceedings of the
        First Symposium on Marine Bio-Acoustics, Bimini", W. N.
        Tavolga, ed., Pergamon Press, New York.
Wood, F. G., 1978, The cetacean stranding phenomenon: an hypothesis,
        in: "Report on the Marine Mammal Stranding Workshop",
        N.T.I.S., Washington, D.C.

PERFORMANCE OF AIRBORNE ANIMAL SONAR SYSTEMS:

I. MICROCHIROPTERA

Hans-Ulrich Schnitzler and O'Dell W. Henson, Jr.

Fachbereich Biologie, Philipps-Universität
355 Marburg/Lahn BRD
Department of Anatomy, U. of North Carolina
Chapel Hill, N.C. 27514   U.S.A.

INTRODUCTION

Animals living in the dark are faced with the problem that
vision is only of limited use for orientation in the environment
and for the detection and identification of relevant targets. In
order to cope with this situation, some animals, during evolution,
developed eyes which were especially adapted for good visual efficiency at low light intensities whereas others improved the remaining
sensory systems.

Echolocation is one of the adaptations that has evolved for a
life in the dark. Echolocating animals emit sounds and analyze the
returning echoes in order to detect the presence and character of the
reflecting targets. In contrast to other sensory modes, echolocation
is an active system which uses its own controlled energy beam to
illuminate interesting parts of the environment and to probe object
characteristics.

Of the numerous nocturnal birds, only the cave-dwelling oil
bird and cave swiftlets use echolocation. In mammals, echolocation
systems of low efficiency have been reported for some species of
rodents, insectivores and Megachiroptera, whereas Microchiroptera
and cetaceans use very efficient and sophisticated systems.

In the first part of this article on airborne animal sonar
systems we shall describe the operational principles and functions
of the highly developed echolocation systems of Microchiroptera. In
the second part, we shall summarize the rather poor echolocation

performance of other animals which use airborne sonar, including
Megachiroptera. Reviews dealing with these topics have been
published by Griffin (1958), Schnitzler (1973b; 1978), Airapetianz
and Konstantinov (1970; 1974) and Novick (1977).

INFORMATION AVAILABLE FROM THE ECHO

Bats analyze the acoustical parameters of their orientation
sound echoes in order to obtain information on the features of
targets in their surroundings. Echolocation is capable of delivering the following information.

Detection: For the detection of a target, the bat has to make
the decision whether an echo of its own orientation sound is present
or not. However, it is hard to conceptualize detection independent
of further information extraction.

Range: The time lag between sound emission and echo reception
is a measure of the distance between bat and target. The more
exact the measurement of this travel time, the more accurate is
the range measurement.

Angular direction: The angle at which the wavefront of an echo
returns to the bat determines the direction of a target. Bats use
two ears, which act as directional antennae, to measure the angle
of the echoes received.

Relative velocity: The relative velocity between bat and
target is encoded in Doppler frequency shifts contained in the
echoes. However, it can also be determined by consecutive range
measurements.

Target oscillations: Fluttering movements of the target, e.g.
insect wingbeats, produce amplitude and frequency modulations in
the echoes owing to changes in the exposed surface area of the target
and due to Doppler shifts.

Target size: The amplitude of an echo can be used as a rough
size indicator for targets. However, this relationship is not always
reliable. A good determination of target size is dependent on the
wavelengths of the signals and is only possible if an echolocation
system can resolve dimensions which are significantly less than
the target dimensions (Skolnik, 1970).

Target shape and surface properties: The illumination of a
target with sound leads to a three-dimensional, scattered field of
echoes. The information on the shape and surface properties of the
target is encoded in the complex spectral composition and temporal
structure of the whole echo field. The analysis of the field

section which returns in the bat's direction gives limited information with regard to the shape and the surface properties of the
target itself. More information may be gained if the target is
probed from different directions.

An echolocation signal that is ideal for the extraction of all
these types of information does not exist. This becomes evident,
for example, when one compares the restraints for range and velocity
determination. For high accuracy in range measurement, broadband,
short signals are needed whereas long signals of constant frequency
allow great precision in velocity determinations. The echolocation
signals of bats are therefore compromises adapted to the special
information needs of each species.

INTERFERENCES IN ECHOLOCATION

Echolocating bats may have problems in characterizing their
targets owing to the following interference factors.

Noise: If a target is very small or far away, the echo may
be so weak that it is buried in external and/or internal noise.
Higher signal energy or an improved noise rejection of the auditory
system may help to overcome this problem.

Clutter: Echoes from reflectors other than the interesting
target are called clutter. In order to obtain maximal information
regarding the interesting target it is necessary to separate the
target echo from the clutter. A proper signal design, high
directionality of the echolocation system and special strategies
in information processing may help to overcome this problem.
Extreme clutter problems arise in fog (Pye, 1971) which scatters
the orientation sounds in all directions and thus makes echolocation
difficult to impossible.

Signals from other bats: If other bats are close to an echolocating bat, it will receive their signals and the echoes from
their signals. In the field, interference from other bats may not
be of any significance owing to the high degree of directionality
of the echolocation systems and the rather low bat density. However
in roosts, or when leaving the roosts, high bat densities are
possible. It may be that in such cases, bats use their highly
developed spatial memory (Neuweiler and Möhres, 1967) to overcome
this problem.

Frequency dependence of directionality: The directionality of
sound emission and hearing is frequency dependent. Echoes from
different directions therefore have a different spectral composition.
In order to overcome this effect, which alters all information based

on spectral cues, bats would have to take the spectral changes typical for each direction into account. Another strategy would be to use spectral cues only after a target has been fixated.

Signal changes due to atmospheric influences: The atmospheric attenuation of sound depends on temperature, humidity, composition of the air and frequency of the propagated sound. High frequencies are attenuated more than low frequencies and high humidity is more effective in attenuation than low humidity (Griffin, 1971). Tropical conditions therefore lead to marked changes in the spectral composition of echoes. A countermeasure would be for bats to recalibrate their echolocation systems to every atmospheric condition.

An echolocating signal ideal for avoiding all these interference factors is theoretically impossible. The echolocation signals of different bats are therefore not only adapted to the information needs but also to the special interference problems encountered by each species.

INFORMATION GATHERING PROBLEMS IN BATS

The more than 800 different species of echolocating bats live in various ecological situations. Apart from problems common to all bats, such as finding their way and their roosts and avoiding obstacles in the dark, the different species must also solve different problems. Those that hunt insects in the open must have different strategies than those hunting in areas full of obstacles or those that hunt stationary insects. The bats that feed on fruit, pollen, nectar, small vertebrates, surface fishes or on blood must also solve a wide variety of different information gathering and interference problems. As a result, the emitter and receiver of the echolocation systems in different species of bats, i.e. the sound emission apparatus, the peripheral auditory receptors and the information processing areas of the auditory system, have been adapted during evolution for an optimal gathering of relevant information.

A comparison of the physical characteristics of the orientation sounds found in different situations (e.g. when looking for food or avoiding obstacles) with signal detection theories developed for radar and sonar systems, allows conclusions to be made regarding the information collection preferences of the echolocation systems in different species.

Neurophysiological studies with stimuli which simulate natural echolocative signal parameters help to reveal the information processing principles of the auditory system. Owing to the fact that behaviorally relevant information can be rather clearly defined

in bats, they are ideal experimental animals for the investigation
of neural information processing.

TYPES OF ECHOLOCATION SIGNALS

The echolocation signals of different bats have been developed,
through evolutionary processes, as optimal adaptations
for gathering behaviorally relevant information and suppressing
unwanted interference. The evolved signals were developed with
differences in frequency structure, sound pressure level and
duration.

The basic echolocative signals of bats consist of frequency
modulated (fm) components alone or a combination of constant frequency (cf) components and fm components. The fm components are
rather short, with durations below 5-10 msec, whereas cf components
lasting from a few msec up to more than 100 msec have been found.
Apart from the pure fm sounds the most frequently observed combinations are the long cf-fm, the short cf-fm and the fm-short-cf sounds.
Further variations in the frequency structure of echolocation signals
can be due to different harmonic content.

The determination of the sound pressure levels of bat sonar
signals is very difficult owing to the highly directional sound
emission patterns and the high directionality of the ultrasonic
microphones used to pick up the sounds. Nevertheless it is
evident that different species of bats produce signals with different
sound pressure levels. Griffin (1958) stated, as a general rule,
that bats which hunt for small targets (e.g. insects) produce loud
sounds whereas bats which echolocate large targets (e.g. fruit or
small vertebrates) use faint signals (whispering bats).

Variations in structure, duration and intensity are not only
found in the signals from different species but also in sounds
used by one species under different situations. These differences
are also interpreted as adaptations to changing information-gathering
and interference problems. Only sounds recorded under similar
conditions should therefore be compared.

Enough detailed information for such a comparison is only
available for a few species. Since another review at this conference will deal with the structure of echolocation sounds of
bats in detail, we will only give an example for the frequency
composition of free flight signals in nine bats from nine different
families (fig. 1) and an example for situation dependent variation
of signals in one species (fig. 19).

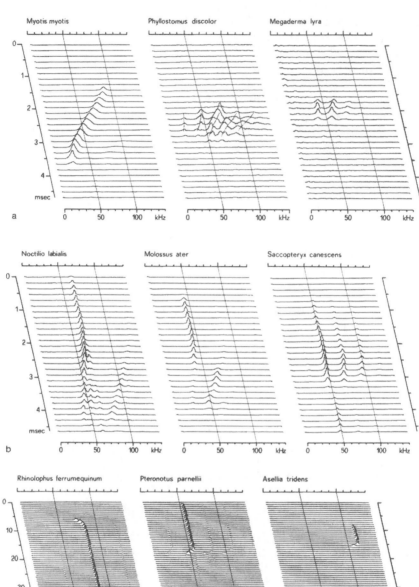

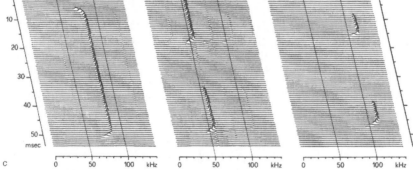

Fig. 1.   Real-time-spectrograms of free flight orientation sounds of nine species of bats from different families (from Schnitzler, 1978).

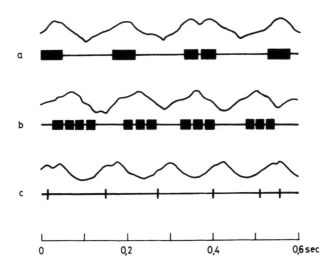

Fig. 2. Correlation between respiratory cycle and sound emission in resting Rhinolophus ferrumequinum (a), Rhinolophus euryale (b) and Phyllostomus hastatus (c) (a and b adapted from Schnitzler, 1968; c from Suthers, Thomas and Suthers, 1972).

PATTERNING OF ECHOLOCATION SIGNALS

Comparative studies of bat signal patterns reveal that different species arrange their sounds in a rather similar way in comparable situations.

All echolocating bats must adjust their pulse emission to the respiratory cycle. Recordings of the flank movements of resting rhinolophids (Schnitzler, 1968) demonstrated that, during one respiratory cycle, one pulse or a group of several pulses is emitted (fig. 2 a, b). Similar results were found in other bats by monitoring their exhalations and inhalations with a thermistor (Suthers, Thomas and Suthers, 1972; Roberts, 1972) (fig. 2c).

In flight, different species of bats also show a more or less distinct tendency to emit single pulses or groups of pulses. This was described for Myotis lucifugus (Grinnell and Griffin, 1958), rhinolophids (Schnitzler, 1967; 1968), Pteronotus parnellii (Schnitzler, 1970a) and hipposiderids (Pye and Roberts, 1970; Gustafson and Schnitzler, in press). Films of different species of bats flying against headwinds in a experimental wind tunnel showed that pulse emission is also correlated with wingbeat (Schnitzler 1971). Species of four different families emitted either a single pulse or a group of pulses per wingbeat (fig. 3 a-e). The

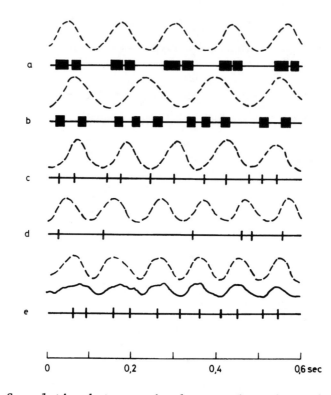

Fig. 3. Correlation between wing beat cycle and sound emission in
Rhinolophus ferrumequinum (a), Rhinolophus euryale (b), Myotis
lucifugus (c), Carollia perspicillata (d) and Phyllostomus hastatus (e) (a - d adapted from Schnitzler, 1971; e from Suthers,
Thomas and Suthers, 1972). The additional curve in e describes
the in and outward air flow during respiration.

correlation between sound emission and wing beat in flying bats led
to the suggestion that flying bats make one wingbeat per respiratory
cycle and, in so doing, either emit a single pulse or a group of
pulses (Schnitzler, 1971). Suthers, Thomas and Suthers (1972) confirmed this with flying Phyllostomus hastatus (fig. 3 e).

The number of pulses which are produced in one respiratory
cycle is dependent on the bat's orientation situation.

In resting bats, the number of pulses per respiratory cycle
is determined by their interest in the surroundings. Undisturbed
Rhinolophus ferrumequinum e.g. produce single pulses at every
exhalation at a rate of 4 - 7 pulses/sec. If they are disturbed or
become interested in a special target, groups with 2-8 pulses are
emitted. Rhinolophus euryale and Asellia tridens echolocating a
mealworm offered with forceps produce up to 10 pulses per group

(Schnitzler 1968; Gustafson and Schnitzler, in press). Resting
Phyllostomus hastatus produce either one pulse or groups of 2 - 4
pulses per respiratory cycle (Suthers, Thomas and Suthers, 1972).
The cited pulse repetition rates of 5 - 10 pulses/sec in resting
Myotis lucifugus (Grinnell and Griffin, 1958) also suggests the
emission of a single pulse per respiratory cycle. The rather sparse
data on echolocation behavior in resting bats reveal the following
tendencies: resting bats make approximately 4 - 10 respiratory
cycles/sec and, depending upon their interest in a target, per
cycle single sounds or groups of up to 10 sounds can be produced.

In flight different species of bats produce similar pulse
patterns. The three behavioral categories used by Griffin, Webster
and Michael (1960) to describe the pulse patterns of hunting vespertilionids can therefore be generalized for all bats. A "free"
or search flight category occurs when bats fly in the open in order
to search for food. When closing in on a target the pulse patterns
change. In this situation Griffin, Webster and Michael (1960)
distinguished between an approach phase and a terminal phase as
the animal neared the target.

Under laboratory conditions, a bat is considered as being in
free flight if there is no detectable target or obstacle in its
flight path within a range of about 2 m. In this situation
rhinolophids (Schnitzler, 1968), Pteronotus parnellii (formerly
Chilonycteris rubiginosa) (Schnitzler, 1970 a, b) and Phyllostomus
hastatus (Suthers, Thomas and Suthers, 1972) emit either single
sounds or groups of two sounds at a rate of approximately 8 - 12
per second (fig. 3e). The group duration, defined as the time
interval between the first sounds of consecutive groups, is normally
in the range of 80 - 120 msec. It is highly probable that these
free flight characteristics can be generalized for all other bats
(fig. 4). However, in bats with short pulses, the arrangement
of pulses into groups is not always clearly evident. When flying
outdoors in the open, most bats probably emit only one pulse per
wingbeat and respiratory cycle with a repetition rate which may
be lower than the 8 - 12 per second found in the laboratory.
Griffin (1958), for example, reports for high flying Eptesicus
fuscus a rate of only 4 pulses/sec.

A bat enters the approach phase when it produces more pulses
per group than in free flight (fig. 4). This increase in pulse
repetition rate indicates that a bat has detected and responded
to a specific target. The number of pulses per group increases
proportionally with decreasing distance to the target. The duration of the approach phase is situation dependant. When detecting
a small, unknown target, e.g. an insect, the approach phase only
contains 1 - 3 groups of sounds, while bats passing a well known

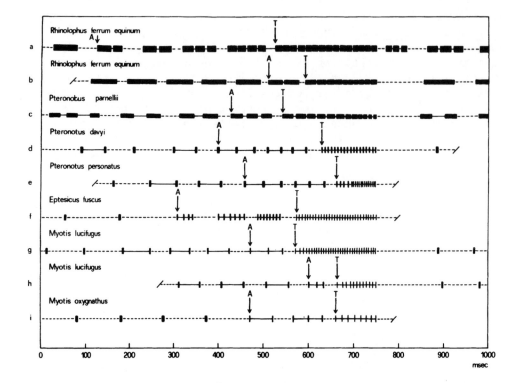

Fig. 4. Sound patterns of different species of bats when approaching a target. Pulses connected with a solid line belong, according to our interpretation, to a pulse group emitted per wingbeat and respiratory cycle. In (a) the bat was approaching a wire obstacle, in (b, c, d, e, h and i) the bats were persuing an insect and in (f and g) a ballistic target. The data are adapted from Schnitzler 1973b (a), Webster, 1967a (b), Novick and Vaisnys, 1964 (c), Novick, 1963 (d), 1965 (e), Griffin, 1962 (f), Webster as cited in Cahlander, 1967 (g), Griffin, Webster and Michael, 1960 (h) and Konstantinov, 1969 (i). Note the difference in the duration of the approach and terminal phases as found in Rhinolophus (a and b) and in Myotis (g and h). (A) determines the start of the approach phase and (T) the start of the terminal phase.

obstacle or landing on a familiar landing site produce up to 8 groups (fig. 4a, b).

Films of landing Rhinolophus ferrumequinum and Pteronotus parnellii demonstrated that the 1:1 correlation between wingbeat and arrangement of sounds in groups is also maintained during the approach phase (Schnitzler 1970 a,b). A similar arrangement of sounds in other species suggests that, in the approach phase, all bats

produce groups of sounds every 80 - 120 msec, these being correlated
with wingbeat and respiratory cycle.

It is not clear whether the start of the approach phase is always an indicator for a directly preceding initial detection of a
target. It can be calculated that, in the case of big targets, the
detection distance should be greater than the response distance,
characterized by the switching from free flight mode to the approach
phase. The bats appear to respond to targets only when they are
nearer than 2 - 3 m. For small targets, however, where the detection
occurs below this distance it is likely that detection is followed
immediately by the start of the approach phase (Grinnell and Griffin, 1958).

The terminal phase begins when a bat approaches the target
closely. It is characterized by a relatively long group of sounds
with rapidly repeated short pulses (fig. 4). The duration of the
terminal group of pulses may be twice as long as that of groups emitted in free flight and during the approach phase. The number of
pulses emitted in the terminal group depends on the echolocation
situation. Webster and Brazier (1965) found that, during the capture
of projected mealworms by Myotis lucifugus, the duration of the final
pursuit determines the duration of the terminal group. In extended
pursuits, where a bat has misjudged the position or the path of a
target, the terminal phase is prolonged and even a second terminal
group is sometimes emitted. During insect pursuit by Myotis
lucifugus, the terminal group contains 3 - 20 pulses (Griffin,
Webster and Michael, 1960) (fig. 4). In obstacle avoidance tests,
the number of pulses in the terminal group is reduced with decreasing
wire diameter (Griffin and Grinnell, 1958; Schnitzler, 1967, 1968)
(fig. 5).

The films of Webster and Griffin showing Myotis lucifugus
catching insects show that, during the terminal group, several
wingbeats are made which bring the bat into the final interception
position. For the terminal phase, therefore, the general rule of one
wing beat and one group of sounds per respiratory cycle, is not valid.

In the approach and terminal phases, the intervals between
pulses within a group also become reduced as the number of pulses per
group increases. The factor that determines the minimum interpulse
interval seems to be the distance to the target. In landing Pteronotus parnellii the duration of the interpulse interval is always
maintained at a length which allows the echo of the target to return
before the next pulse is emitted (Schnitzler, 1970a).

The approach to a target is also characterized by a more or
less pronounced decrease in the bat's flight speed. This was demonstrated for Lasiurus, Eptesicus and Myotis during insect catching

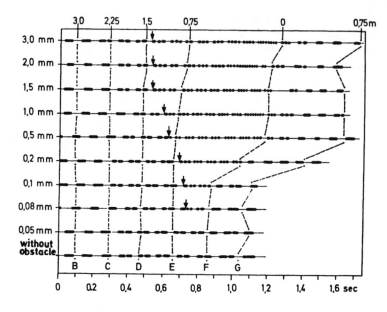

Fig. 5. Sound patterns of <u>Rhinolophus ferrumequinum</u> in obstacle experiments with different wire diameters. The arrows indicate the start of the approach phase. At 0 m the bats passed the obstacle (adapted from Schnitzler, 1968).

(Webster and Brazier, 1968) and also for obstacle avoidance and landing in rhinolophids (Schnitzler 1967, 1968) (fig. 5). This decrease in flight speed, which occurs together with the previously described increase in pulse repetition rate, results in a rapid increase in pulse density (number of pulses emitted per distance flown) when a bat closes in on a target (Schnitzler, 1967).

In all bats, the terminal group of pulses is often followed by a long interval with no sound emission, this being at least as long as individual group durations in free flight or during the approach phase (fig. 4). After this silent period, which may be a respiratory cycle without sound emission, the bats switch back to the pulse pattern of free flight with single pulses, or groups with two pulses.

In areas where <u>Myotis</u>, <u>Eptesicus</u> and <u>Lasiurus</u> have encountered targets in former flights, the pulse patterns are characterized by an increase in pulse repetition rate even if no targets are present (Webster and Brazier, 1968). This kind of 'target expectancy' has also been found in <u>Rhinolophus ferrumequinum</u> (Schnitzler, 1968) and <u>Asellia tridens</u> (Gustafson and Schnitzler, in press) accustomed to flying through obstacles. Trained <u>Rhinolophus ferrumequinum</u>, for

instance, emitted two pulses per wingbeat within a distance
of about 2 m in front of the place where the obstacle had previously
been positioned (fig. 5). Untrained bats, on the other hand, showed
typical free flight behavior with single pulses in the same situation.

DURATION OF ECHOLOCATION SIGNALS

Different species of bats show wide variation with respect to
their pulse duration but basic similarities in the way they change
this parameter in different situations.

In resting bats, sound duration is maximal in single pulses
and becomes reduced with an increase in the number of pulses per
group. In all bats, the longest pulses are normally emitted in free
flight. During the approach and terminal phases, the pulses are
progressively shortened as the number of pulses per group increases.
The shortest pulse durations are always found in the terminal phase.
In Pteronotus davyi, Pteronotus personatus (formerly Chilonycteris
psilotis), Pteronotus parnellii (Novick 1963, 1965; Novick and
Vaisnys, 1964), Noctilio leporinus (Suthers, 1965), Myotis lucifugus
(Cahlander, McCue and Webster, 1964) and Rhinolophus ferrumequinum
(Schnitzler, 1968) there is an approximately linear relationship
between the reductions in pulse durations and the target distance.

Under laboratory conditions, the size of the flight room probably
influences sound duration in free flight. Griffin (1953) reports
that Eptesicus fuscus flying in the open produce pulses of
10 - 15 msec whereas in the laboratory pulses with only 2 - 4 msec
durations were recorded. In Noctilio leporinus flying outdoors pulse
durations of 11.1 - 16.7 msec are produced but in a flight cage, the
pulse durations are 5.9 - 9.11 msec.

If the echo delay is shorter than the pulse duration, the returning
echoes overlap the emitted pulse. In all vespertilionid
bats studied so far, the pulses are kept so short that no overlapping
occurs (Webster and Brazier, 1968). This is probably the case in all
bats emitting pure fm sounds. When catching mealworms Myotis lucifugus
show, in the approach and terminal phases, a strong tendency to
maintain the pulse durations at about half the echo delay (Cahlander,
McCue and Webster, 1964). In bats which produce cf-fm sounds, the
returning echoes usually overlap the outgoing sounds. Detailed
studies of this problem in Rhinolophus ferrumequinum (Schnitzler,
1968) and in Pteronotus parnellii (Schnitzler, 1970 a,b) reveal,
however, that a decrease in the duration of the fm component during
the approach and terminal phases excludes any fm component pulse-
echo overlap. When the fm component of cf-fm sounds is considered
by itself, it emerges that this component is modified in a way
similar to that of the fm sounds in pure fm bats.

DETECTION OF TARGETS

The detection limits of echolocation in bats have been studied
by measuring obstacle avoidance skills with vertically and horizontally stretched wires. The ability to detect the wires is usually
measured as the percentage of through flights without collisions.
The data for different species are not fully comparable owing to
the different methods used in the experiments. For example, a bat
which has been trained to fly through obstacles will attain a better
avoidance score than an untrained bat forced to fly the same route.
The results may also be influenced by differences in the distance
ratios of the wires and the wingspans of the animals being tested.
In order to obtain comparable data we suggest that, in future experiments, trained bats should be enticed to fly through the obstacles
with a food reward and a 2:3 ratio of distance between wires and
wingspan should be maintained.

A score above the random avoidance level was reached at a
wire diameter of 0.26 mm in Myotis lucifugus (Curtis, 1952), of
0.175 mm in Carollia perspicillata, Glossophaga soricina and Artibeus jamaicensis (Griffin and Novick, 1955), of 0.19 mm in Macrotus
mexicanus (Griffin and Novick, 1955), of 0.2 mm in Myotis oxygnathus
and Plecotus auritus (Konstantinov, Sokolov and Stosman, 1967) and
of 0.21 mm in Noctilio leporinus and Pizonyx vivesi (Suthers, 1967).
The avoidance capacities of some other species are even better. The
detection limit of Rhinolophus ferrumequinum is between 0.08 and
0.05 mm (Schneider and Möhres, 1960; Schnitzler, 1967, 1968).
Rhinolophus euryale and Rhinolophus mehelyi are able to detect wires
with diameters of only 0.05 mm (Schnitzler, 1968; Konstantinov,
Sokolov and Stosman, 1967). Megaderma lyra can avoid wires of 0.06 –
0.08 mm (Möhres and Neuweiler, 1966). In Asellia tridens the avoidance score drops below the chance level when the wire diameter is
between 0.065 and 0.05 mm (Gustafson and Schnitzler, in press).

These results demonstrate that wires with diameters far below
the orientation sound wavelength can be detected. In rhinolophids
the minimal wire diameter was 50 – 60 times smaller than the
wavelength of the pulse cf component.

In the section on pulse patterns it was noted that with small
targets a change from the free flight to the approach phase is a
definite indication that a bat has just detected a target. The
analysis of the sound sequences produced during flights through
obstacles therefore allows the determination of the response distance
and also an evaluation of the detection distance. In addition, the
response to a target is very often characterized by a decrease in
pulse duration and flight speed (figs. 4 and 5). Response and detection distances have been determined for Myotis lucifugus (Grinnell
and Griffin, 1958), Rhinolophus ferrumequinum and Rhinolophus

euryale (Schnitzler, 1968), Myotis oxygnathus (Golubkov, Konstantinov and Makarov, 1969) and Rhinolophus ferrumequinum and Rhinolophus mehelyi (Sokolov, 1972). All these measurements revealed a general tendency for the detection distance to diminish with decreasing wire diameter. A rather rough comparison of the experimental data from different species indicates that wires with diameters of about 2 mm are detected at distances of approximately 2 m, whereas the detection range for 0.2 mm diameter wires is about 1 m.

Griffin (1958) studied the echolocation behavior of free-flying Eptesicus fuscus in the field. He estimated that the bats detect 1 cm diameter objects thrown into their flight path at a distance of about 2 m. Webster and Brazier (1965) investigated the detection abilities of Myotis lucifugus trained to catch airborne targets such as mealworms and small spheres. They used two lines of evidence to decide whether a bat had detected and responded to a target. Apart from the increase in pulse repetition rate at the start of the approach phase, they also found that, after detection, the bats turned their heads toward a target and if they intended to explore the target further they directed the flight path to the target or to the place where they expected to intercept it. Using both criteria, they determined that the response to spheres 2.1 mm in diameter occurred at distances of about 60 cm and initial detection occurred at about 90 cm. With 4.2 mm diameter spheres the response range was between 90 - 105 cm and the detection range was judged to be 120 - 135 cm. With spheres of greater diameter, however, the response point was not further away although there should have been an increase in echo strength.

These findings and similar ones by Grinnell and Griffin (1958) and Schnitzler (1968) in bats avoiding obstacles raise the question as to whether objects large enough to reflect audible echoes are also processed beyond the maximal response range indicated by the pulse patterns. The observations of Webster and Brazier (1965), in which bats turned away from a volley ball thrown into their flight path at a distance of about 3 - 4 m without showing a change in the pulse pattern suggest that these bats are able to detect targets beyond the indicated range. Similar conclusions can be made in the case of rhinolophids which respond with a fright response to big obstacles moving towards them at distances of several meters. However, it is still possible that bats use a form of time or attention gate which is mainly open for echoes coming back from targets within a maximum response range of about 2 m. A time or attention gate for the near field in front of a bat would allow the bats to concentrate on a few near-by targets and prevent overloading of the echolocation system with echoes from more distant targets.

Many species of bats collide with objects placed in an accustomed flight path which, over a long period of time had

previously been obstacle-free. Although strong echoes returning from
these newly placed obstacles must be present, they are not detected
and avoided by the bats. Griffin called this situation, where
obviously present information is not used, the "Andrea Doria effect".
It has been shown by Neuweiler and Möhres (1967), that bats can
establish a very precise three-dimensional picture of the environment. We suggest that this spatial memory completely dominates the
echolocation system in cases where bats do not avoid new obstacles
in a well known environment. This raises the question as to
whether in this situation the "attention gate" is closed for obstacle information but open for prey detection. In this case bats,
after having established their spatial memory, would disregard
all echo information returning from the environment and only concentrate on the targets in which they are interested. Further experiments must be conducted to study and solve this question.

If one knows the detection range and the sound pressure level
of the emitted sounds, it is possible to calculate the sound pressure
level of the echoes used for detection of targets with known reflective properties. Griffin (1958) estimated that, in Eptesicus
fuscus, the echoes return with a SPL of 17 dB (relative to 0.0002
dyne/cm^2) from a spherical object detected at about 2 m distance.
For Myotis lucifugus catching fruitflies a detection level of about
30 dB at a range of 50 cm, and 15 - 20 dB at 100 cm was assumed
(Griffin, Webster and Michael, 1960). In bats avoiding obstacles
a detection level of 23 - 28 dB was calculated in the same
way for Myotis lucifugus (Griffin, 1958), of 25 - 34 dB for
Rhinolophus ferrumequinum (Sokolov, 1972) and 9.2 - 21.6 dB in
Myotis oxygnathus (Airapetianz and Konstantinov, 1974).

Novick suggested another criterion for the detection of targets
in mormoopids and other bats with cf-fm sounds. He assumed that detection first occurred when a returning echo overlapped the outgoing
pulse (Novick, 1963, 1965, 1971, 1973, 1977; Novick and Vaisnys,
1964). In Pteronotus personatus whose pulse durations are 4 msec
in free flight, the detection distance would be about 0.69 m
whereas Pteronotus parnellii with 20 msec sounds should detect
targets at 3.5 m. In Rhinolophus ferrumequinum, with its free
flight pulses of more than 50 msec duration, the detection should
occur at distances greater than 8.5 m. Since mormoopids produce
the same basic fm sound patterns in their free flight, approach and
terminal phases as other bats, it may be that the detection mechanisms in these bats do not differ from those of all other species.

DIRECTIONAL SENSITIVITY OF ECHOLOCATION SYSTEMS

The likelihood that a target attracts the attention of a bat
decreases the more laterally the target is positioned relative to
the flight path (Griffin, 1958). This observation can be explained
by the directionality of sound emission and of hearing. The more
lateral a target, the fainter are the echoes heard by the bats.

Measurements to determine the directional emission pattern have
been made in several species of bats including Myotis lucifugus
(Griffin, 1958; Griffin and Hollander, 1973; Shimozawa, Suga, Hand-
ler and Schuetze, 1974), Myotis oxygnathus (Airapetianz and Kon-
stantinov, 1974), Myotis grisescens (Shimozawa, Suga, Handler and
Schuetze, 1974), Eptesicus fuscus (Simmons, 1969; Griffin and
Hollander, 1973), Megaderma lyra (Möhres and Neuweiler, 1966),
Pteronotus parnellii (Simmons, 1969) and Rhinolophus ferrumquinum
(Möhres, 1953; Schnitzler, 1968; Sokolov and Makarov, 1971;
Schnitzler and Grinnell, 1977).

In most cases, only a few angles in the horizontal and vertical
axes were tested, but the results have demonstrated that the emission
patterns are highly directional and frequency dependent. Some of
these measurements show that the radiation patterns are determined
by the structure of the nose leaf in bats which emit sounds through
the nostrils and by the form of the mouth and lips in bats which emit
sounds through the mouth. Complete measurements of the sound field
are only available for 3 frequency components of the fm bat Myotis
grisescens (Shimozawa, Suga, Handler and Schuetze, 1974) and of the
cf components of the cf-fm sounds of Rhinolophus ferrumequinum
(Schnitzler and Grinnell, 1977) (fig. 6B). Contour maps of the
emission pattern of these bats reveal sound fields with a strong
asymmetry in the vertical axis. In Myotis grisescens, the main
beam was emitted 5 - 10° downward from the eye-nostril line, whereas
Rhinolophus emitted the main energy perpendicular to the plane of
the horseshoe-shaped nose leaf; this corresponds to the normal
flight direction in horizontal flight. In both species rather
wide and prominent downward-pointing side lobes were found.

The likelihood of target detection at different angles cannot
be judged from the directionality pattern of sound emission alone.
For instance, the presence of a downward pointing side lobe has led
to the erroneous conclusion, that Rhinolophus has a form of ground
sonar for monitoring the flight altitude (Sokolov and Makarov, 1971).
However, to obtain a complete picture of the directional sensitivity
of an echolocation system, the directionality of the hearing system
also has to be taken into account.

The directionality of hearing has been measured in several
species using neural responses as a criterion. This has the

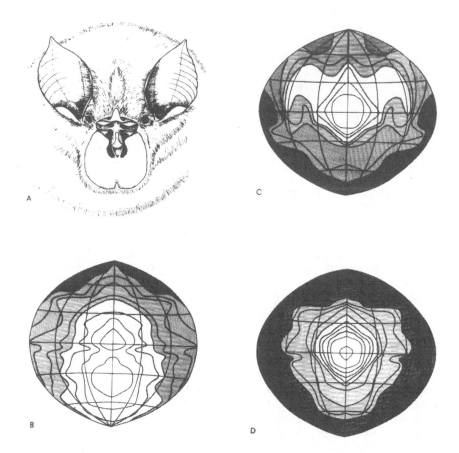

Fig. 6. Directional sensitivity of echolocation in <u>Rhinolophus</u>
<u>ferrumequinum</u>. (A) head of the Greater Horseshoe Bat as seen from
the direction perpendicular to the plane of the horseshoe. (B) contour map of the sound field in front of a bat. (C) contour map of
the behaviorally determined directionality of hearing. (D) contour
map of the directionality of the whole echolocation system as calculated by superimposing (B) and (C). Lines in the contour maps are
separated by 3 dB. Parallels of latitude and meridians of longi-
tude are shown at 30° intervals (adapted from Schnitzler and
Grinnell, 1977, and Grinnell and Schnitzler, 1977). Note in (D)
the symmetry of the pattern and the steep fall-off of echo effectiveness in all directions relative to the line of flight.

advantage that the effectiveness of sounds from any angle can be
quickly measured. However, all these measurements have been made
on one side of the brain only and do not provide a good means of
assessing the capabilities of the hearing system as a whole.
A method has been developed which allows the determination
of behavioral hearing directionality in Rhinolophus ferrumequinum
(Grinnell and Schnitzler, 1977). Contour plots of the hearing
field of Rhinolophus reveal an asymmetry in the vertical axis
with a sharp drop in sensitivity beneath the bat (fig. 6C). When
the data on hearing are combined with the data on sound emission,
a contour map of the sensitivity of the whole echolocation system
can be constructed (fig. 6D). This map shows rather radially
symmetrical contour lines with practically no side lobes. The
sensitivity is reduced to half (- 6 dB) at about 15 degrees lateral
to an axis perpendicular to the plane of the noseleaf. The
prominent downward side lobes of the emission pattern therefore
do not increase the chances of ground echo detection, since hearing
sensitivity decreases rapidly at this angle.

The measurements in Rhinolophus ferrumequinum reveal that the
asymmetries found in sound emission and hearing compensate each
other. This results in a radially symmetrical and highly beamed
sensitivity curve. If in Rhinolophus the chance to detect an echo
only depends on echo strength the echoes returning at an angle 15
degrees lateral to the flight path should be about 6 dB louder and
those from 30 - 40° approximately 20 dB louder than echoes returning
from straight ahead to have the same chance of detection.

Unfortunately we only have complete measurements on directional
sensitivity in one species. Further experiments have to be conducted in order to determine whether other species have similar radially
symmetrical characteristics.

FIXATION, TRACKING AND CAPTURE OF TARGETS

Search Volume

The highly directional sensitivity of echolocation in bats
suggests that their angular search volume may be rather limited
and restricted to a cone-shaped area which extends outward from
the mouth. Griffin, Webster and Michael (1960) have studied this
problem in Myotis lucifugus during insect-catching in the laboratory.
They estimated that Myotis is able to detect fruit flies in a spatial
cone of about 120° diameter in front of the head at detection distances between 27 - 83 cm. There was, however, little correlation
between angular position and detection distance. Webster and
Brazier (1965) describe a case where Myotis was even able to detect
a moth roughly 90° to one side and about 45 cm away.

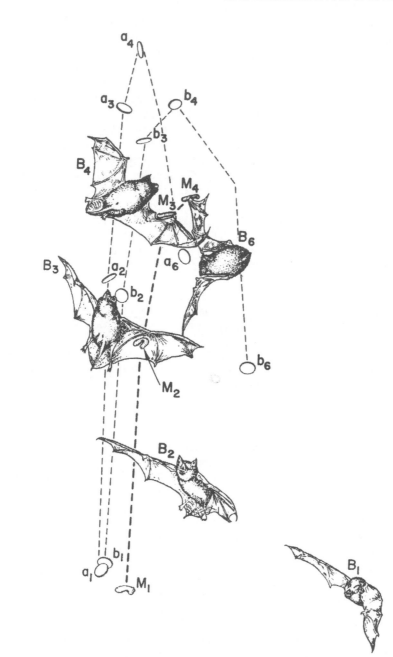

Fig. 7. Myotis lucifugus selecting a mealworm out of a cluster of three discs. Image 5 and the third disc are omitted. Note the precise head aim of the bat (adapted from Webster and Durlach, 1963).

Fixation

Immediately after detection, flying <u>Myotis lucifugus</u>,
<u>Eptesicus fuscus</u> and <u>Rhinolophus ferrumequinum</u> turn toward the
target and orient their head so as to bring the target into their
echolocation system's region of maximal sensitivity (Webster, 1963
a, b, 1967 a,b; Webster and Brazier, 1965, 1968). Webster (1967a)
estimated that this initial fixation of the target occurs within
about 1/10th sec. The precise aiming of the bat's mouth and ears
toward the target indicates that they are able to pinpoint it
exactly. From photographs of <u>Myotis lucifugus</u> catching insects
Webster and Brazier (1965) judged the accuracy of head aim to be
+ 5 degrees. Fig. 7 gives an example of the excellent head aim in
a <u>Myotis lucifugus</u> selecting a mealworm from a cluster of targets.

Tracking

The tracking of targets, i.e. the ability to determine the
pathways of targets and to predict their future position, has been
studied in <u>Myotis lucifugus</u> <u>Eptesicus fuscus</u>, <u>Lasiurus borealis</u>
and <u>Rhinolophus ferrumequinum</u> (Webster, 1963 a,b, 1967 a,b,
Webster and Brazier, 1965, 1968).

The analysis of the bat's flight path with respect to that of
a target reveals that bats are able to evaluate the course of
targets and to make predictions regarding their future position.
Experiments with ballistic targets (spheres, discs, etc.) demonstrate that trajectory evaluation contains a strong learning component. Bats initially misjudge the vertical position of targets
and try to capture objects at a position several cm above the
real one. They are probably not able to appreciate acceleration
due to gravity. After a few trials, however, they learn to predict
the trajectories so precisely that they no longer need to correct
their flight path over the last half meter before target capture.

Since flying insects are capable of maneuvering, tracking is
therefore far more difficult than in the case of ballistic targets,
especially when the insect is a moth making evasive movements.
Nevertheless remarkable predictions of flight paths have been
observed. It may be that in this situation, bats make a type of
probabilistic estimation of the future target position. Well-timed
and sudden changes in flight direction, however, commonly result
in successful evasion of bats by insects.

The tracking behaviour of <u>Myotis</u>, <u>Eptesicus</u> and <u>Rhinolophus</u>
is characterized by quick and precise following of the target with
the head. Such precise head orientation was found at distances
up to about one meter in laboratory tests with <u>Myotis</u> and up to two
meters or more in <u>Eptesicus</u> hunting in the field. Even in situations

where the bats were no longer capable of bringing their bodies into
an interception position, the head aim still indicated that they had
located the target. A different tracking behavior was found in
Lasiurus borealis; this bat can accurately predict the path of its
prey without the precise head-following observed in other bats.

Capture Techniques in Bats

The following movements of the head, characteristic of target
flight-path evaluation, cease at close range and the final interception is made by simple deduction. The following target-capture
techniques for insects, projected mealworms or other ballistic
targets have been described for Myotis lucifugus, Eptesicus fuscus,
Lasiurus borealis and Rhinolophus ferrumequinum (Webster, 1962, 1963
a,b, 1967 a,b; Webster and Griffin, 1962; Webster and Durlach, 1963;
Webster and Brazier, 1965, 1968).

Mouth catch: In Myotis it was observed that small and slow
flying insects, such as fruit flies, may sometimes be seized directly with the mouth. That means that the insect must have been
localized with extremely high accuracy.

Tail membrane catch: Myotis, Eptesicus and sometimes Lasiurus
direct their flight path to the interception point and form a
pouch with the interfemoral membrane by forward flexion of the
hind limbs and tail. After the target is caught in this pouch, the
bat bends forward and seizes it with the jaws. This kind of catching
behavior was observed with insects as well as with projected mealworms and other ballistic targets.

Wing scoop catch: If the target is out of reach of the tail
membrane pouch, bats sometimes reach out for the target with the
wing, shovel it into the pouch and grasp it with the mouth. Even
targets which can be reached only with the tip of the wing may be
caught in this way. In Rhinolophus it is probable that insects
intercepted with the wing are brought to the mouth directly without
the use of the interfemoral membrane pouch.

Somersault catch: Lasiurus uses all the capture techniques
found in Myotis and Eptesicus and with an even higher predictive
skill. Sometimes, however, these bats fly underneath the target
and make a backward somersault by swinging the tail membrane
around in a complete 360° circle. During this action they
produce a form of funnel with the wings leading to the tail
membrane pouch. The bats come out of this maneuver with the target
in the pouch. After reorientation, they reach into the pouch and
seize the prey.

The use of the tail membrane and the wings for the capture of flying insects serves to increase the area within which a bat can seize its prey (Webster and Griffin, 1962).

The high capture efficiency observed in the different species indicates that the bats know exactly where their targets are. Webster (1963) estimated that targets are sometimes localized and intercepted with an accuracy of $\pm$ 0.5 cm. This means that bats are able to pinpoint their prey in space within about one cubic centimeter. The speed of action is also astonishing. The complete sequence with detection, initial fixation, tracking and capture may only last 300 - 500 msec.

The rate at which bats catch insects has been studied in Myotis lucifugus hunting insects in the open (Gould, 1955, 1959). During a 2369 sec observation time, the Myotis, on an average, pursued an insect every 3 sec and probably caught every second one. Similar results were presented by Griffin, Webster and Michael (1960). They estimated that, in the laboratory, Myotis caught fruit flies with an average rate of one every hour seconds.

DETERMINATION OF RANGE

Hartridge (1945 a,b) suggested that bats determine the range of targets by measuring the time between sound emission and echo reception. Griffin (1944) had already anticipated this principle by comparing echolocation with radar and sonar systems.

Beat Note Hypothesis

Pye (1960, 1961a, 1961b, 1963) and Kay (1961a, 1961b, 1962) suggested that in fm bats, range information might be contained in beat notes arising between the outgoing sound and the overlapping echo. Such a beat frequency would decrease linearly with decreasing range. Cahlander, McCue and Webster (1964), however, demonstrated that, in Myotis lucifugus during the approach and terminal phases, the sound duration decreases almost linearly with decreasing target range and in such a way that no overlap occurs. The beat note hypothesis can also be rejected for other species since the fm sounds in other pure fm bats and also the fm components in cf-fm bats do not overlap with their echoes in the approach and terminal phases.

Accuracy of Target Range Determination

When judging the ability of bats to measure distances by echolocation, it is important to distinguish between accuracy of range measurement and range resolution. Accuracy describes the acuity

with which the range of a single target can be measured whereas
resolution refers to the ability to distinguish between two or more
closely spaced targets.

In experiments to investigate range measurement performance
(Simmons, 1968, 1971, 1973; Simmons and Vernon, 1971; Airapetianz
and Konstantinov, 1974) bats had to decide which of two similar
targets offered in different directions and at different absolute
ranges was the closer. The discrimination performance was measured
as the percentage of correct responses and the threshold was set at
75%. In these experiments, the angle between the targets was so
great that the bat had no difficulty in resolving them. While
measuring and discriminating the distances, the bats scanned from
one target to the other, thus making it highly probable that determination of the target range was made successively and not simultaneously. The discrimination threshold is therefore a measure for
the accuracy of range measurement and not of range resolution.

Similar experiments have been conducted with Eptesicus fuscus,
Phyllostomus hastatus (both fm bats), Pteronotus suapurensis (short
cf/fm) and Rhinolophus ferrumequinum (long cf/fm); the bats were
trained to discriminate the range of triangular targets offered at
an angle of 40° to each other (Simmons, 1968, 1971, 1973; Simmons
and Vernon, 1971). In other experiments Myotis oxygnathus (fm bat)
and Rhinolophus ferrumequinum were trained to discriminate the
range of cylinders or corner reflectors offered at an angle of 13°
(Airapetianz and Konstantinov, 1974). Table 1 summarizes the results
of all bats tested to date and fig. 8 describes the discrimination
performance of Eptesicus fuscus. In general, range discrimination
performance diminishes with a decrease in the band widths of the
echolocation signals used for discrimination. In the bats tested,
thresholds between 8 – 45 mm were found. With a two channel target
simulator (two loudspeakers transmitting bat signals as echoes back to
the bat) where similar echoes differing only in arrival time were
presented, it was demonstrated that Eptesicus discriminates differences in echo arrival times with the same accuracy as comparable
difference in range (Simmons, 1973). This suggests that Eptesicus
and probably all other bats use time differences between signals for
the determination of ranges.

An equivalent discrimination performance was reached with the
two-channel simulator when the echoes were presented successively
instead of simultaneously (Simmons and Lavender, 1976). This also
indicates that absolute ranges (time delays) are compared and that
the range discrimination threshold is a measure for the accuracy of
range determination and not for range resolution.

An astonishing result of Simmons' measurements is that the discrimination threshold hardly changes when the targets are offered

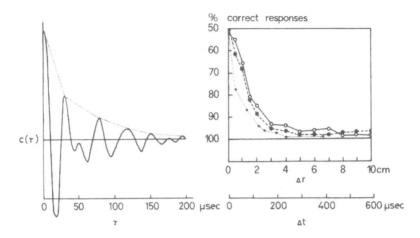

Fig. 8. Left: Autocorrelation function with envelope of an orientation sound of Eptesicus fuscus produced during a range discrimination experiment. Right: Average performance of 8 Eptesicus fuscus
in a range discrimination experiment (solid line) compared with a
predicted performance curve (dashed line) which was derived from the
envelope curve of the autocorrelation function (dotted line). In
order to obtain the prediction curve the envelope curve was corrected
for the bats' head movements during discrimination (adapted from
Simmons, 1968).

at different ranges (table 1). In Eptesicus, for example, the discrimination threshold was 12 mm and 14 mm when the target distance
was 30 cm and 240 cm respectively. This indicates that range determination is independent of echo intensity since echoes returning from
240 cm must be far less intense than echoes from targets 30 cm away.
Airapetianz and Konstantinov (1974) also discovered no difference
in the range performance of Rhinolophus when the echo intensities
differed between 71 dB and 64 dB. In Myotis, however, a threshold
of 8 mm was measured with echo intensities of 25 dB, whereas
only 23 mm was reached with 18 dB echoes.

Prediction of Range Determination Accuracy

     Strother (1961, 1967) pointed out that the signals of fm bats
are rather similar to the signals used in chirp or pulse compression
radar systems. He therefore assumed that echo processing in bats
might be comparable to that in the matched filter receiver of a
pulse compression radar system; in these radar systems the returning

Table 1. Range discrimination threshold (dr) and corresponding
time measuring errors (dt) of different species of bats using
signals of different bandwidth (b) measured with targets positioned
at range (r). In one Eptesicus the measuring error was determined
directly with a two target simulator. The angle between targets
was 40° in the experiments of Simmons and 13° in the experiments
of the Russian scientists. The different thresholds obtained for
Myotis oxygnathus were explained by differences in echo intensity
(18 dB at dr = 43 mm and 25 dB at dr = 8mm).

Species	dr	dt	b	r	Author
Eptesicus fuscus	13 mm	75µsec	25 kHz	60 cm	Simmons 1971
Eptesicus fuscus	12 mm	70µsec	25 kHz	30 cm	Simmons 1973
Eptesicus fuscus	14 mm	81µsec	25 kHz	240 cm	Simmons 1973
Eptesicus fuscus	-----	60µsec	25 kHz	30 cm	Simmons 1973
Phyllostomus hast.	12 mm	70µsec	35 kHz	60 cm	Simmons 1971
Phyllostomus hast.	12 mm	70µsec	35 kHz	120 cm	Simmons 1973
Pteronotus suapur.	15 mm	87µsec	22.5kHz	30 cm	Simmons 1973
Pteronotus suapur.	17 mm	98µsec	22.5kHz	60 cm	Simmons 1973
Rhinolophus ferr.	25 mm	145µsec	15 kHz	30 cm	Simmons 1973
Rhinolophus ferr.	41 mm	240µsec	15 kHz	100 cm	Airapet.et al.1974
Myotis oxygnathus	8 mm	46µsec	85 kHz	100 cm	Airapet.et al.1974
Myotis oxygnathus	23 mm	133µsec	85 kHz	100 cm	Airapet.et al.1974

echo is cross-correlated with the transmitted signal by passing the
echo through a filter matched to the signal.

Various aspects of such optimal filtering in bats have been discussed in detail by Cahlander (1963, 1964, 1967), van Bergejk
(1964), McCue (1966, 1969), Altes and Titlebaum (1970), Altes
(1973, 1975, 1976), Glaser (1971a, 1971b, 1974) and Beuter (1977).
From all these theoretical studies, the only result indicating that
bats might use optimal filters was that the echolocation signals
of Myotis lucifugus would be optimally Doppler tolerant if they were
processed in a matched filter receiver.

A matched filter receiver (ideal receiver, optimal filter receiver) concentrates the signal energy within a small time interval. This compression increases the signal-to-noise ratio at the
filter output and allows an optimum estimate of the pulse arrival
time. Under the assumption that the postulated optimal filter in
bats is matched to the waveform of the signals used, the efficiency
of the receiver in compressing signals can be judged by comparing

the cross-correlation function as derived from the echo and the
emitted signal. Simmons therefore assumed in his range determination
experiments that the cross-correlation function reflects the ambiguity encountered by an optimal filter receiver in estimating the
arrival time of echoes. With the high frequencies used by bats, it
is unlikely that phase information can be extracted by the auditory
system. A postulated matched filter receiver would therefore be a
noncoherent receiver, i.e. the fine structure of the cross-correlation function is not available and only the envelope may be used to
make predictions regarding range measurement accuracy. Since the
echoes in the discrimination experiments are similar to the emitted
signals, the envelope of the cross-correlation function can be conveniently approximated to the envelope of the autocorrelation
function of the emitted signal.

In fig. 8, a predicted performance curve is derived from the
envelope of an echo's autocorrelation function in Eptesicus. It
is compared with the average range discrimination performance of
8 bats. After a correction of the prediction curve, necessary
to compensate for the bat's head movements, the measured performance
curve shows good correspondence with the corrected prediction curve.
Such a close fit between predicted curve and measured performance
was also found in Phyllostomus hastatus (fm) and Pteronotus
suapurensis (short cf-fm) when the entire sounds were used for
autocorrelation function derivation. In Rhinolophus (long cf-fm),
however, only the autocorrelation function of the fm component
produces a corresponding prediction curve.

These results led Simmons (1973) to suggest that there is a
neural equivalent of a matchedfilter, an ideal sonar receiver
which functionally cross-correlates a replica of the outgoing sound
with the returning echo to detect the echo and to determine the
arrival time. Whether this experimental approach allows this
far-reaching conclusion is a matter of conjecture.

The accuracy of time and therefore also of range measurements
in a matched filter receiver depends on the effective bandwidth (b)
of the signal and also on the signal energy and the noise conditions.
In the case of echolocation signals, the spectral bandwidth (b) is
a good approximation of the effective bandwidth (Glaser, 1974). The
maximum-peak-signal-to-mean-noise (power) ratio at the output of a
matched filter is equal to 2E/No where E is the signal energy and
N_0 the noise power per cycle of bandwidth at the input. The error
of accuracy in time measurements δ_t for a coherent matched filter
receiver (Skolnik, 1962) was calculated as:

$$\delta_t = \frac{1}{\beta \ 2E/N_0}$$

This formula was used to predict time (range) error for the signals with bandwidths between 10 - 100 kHz and for different $2E/N_0$ ratios (fig. 9). If we assume constant noise conditions, the following prediction can be made for a matched filter receiver which uses phase information: The time error increases with decreasing echo energy and decreasing echo bandwidth. This can also be generalized for a matched filter receiver with no phase information; in this case, the time error must be greater, since the "fine grain" information of the cross-correlation function is lost, and only the envelope is available. However, the general concept that time error depends on echo energy and bandwidth is the same.

In order to decide whether or not bats use a matched filter receiver, the experimental data have to be compared with the predictions made for such a receiver.

One of Simmons' most important results is that the measured discrimination thresholds do not depend on absolute target range and therefore not on echo intensity. In Eptesicus, for instance, the energy of the echoes returning from 240 cm must have been far below that of echoes from 30 cm. Although great differences in the $2E/N_0$ ratio must have been present, the measured accuracy of time (range) determination was approximately the same at both distances. This is in clear contradiction to the behavior of a matched filter receiver, suggesting that it is improbable that bats use optimal filtering.

For a second comparison, the experimental results of the range discrimination experiments in different species of bats are shown in fig. 9. At first glance one would assume that the measured time errors of different species decrease with increasing bandwidth in a predictable way i.e. parallel to a line of equal $2E/N_0$ ratio. However, the energy of the echoes returning in the discrimination experiments is very different in different bats. Simmons (1973), for example, measured different signal durations and different sound pressure levels of echoes ranging from 0.005 to 0.1 N/m^2. As the noise conditions were approximately the same for all bats, the differences in echo energy also mean differences in the $2E/N_0$ ratios. The experimental data from different species are therefore not comparable and the observed tendency of decreasing time error with increasing bandwidth cannot be used to conclude that bats use a matched filter receiver.

Another argument against optimal filtering is the high variability of sound emissions in different echolocative phases and in different species of bats. Should a matched filter receiver be postulated, one would then have to assume that each sound had its own matched filter.

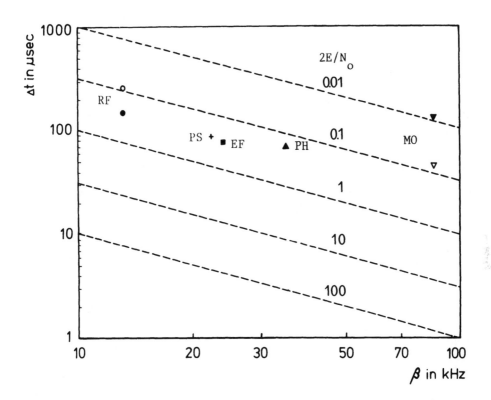

Fig. 9 Diagram which predicts for a coherent matched filter receiver the time measuring error for signals of different effective
bandwidth (β) and different $2E/N_0$ ratios. For comparison, experimental data on time measuring errors have also been included; these
were derived from range discrimination experiments with different
species. (RF) Rhinolophus ferrumequinum, (PS) Pteronotus suapurensis, (EF) Eptesicus fuscus, (PH) Phyllostomus hastatus, (MO)
Myotis oxygnathus. For further information see section on "prediction of range determination accuracy" and table 1.

     In summary, one can state that bats determine the distances of
targets separated in angle by measuring the delay between the outgoing sound and its returning echo. In bats with long cf-fm sounds
probably only the fm component is used for range determination
(Schnitzler, 1968; Simmons, 1973). This information processing in
the time domain allows a range determination accuracy of approximately 1 - 2 cm, this corresponds to a time accuracy of about
60 - 120μsec. The high degree of accuracy found suggests that some
kind of filtering improves the arrival time estimate. No experiment
to date, however, has been able to prove that this filtering is
optimal, as should be the case for a matched filter receiver. It

is probable that the range discrimination data can be explained on the basis of suboptimal receiver types since the experiments were performed under relatively favorable noise conditions.

Range Resolution

Range resolution is the ability to distinguish between two closely spaced targets. These two targets may be simulated by a single target reflecting two wavefronts with a variable time delay between them, as, for instance, a planar target with holes of variable depth. In this case, the range resolution performance could be measured as a percentage of correct choices for the two-wavefront-target when discriminating it from a one-wavefront-target (flat plate). Such an experiment has yet to be conducted. From discrimination experiments with two-wavefront-targets made in another context (Simmons et al., 1974; Habersetzer, Vogler and Leimer, personal communication), it can be assumed that a bat's range resolution threshold is approximately 1 mm or less. How can this high degree of resolution be explained?

The echoes from a two-wavefront-target are distinctly different from the echoes of a one-wavefront-target. The frequency spectrum of the one-wavefront-target corresponds well with the spectrum of the emitted sound. The spectrum of the two-wavefront-target, however, has distinct minima which change their position in relation to the time delay (hole depth) between the two wavefronts (fig. 18). These minima reflect the interference patterns between the overlapping echoes returning from the two reflecting planes of the two-wavefront-target. Simmons et al. (1974) demonstrated that bats are able to use such spectral cues to discriminate between two targets, each of which produced two wavefronts but with different wavefront delays between them. In this experiment, a discrimination threshold of 1 mm or less was measured. It is therefore highly probable that this type of information processing in the frequency domain could also be applied in the range resolution experiment suggested above and that the range resolution threshold would be approximately the same or even better.

DETERMINATION OF ANGULAR DIRECTION

Acoustical Cues

For the determination of target direction, the angle at which the echo's returning wavefront arrives at the bat has to be measured. For this measurement, the bats collect information with their two ears which act as directional antennae. Such a system could deliver information on interaural phase, time, and sound pressure differences as well as monaural cues. Flying bats could

also use the angular dependence of the Doppler shift as additional
information for the directional localization of nonmoving targets
(Schnitzler, 1968). It is probable that various species of bats
use the available information in different ways. The following
suggestions have been forwarded to explain directional determination by bats.

Binaural Comparison of Interaural Differences in Arrival Time

The azimuth of target direction is encoded by differences
in the phase and arrival times of echoes. Phase differences,
however, do not allow unique direction determination, since the
half wavelength of ultrasonic sounds is shorter than the bat's
interaural distance. It is also improbable that phase information is preserved in the auditory system at frequencies above
5 kHz. It is therefore suggested that phase information is not
used for the determination of horizontal target angles.

The interaural differences in arrival time depend on the distance between the two ears. This distance is relatively short in
bats (8 - 22 mm). For reasonable accuracy in angle determination,
the barely detectable interaural time difference should also be
very small. Shimozawa et al., (1974) calculated for Myotis
grisescens that the interaural time difference changes 0.45 µsec
per degree. Behavioral data suggest that bats are able to determine
direction with an accuracy of about 5° (see below). Depending on
the interaural distance in different species of bats, the barely
detectable time difference would be in the range of 3 - 5 µsec.
This is comparable to the smallest detectable interaural time
difference of 5 µsec measured in man (cited in Shimozawa et al.,
1974). The fact that bats might be capable of using the interaural
time difference for the determination of a target's azimuth should
therefore still be considered a possibility. It is, however, more
probable that the interaural differences in the sound pressure levels
of different frequency components are the more important cues.

Binaural Comparison of Different Echo Frequency Components

Grinnell and Grinnell (1965) demonstrated for Myotis lucifugus
and Plecotus townsendii with collicular evoked potentials that the
directionality of hearing is frequency dependent and that the directionality increases with frequency. They found that the measured
directionality depends mainly on the structure of the external ear
and demonstrated that every angle in space is uniquely characterized
by a family of binaural ratios of echo spectral components of different frequencies. They therefore suggested that bats with fm
sounds measure the binaural ratios of different frequency
components in the echo and use this information for determination of
direction. The conditions for this form of angle determination are

orientation sounds with different frequency components and a change
of ear directionality with different frequencies. These are not
only found in fm bats but also in bats with cf-fm sounds such as
Pteronotus and Saccopteryx (Grinnell, 1970; Grinnell and Hagiwara,
1972). Different frequencies not only occur in the fm components
but also in the harmonics of the cf components of the signals.

Binaural Comparison of cf Components in the Echo at Different
Ear Positions

    Bats with long cf-fm sounds such as rhinolophids, hipposiderids
and also Pteronotus parnellii often concentrate most of their sound
energy in the cf part. The terminal fm sweep is short and covers
only a small frequency range. This raises the question as to whether
these bats can use their cf part for the determination of angular
directions. The answer to this question might be determined if an
explanation could be found for the remarkable ear movements used by
bats with long cf-fm sounds (Rhinolophids: Schneider and Mohres,
1960; Griffin, Dunning, Cahlander and Webster, 1962; Pye, Flinn and
Pye, 1962; Webster, 1967a; Gorlinsky and Konstantinov, 1978; Hipposiderids: Pye and Roberts, 1970; Pteronotus parnellii: Schnitzler,
1970a). In order to explain the determination of direction with
cf components, Schnitzler (1973b) suggested that the ear movements
during echo reception change the binaural SPL ratio of an echo's
cf component in a way typical for each target direction. Since the
bats control their ear movements, they should be able to decode the
information concerning target direction contained in the changing
binaural SPL ratios. The dependence of hearing directionality on
the position of the external ear measured with collicular evoked
potentials in Rhinolophus (Neuweiler, 1970) supports this hypothesis.
Further evidence for the importance of ear movements in the determination of direction was given by Gorlinsky and Konstantinov (1978).
They showed that ablation or partial immobilization of the pinnae of
Rhinolophus prevents vertical and diminishes horizontal localization
of long cf signals played over a loudspeaker.

Directional Information from Doppler Shifts

    In flying bats, the echo frequency is higher than the emission
frequency owing to Doppler shifts caused by the relative velocity
between bat and target. For stationary targets, the relative velocity is dependant on the angle between flight path and target. The
Doppler shifts decrease with the cosine of the increasing target
angles. It may be possible that, in bats with cf-fm sounds, Doppler
shifts might be a cue for a rough determination of target angle
(Schnitzler, 1968). In Rhinolophus flying 4.1 m/sec the maximal
Doppler shift is 2000 Hz, at 30° it is still 1732 Hz, at 60° 1000 Hz
and at 90° it would be 0 Hz. This cue would only be of aid in flight
and with stationary targets. It is therefore highly probable that it

may simply be an additional cue which can be used together with other mechanisms for determination of direction.

Discrimination Experiments

Peff and Simmons (1972) trained Eptesicus fuscus and Phyllostomus hastatus to discriminate the horizontal angle between a spherical target offered at 0° and a fixed target at 19° from the angle between the target at 0° and a movable target offered at variable angles between 37° and 19°. The bats were able to discriminate the two angles when the variable angle was 6° - 8° larger than the fixed one in Eptesicus fuscus and 4° - 6° larger than in Phyllostomus hastatus (fig. 10). Airapetianz and Konstantinov (1974) determined the angular resolution in the following way: they measured the ability of Rhinolophus ferrumequinum to discriminate a pair of metal rods offered at an angle below the angular resolutation (24') from a pair of similar rods offered at variable angles between 7° 50' and 2° 58'. The bats reached the threshold level of 75% correct responses at an angle of 4° 30' (fig. 10).

Monaural Versus Binaural Localization of Targets by Echolocation

Several experiments have been made to determine the influence of auditory input attenuation for one ear on a bat's ability to localize targets. All these experiments were made under the assumption that a decrease of localization acuity depends on a reduced ability to determine direction. Griffin and Galambos (1941) showed that monaural earplugs reduced the ability to avoid wire obstacles in Myotis lucifugus Konstantinov, Sokolov and Stosman (1967) found that, after unilateral perforation of the tympanic membrane, a reduction in obstacle avoidance ability occurred in Plecotus auritus and Rhinolophus mehelyi but no effect was observed in Myotis oxygnathus. Möhres (1953) stated that obstacle avoidance in Rhinolophus ferrumequinum is not influenced by plugging one ear. Further experiments with Rhinolophus ferrumequinum (Flieger and Schnitzler, 1972), however, showed that the obstacle avoidance score was reduced from a normal value of 90% down to 60% when a monaural earplug having a measured attenuation of 15 - 20 dB was inserted. If the second ear was plugged in the same way, these bats reached 90% again. This fact demonstrates that both ears must be provided with a symmetrical input for optimal target localization.

These data show that monaural echo attenuation reduces the obstacle avoidance performance in nearly all species of bats tested to date. This result and the restoration of the localization acuity in Rhinolophus ferrumequinum with symmetrical attenuation in both ears, leads to the conclusion that, in bats, the interaural comparison of echo intensity plays an important role in angular direction determination.

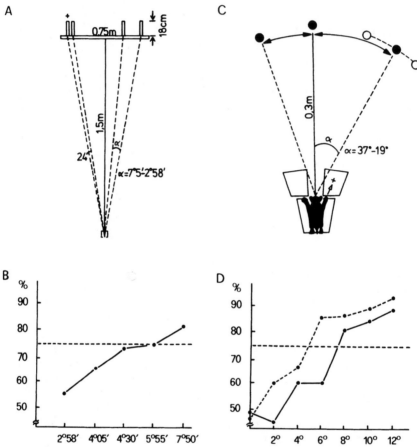

Fig. 10. Experimental set-up and results of discrimination experiments for the determination of the angular resolution threshold in Rhinolophus ferrumequinum (A and B) and of the angle measuring accuracy in Eptesicus fuscus (solid lines) and Phyllostomus hastatus (dashed lines) (C and D). Adapted from Airapetianz and Konstantinov, 1974 (A and B) and from Peff and Simmons, 1972 (C and D).

Localization of Sound Sources by Passive Hearing

The ability of bats to localize ultrasonic sound sources by passive hearing offers additional information on the acoustical cues required for the determination of angular direction.

Konstantinov, Stosman and Gorlinsky (1973) trained Rhinolophus ferrumequinum and Myotis sp. to discriminate which of two loud-speakers positioned at various angles (21° - 2°) 2.7 m away trans-mitted ultrasonic signals of constant frequency and with a sound duration of 5 msec. Both bats learned to make correct right-left decisions with large angles. The 75% level of correct responses was reached at an angle of 4° in Rhinolophus ferrumequinum and of 5° in Myotis sp. (at 83.5 and 40 kHz respectively). In further experiments the localization ability of Rhinolophus was also measured in the vertical plane (Gorlinsky and Konstantinov, 1978). With a signal duration of 50 msec and a frequency of 81 kHz, the bats reach-ed the 75% discrimination level in both planes at angles of about 4°. With frequency modulated signals which imitated the terminal sweep of the echolocation sounds (2.5 msec duration, sweep of 81 to 62 kHz) the threshold angle of 19° in the vertical plane was far greater than the angle of 6° in the horizontal plane. The plugging of one ear prevented the discrimination of long cf signals in both planes even at angles of 25°. Partial immobilization or ablation of the pinnae prevented the vertical and reduced the horizontal discrimination ability with long cf signals at an angle of 25°. These data support the hypothesis (Schnitzler, 1973b) that, in Rhinolophus, direction determination is based upon evaluation of successive interaural in-tensity differences caused by the ear movements.

DETERMINATION OF RELATIVE VELOCITY BETWEEN BAT AND TARGET AND DOPPLER SHIFT COMPENSATION

Relative Velocity and Doppler Shifts

The relative velocity between bat and target can be determined either by measuring the decrease in distance between consecutive signals or by evaluating the Doppler shifts in echoes. No experi-ments have been conducted to date to determine a bat's efficiency in measuring relative velocity by range tracking. We will there-fore concentrate in this chapter only on the Doppler shift evalua-tion.

Doppler shifts originate when a sound source or a receiver moves when transmitting or receiving sounds. In the case of a bat flying with a velocity v_B toward a target, which moves with a velocity v_T toward the bat, the echo frequency f_e can be calculated from the transmission frequency f_t (Schnitzler, 1968):

$$f_e = f_t \frac{c + v_B}{c - v_B} \times \frac{c + v_T}{c - v_T}$$

If v_B and v_T are small compared to the velocity of sound (c) a good approximation for f_e is

$$f_e = f_t \frac{2 v_B + 2 v_T}{c}$$

As $v_B + v_T$ correspond to the relative velocity v_R between bat and target, f_e can be expressed as

$$f_e = f_t + f_t \frac{2 v_R}{c}$$

The frequency shift caused by Doppler shifts is therefore

$$f_e - f_t = f_t \frac{2 v_R}{c} = \Delta f$$

The Doppler shift thus increases linearly with frequency and relative velocity so that the ability of a bat to measure relative velocity is directly determined by its ability to measure the Doppler shifts in echoes.

Accuracy of Frequency Measurement

Frequency cannot be measured with infinite accuracy as all filters have finite time constants and bandwidths. A rule of thumb in technical frequency analysis is that, with decreasing time for analysis, the bandwidth of analysis filters must be increased so that

$$\Delta f \times \Delta t \approx 1$$

This means that the accuracy is already limited by the time a signal is available for analysis. Accuracy is limited in addition by noise and by the bandwidth of the analysis filters even when the analysis time is long. If we use the same rule of thumb for bats, we can assume for low noise levels that the accuracy in frequency determination is limited by the characteristics of the neuronal filters and by the time each individual filter is excited.

In cf signals, the excitation time corresponds to the duration of the echolocation signals (T). The theoretical upper limit of frequency accuracy is therefore only limited by the duration of the signal:

$$\Delta f \approx \frac{1}{T}$$

This theoretical limit cannot be attained when the bandwidth of the filtering neurons is larger than the predicted value.

In fm bats which use signals with bandwidths (B_s) larger than the bandwidth of the neurons (B_n) the signal sweeps through the neuronal filters. The signal duration therefore has to be reduced by the ratio B_n/B_s in order to determine the excitation time. The theoretical upper accuracy limit can thus be expressed as

$$\Delta f \approx \frac{1}{T} \times \frac{B_s}{B_n}$$

It must be emphasized that these calculations are made using a "rule of thumb" to obtain approximate predictions for the frequency measurement accuracy. It may well be that bats improve the accuracy of frequency measurement by additional informaion processing. There is no experimental evidence, however, to prove that this is the case.

From these considerations, it can be concluded that long signals of constant frequency are especially suited for Doppler shift evaluation whereas short fm signals with large sweeps are far less effective. Theoretically, frequency changes of 20 Hz could be detected with a cf sound of 50 msec duration.

Therefore it is probable that bats with long cf-fm sounds, such as rhinolophids, hipposiderids and the mormoopid bat Pteronotus parnellii, are very sensitive to Doppler shifts whereas bats using pure fm sounds do not avail themselves of Doppler information. This conclusion is supported by neurophysiological studies. In bats with long cf-fm sounds, many neurons are extremely sharply tuned, thus allowing good frequency determination whereas in fm bats, only widely tuned neurons have been found.

No systematic experiments have been made to date to investigate the accuracy of frequency determination in bats. In Rhinolophus ferrumequinum, however, two factors suggest that these bats are able to determine frequencies with an accuracy of 30 - 60 Hz: in the first place during Doppler compensation they have the ability to maintain the echo frequency in the cf component of consecutive sounds with a standard deviation of only 30 - 40 Hz; secondly, they have the ability to discriminate echoes with sinusoidal frequency modulations from non-modulated ones when the modulation depth is 60 Hz

(Schnitzler, 1978). Since the frequency of the cf component is
approximately 83 kHz, one can calculate that relative velocities
of 6 - 12 cm/sec could be detected by evaluating Doppler shifts.

Everyone who has worked with rhinolophids knows that these bats
are very sensitive to Doppler shifts. Even slow movements of the
hand or of a large target toward the bat cause a typical fright
response, characterized by an increase in scanning movements of the
head and a retraction of the body on the hind legs. More rapid movements immediately result in the bats flying off. Movements away from
the bat evoke no response at all. This indicates that the Doppler
shifts from large objects moving toward a horseshoe bat have a warning function.

Beat Note Hypothesis

In rhinolophids, hipposiderids, and Pteronotus parnellii the
cf component is so long that returning echoes often overlap the
transmitted signals. In the case of relative movements between bat
and target, the bats would hear two different frequencies, the
transmission frequency and the Doppler shifted echo frequency. Pye
(1960, 1961a, 1961b, 1963) and Kay (1961a, 1961b, 1962) suggested
therefore that cf bats might hear the frequency difference as low
frequency beat notes produced by a multiplicative process. This beat
frequency would correspond to the Doppler shifts and therefore allow
a precise determination of the relative velocity between bat and
target. Since bats have high auditory thresholds in the range of
possible beat frequencies (Neuweiler, Schuller and Schnitzler, 1971;
Long and Schnitzler, 1975) and since there are many results indicating that bats process their echo frequencies directly in the ultrasonic frequency range, this hypothesis is now only of historical
interest.

It should be noted, however, that the overlapping of signals
with two frequencies (f_1, f_2) sometimes produce amplitude modulations
with a modulation frequency ($f_2 - f_1$) and these modulations are
capable of evoking synchronized responses in the auditory system
(Grinnell, 1970; Suga, Simmons and Jen, 1975; Vasiliev, 1976).

Doppler Shift Compensation in Free Flight

At positive relative velocities between bat and target the echo
frequency is always higher than the transmission frequency owing to
Doppler shifts. For instance, in Rhinolophus ferrumequinum flying
with a speed of 4.1 m/sec toward a stationary target the difference
between transmission and echo frequency is as high as 2 kHz.

Laboratory studies with long cf-fm bats such as the rhinolophids
Rhinolophus ferrumequinum and Rhinolophus euryale (Schnitzler, 1968),

the mormoopid bat Pteronotus parnellii (Schnitzler 1970a) and the
hipposiderid bat Asellia tridens (Gustafson and Schnitzler, in
press) revealed that in flight, these bats lower the frequency of
the cf component in such a way that the Doppler shifts are compensated and the echo frequency is kept constant within a small frequency band about 200 Hz above the average cf frequency of resting
bats (resting frequency). This behavioral response to Doppler shifts
was termed "Doppler shift compensation". Russian scientists recently
repeated the experiments with flying Rhinolophus ferrumequinum and
confirmed the presence of Doppler shift compensation (Konstantinov,
Makarov, Sokolov and Sanotskaya, 1976; Konstantinov, Makarov and
Sokolov, 1978).

In order to decide whether horsehoe bats compensate Doppler
shifts in a control system or if they derive the necessary downward frequency shift from measurement of the air speed (perhaps with
their vibrissae), two crucial experiments with bats flying in a
wind tunnel or in a He-O_2 gas mixture were made (Schnitzler, 1973a).
The following results proved that bats use a feedback control system
to hold the echo frequency constant at a reference frequency. Bats
flying in head winds in a wind tunnel lowered the transmission
frequency to compensate for the Doppler shifts caused by the ground
speed and not with reference to the far higher air speed. In the
He-O_2 mixture, the bats also made no use of the flight speed information and compensated for the Doppler shifts with reference to the
different sound velocity in the gas mixture.

Bats also compensate for Doppler shifts originating from other
sources. A pendulum moving in front of a stationary horseshoe bat
will evoke the Doppler compensating response (Schnitzler, 1968).
Compensation was also found in Rhinolophus ferrumequinum sitting on
a moving platform (Sokolov and Lipmanova, 1977; Konstantinov, Makarov
and Sokolov, 1978) and Pteronotus parnellii sitting on a pendulum
(Henson, poster in this conference) or on a moving platform (O'Neill,
personal communication).

Doppler Shift Compensation with Simulated Doppler Shifts

With free flying bats, the performance of the feedback control
system for Doppler shift compensation cannot be investigated sufficiently well. In Rhinolophus, however, electronically simulated
Doppler shifts evoked the Doppler compensation response. It is thus
possible to determine the compensation range (Schuller, Beuter and
Schnitzler, 1974) and the dynamic properties of the compensation
system by this means (Schuller, Beuter and Rübsamen, 1975). Simmons
(1974) using a different method to simulate Doppler shifts in the
echoes also evoked Doppler compensation in a horseshoe bat.

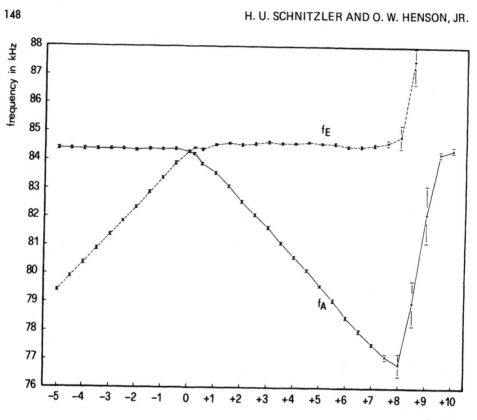

Δf in kHz

Fig. 11. Dynamic range of the Doppler shift compensation system of Rhinolophus ferrumequinum. For the cf components, the transmitted frequency (f_A - solid line) and the echo frequency (f_E - dashed line) was measured for simulated positive and negative Doppler shifts (Δf). Each point corresponds to the average frequency of 200 cf components with standard deviation of the mean. Adapted from Schnitzler (1978).

In the Doppler simulation experiments, bats were held in a restrainer in front of a microphone and a loudspeaker. The transmitted sounds were picked up with the microphone and electronically shifted in frequency upward or downward with the moving target simulator. The shifted signals were played back to the bats as artificial echoes thus simulating relative movements toward or away from the bat. Some bats responded to these simulated target movements by compensating positive Doppler shifts without paying attention to the fact that range did not change in this situation.

Fig. 11 gives an example for experiments with Rhinolophus ferrumequinum in which positive and negative Doppler shifts were simulated over a wide range in order to determine the entire compensation ability.

With positive frequency shifts, horseshoe bats lower the transmission frequency (f_A) of their pulse cf component so that the echo frequency (f_E) is kept constant at approximately the reference frequency of the feedback control system. In Fig. 11 the reference frequency is about 300 Hz above the average cf produced when there is no frequency shift (resting frequency). In other species frequency differences of 50 - 300 Hz between resting and reference frequency have been observed (Schuller, Beuter and Schnitzler, 1974). The horseshoe bat in our example (fig. 11) was able to compensate for frequency shifts up to 8 kHz. At higher Doppler shifts, compensation was incomplete and ceased above 10 kHz. The compensation range for positive frequency shifts differs from bat to bat but normally lies between 4 - 8 kHz. The normal flight speed of horseshoe bats is seldom higher than 5 m/sec (Schnitzler, 1971). The Doppler shifts this brings about do not exceed 2.5 kHz. Therefore in flying horseshoe bats sufficient compensation range remains to compensate for additional Doppler shifts caused by targets moving toward the animal.

Negative Doppler shifts simulating relative movement away from a bat are not compensated for. In this case, the transmission frequency is similar to the resting frequency. This behavior might be of biological value since targets which move away do not indicate danger or, in the case of prey, are rather difficult to catch since the maximum flight speed of Rhinolophus ferrumequinum is low (Schnitzler, 1971).

An astonishing result of the compensation experiments is the high precision with which the echo frequency is kept constant. When compensating positive Doppler shifts, some horseshoe bats reach standard deviations of only 30 - 40 Hz. This indicates that the accuracy of frequency determination must be equally good.

Insects changing their flight paths can cause rapid changes in the relative velocity between a bat and its prey. It is therefore of interest to know how bats respond to changing Doppler shifts. The dynamic properties of Rhinolophus ferrumequinum's Doppler compensation system has been studied in detail by Schuller, Beuter and Rübsamen (1975). Fig. 12 summarizes some of their results. In their experiments, horseshoe bats had to compensate for positive frequency shifts changing sinusoidally between +1 and +3 kHz with different modulation frequencies.

At modulation frequencies below 0.1 Hz, the compensation system is fast enough to compensate for the simulated Doppler shifts completely by lowering the transmission frequency for each signal (fig. 12 A). Therefore the echo frequency is always kept constant at the reference frequency (fig. 12B). At a modulation frequency of 0.25 Hz the compensation system cannot respond quickly enough to compensate for the Doppler shifts for every sound. With increasing

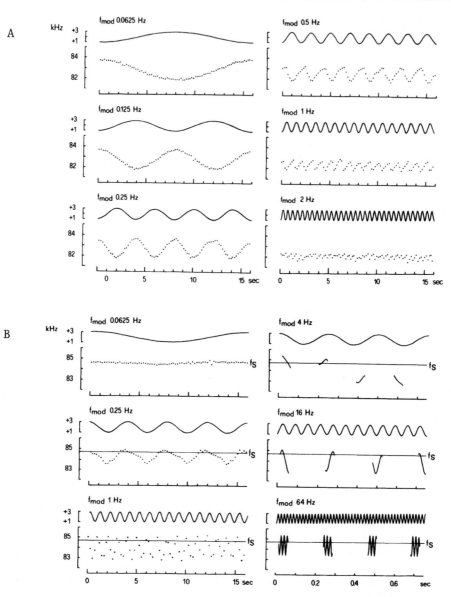

Fig. 12. Transmission frequency (A) and echo frequency (B) of the cf components of echolocation sounds in Rhinolophus ferrumequinum when compensating for simulated Doppler shifts changing sinusoidally between +1 and +3 kHz at different modulation frequencies (f_{mod}). In (A) and in the left half of (B) each point corresponds to the frequency of one cf component. In the right half of (B) the time base is changed to make the frequency structure of each individual echo visible. In (B) the echo frequency can be compared with the reference frequency (f_s). Part (B) adapted from Schnitzler (1978).

modulation frequencies only the highest Doppler shifts are completely
compensated for since the feedback system is faster when compensating
for increasing rather than for decreasing Doppler shifts. At modulation frequencies above 2 kHz, the transmission frequency remains at
the lower end of the corresponding compensation range the entire time
(fig. 12 A). In situations where the modulation frequency is so
high that one or more cycles fall within one echo (f_{mod}> 20 Hz), the
transmission frequency is also kept at the lower end of the compensation range so that the highest frequency components in the echoes
stay near the reference frequency.

The experiments with Doppler compensating bats have revealed
that these bats regulate their transmission frequency in order to
adjust the cf component of the echo to a specific reference
frequency. This raises the question as to whether the receiver,
i.e. the hearing system, shows a corresponding adaptation to the
reference frequency.

Hearing in Doppler Compensating Bats

Auditory thresholds of bats with fm signals are widely tuned
over the whole frequency range of their emissions. In bats with
long cf-fm sounds and Doppler shift compensation, however, a different threshold curve with a small minimum in the range of the reference frequency occurs. The hearing system of these bats is
further characterized by numerous neurons with extremely sharp
tuning and best frequencies near the reference frequency. These
specializations have been studied with neurophysiological methods
in rhinolophids (Neuweiler, 1970; Neuweiler, Schuller and Schnitzler,
1971; Schnitzler, Schuller and Neuweiler, 1971; Suga, Neuweiler and
Möller, 1976; Schnitzler, Suga and Simmons, 1976; Neuweiler and
Vater, 1977; Möller, Neuweiler and Zöller, 1978; Ostwald, 1978;
Airapetianz and Vasiliev, 1970, 1971; Vasiliev, 1971, 1975, 1976;
Vasiliev and Andreeva, 1971; Vasiliev and Timoshenko, 1973), in
Pteronotus parnellii (Grinnell, 1967, 1970; Pollak, Henson and
Novick, 1972; Suga, Simmons and Shimozawa, 1974; Suga, Simmons and
Jen, 1975; Suga and Jen, 1977; Suga, 1978; Suga and O'Neill, 1978)
and in hipposiderids (Grinnell and Hagiwara, 1972). The sharply
tuned threshold minumum near the reference frequency has also been
found in the behavioral audiogram of Rhinolophus ferrumequinum
(Long and Schnitzler, 1975).

Studies on the organization of the auditory cortex in Pteronotus parnellii (Suga and Jen, 1976; Suga, 1977; Suga, O'Neill and
Watanabe, 1978) and Rhinolophus ferrumequinum (Ostwald, 1978, and
poster at this conference) revealed, as a further adaptation, a disproportionate tonotopical organization with large areas specialized
for the processing of the cf and fm components of the echoes.

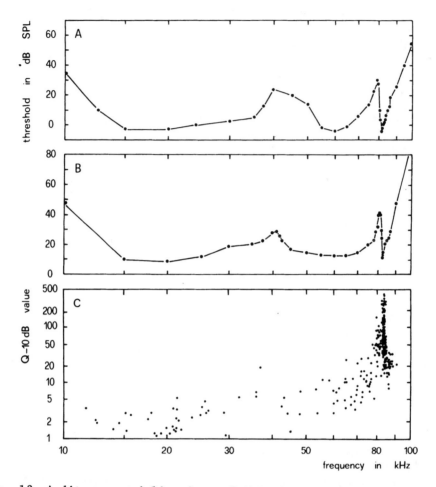

Fig. 13. Auditory specializations of Rhinolophus ferrumequinum.
(A) Behavioral audiogram (adapted from Long and Schnitzler, 1975);
(B) evoked potential threshold as measured in the colliculus
inferior (adapted from Schnitzler, 1978); (C) Q-values of primary
auditory cortex neurons (adapted from Ostwald, poster at this
conference).

    Details on information processing in the auditory system of
Doppler compensating bats will be given in other reviews at this
conference. Therefore only a few basic data on hearing in Rhinolophus ferrumequinum, to understand the performance of such a specialized echolocation system, are summarized in fig. 13.

    In Rhinolophus ferrumequinum, the behavioral audiogram (Long
and Schnitzler, 1975) is similar to the evoked potential threshold

curve (Neuweiler, 1970; Neuweiler, Schuller and Schnitzler, 1971).
Both curves are characterized by a narrowly tuned threshold minimum
in the reference frequency range. This minimum is separated from a
wide tuned area of low threshold which partly covers the frequency
range of the terminal fm sweeps (83 - 68 kHz) in the bat's cry. The
two areas where the threshold minimum is low are separated by an
insensitive region where the threshold is 25 - 30 dB higher. Another insensitivity peak occurs at the frequency of the orientation
sound's suppressed first harmonic (about 41.5 kHz) and this
separates the fm area from another area of low threshold which corresponds to the main frequency range of communication sounds. Below
15 kHz, the threshold is relatively high, indicating that horseshoe
bats have poor hearing at low frequencies.

The ability of the auditory system to measure frequencies is
determined by the sharpness of the tuning curves of single neurons.
The degree of tuning can be characterized by the Q_{10}-dB value (ratio
of best frequency to bandwidth at 10 dB above minimum threshold). The
sharper the tuning, the higher the Q_{10}-dB value and the better the
frequency resolution. In fig. 13 C the Q_{10}-dB values of auditory
neurons from the cf-processing area in the cortex of Rhinolophus
(Ostwald, poster at this conference) are compared with the auditory
threshold curves. This comparison shows that at frequencies below
the threshold minimum at the reference frequency, the Q-values are
low, indicating wide tuning. Similar values were also found in
fm bats (Suga, Neuweiler and Möller, 1976). However, near the reference frequency and a few kHz above it, the neurons are extremely
sharply tuned, reaching Q-values up to 400. This suggests a very
high accuracy in frequency measurement for this frequency range.

The described specializations of transmitters and receivers in
the echolocation systems of Doppler compensating bats are interpreted as adaptations for the detection and evaluation of movements.

It has already been pointed out that long cf-fm sounds are
advantageous for the determination of Doppler shifts and therefore
probably for the measurement of relative velocity between bat and
target. Doppler shift compensation and the sharply tuned threshold
minimum near the reference frequency may improve the determination
of the frequency differences between transmitted sounds and overlapping echoes. Faint echoes are adjusted in a frequency range of
low auditory threshold whereas loud outgoing sounds fall in frequency
ranges where auditory thresholds are high. The masking effect of the
loud transmitted sounds is therefore reduced (Neuweiler, 1970).

A further advantage of Doppler compensation combined with a
sharply tuned auditory system is that, in a situation where a bat
compensates for echoes from a target moving toward it, all echoes
from stationary or slowly moving targets are rejected since they

have lower frequencies and therefore fall in ranges of higher
auditory thresholds (Schnitzler, 1973 a, b; Simmons, 1974).

Long cf component sounds also have the potential to deliver
directional cues since they allow the measurement of the Doppler
shifts which are reduced laterally in accordance with the cosine of
target angle (Schnitzler, 1968).

The most important advantage of echolocation in Doppler compensating bats is that amplitude and frequency modulations in
echoes, brought about by the wingbeats of insects, can be detected.
This advantage is so important that it will be discussed in a
separate section.

DETECTION AND EVALUATION OF FLUTTERING TARGET MOVEMENTS

Hypothesis

The beating wings of insects frequency modulate and amplitude
modulate cf signals and insects are the prey of all Doppler-shift-
compensating bats studied to date. After the discovery of Doppler
shift compensation it was therefore suggested that these bats keep
the echo frequency constant at the reference frequency in order
to maintain a constant carrier frequency for such frequency and
amplitude modulations (Schnitzler, 1970a, 1970b). These modulations could aid in discriminating echoes from fluttering and nonfluttering targets. Since different species of insects cause
different modulations, it is even possible that bats use these
cues to classify or identify their insect prey (Schnitzler, 1978).

Behavioral Experiments

In order to answer the question as to whether bats utilize frequency modulations caused by fluttering target movements, the following experiment was conducted (Schnitzler, 1978; Schnitzler and
Flieger, in prep.). Several horseshoe bats were trained to discriminate an oscillating from a motionless target. By controlling the
oscillation amplitude and the oscillation frequency of the target,
frequency modulations of defined modulation depth (Δf) and modulation frequency (f_{mod}) were produced. It was therefore possible
to determine, for different modulation frequences, the threshold
modulation depth just necessary to discriminate the oscillating
target from the motionless target with a percentage of 75% correct
choices (fig. 14).

At oscillation rates of 30 Hz and below, the bats required a
modulation depth of at least 60 Hz in order to discriminate. Since
modulation depth corresponds to the difference between highest and

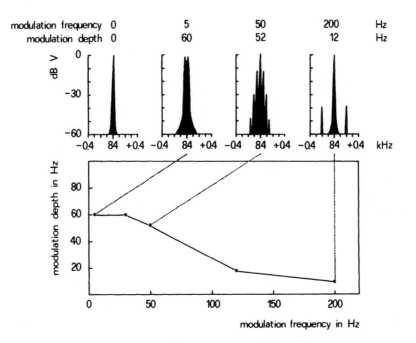

Fig. 14. Lower part: Performance of <u>Rhinolophus ferrumequinum</u> when discriminating a target oscillating at different rates (or modulation frequencies) from a nonoscillating target on the basis of Doppler shifts caused by the oscillations. The threshold curve indicates the echo modulation depth (maximal frequency difference in the sinusoidally modulated echoes) necessary to make the discrimination with 75% correct choices. Upper part: Comparison of echo spectra simulating threshold situations with the spectrum of a nonmodulated echo. For the simulation a carrier frequency of 84 kHz was used (adapted from Schnitzler, 1978).

lowest frequency in the sinusoidal changing echo frequency, this result suggests that bats have a frequency resolution of about 60 Hz in this situation. This value is comparable to the frequency determination accuracy of 30 – 40 Hz derived from Doppler shift compensation experiments. At oscillation rates above 30 Hz, however, the discrimination threshold decreases, finally reaching a value of only 12 Hz when the target's oscillating frequency is 200 Hz. This finding cannot be explained on the basis of the frequency discrimination performance of horseshoe bats.

A comparison of the spectra of echoes used at different modulation frequencies for threshold discrimination may help to explain these results (fig. 14). The unmodulated carrier frequency has a small single peak in the spectrum. At low modulation frequencies,

this peak oscillates up and down at a rate sufficient for the bats
to make normal frequency discriminations. A time averaged spectrum
of this situation shows a peak which is as wide as the modulation
depth. At oscillation frequencies above 50 Hz, the spectra are
characterized by side bands. The height of the side bands determined
by the modulation depth and the distance between the bands corresponds to the modulation frequency. Although modulation depth is
only 12 Hz at an oscillation rate of 200 Hz, the side bands are
clearly visible in the spectra and must be audible to the animals.
It was therefore suggested that, at low oscillation rates, bats
measure the frequency changes in the echoes directly, whereas the
high sensitivity to frequency modulations at high oscillation rates
can be explained on the basis of side band detection (Schnitzler,
1978). A similar mechanism is known in humans where tones with
high modulation rates and small modulation depth sound "rougher"
than pure tones.

Behavioral discrimination experiments support the hypothesis
that Doppler compensating bats utilize frequency modulations in
their echoes to obtain information on oscillating target movements.
These experiments proved, in addition, that this detection system
is extremely sensitive, e.g. a target oscillating with a frequency
of 40 Hz only requires an oscillation amplitude of 0.3 mm in order
to be distinguished from a motionless target.

The Doppler shifts produced by the wing movements of insects are
far above the discrimination thresholds measured in Rhinolophus
ferrumequinum, and it must be concluded that the frequency modulations in echos are detectable by Rhinolophus and other Doppler compensating bats.

The fact that wing movements are very important for the detection of insects can be derived from observations on horseshoe bats
hunting insects in a flight room. Only insects with beating wings
were pursued whereas moths whose wings were not moving did not
evoke the bats' interest even when they were in the acoustical beam
of the transmitted sounds (Schnitzler and Hollerbach, unpublished).
Russian scientists made similar observations with rhinolophids
(Sokolov, 1972; Airapetianz and Konstantinov, 1974) but interpreted
their results in a different way. They assumed that the bats hear
the wing movements of the moths and find their prey by passive acoustical localization. That the hearing of the wing sounds might act as
an additional cue for the detection of insects cannot be excluded.
We, however, suggest that the main information on flying insects is
derived from the frequency and amplitude modulations in the echoes,
since horseshoe bats still pursued fluttering insects when the sounds
produced by their wings were completely masked for a human listener,
by the loud noise of a ventilation fan (Schnitzler and Hollerbach,
unpublished).

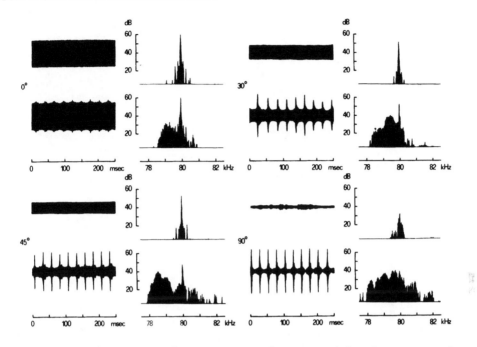

Fig. 15. Oscillograms and time averaged spectra of echoes returning from flying and nonflying sphingid moths Daphnis nerii when oriented at different angles in the acoustical beam of a loudspeaker transmitting an 80 kHz signal (adapted from Schnitzler, 1978).

The importance of wing beats has also been studied in Pteronotus parnellii by Goldman and Henson (1977). This species only pursued insects with flapping wings and it routinely attacked a stationary mechanical insect model when its wing-like parts were moving at rates of 15 Hz or higher. They found, in addition, that Pteronotus rejects some species of insects, thus demonstrating selective prey acquisition.

Echo Analysis

For the description of amplitude and frequency modulation caused by wing beats we have analyzed echoes from flying insects. Several species of moths were tethered in the beam of an ultrasonic loudspeaker which transmitted a carrier frequency of 80 kHz. The echoes returning from the insects were picked up with a microphone positioned near the loudspeaker, recorded on tape and afterwards analyzed (Schnitzler, 1978; Schnitzler and Flieger, in prep.). Figs. 15 and 16 summarize results from a sphingid moth (Daphnis nerii) flying and resting when oriented at different angles to the loudspeaker.

Echoes returning from a resting moth had the highest sound

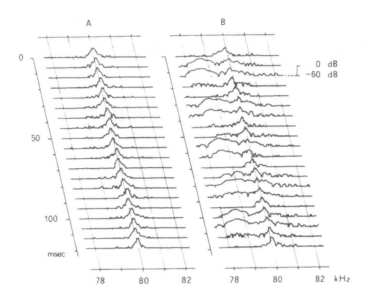

Fig. 16. Real-time spectrograms of echoes returning from a nonflying
(A) and a flying (B) <u>Daphnis nerii</u> when oriented at an angle of
45° (adapted from Schnitzler, 1978).

pressure level (SPL) when the moth was oriented with the head directly toward the loudspeaker (60 dB re $2x10^{-5}N/m^2$ at 0°). When the moth
was turned to the side, the echo SPL decreased and only reached
35 dB at an orientation of 90°. At 0° the flapping wings caused only
minor amplitude modulations whereas at 90°, a modulation ratio of
30 dB was reached. The number of amplitude peaks per second indicated
a wing beat frequency of 38 - 40 Hz.

Frequency spectra averaged over several beats allow the degree
of frequency changes caused by the wing beats in the echoes to be
determined. The spectra of echoes from resting moths show a single
small peak which decreases in height at lower echo intensities. In
flying moths, the frequency and amplitude modulations in the echoes
are represented by sidebands in the spectra. These sidebands are
asymmetrical, the main energy being normally concentrated below the
carrier frequency. The width of the sidebands and their fine structure differ for the moth's varying angles of orientation. At 30°
the lower sideband is approximately 1.5 kHz wide whereas at 90°,
widths of more than 2 kHz are reached.

Comparative studies with echoes from different species of moths also reveal species specific characteristics in amplitude and spectral cues. In this case, moths with large wings cause larger amplitude modulation ratios and wider sidebands than smaller moths. Also differences in wing beat frequencies were found.

Time averaged spectra do not show the time relationship between frequency changes in the spectrum and wing beat rhythm. In order to obtain this information, real time spectra with time intervals of 6.25 msec were made (fig. 16). These spectra indicate asymmetrical spectrum broadening in rhythm with the wing beat.

All behavioral experiments conducted to date demonstrate that echo modulations can certainly be used to discriminate fluttering from nonfluttering targets. Several neurophysiological approaches give additional support to this conclusion.

Schuller (1972) demonstrated, with evoked potential measurements, that horseshoe bats are very sensitive to small linear frequency modulations and Johnson, Henson and Goldman (1974) showed for Pteronotus parnellii that frequency modulations are encoded in cochlear microphonic potentials. Experiments in Rhinolophus ferrumequinum with sinusoidally amplitude modulated (Andreeva and Vasiliev, 1977; Schuller, 1979) and frequency modulated stimuli (Ostwald, 1978; Schuller, 1979) and with frequency modulations in Pteronotus parnellii (Suga and Jen, 1977) revealed a synchronization of modulation frequency and neuronal responses.

When stimulated with natural moth echoes reproduced with a laboratory computer, Ostwald (poster at this conference) demonstrated that neurons in the cf-processing area of Rhinolophus ferrumequinum respond in synchronization with the wing beat (fig. 17). This indicates that Doppler compensating bats are not only able to discriminate between echoes from fluttering insects and stationary targets but also to measure the wing beat frequency of their prey. This makes it possible that bats are capable of categorizing their prey according to differences in wing beat frequency.

The fine structure of moth echoes differs in various species but also depends on the angular orientation of the moth in the acoustical beam. If bats are capable of learning to discriminate the acoustical images of different species and of recognizing an individual species oriented at different angles, they would have additional cues for the classification or even the identification of their prey. This could be an explanation for the selective prey acquisition found in Pteronotus parnellii (Goldman and Henson, 1977). Further experiments are in progress to investigate this question.

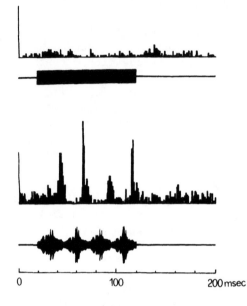

Fig. 17. Neuronal responses recorded in the cf-processing area of the primary auditory cortex of <u>Rhinolophus ferrumequinum</u> when stimulated with a pure cf tone and an echo of a flying insect. The insect echo was reproduced with a laboratory computer using a carrier frequency equal to the frequency of the pure tone. Note the synchronization between neuronal responses and wing beat frequency (adapted from Ostwald, poster at this conference).

Feeding Behavior of Doppler Compensating Bats

Laboratory studies on the feeding behavior of <u>Rhinolophus ferrumequinum</u> have demonstrated that these bats only pursue moths which are flapping their wings, and pursuit occurs even when the insects fly in very confined spaces or sit on surfaces (Schnitzler and Hollerback, unpublished). Goldman and Henson (1977) observed similar behavior in <u>Pteronotus parnellii</u>.

Under natural conditions, rhinolophids usually forage near the ground or close to the branches of low trees, bushes and along walls (Eisentraut, 1950; Brosset, 1966; Wallin, 1969). It is even possible that they take insects from the foliage of trees (Griffin and Simmons, 1974) or from the ground (Blackmore, 1964). <u>Rhinolophus</u>

aethiops has also been observed hanging on a tree branch and hawking moths like a flycatcher by taking short flights and returning to the same vantage point (Shortridge, 1934).

Pteronotus parnellii have been observed foraging mainly within 3.5 m of the ground under a dense canopy of foliage (Bateman and Vaughan, 1974).

The foraging behavior of Hipposideros commersoni is described in an excellent paper by Vaughan (1977). These bats mainly forage in riverine vegetations using vantage points some 6 m above the ground. From these points they scan the environment with echolocation signals and hunt large beetles in the "flycatcher style", thereby demonstrating high precision in trajectory evaluation.

Since Hipposideros is very selective in its choice of prey, one can suggest that this species must have a highly developed target discrimination ability (Vaughan, 1977). The same can be concluded for Pteronotus, which also showed preferences in prey acquisition (Goldman and Henson, 1977). In rhinolophids, no observations concerning the selection of prey have been made to date.

These observations indicate that Doppler compensating bats often hunt in areas where there are many additional stationary targets (bushes, trees, ground). They therefore have to detect the insect echoes amongst heavy background clutter. The extreme specialization for the detection of echoes which are modulated by the insects' wing beats can be interpreted as adaptations to overcome this problem. Another point arising from these observations is that Pteronotus parnellii and Hipposideros commersoni are selective in their prey acquisition. This high degree of target discrimination can be explained by the ability of Doppler compensating bats to measure the wing beat frequency of their prey and maybe even to evaluate additional cues in the species specific fine structure of the modulated insect echoes.

The long cf-fm sounds of Doppler compensating bats are also of advantage in hunting insects in the "flycatcher style". The high duty cycle increases the chances of detecting an insect when scanning the environment and the high echo energy of the cf component allows the detection of insects over long ranges.

Since the feeding behavior of Doppler compensating bats from different families is similar, we suggest that echolocation systems with long cf-fm sounds, Doppler shift compensation and specialization of the hearing system for the evaluation of modulated echoes, can be interpreted as an evolutional adaptation allowing the capture of insects in a specialized ecological niche.

DETERMINATION OF OTHER TARGET PROPERTIES

Targets encountered by bats under natural conditions are characterized by many more features than range, angular direction, relative velocity and oscillatory movements. The additional target properties of size, shape, surface properties and resonance must also be encoded in the echo structure. This raises the question as to whether bats are able to evaluate these properties in order to classify or identify targets.

Discrimination experiments have been used to test the ability of some bats to discriminate between different target features. In some of these experiments an attempt was made to explain the discrimination performance on the basis of differences in echo structure.

Discrimination of Targets Differing in Size

In discrimination experiments with cylinders of different diameters a level of 75% correct choices was reached at echo intensity differences of 1 dB in Noctilio leporinus, of 2 dB in Myotis oxygnathus and of 4 - 5 dB in Rhinolophus ferrumequinum (Konstantinov, 1970; Airapetianz and Konstantinov, 1974). In Noctilio the discrimination threshold of 1 dB corresponds well to the intensity discrimination threshold of 1 - 2 dB found in an experiment where these bats had to discriminate one wire from two wires (Suthers, 1965). In further experiments with Rhinolophus, a discrimination threshold of 2.4 - 4.4 dB with spherical targets (Airapetianz and Konstantinov, 1974) and of 2.8 - 3.2 dB with circular planar targets (Fleissner, 1974) was measured. In Eptesicus fuscus an echo intensity difference of 1.5 - 3 dB was necessary to discriminate triangular targets of different size (Simmons and Vernon, 1971).

Discrimination of Targets Differing in Shape and Material

Dijkgraaf (1957) had previously demonstrated that Plecotus auritus can discriminate a disc from a cross, both of which had the same surface area. Airapetianz and Konstantinov (1965) successfully trained the same species to discriminate between a cube, pyramid and cylinder of equal volume. Myotis oxygnathus learned to discriminate triangles, squares and circles with equal surface areas (Konstantinov and Akmarova, 1968). Myotis oxygnathus and Rhinolophus ferrumequinum were able to discriminate a square with smooth edges from squares with serrated edges, and a square with holes in it from concave and convex squares (Airapetianz and Konstantinov, 1970). Vampyrum spectrum learned to discriminate a sphere from a prolate spheroid. The targets were designed so as to produce echoes of approximately equal overall amplitude (Bradbury, 1970). In this experiment, one bat appeared to be using differences in echo spectrum

and another differences in the overall intensity as discrimination
cues.

Konstantinov et al. (1973) tested the ability of Myotis
oxygnathus and Rhinolophus ferrumequinum to discriminate a regular
lattice of 7 metal rods (150 mm high and 3 mm in diameter) positioned along a 150 mm horizontally placed plank from a lattice with fewer
or more rods arranged on a plank of the same length. Myotis was
able to distinguish between the two targets with 75% correct choices
when the variable target contained only one more or one less rod.
Rhinolophus required a change of at least three rods for the same
discrimination. The discrimination performance in this experiment
was explained by the bats' ability to discriminate the different
interference patterns of the echoes returning from the different
targets (Malashin, 1970).

In discrimination experiments with similar targets of
different material it was found that Eptescus serotinus can distinguish between targets covered in the one case with glass and in
the other with velvet (Dijkgraaf, 1946). Myotis oxygnathus and
Rhinolophus ferrumequinum were able to discriminate an aluminum
plate from similar plates made from plexiglass, plywood and textolite
but not from brass and iron plates. The echo cues used for
discrimination in this experiment have not yet been identified.
The possibility of differing resonances causing differences in the
echo structure has been discussed by Airapetianz and Konstantinov
(1974).

Discrimination of Mealworms from Other Targets

Bats which have learned to catch mealworms projected into
their flight path are also able to discriminate these from other
targets (Webster, 1963a, 1963b, 1967a, 1967b; Webster and Durlach,
1963; Webster and Brazier, 1965, 1968; Griffin, Friend and Webster,
1965; Griffin, 1967). Several species of vespertilionids learned
to select and catch mealworms out of a cluster of discs of comparable
size (fig. 7). Experienced bats reached such a high discrimination
level that they caught more than 90% of mealworms and avoided
85 - 90% of the discs. The investigation of the target echo fine
structure led to the suggestion that the bats might be able to recognize spectral differences in the echoes of mealworms and discs
(Griffin, Friend and Webster, 1965; Griffin, 1967).

Discrimination of Targets with Different Depth Structure

In point targets, the returning echoes arrive as one wavefront
at the bat's ear. Three dimensional targets with several reflecting
surfaces return several echoes delayed temporally according to the
depth differences. Interferences between time delayed echoes create
a complex echo.

Simmons et al. (1974) designed a crucial experiment to determine whether bats are able to evaluate information contained in such
complex echoes and use it for the discrimination of targets differing in their depth structure. They trained Eptesicus fuscus to
discriminate a plate with holes of 8 mm depth from other plates
with hole depths between 6.5 - 8 mm. At depth differences below
0.6 - 0.9 mm, the bats were no longer able to discriminate between
the targets.

The analysis of echoes from different targets showed that the
interference between the echoes from the two reflecting surfaces
of each target caused differences in the echo spectra, these being
related to the hole depth (fig. 18). As the overall intensity of
the echoes differed only by 0.1 - 0.5 dB it was suggested that the
bats used the spectral differences and not the overall intensity
for discrimination. Similar experiments with Myotis myotis and
Megaderma lyra yielded comparable results (Habersetzer, Vogler and
Leimer, personal communication).

Many targets are structured in depth and therefore deliver
multiple echoes interfering with each other. The resulting spectral
cues are available for the information evaluation in the frequency
domain. In discrimination experiments, the observation that bats are
able to analyze and to learn to use this spectral information,
allows the conclusion that it is at least possible that they are
able to classify or even to identify targets on the basis of
spectral cues.

Under natural conditions, this identification mechanism may
frequently not be necessary. For instance, in bats hunting in the
open, it might be a good strategy to pursue each detected target
since the chance that the target is not an insect is remote. This
could explain why outdoor hunting Eptesicus pursue pebbles thrown
into their flight path when hunting in the open but learn to avoid
similar targets when they are offered together with mealworms in the
laboratory.

As a general rule it can be stated that fm signals of wide
bandwidth are useful for target discrimination and the echo
spectral cues may be a means of identification. Another result
of the experiments described above is that learning plays an
important role in target discrimination.

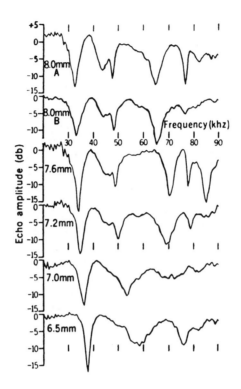

Fig. 18. Echo spectra of targets used in the hole-depth discrimination experiment. The targets were exposed to a simulated wideband signal. Amplitudes of the target spectra are expressed in dB relative to those of a smooth target (adapted from Simmons et al., 1974).

BATS' RESISTANCE TO INTERFERENCE

Under natural conditions, bats have the problem of separating echoes from targets they are interested in from background echoes (clutter), and of detecting their echoes in the presence of other noise. This problem is especially difficult to solve when the intensity of the useful echo is weak in comparison to the intensity of the interference signals. In order to overcome such interference, echolocation systems have evolved effective adaptations for the

gathering of biologically relevant information and for effectively
operating in the presence of interference encountered under
natural conditions.

Resistance to Noise Jamming

In Plecotus townsendii, the ability to resist broad band noise
jamming was measured as an obstacle avoidance score reached at
different noise levels (Griffin and Grinnell, 1958; Griffin, McCue
and Grinnell, 1963); with low intensity noise, the bats performed
normally. With high intensity noise, the bats were reluctant to
fly. After dropping them into the air, the most skillful animals
still performed far above chance level. In these experiments,
the bats emitted signals similar to those produced without noise.
It was therefore possible to determine the detection range
and to calculate, from transmitted sound energy and wire diameter,
the echo energy at detection. The calculated data were then compared with predictions made under the assumption that the bats
might use an optimal filter receiver. This comparison indicated
that the bats performed even better than predicted.

The explanation for this astonishing result was that the bats
reduced noise jamming by approaching the wire obliquely, so that
echoes and noise reached the ears from different directions. They
used the directionality of their hearing system to suppress the
unwanted interference.

That hearing directionality is a very effective countermeasure
for the rejection of masking noise has also been demonstrated in
Rhinolophus ferrumequinum in experiments to measure behavioral
auditory directionality (Grinnell and Schnitzler, 1977) (fig. 6).
For example, a noise source masking a band of $\pm$ 3 kHz around the
signal frequency, positioned 30° laterally, required a 9 dB higher
SPL than one positioned directly ahead. An additional result of the
experiments with Plecotus is that the amplitude of the signals - at
least in hand-held bats - changed with the noise level.

Russian scientists studied the ability of Myotis oxygnathus to
discriminate a cube from a cone or cylinder in the presence of noise.
With increasing noise levels, the signal durations were increased
from 1.5 - 2 msec (no noise) to 5 - 6 msec (Airapetianz and
Konstantinov, 1974).

The ability of Myotis oxygnathus and Rhinolophus ferrumequinum
to detect a small sphere in the presence of wideband and bandlimited
noise has also been tested (Konstantinov, Movtchan and Makarov, 1973;
Makarov, 1973). Both species increased the intensity of their signals with increasing noise levels. The experiments with small-band

jamming noise in <u>Rhinolophus</u> revealed that the masking effect was
stronger when the fm-sweeps of the echoes were masked than when
the long cf components were jammed.

Simmons, Lavender and Lavender (1978) tested the range discrimination performance of <u>Eptesicus fuscus</u> under controlled noise
levels and analyzed the sounds used for discrimination. At noise
levels of -70 to -10dB (re-0.1 N/m^2) the bats showed a steady performance with about 80% correct responses in the discrimination of
a 2 cm target range difference. At 0 dB the discrimination performance declined to less than 60% correct responses and at +10 dB
the bats flew away.

Under quiet conditions and up to -50 dB, the target ranging
signals were 1 - 2 msec in duration and frequency modulated with a
strong second harmonic (fig. 19). With increasing noise level the
signal amplitude was increased as much as 6 to 10 dB and sound
durations up to 8 msec were obtained. Most of the increase in
sound duration was due to the addition of a nearly constant frequency tail, thus forming a fm short-cf signal (fig. 19).

When comparing the results of the different noise jamming
experiments the following tendencies are apparent: bats avoid
noise jamming by using the directionality of their echolocation
system so as to separate the useful echo from the unwanted interference. They also improve the signal to noise ratio when
increasing the amplitude and the duration and thus the energy
of the transmitted signal.

Clutter Rejection

Many bats can hunt insects in situations where there must be
many echoes from nearby objects. They even learn to identify and
selectively catch one mealworm out of a cluster of other ballistic
targets (fig. 7). For the explanation of this performance, we must
assume that they can recognize and separate useful echoes from
interfering clutter.

For Doppler-compensating bats we have already shown that the
utilized echoes from flying insects are characterized by typical
frequency and amplitude modulations caused by wing beats and
that a specialized hearing system is used for the selective
evaluation of this information.

In other bats, the separation of useful echoes from clutter
may be possible because of other typical echo features probably
contained in the echo spectrum.

Range (in noise)

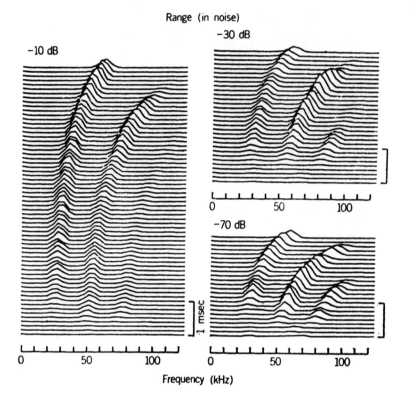

Fig. 19. Frequency structure of <u>Eptesicus fuscus</u> target-ranging signals used in different noise interferences. Noise in dB relative to a SPL of $0.1 \ N/m^2$ (adapted from Simmons, 1978).

After a bat has made the decision that an echo should be utilized, it can use the echolocation system's high directional sensitivity to focus on the target in such a way that the echoes from cluttering targets are excluded. That bats actually use this strategy to separate a target out of clutter, has been demonstrated for <u>Myotis lucifugus</u>, <u>Lasiurus borealis</u>, <u>Eptesicus fuscus</u> and <u>Rhinolophus ferrumequinum</u> (Webster and Brazier, 1965, 1968).

EVOLUTIONARY ADAPTATION OF ECHOLOCATION TO FEEDING HABITS

Echolocation systems have evolved as effective adaptations for gathering behaviorally relevant information and suppressing unwanted interference. All bats have similar problems in finding their way and in avoiding obstacles, but owing to the wide variations in diet, there are very different problems in finding food. As a result, the differences found in their echolocation systems have very probably been developed under selective pressure as a means of finding the species specific nourishment.

The relationship between echolocation, feeding behavior and the acoustical environment in which bats seek their food has been discussed by many authors. The most recent publications in this field come from Novick (1977), Neuweiler (1977), Schnitzler (1978) and Simmons, Fenton and Farell (1979). In these publications, different species of bats have been mainly categorized according the types of echolocation signals present e.g.   long cf-fm, short cf-fm, fm-short cf and pure fm bats. This raises the question as to whether such a categorization according to sound structure alone reflects the different types of feeding behavior found in the different species.

Theoretically cf components allow the evaluation of Doppler shifts and improve the chances of detecting faint echoes, but for range measurement, evaluation of target fine structure and clutter rejection on the basis of spectral cues, they are not as effective as fm components. The use of different signal types in different species and also the change from one type to another within one species therefore reflects the use of different information gathering strategies.

On the basis of the data presented in this review, we can roughly categorize three different main types of echolocation systems, these corresponding to three different main types of feeding behavior.

Echolocation Systems of Insect Hunting Bats with Specializations for the Evaluation of Doppler Shifts

Bats with this type of echolocation system utilize loud cf-fm sounds; they compensate for Doppler shifts and have hearing systems specialized for the evaluation of frequency and amplitude modulations in echoes. This allows the detection and probably also the classification of fluttering insects even in dense clutter. The high signal energy of the cf component also allows the detection of insects over long ranges, thus making it possible to hunt for insects in the "flycatcher style". The basic frequency structure of

the echolocation signals remains similar in all orientation situations. Only the sound duration is changed. The terminal fm sweeps
are necessary for accurate range measurement, especially during
the approach and terminal phases. Ear movements are important for
the determination of angular direction.

All rhinolophids, probably all hipposiderids and Pteronotus
parnellii belong to this group. Further experiments must be conducted in order to determine whether some of the emballonurids
which produce cf-fm signals are also specialized for the evaluation of Doppler shifts.

Echolocation Systems of Insect Hunting Bats without Specializations
for the Evaluation of Doppler Shifts

This category includes bats which forage in open spaces and
species which occasionally or usually pursue insects in the vicinity
of obstacles. The echolocation signals of this group are of high
intensity. When searching for food, a wide variety of signal types
such as pure cf, short cf-fm, fm-short cf, shallow and steep fm
with and without harmonics have been recorded.

Theoretically, the cf components used in search flight by these
bats are as useful for the evaluation of Doppler shifts as those
in Doppler compensating bats. In most cases, however, the cf components are shorter in duration than those of the first group,
and the hearing systems of these bats are not specialized for the
evaluation of Doppler information. The ability to utilize Doppler
shifts is therefore almost certainly far less well developed here
than in Doppler compensating bats.

These bats, however, have the advantage that the high signal
energy of the cf components or of the shallow sweeps remains concentrated for the entire duration within the bandwidth of individual neuronal filters. The chance of detecting faint insect echoes
is therefore much improved (Grinnell and Hagiwara, 1972). The
use of such components of high signal energy thus reflects an
information gathering strategy oriented toward the detection of
faint echoes.

Pure fm signals, with and without harmonics, are usually emitted by these bats when they are close to targets and it has been
demonstrated through discrimination experiments that bats from this
group can learn to utilize spectral cues for target discrimination.
They also use the directionality of their echolocation system as
well as spectral cues to separate useful echoes from clutter.

The auditory sensitivity of the bats in this group is usually widely tuned over the whole frequency range of the echolocation sounds. The hearing systems have no specializations for the evaluation of Doppler shifts in the frequency range of the cf components or the shallow fm sweeps.

Some species use a wide variety of signal types. This demonstrates high adaptiveness to different information gathering problems. Species from different families belong to this group, varying from molossids hunting for insects in the open far away from obstacles to emballonurids which pursue insects in the forest. Further studies on echolocation behavior, feeding habits and hearing in different species may allow a further subdivision of this category.

Echolocation Systems of "Surface Bats"

Nycterids, megadermatids, Antrozous, Macrotus and Plecotus belong to a group specialized for preying on insects, arthropods or small vertebrates resting on surfaces (gleaning bats) and also bats which feed on fruit, pollen and nectar (phyllostomatids) or on blood (desmodontids). All these "surface bats" have the problem that the echoes from the targets in which they are interested are buried in dense clutter.

The echolocation signals of these bats are characterized by low intensities, short durations and shallow sweeps with many harmonics. These allow a good evaluation of target fine structure on the basis of spectral cues. This may aid in resolving the useful target echoes from the dense background clutter.

Several observations indicate that in "surface bats" nonecholocation cues may be used alone or as additional help to identify or localize food.

For instance, all "gleaning bats" have large ears which can be turned toward sound sources. This suggests a highly developed localization acuity due to passive hearing. In Megaderma lyra it has even been demonstrated (Fiedler, personal communication) that they switch off the emission of echolocation signals and hunt mice solely by passive hearing.

The musky odor of all flowers visited by nectar bats also suggests that smell plays an important role in the localization of these food sources. Another orientation aid may be the vision found in some species of this group.

There are probably many species of bats which do not fit completely into one of the proposed categories, e.g. all fm bats which

hunt for flying insects but sometimes also catch insects on the
ground. We hope that further field observations on echolocation
and feeding behavior in more species of bats will allow us to gain
a better understanding of the importance of echolocation as an
evolutionary adaptation for information gathering problems in bats.

ACKNOWLEDGEMENTS

The research yielding new data published for the first time in
this review has been supported by the Deutsche Forschungsgemein-
schaft Grant No. Schn 138/7-9 and NIH Grant NS 12445.

We thank H.-H. Gŏrs, U. Heilmann, W. Hollerbach, D. Menne, J.
Ostwald, A. Rassa, S. Riediger, M.-L. Zander and M.M. Henson for
critical discussions and technical assistance. Without their
patience and valuable help this review would not have been
finished on time.

REFERENCES

Airapetianz, E. Sh. and Konstantinov, A.I., 1970, Echolocation in
     Nature. Nauka, Leningrad (in Russian).
Airapetianz, E. Sh. and Konstantinov, A.I., 1970, Echolocation in
     Animals. Nauka, Leningrad. English Translation. Israel Program
     of Scientific Translations, Jerusalem, 1973.
Airapetianz, E. Sh. and Konstantinov, A.I., 1974, Echolocation in
     Nature. Nauka, Leningrad (in Russian).
Airapetianz, E. Sh. and Konstantinov, A.I., 1974, Echolocation in
     Nature. Nauka, Leningrad. English Translation. Joint Publi-
     cations Research Service, No. 63328, 1000 North Glebe Road,
     Arlington, Virginia 22201.
Airapetianz, E. Sh. and Konstantinov, A.I., 1965, The problem of
     the role of echolocation in spatial analysis by bats. Bionika,
     p. 334, Nauka, Moscow (in Russian).
Airapetianz, E. Sh. and Vasiliev, A.G., 1970, The characteristics of
     the evoked responses in the auditory system of bats to ultra-
     sonic stimuli of different fill frequency. Sechenov. Physiol.
     J. $\underline{56}$, 1721-1730 (in Russian).
Airapetianz, E. Sh. and Vasiliev, A.G., 1971, On neurophysiological
     mechanism of the echolocating apparatus in bats (frequency
     parameters). Internat. J. Neuroscience $\underline{1}$, 279-286.
Altes, R.A., 1973, Some invariance properties of the wide band
     ambiguity function. J. Acoust. Soc. Am. $\underline{53}$, 1154-1160.
Altes, R.A., 1975, Mechanism for aural pulse compression in mammals.
     J. Acoust. Soc. Am. $\underline{57}$, 513-515.
Altes, R.A., 1976, Sonar for generalized target description and its

similarity to animal echolocation systems. J. Acoust. Soc. Am. 59, 97-105.

Altes, R.A. and Titlebaum, E.L., 1970, Bat signals as optimally tolerant waveforms. J. Acoust. Soc. Am. 48, 1014-1020.

Andreeva, N.G. and Vasiliev, A.G., 1977, Responses of the superior olivary neurons to amplitude modulated signals. Sechenov. Physiol. J. 63, 496-503 (in Russian).

Bateman, G.C. and Vaughan, T.A., 1974, Nightly activities of mormoopid bats. J. Mammal. 55, 45-65.

van Bergeijk, W.A., 1964, Sonic pulse compression in bats and people: a comment. J. Acoust. Soc. Am. 36, 594-597.

Beuter, K.J., 1977, Optimalempfängertheorie und Informationsverarbeitung im Echoortungssystem der Fledermäuse. In: Kybernetik 1977, G. Hauske and E. Butenandt, eds., pp. 106-125, München-Wien, Oldenbourg.

Blackmore, M., 1964, Order Chiroptera. In: The handbook of British mammals, N.H. Southern, ed., pp. 224-249, Blackwell Scientific Publication, Oxford.

Bradbury. J., 1970, Target discrimination by the echolocating bat, Vampyrum spectrum. J. Exp. Zool. 173, 23-46.

Brosset, A., 1966, La biologie des chiroptères. Masson et Cie., Paris.

Cahlander, D.A., 1963, Echolocation with wide-band waveforms. Thesis deposited in the library of MIT, Boston, Mass.

Cahlander, D.A., 1964, Echolocation with wide-band waveforms: Bat sonar signals. MIT Lincoln Lab. Techn. Rep. No. 271, AD 605322.

Cahlander, D.A., 1967, Discussion of Batteau's paper. In: Animal Sonar Systems, vol. II, R.-G. Busnel, ed., pp. 1052-1081. Laboratoire de Physiologie acoustique, Jouy-en-Josas.

Cahlander, D.A., McCue, J.J.G. and Webster, F.A., 1964, The determination of distance by echolocating bats. Nature 201, 544-546.

Curtis, W.E., 1952, Quantitative studies of echolocation in bats (Myotis 1. lucifugus); Studies of vision in bats (Myotis 1. lucifugus and Eptesicus f. fuscus) and quantitative studies of vision in owls (Tyto alba practincola). Thesis deposited in the library of Cornell Univ., Ithaca, N.Y.

Dijkgraaf, S., 1946, Die Sinneswelt der Fledermäuse. Experientia 2, 438-448.

Dijkgraaf, S., 1957, Sinnesphysiologische Beobachtungen an Fledermäusen. Acta Physiol. Pharmacol. Neerlandica 6, 675-684.

Eisentraut, M., 1950, Die Ernährung der Fledermäuse. Zoologische Jahrbücher 79, 115-177.

Flieger, E. and Schnitzler, H.-U., 1973, Ortungsleitungen der Fledermaus Rhinolophus ferrumequinum bei ein- und beidseitiger Ohrverstopfung. J. Comp. Physiol. 82, 93-102.

Fleissner, N., 1974, Intensitätsunterscheidung bei Hufeisennasen

(Rhinolophus ferrumequinum). Staatsexamensarbeit vorgelegt am
   Fachbereich Biologie der Universität Frankfurt/Main.

Glaser, W., 1971a, Eine systemtheoretische Interpretation der
   Fledermausortung. Studia Biophysica 27, 103-110.

Glaser, W., 1971b, Zur Fledermausortung aus dem Gesichtspunkt der
   Theorie gestörter Systeme. Zool. Jb. Physiol. 76 209-229.

Glaser, W., 1974, The hypothesis of optimum detection in bats
   echolocation. J. Comp. Physiol. 94, 227-248.

Goldman, L.J. and Henson, O.W. Jr., 1977, Prey recognition and
   selection by the constant frequency bat, Pteronotus p.
   parnellii. Behav. Ecol. Sociobiol. 2, 411-419.

Golubkov, A.G., Konstantinov, A.I. and Makarov, A.K., 1969,
   Effective range and sensitivity of the microchiropteran echolocation system. TR. LIAP 64, 117-127 (in Russian).

Gorlinsky, I.A. and Konstantinov, A.I., 1978, Auditory localization
   of ultrasonic source by Rhinolophus ferrumequinum. Kenya
   National Acad. for Advancement of Arts and Science 145-153.

Gould, E., 1955, The feeding efficiency of insectivorous bats.
   J. Mammal 36, 399-407.

Gould, E., 1959, Further studies of the feeding efficiency of bats.
   J. Mammal. 40, 149-150.

Griffin, D.R., 1944, Echolocation by blind men, bats and radar.
   Science 100, 589-590.

Griffin, D.R., 1953, Bat sounds under natural conditions, with
   evidence for the echolocation of insect prey. J. Exp. Zool.
   123, 435-466.

Griffin, D.R., 1958, Listening in the dark. Yale University Press,
   New Haven.

Griffin, D.R., 1962, Echoortung der Fledermause, insbesondere beim
   Fangen fliegender Insekten. Naturw. Rundsch. 15, 169-173.

Griffin, D.R., 1967, Discriminative echolocation in bats. In: Animal
   Sonar Systems, vol. I, R.-G. Busnel, ed., pp. 273-299. Laboratoire de Physiologie acoustique, Jouy-en-Josas.

Griffin, D.R., 1971, The importance of atmospheric attenuation for
   the echolocation of bats (Chiroptera). Anim. Behav. 19, 55-61.

Griffin, D.R., Dunning, D.E., Cahlander, D.A. and Webster, F.A.,
   1962, Correlated orientation sounds and ear movements of horseshoe bats. Nature 196, 1185-1186.

Griffin, D.R., Friend, J.H. and Webster, F.A., 1965, Target
   discrimination by the echolocation of bats. J. Exp. Zool.
   158, 155-168.

Griffin, D.R. and Galambos, R., 1941, The sensory basis of obstacle
   avoidance in flying bats. J. Exp. Zool. 86, 481-506.

Griffin, D.R. and Grinnell, A.D., 1958, Ability of bats to discriminate echoes from louder noise. Science 128, 145-147.

Griffin, D.R., and Hollander, P., 1973, Directional patterns of
   bats' orientation sounds. Period. Biol. 75, 3-6.

Griffin, D.R., McCue, J.J.G. and Grinnell, A.D., 1963, The resistance of bats to jamming. J. Exp. Zool. 152, 229-250.

Griffin, D.R., and Novick, A., 1955, Acoustic orientation of
    neotropical bats. J. Exp. Zool. 130, 251-300.
Griffin, D.R., and Simmons, J.A., 1974, Echolocation of insects by
    horseshoe bats. Nature 250, 731-732.
Griffin, D.R., Webster, F.A. and Michael, C.R., 1960, The
    echolocation of flying insects by bats. Anim. Behav.
    8, 141-154.
Grinnell, A.D., 1967, Mechanism of overcoming interference in
    echolocating animals. In: Animal Sonar Systems, Vol. I.
    R.-G. Busnel, ed., pp. 451-481. Laboratoire de Physiologie
    acoustique, Jouy-en-Josas.
Grinnell, A.D., 1970, Comparative auditory neurophysiology of
    neotropical bats employing different echolocation signals.
    Z. vergl. Physiol. 68, 117-153.
Grinnell, A.D. and Grinnell, V.S., 1965, Neural correlates of
    vertical localization in echolocating bats. J. Physiol.
    181, 830-851.
Grinnell, A.D. and Griffin, D.R., 1958, The sensitivity of
    echolocation in bats. Biol. Bull. 114, 10-22.
Grinnell, A.D. and Hagiwara, S., 1972, Adaptations of the auditory
    nervous system for echolocation. Studies of New Guinea bats.
    Z. vergl. Physiol. 76, 41-81.
Grinnell, A.D. and Schnitzler, H.-U., 1977, Directional sensitivity
    of echolocation in the horseshoe bat, Rhinolophus ferrumequinum.
    II. Behavioral directionality of hearing. J. Comp. Physiol.
    116, 63-76.
Gustafson, Y. and Schnitzler, H.-U., Echolocation and obstacle
    avoidance in the hipposiderid bat, Asellia tridens. J. Comp.
    Physiol., in press.
Hartridge, H., 1945a, Acoustic control in the flight of bats. Nature
    156, 490-494.
Hartridge, H., 1945b, Acoustic control in the flight of bats. Nature
    156, 692-693.
Johnson, R.A., Henson, O.W., Jr. and Goldman, L.R., 1974, Detection
    of insect wing beats by the bat, Pteronotus parnellii. J.
    Acoust. Soc. Am. 55, 53.
Kay, L., 1961a, The orientation of bats and men by ultrasonic echo-
    location. Brit. Commun. Electron. 8, 582-586.
Kay, L., 1961b, Perception of distance in animal echolocation.
    Nature 190, 361.
Kay, L., 1962, A plausible explanation of the bats echolocation
    acuity. Anim. Behav. 10, 34-41.
Konstantinov, A.I., 1969, The relationship of auditory perception
    and echolocation during detection and catching of
    insects by bats (Vespertilionidae). J. Evol. Biochem.
    Physiol. 5, 566-572 (in Russian).
Konstantinov, A.I., 1970, A description of the ultrasonic location
    system of piscivorous bats. Tez. Dokl. 23-y Nauchno-Tekhn.
    Konf. LIAP, Leningrad, 51-52.

Konstantinov, A.I. and Akhmarova, N.I., 1968, Discrimination
    (analysis) of target by echolocation in Myotis oxygnathus.
    J. Biol. Sci., Moscow Univ. 4, 22-28 (in Russian).
Konstantinov, A.I., Makarov, A.K. and Sokolov, B.V., 1978, Doppler-
    pulse sonar system in Rhinolophus ferrumequinum. Kenya National
    Acad. for Advancement of Arts and Science, 155-163.
Konstantinov, A.I., Makarov, A.K., Sokolov, B.V. and Sanotskaya,
    N.N., 1976, Physiological mechanism of the use of the Doppler
    effect in echolocation by the bat Rhinolophus ferrumequinum.
    J. Evol. Biochem. Physiol. 12, 413-418, (in English) and
    Zh. Evol. Biokhim. Fiziol. 12, 466-472 (1976) (in Russian).
Konstantinov, A.I., Malashin, V.I., Makarov, A.K., Kozlova, N.A.,
    Sokolova, N.N. and Chilingiris, V.I., 1973, The range and
    the principles of echolocation recognition in bats. Vopr.
    Sravn. Fiziol. Analizatorov No. 3 Izd-vo LGU (in Russian).
Konstantinov, A.I., Movtchan, V.N. and Makarov, A.K., 1973, Influ-
    ence of band limited noises on the efficiency of echolocation
    detection of target by Rhinolophus ferrumequinum. Period.
    Biol. 75, 7-11.
Konstantinov, A.I., Sokolov, B.V. and Stosman, J.M.A., 1967, Com-
    parative research on bat echolocation sensitivity. DAN SSSR
    175, 1418 (in Russian).
Konstantinov, A.I., Stosman, I.M. and Gorlinskiy, I.A., 1973,
    Characteristics of the directionality of reception and precision
    of localization of ultrasound by echolocating bats. Ref. Dokl.
    IV Vses. Konf.po.Bionike 4, 5156.
Long, G.R. and Schnitzler, H.U., 1975, Behavioral audiograms from
    the bat Rhinolophus ferrumequinum, J. Comp. Physiol. 100,
    211-220.
Makarov, A.K., 1973, Acoustic methods and apparatus for studying
    ultrasonic location in animals. Vopr. Sravn. Fiziol. Analiza-
    torov, No. 3, Izd-vo LGU, 77-86 (in Russian).
Malashin, V.I., 1970, Theoretical and experimental research on
    frequency-modulated signals reflected from objects being
    recognized by bats. Tez. Dokl. 23-y Nauchno-Teckn. Konf.
    LIAP, Leningrad, p. 57 (in Russian).
McCue, J.J.G., 1966, Aural pulse compression by bats and humans.
    J. Acoust. Soc. Am. 40, 545548.
McCue, J.J.G., 1969, Signal processing by the bat, Myotis lucifugus.
    J. Auditory Res. 9, 100107.
Möhres, F.P., 1963, Über die Ultraschallorientierung der Hufeisen-
    nasen (Chiroptera Rhinolophinae). Z. vergl. Physiol. 34,
    547-588.
Möhres, F.P. and Neuweiler, G., 1966, Die Ultraschallorientierung
    der Großblatt-Fledermäuse (Chiroptera-Megadermatidae). Z.
    vergl. Physiol. 53, 195-227.
Möller, J., Neuweiler, G. and Zöller, H., 1978, Response character-
    istics of inferior colliculus neurons of the awake cf-fm bat
    Rhinolophus ferrumequinum. J. Comp. Physiol. 125, 217-225.

Neuweiler, G., 1970, Neurophysiologische Untersuchungen zum Echo-
    ortungssystem der Großen Hufeisennase Rhinolophus ferrumequinum.
    Z. vergl. Physiol. 67, 273-306.
Neuweiler, G., 1977, Recognition mechanisms in echolocation of bats.
    In: Processing of complex acoustic signals T.H. Bullock, ed.,
    pp. 111-126. Dahlem Konferenzen, Berlin.
Neuweiler, G. and Möhres, F.P., 1967, Die Rolle des Ortungsgedächt-
    nisses bei der Orientierung der Grossblatt-Fledermaus Megderma
    lyra. Z. vergl. Physiol. 57, 147-171.
Neuweiler, G., Schuller, G. and Schnitzler, H.-U., 1971, On- and
    off- responses in the inferior colliculus of the greater horse-
    shoe bat to pure tones. Z. vergl. Physiol. 74, 57-63.
Neuweiler, G. and Vater, M., 1977, Response patterns to pure tones
    of cochlear nucleus units in the CF-FM bat Rhinolophus ferrum-
    equinum, J. Comp. Physiol. 115, 119-134.
Novick, A., 1963, Pulse duration in the echolocation of insects by
    the bat, Pteronotus. Ergebn. Biol. 26, 21-26.
Novick, A., 1965, Echolocation of flying insects by the bat
    Chilonycteris psilotis. Biol. Bull. 128, 297-314.
Novick, A., 1971, Echolocation in bats: some aspects of pulse
    design. Amer. Scientist 59, 198-209.
Novick, A., 1973, Echolocation in bats: a zoologist's view. J.
    Acoust. Soc. Am. 54, 139-146.
Novick, A., 1977, Acoustic orientation. In: Biology of bats,
    Vol. III, W.A. Wimsatt, ed., pp. 73-287.
Novick, A. and Vaisnys, J.R., 1964, Echolocation of flying insects
    by the bat, Chilonycteris parnellii. Biol. Bull. 127, 478-488.
Ostwald, J. 1978, Tonotopical organisation of the auditory cortex in
    the cf-fm bat Rhinolophus ferrumequinum. Verh. Dtsch. Zool.
    Ges., Gustav Fischer Verlag.
Peff, T.C. and Simmons, J.A., 1972, Horizontal-angle resolution
    by echolocating bats. J. Acoust. Soc. Am. 51, 2063-2065.
Pollak, G., Henson, O.W. and Novick, A., 1972, Cochlear microphonic
    audiograms in the "pure tone" bat, Chilonycteris parnellii
    parnellii. Science 176, 66-68.
Pye, J.D., 1960, A theory of echolocation by bats. J. Laryngol.
    Otol. 74, 718-729.
Pye, J.D., 1961a, Echolocation by bats. Endeavour 20, 101-111.
Pye, J.D., 1961b, Perception of distance in animal echolocation.
    Nature 190, 362-363.
Pye, J.D., 1963, Mechanism of echolocation. Ergebn. Biol. 26,
    12-20.
Pye, J.D., 1971, Bats and fog. Nature 229, 572-574.
Pye, J.D., Flinn, M. and Pye, A., 1962, Correlated orientation
    sounds and ear movements of horseshoe bats. Nature 196,
    1186-1188.
Pye, J.D. and Roberts, L.H., 1970, Ear movements in a hipposiderid
    bat. Nature 225, 285-286.

Roberts, L.H., 1972, Correlation of respiration and ultrasound
    production in rodents and bats? J. Zool. Lond. 168,
    439-449.
Schneider, H. and Möhres, F.P., 1960, Die Ohrbewegungen der
    Hufeisenfledermäuse (Chiroptera, Rhinolophidae) und
    der Mechanismus des Bildhörens. Z. vergl. Physiol. 44,
    1-40.
Schnitzler, H.-U., 1967, Discrimination of thin wires by flying
    horseshoe bats (Rhinolophidae). In: Animal Sonar Systems,
    Vol. I. R.G. Busnel, ed., pp. 69-87. Laboratoire de
    Physiologie acoustique, Jouy-en-Josas.
Schnitzler, H.-U., 1968, Die Ultraschall-Ortungslaute der
    Hufeisen Fledermäuse (Chiroptera-Rhinolophidae) in verschiedenen
    Orientierungs-situationen. Z. vergl. Physiol. 57, 376-408.
Schnitzler, H.-U., 1970a, Echoortung bei der Fledermaus
    Chilonycteris rubiginosa. Z. vergl. Physiol. 68, 25-39.
Schnitzler, H.-U., 1970b, Comparison of the echolocation behavior
    in Rhinolophus ferrumequinum and Chilonycteris rubiginosa.
    Bijdr. Dierk. 40, 77-80.
Schnitzler, H.-U., 1971, Fledermäuse im Windkanal. Z. vergl. Physiol.
    73, 209-221.
Schnitzler, H.-U., 1973a, Control of Doppler shift compensation in
    the greater horseshoe bat, Rhinolophus ferrumequinum. J. Comp.
    Physiol. 82, 79-92.
Schnitzler, H.-U., 1973b, Die Echoortung der Fledermäuse und
    ihre hörphysiologischen Grundlagen. Fortschr. Zool.
    21, 136-189.
Schnitzler, H.-U., 1978, Die Detektion von Bewegungen durch
    Echoortung bei Fledermäusen. Verh. Dtsch. Zool. Ges., p. 16-33,
    Gustav Fischer Verlag, Stuttgart.
Schnitzler, H.-U. and Grinnell, A.D., 1977, Directional sensivity of
    echolocation in the horseshoe bat, Rhinolophus ferrumequinum.
    I. Directionality of sound emission. J. Comp. Physiol. 116,
    51-61.
Schnitzler, H.-U., Schuller, G. and Neuweiler, G., 1971, Antworten
    des Colliculus inferior der Fledermaus Rhinolophus euryale auf
    tonale Reizung. Naturw. 58, 617.
Schnitzler, H.-U., Suga, N. and Simmons, J.A., 1976, Peripheral
    auditory tuning for fine frequency analysis by the cf-fm
    bat, Rhinolophus ferrumequinum. III. Cochlear microphonics
    and N1-response. J. Comp. Physiol. 106, 99-110.
Schuller, G., 1972, Echoortung bei Rhinolophus ferrumequinum mit
    frequenzmodulierten Lauten. J. Comp. Physiol. 77, 306-331.
Schuller, G., 1979, Coding of small sinusoidal frequency and
    amplitude modulations in the inferior colliculus of "cf-fm"
    bat, Rhinolophus ferrumequinum. Exp. Brain Res. 34, 117-132.
Schuller, G., Beuter, K. and Schnitzler, H.-U., 1974, Response to
    frequency shifted artificial echoes in the bat Rhinolophus
    ferrumequinum. J. Comp. Physiol. 89, 275-286.

Schuller, G., Beuter, K. and Rübsamen, R., 1975, Dynamic properties
    of the compensation system for Doppler shifts in bat,
    Rhinolophus ferrumequinum. J. Comp. Physiol. 97, 113-125.
Shimozawa, T., Suga, N., Hendler, P. and Schuetze, S., 1974,
    Directional sensitivity of echolocation system in bats producing
    frequency-modulated signals. J. Exp. Biol. 60, 53-69.
Shortridge, G.C., 1934, The mammals of south west Africa. Heinemann,
    London.
Simmons, J.A., 1968, Depth perception by sonar in the bat Eptesicus
    fuscus. Ph.D. Dissertation, Princeton Univ., Princeton.
Simmons, J.A., 1969, Acoustic radiation patterns for the echolocating
    bats Chilonycteris rubiginosa and Eptesicus fuscus. J. Acoust.
    Soc. Am. 46, 1054-1056.
Simmons, J.A., 1971, Echolocation in bats: Signal processing of
    echoes for target range. Science 171, 925-928.
Simmons, J.A., 1973, The resolution of target range by echolocating
    bats. J. Acoust. Soc. Am. 54, 157-173.
Simmons, J.A., 1974, Response of a Doppler echolocation system in the
    bat, Rhinolophus ferrumequinum. J. Acoust. Soc. Am. 56,
    672-682.
Simmons, J.A., Fenton, M.B. and O'Farrell, M.J., 1979, Echolocation
    and pursuit of prey by bats. Science 203, 16-21.
Simmons, J.A. and Lavender, W.A., 1976, Representation of target
    range in the sonar receivers of echolocating bats. J. Acoust.
    Soc. Am. 60 (suppl. 1), S5.
Simmons, J.A., Lavender, W.A. and Lavender, B.A., 1978, Adaptation
    of echolocation to environmental noise by the bat Eptesicus
    fuscus. Kenya National Academy for Advancement of Arts and
    Science 97-104.
Simmons, J.A., Lavender, W.A., Lavender, B.A., Doroshow, C.F.,
    Kiefer, S.W., Livingston, R., Scallet, A.C., and Crowley, D.E.,
    1974, Target structure and echo spectral discrimination by
    echolocating bats. Science 186, 1130-1132.
Simmons, J.A. and Vernon, J.A., 1971, Echolocation: discrimination
    of targets by the bat Eptesicus fuscus. J. Exp. Zool. 176,
    315-328.
Skolnik, M.I., 1962, Introduction to radar systems. McGraw-Hill,
    New York.
Skolnik, M.I., 1970, Radar handbook. McGraw-Hill, New York.
Sokolov, B.V., 1972, Interaction of auditory perception and
    echolocation in bats Rhinolophidae during insect catching.
    Vestn. Leningr. Univ. Ser.Biol. 27, 96-104 (in Russian).
Sokolov, B.V. and Makarov, A.K., 1971, Direction of the ultrasonic
    radiation and role of the nasal leaf in Rhinolophus
    ferrumequinum. J. Biol. Sci. Moscow Univ. 7, 37-44 (in
    Russian).
Sokolov, B.V. and Lipmanova, E.E., 1977, Echolocational estimation
    of the rate of motion of horseshoe bats. Vestn. Leningr. Univ.
    Biol. 3, 95-103 (in Russian).

Strother, G.K., 1961, Note on the possible use of ultrasonic pulse
    compression by bats. J. Acoust. Soc. Am. 33, 696-697.
Strother, G.K., 1967, Comments on aural pulse compression in bats
    and humans. J. Acoust. Soc. Am. 41, 529.
Suga, N., 1977, Amplitude spectrum representation in the Doppler-
    shifted-cf processing area of the auditory cortex of the
    mustache bat. Sicence 196, 64-67.
Suga, N., 1978, Specialization of the auditory system for reception
    and processing of species specific sounds. Fed. Proc. 37,
    2342-2354.
Suga, N. and Jen, P., 1976, Disproportionate tonotopic representation
    of processing cf-fm sonar signals in the mustache bat auditory
    cortex. Science 194, 542-544.
Suga, N. and Jen, P., 1977, Further studies on the peripheral auditory
    system of cf-fm bats specialized for fine frequency analysis
    of Doppler shifted echoes. J. Exp. Biol. 69, 207-232.
Suga, N., Neuweiler, G. and Moeller, J., 1976, Peripheral
    auditory tuning for fine frequency analysis by the cf-fm bat
    Rhinolophus-ferrumequinum. IV. Properties of peripheral
    auditory neurons. J. Comp. Physiol. 106, 111-125.
Suga, N. and O'Neill, W.E., 1978, Mechanism of echolocation in bats.
    TINS 1, 35-38.
Suga, N., O'Neill, W.E. and Manabe, T., 1978, Cortical neurons
    sensitivity to combinations of information-hearing elements of
    biosonar signals in the mustache bat. Science 200, 778-781.
Suga, N., Simmons, J.A. and Jen, PH.-S., 1975, Peripheral
    specialization for fine frequency analysis of Doppler-shifted
    echoes in the auditory system of the CF-FM bat Pteronotus
    parnellii. J. Exp. Biol. 63, 161-192.
Suga, N., Simmons, J.A. and Shimozawa, T., 1974, Neurophysiological
    studies on echolocation systems in awake bats producing cf-fm
    orientation sounds. J. Exp. Biol. 61, 379-399.
Suthers, R.A., 1965, Acoustic orientation by fish-catching bats.
    J. Exp. Zool. 158, 319-348.
Suthers, R.A., 1967, Comparative echolocation by fishing bats.
    J. Mammal. 48, 79-87.
Suthers, R.A., Thomas, S.P. and Suthers, B.J., 1972, Respiration,
    wing beat and ultrasonic pulse emission in an echolocating bat.
    J. Exp. Zool. 56, 37-48.
Vasiliev, A.G., 1971, Characteristics of electric responses of the
    cochlear nuclei in Vespertilionidae and Rhinolophidae to
    ultra-sonic stimuli with different fill frequency. Neuro-
    physiologica 4, 379-385 (in Russian).
Vasiliev, A.G., 1975, Characteristics of unit responses of the
    cochlear nuclei of bats Rhinolophidae to single and
    paired ultrasonic stimuli. Neurophysiology 7, 195-199
    (English translation).
Vasiliev, A.G., 1976, Characteristics of the responses of neurons in

the superior olive of bats in response to single and paired
ultrasonic stimuli. Neirofiziologiya 8, 30-38 (in Russian).

Vasiliev, A.G. and Andreeva, N.G., 1971, Characteristics of the
electric responses of medial geniculate body of Vespertilionidae
and Rhinolophidae to ultrasonic stimuli with different fill
frequency. Neirofiziologiya 3, 138-144 (in Russian).

Vasiliev, A.G. and Timoshenko, T.E., 1973, Characteristics of
electric responses of superior olivary complex in
Vespertilionidae and Rhinolophidae bats to ultrasonic
stimuli with different fill frequency. Neirofiziologiya
5, 33-39 (in Russian).

Vaughan, T.A., 1977, Foraging behavior of the giant leaf-nosed bat
Hipposideros commersoni. East Afr. Wildl. J. 15, 237-250.

Wallin, L., 1969, The Japanese bat fauna. Zool. Bidr. Uppsala
37, 408-413.

Webster, F.A., 1962, Mobility without vision by living creatures
other than man (with special reference to the insectivorous
bats). In: Proc. Mobil. Res. Conf. A.F.B., New York, J.W.
Linser, ed., pp. 110-127.

Webster, F.A., 1963a, Active energy radiating systems: the bat and
ultrasonic principles II, acoustical control of airborne interceptions by bats. Proc. Int. Congr. Tech. and Blindness A.F.B.,
New York 1, 49-135.

Webster, F.A., 1963b, Bat-type signals and some implications. In:
Human factors in technology, Bennett, Degan and Spiegel, eds.,
pp. 378-408, McGraw-Hill.

Webster, F.A., 1967a, Some acoustical differences between the bats
and men. Proc. of the International Conference on Sensory
devices for the blind. St. Dunstan's London, 63-87.

Webster, F.A., 1967b, Interception performance of echolocating bats
in the presence of interference. In: Animal Sonar Systems,
Vol. I, R.G. Busnel, ed., pp. 673-713. Laboratoire de
Physiologie acoustique: Jouy-en-Josas.

Webster, F.A. and Brazier, O.B., 1965, Experimental studies on
target detection, evaluation and interception by echolocating
bats. Aerospace Medical Res. Lab., Wright-Patterson Air Force
Base, Ohio, AD 628055.

Webster, F.A. and Brazier, O.G., 1968, Experimental studies on
echolocation mechanisms in bats. Aerospace Medical Res. Lab.,
Wright-Patterson Air Force Base, Ohio, AD 673373.

Webster, F.A. and Durlach, N.I., 1963, Echolocation systems of the
bat. MIT Lincoln Lab. Rep. No. 41-G-3, Lexington, Mass.

Webster, F.A. and Griffin, D.R., 1962, The role of the flight membranes in insect capture in bats. Anim. Behav. 10, 332-340.

PERFORMANCE OF AIRBORNE BIOSONAR SYSTEMS:

II. VERTEBRATES OTHER THAN MICROCHIROPTERA

O. W. Henson, Jr. and H.-U. Schnitzler

University of North Carolina at Chapel Hill, and

Philipps-Universität, Lahnberge

INTRODUCTION

Biosonar is a means by which some vertebrates gain information about obstacles and prey or other aspects of their environment. It occurs in a variety of forms and it is accomplished by
the emission of brief signals and the subsequent perception and
analysis of echoes. It allows animals living under conditions of
limited visibility or with limited vision to obtain information
that may not be available or fully appreciated through other
sensory systems. Among terrestrial vertebrates the most sophisticated biosonar systems are encountered in the Microchiroptera
and indeed their lives are dependent upon their acoustic orientation capacities. By listening to echoes they can assess the
distance, direction, size, texture, shape and movement of obstacles and/or prey. The performance of the refined biosonar systems
in these animals is the subject of the second part of this review.
The first part will examine the echolocative capacities of terrestrial vertebrates whose lives are much less dependent on biosonar
and which have less sophisticated biosonar systems than
Microchiroptera.

Our knowledge of the use of biosonar by various animals is
based on behavioral experiments. In these experiments biosonar
signals are usually monitored in relation to the performance of
specific tasks and the performance of that task is quantified with
the system intact and experimentally altered. It is through such
experiments that biosonar has been convincingly demonstrated in
certain cave dwelling birds, cave dwelling Megachiroptera of
the genus Rousettus and in some of the Insectivora. In addition

it has been demonstrated that certain animals, such as rats and
man, may utilize biosonar in the absence of vision.

BIRDS

1.   Oilbirds (Steatornis caripensis)

Echolocation by cave dwelling oilbirds was first described
by Griffin (1953) 25 years ago and since that time no additional
studies have been undertaken. In complete darkness the oilbirds
make loud clicking sounds which are clearly audible to man at a
considerable distance. Signal durations are so short that only a
few waves may be apparent on the display of an oscilloscope. The
signals have not been analyzed with modern signal processing
devices, but the major energy appears to be in the 6-10 kHz band.
Evidence strongly indicates that these birds use their biosonar
systems primarily, if not entirely, for the detection of the
rock walls within the caves where they roost. When placed in a
dark experimental chamber the birds immediately begin to emit
their echolocative signals and they can avoid direct collisions
with the walls of the chamber. If their ears are plugged, however, they typically collide with the first wall they encounter;
when the ears are subsequently unplugged they regain their ability
to avoid the walls with their previous skill level. Griffin noted
that the birds were unable to avoid an electric light cord which
was in their flight path, but no obstacle avoidance experiments
have been carried out with these birds.

2.   Cave swiftlets (Collocalia)

Echolocation has been demonstrated in some but not all
representatives of the genus Collocalia. Those that echolocate
have been shown to produce click-like sonar signals with brief
durations and frequencies primarily in the 4-7.5 kHz band (Novick,
1959; Medway, 1959, 1967; Cranbrook and Medway, 1965; Griffin and
Suthers, 1970; Fenton, 1975).

There have been three reports dealing with the ability of
Collocalia to avoid obstacles. Medway (1967) studied the ability
of C. esculenta and C. fuciphaga to avoid walls and a barrier of
vertically placed wooden rods 15 cm apart, each with a cross-
sectional area of 1 cm^2. On the basis of these studies he concluded that C. esculenta do not echolocate. On the other hand
C. fuciphaga was able to use biosonar with sufficient skill to
detect and avoid walls in total darkness; their ability to detect
the wooden rods was limited.

Griffin and Suthers (1970) carried out experiments to test
the obstacle avoidance skills of C. vankorensis granti. Cylinders
were placed horizontally 40 cm apart and attached to a frame which
could be shifted in position. In addition, three rows of vertical

wires were hung from the ceiling of the chamber. The obstacles
used were 2 mm diameter plastic covered wires, 6.3 mm diameter iron
rods and 8 mm diameter plastic tubing. These experiments were the
first to seek information on the size range of objects which could
be detected and successfully avoided. The four birds tested avoided
the 6.3 mm diameter rods an average of 80% of the time (74%-92%
range); the 8 mm diameter plastic tubing 67% of the time (58%-80%
range); and the 2 mm diameter wires only 43% of the time (31%-50%
range). The reason the 6.3 mm rods were avoided better than the
8 mm tubing is not clear.

Fenton (1975) assessed the echolocative skills of C.
hirundinacea by constructing an obstacle course 10 M inside a mine
tunnel where the birds normally roosted. The obstacles were 10 mm
diameter wooden dowels, 1.5 M in length; 4 mm diameter pieces of
aluminum, 0.9 M in length; and 1.5 mm diameter copper wire strips,
0.9 M in length. Longer obstacles were not necessary since C.
hirundinacea invariably flew within 0.5 M of the tunnel ceiling.
Flight speed through the obstacle course was not known but as
the birds flew into the tunnel the average flight speed was about
10 M/sec. The data on avoidance were as follows:

obstacle diameter	percent misses	conditions
10 mm	91	flying out of mine
10 mm	73	flying into mine
4 mm	64	flying out of mine
4 mm	67	flying into mine
1.5 mm	62	flying out of mine
1.5 mm	55	flying into mine

From the published data it appears that some birds of a given
species are more skilled than others in avoiding obstacles; the
most skilled birds were usually able, at better than chance
levels, to avoid obstacles with diameters ranging from 4 mm to
10 mm.

MAMMALS

1. Shrews
    The utilization of echolocation by some species of shrews was
first demonstrated by Gould, Negus and Novick (1964). Shrews were
trained to locate and jump onto a small ($12.7mm^2$) platform 55-110 mm
below the rim of a circular disc. Experiments were carried out with
the masked shrew (Sorex cinereus), the Northern water shrew (Sorex
palustris), the wandering shrew (Sorex vagrans) and the short-tailed

shrew (<u>Blarina brevicauda</u>). It was shown that biosonar signal emission was associated with searching for the platform. The sonar signals were of low intensity and thus difficult to record. They have been described as "pure" tones but they have not been carefully scrutinized with sophisticated equipment. Under normal conditions the shrews found the platform easily but when their ears were plugged they were either unable to find it or found it with considerable difficulty.

Buchler (1976) made more extensive studies on <u>Sorex vagrans</u>. By testing the ability of these shrews to detect a 15 cm^2 barrier, set at various distances along one limb of a Y-maze, he determined that the greatest distance at which detection took place was about 65 cm. By varying the size of the barrier he found that the smallest barrier that could be detected at a distance of 20 cm was a 3.5 cm square (12.25 cm^2 area). Buchler was able to train the shrews to discriminate between a solid barrier with a 4 cm square hole and a similar barrier with no hole, when the barriers were at distances ranging from 5 to 30 cm. On the basis of these and other experiments it was concluded that: (1) the sonar system of the wandering shrew is crude and operates only over a limited range; (2) it is unlikely that it is used for prey identification or detection; and (3) once an animal becomes familiar with an area, spatial memory replaces echolocation as a means of orientation.

2.   Tenrecs
Tenrecs (<u>Hemicentetes semispinosus</u>, <u>Echinops telfairi</u> and <u>Microgale dobsoni</u>) have also been studied with a disc-platform apparatus similar to that used for studying shrews (Gould, 1965). The animals were trained to search along the rim of a 45.7 cm diameter disc and locate a 10 cm^2 aluminum platform which was 11 cm below the disc and thus beyond the animals' reach. They were rewarded when they successfully dropped onto the platform and followed a ramp to the reward box. The platform was boiled in potassium permanganate to minimize olfactory cues and its position was changed before each trial. The experiments were carried out in complete darkness; two animals were also blindfolded but this had no effect on their performance. In six <u>Hemicentetes</u> short (10-12 mm) tubes with 4 mm inside diameters were glued into the external acoustic meatus of each ear so that the echolocative skills could be tested with both open and plugged tubes. When the tubes were open all of the animals were able to locate and drop onto the platform, but when the tubes were plugged with cotton, only two of six animals were able to find the platform. How these two animals succeeded in locating the platform is not known. Although the ear plugging experiments were done only with <u>Hemicentetes</u>, one <u>Echinops</u> and one <u>Microgale</u> were judged to utilize echolocation because they were able to find the platform and because they, like <u>Hemicentetes</u>, made clicking noises when searching for the platform. <u>Centetes ecaudatus</u>

were also observed to make clicking sounds and it is presumed that
they also echolocate. Of all the tenrecs studied by Gould,
Centetes made the loudest noises and these were said to be
audible to an observer at a distance of seven meters. According to
Gould, the sounds produced by Centetes were made by pressing the
tongue against the palate and then lowering it; the source of the
clicks produced by the other tenrecs was not clearly established.
The click rate was found to vary from animal to animal but in
general was slow, like that of shrews; the signal durations were
very short, 0.1-1.8 msec, and the main frequency components were
in the 5-17 kHz frequency band. Gould noted that Echinops
produce clicking noises while climbing in shrubbery and all of the
other tenrecs that he studied made clicking noises while exploring
strange places. His observations suggest that echolocation is used
under natural conditions for gaining general rather than very
specific information about the surroundings.

3.   Rousettus
      Three species of Rousettus (R. aegypticus, R. amplexicaudatus
and R. seminudus) have been studied to date and these are the only
Megachiroptera known to echolocate. Among Chiroptera they are
unique in their utilization of tongue clicks for echolocative
purposes (Kulzer, 1956, 1960; Möhres and Kulzer, 1956; Möhres, 1956;
Novick, 1958). The clicking sounds of these bats vary from about
2 to 4.5 msec in duration but only the initial 0.3-1.0 msec portion
of each click has a high amplitude. Different studies report that
the fundamental frequency is in the 10 to 17 kHz range with ultrasonic harmonic components extending up to 100 kHz. Certain harmonics
are emphasized, however, (ca 13, 26 and 38 kHz) and this has been
well demonstrated as being the result of resonance within the buccal
cavity (Roberts, 1975).

      When flying from tree to tree Rousettus make rapid and regular
successions of clicks which are clearly audible to the human ear.
The repetition rate and loudness of the clicks appear to increase
when the bats approach a branch or begin their take off. Our
knowledge of the ability of Rousettus to detect and avoid obstacles,
however, is limited to a single study on one bat (Griffin et al.,
1958). After considerable training this animal was able to avoid
1.0 mm cylindrical obstacles 68% of the time, 1.5 mm obstacles
77% of the time and 3 mm obstacles 85% of the time.

4.   Blinded Laboratory Rats
      Blinded laboratory rats appear to be able to utilize the
echoes of self-generated signals (sniffs, sneezes, clicks and
scratches) in order to gain some appreciation of their environment
(Rosenzweig et al., 1955). This was demonstrated by training
10 rats to choose between one of two elevated runways along which
they could walk toward a source of food; on one or the other runway

a randomly placed sheet of metal blocked the pathway. Seven of the
rats learned to distinguish the open from the blocked runway at a
90% level (18 of 20 trials) when the metal barrier was perpendicular
to the runway; when the barrier was turned at a 45 degree angle to
reduce the echo intensity the rats lost this ability. They were also
unable to choose the proper runway when their ears were plugged. The
sounds produced by the rats were monitored and although rats are
capable of producing ultrasonic signals they did not appear to use
these when performing in the maze. In some trials the rats did not
produce any sounds that could be detected by the observers, yet the
rats made the correct decisions.

It has been suggested that other types of rodents might utilize
echolocation under natural conditions, but the information presented
to date is inconclusive (see Aryapet'yants et al., 1974).

5.   Blind Persons
The amazing ability of some blind persons to detect and avoid
obstacles has long been of interest (Diderot, 1749) and many remark-·
able cases have been described and discussed (Hayes, 1935; Griffin,
1958; Rice, 1967, 1969). A number of experiments have been carried
out on the obstacle detection skills of the blind; all have shown a
high degree of variability and it is clear that some subjects are
much more skilled than others. Some of the most extensive early
experiments were carried out by Heller (1904) who reported that
his blind students could detect a school chart (1 X 1.65 M) at
distances of about 3 to 4 M by noting changes in the sounds of their
own footsteps; he also found that they perceived a tactile sensation
when they were 60-70 cm from the chart. That this tactile sensation
was produced by acoustic signals reaching the ear and that echoper-
ception was the basis of the tactile sensation known as facial vision
was clearly demonstrated by Supa, Cotzin and Dallenbach (1944).
Qualitative assessments of the capacity of blind persons to detect
objects of different sizes and texture and at different distances and
to localize an echo source have been made by Kellogg (1962), Rice,
Feinstein and Schusterman (1965) and Rice (1967, 1969). In these
experiments the subjects were allowed to make whatever natural sound
they wished; this included talking, singing, whistling, hissing,
finger snapping and tongue clicking.

Resolution

Rice (1969) found one subject who had been blind since birth
and who had superior echolocative skills; this individual could
consistently, with 90% accuracy, detect the presence of a disc 50.8 mm
in diameter and he could detect a disc 38 mm in diameter 75% of the
time. The smallest target which could be detected was 27 mm in
diameter. In these tests the targets were located 0.91 M in front
of the subject. Other subjects, labeled "late blind", were judged

to be considerably less skilled than the above individual since
95-100% correct responses were obtained only when the diameter of
the disc was 104 mm or larger.

Distance perception

It is clear that acoustic delay between a signal and its
reflected echo can be appreciated by the human ear and an assessment
of this delay should provide a measure of target distance. Nevertheless, there have apparently been no attempts to test the ability
of blind persons to perceive the distance of targets; that is, on
the basis of echo information to state that an object is subjectively
at a specific range. Supa et al. (1944), however, did test for the
ability to detect barriers at unknown distances; subjects were asked
to approach the barrier as close as possible without touching it and
at this task the subjects were able to stop within several inches
of the barrier. Kellogg (1962) found that blind persons can also
resolve changes in the distance of targets. This was determined by
attaching a flat, 30.48 cm diameter, disc to a device that allowed
its position to be rapidly changed. Judgments were made in relation
to two successive target presentations. One presentation, the
"standard" target, was always positioned at a distance of 60.9 cm
and there were three fixed points where the targets could be positioned closer to the subject and three that were further away. In
these cases the subjects responded by stating whether a given target
was nearer or further away than the preceeding target in a pair of
presentations. Kellogg tested both blind and sighted (blindfolded)
persons; the blind persons were considerably more skilled than the
sighted ones. The data from Kellogg's graph No. 4 can be tabulated
as follows:

Distance of target from subject's face (in cm)	Percentage of time that the target was judged to be closer than the target at the standard distance	
	Blind subjects	Sighted (blindfolded) subjects
30.48	98.0%	56.0%
38.4	97.5	55.5
48.4	84.5	49.5
60.9 standard		
77.7	29.5	47.5
97.2	15.5	12.5
121.0	8.0	33.0

On the basis of these data Kellogg concluded that changes in distance of a little more than 10 cm could be appreciated by his skilled
blind subject and that this person could perceive differences in the

location of two targets better than a person using monocular vision. This conclusion might be further tested by examining the ability of subjects to detect two targets simultaneously, and to align the targets side by side. One also must question whether these experiments were really testing distance perception or simply fine structural analysis of echoes.

Size discrimination

The ability of blind persons to discriminate among targets of different sizes at set distances was also examined by Kellogg (1962). In these experiments each target was set at the same distance and seven different sized plywood discs (painted with a sand-texture paint) were presented. The test subjects were asked whether the target was larger or smaller than a 23.8 cm diameter standard. The targets were always presented in pairs, but their order of presentation was random. In general it was found that the ability to recognize targets that were smaller than the standard was good. One subject could, with 100% accuracy, tell that a 14.7 cm target was smaller than the 23.8 cm target at distances of 30.4 or 45.7 cm from the face. In general the discrimination was not as good with larger sized targets, especially at distances of 30.4 and 45.7 cm from the face. Although the subjects were undoubtedly responding to differences in the physical characteristics of the echoes it is not clear that there was an actual appreciation of size differences.

Texture discrimination

Kellogg (1962) tested the ability of his subjects to discriminate among targets with different surface characteristics. The surfaces of targets were covered with different materials (velvet, denim, plain wood, painted wood, glass, metal) and placed 30.4 cm in front of the subject's face. Each disc was paired with every other disc by the method of paired comparisons. The results indicate that the qualities of the echoes from hard surfaces can be distinguished from those from soft surfaces. In these cases it does not appear that there was a real appreciation of the surfaces but rather that the echoes from the different targets sounded differently.

Shape discrimination

Rice (1966) presented some preliminary data on shape discrimination. Four blind subjects were presented with square, circular and triangular shaped targets of equal area, and after training they were able to discriminate among the three targets with 80% accuracy. One subject was tested to determine whether an auditory concept of shape had been formed but this did not seem to be the case. It is interesting to note that the subject best able to discriminate shape used hissing sounds of long duration rather than clicks.

Rice (1967) noted that some blind persons used scanning movements of their heads while attempting to detect a target. By comparing target detection capacities in subjects with mobile and immobile heads, however, it was clear that head mobility was not necessary for detection. It may have facilitated detection but made little practical difference. Rice also tested performance of blind persons with one ear plugged and compared binaural vs. monaural echolocative capacities for targets at a distance of 30.4 cm. The results indicated that binaural detection was superior to monaural although the capacities were not dramatically different. Binaural and monaural echolocative capacities were also compared when the head was mobile and immobile.

The ability of blind subjects to localize the centers of targets has also been tested (Rice, 1967); this was done by having each subject point his nose at the middle of a target which was changed in position from test to test. Localization ability was compared with varying target size and distance, with and without the subjects' knowledge of results and with and without kinesthetic feedback. The results showed that when a 16.2 cm target was presented at a distance of 0.91 M, the center of the target could be well localized under all conditions; the accuracy of localization could be only slightly improved with verbal or kinesthetic feedback. Since it is possible that the ability to find the center of a target may vary with targets of different diameter, Rice also used targets with 9.1 and 21.3 cm diameters. Target size, however, did not affect the ability of the subject to localize the center nor did moving the target from 0.91 to 1.52 M from the subject change the localization acuity. To test localization acuity further, Rice (1967) asked the subjects to find the left edge of a 16.2 cm target 0.91 M from the subject's ear. On the basis of these experiments Rice concluded that the ability to find the left edge was as good or better than that of finding the center of the target. In these experiments it was not clear whether the accuracy was due to kinesthetic adjustments made by finding the center and moving left, or by actually listening for an edge cue. Experiments were also conducted to evaluate the extent of the arc over which targets could be localized from signals directed straight ahead. Again, the subjects pointed their noses at a 16.2 cm diameter target 0.91 M away after signals were emitted at 0° azimuth. The target was in one of 13 positions in a $\pm 90°$ arc, but the accuracy was best between $\pm 30°$ of the midline. It is interesting to note that the only subjects who could reliably detect targets beyond 45° were those that were congenitally blind. These data support prior studies indicating that blind persons are generally superior to others in their ability to localize sound.

DISCUSSION

It is interesting to compare the echolocative capacities of Rousettus with Collocalia since both of these animals have developed biosonar systems which seem to be designed for general orientation within the dark confines of caves and tunnels. As noted above, experiments with one Rousettus revealed that 1 mm diameter cylindrical obstacles can be avoided 68% of the time, 1.5 mm diameter obstacles 77% of the time and 3 mm diameter obstacles 85% of the time. By contrast, comparable date on Collocalia shows 55-67% avoidance for 1.5 mm diameter obstacles and 64-67% for 4 mm diameter obstacles. It is clear from the studies of Roberts (1975) that sonar signals of Rousettus contain ultrasonic components of significant strength and thus differ from the clicking noises of Collocalia in being much broader in band width. It is also important to note that the hearing capacities of birds at frequencies above 10 kHz, are much inferior to those of Rousettus (Grinnell and Hagiwara, 1972). Thus, it is to be expected that Rousettus should hear echoes from small objects better than Collocalia. One must, however, be cautious in comparing the performance of these animals because the tests performed encompass both echo perception and maneuverability; experimental conditions under which the most critical tests were made were different (laboratory vs. natural roost, trained vs untrained, blindfolded vs. not blindfolded). In addition to comparative aspects of maneuverability other information not known were the different speeds at which the animals approached the obstacles, the distance at which detection took place.

The levels of obstacle detection and avoidance skills in the Microchiroptera are very clearly superior to those of Rousettus and all of the other animals considered here. The biosonar signals of Microchiroptera are much more structured and versatile and certain aspects of the sense of hearing is clearly specialized for echo perception. While Rousettus has difficulty detecting and avoiding 1 mm diameter wires, most Microchiroptera can avoid wires with diameters of 0.12 mm (Griffin, 1958; Konstantinov et al., 1966) and Megaderma and Rhinolophus can avoid 0.06-.08 mm diameter wires (Schnitzer, 1967, 1968).

There is no question but that the echolocative capacities of shrews and tenrecs are crude in comparison with the Microchiroptera, but they must be considered adequate for general orientation. One of the more interesting aspects of biosonar signal emission by these small terrestrial animals is the rate of emission. In Microchiroptera rates of 10-150/sec are often cited while the rate in shrews and tenrecs has typically been expressed in terms of 3 minute intervals. For tenrecs Gould (1965) gives values of 65 clicks in some 3 minute intervals and for shrews 13-16. These slow rates support the idea that the sonar systems of these animals are primarily used for general appreciation of the immediate environment.

It is tempting to compare the signal emission behavior of the
shrews and tenrecs with blind persons. In working with blind
persons, Rice (1969) noted that some of his subjects routinely
made a few tongue click noises and from these they could obtain
accurate information about the presence and location of objects
over a wide range of positions and distances. He referred to this
as an "auditory glance". It is curious that the biosonar capacities
of the blind have not been pursued in more detail. There is much
that might be learned and the principles applied to other animals.

ACKNOWLEDGEMENTS

We wish to thank Dr. M.M. Henson for discussions and technical
assistance in the preparation of this manuscript, and we wish to
acknowledge the support from the Deutsche Forschungsgemeinschraft
(Grant No. Schn 138/7-9) and from the National Institutes of
Health (Grant No. NS 12445).

ADDENDUM

A recent report on oilbirds by M. Konishi and E. I. Knudson
(The oilbird: Hearing and Echolocation, Science, 204, 425-427,
1979) has clarified some of the questions concerning the frequency
composition of the cries, the sense of hearing and echolocative
skills of these animals. They found the sound energy of the click-
like signals to be unevenly distributed in the 1-15 kHz range with
most of the energy in the 1.5-2.5 kHz band. This band corresponds
to the most sensitive range of the oilbirds' hearing as determined
by cochlear microphonic and evoked neural potentials. The echolocative skills were tested by the ability of the birds to avoid
disks with diameters of 5, 10, 20, 25, 30, 35 and 40 cm. The disks
were spaced apart by 5 diameters in a square array consisting of
rows and columns, and the number of disks in each array was adjusted
to maintain a total disk surface of 1256 cm^2. Under these conditions the birds hit the 5 and 10 cm disks as if nothing existed in
their paths. The first signs of avoidance appeared when 20 cm disks
were presented, and all of the five birds tested were able to avoid
the 40 cm disks. In test trials the obstacles were placed in a
familiar passageway. On the basis of these studies the authors
concluded that the role of echolocation in the oilbird is the
detection of relatively large objects.

REFERENCES

Buchler, E. R., 1976, The use of echolocation by the wandering
    shrew (Sorex vagrans). Anim. Behav. 24, 858-873.

Cranbrook, Earl of, and Lord Medway, 1965, Lack of ultrasonic
    frequencies in the calls of swiftlets. Ibis 107, 259.
Diderot, D. Letter on the blind, 1749, in Early Philosophical Works,
    M. Jourdain, Trans. (Open Court Chicago, 1961), pp. 68-141
    (cited from Rice, 1967).
Fenton, M. B., 1975, Acuity of echolocation in Collocalia
    hirundinacea (Aves: Apodidae), with comments on the distribution
    of echolocating swiftlets and molossid bats. Biotropica 7,
    1-7.
Gould, E., 1965, Evidence for echolocation in the Tenrecidae
    of Madagascar. Proc. Amer. Phil. Soc. 109, 352-360.
Gould, E., N.C. Negus and A. Novick, 1964, Evidence for echolo-
    cation in shrews. J. Exp. Zool. 156, 19-38.
Griffin, D. R., 1953, Acoustic orientation in the oilbird,
    Steatornis. Proc. Nat. Acad. Sci. 39, 884-893.
Griffin, D. R. Listening in the Dark, Yale Univ. Press, New Haven,
    1958.
Griffin, D. R. and R. A. Suthers, 1970, Sensitivity of echolocation
    in cave swiftlets. Biol. Bull. 139, 495-501.
Griffin, D. R., A. Novick and M. Kornfield, 1958, The sensitivity
    of echolocation in the fruit bat, Rousettus. Biol. Bull.
    115, 107-113.
Grinnell, A. D. and S. Hagiwara, 1972, Studies on auditory physi-
    ology in nonecholocating bat, and adaptations for echolocation
    in one genus, Rousettus. Z. vergl. Physiol. 76, 82-96.
Hayes, S. P., 1935, Facial vision or the sense of obstacles.
    Perkins Publications, No. 12. Perkins Institution, Watertown
    Mass.
Heller, T. Studien zur Blindenpsychologie. Leipzig, W. Engelmann,
    1904, 136 pp. (cited from Griffin, 1958).
Kellogg, W. N., 1962, Sonar system of the blind. Science 137,
    399-404.
Konstantinov, A. I., B. V. Sokolov and I. M. Stosman, 1966,
    Comparative study of the sensitivity of echo-location in bats.
    Doklady Akad. Nauk SSSR 175, 1418-1421.
Kulzer, E., 1956, Flughunde erzeugen Orientierungslaute durch
    Zungenschlag. Naturwiss. 43, 117-118.
Kulzer, E., 1960, Physiologische und morphologische untersuchungen
    uber die erzeugung der orientierungslaute von slughunden der
    Gattung Rousettus. Z. vergl. Physiol. 43, 231-268.
Medway, Lord, 1959, Echolocation among Collocalia. Nature 184,
    1352-1353.
Medway, Lord, 1967, The function of echonavigation among swiftlets.
    Anim. Behav. 15, 416-420.
Möhres, F. P., 1956, Uber die Orientierung der Flughunde
    (Chiroptera-Pleropididae). Z. vergl. Physiol. 38, 1-29.
Möhres, F. P. and E. Kulzer, 1956, Uber die orientierung der
    flughunde (Chiroptera-Pteropididae). Z. vergl. Physiol. 38,
    1-29.

Novick, A., 1958, Orientation in paleotropical bats. II. Megachi-
    roptera. J. Exp. Zool. 137, 443-462.

Novick, A., 1959, Acoustic orientation in the cave swiftlet. Biol.
    Bull. 117, 497-503.

Rice, C. E., 1967, Human echo perception. Science 155, 656-664.

Rice, C. E., 1969, Perceptual enhancement in the early blind?
    Psychol. Rec. 19, 1-14.

Rice, C. E. and S. H. Feinstein, 1965, The influence of target
    parameters on a human echo-detection task. Proc. Amer.
    Psychol. Assn., p. 65.

Rice, C. E., S. H. Feinstein. and R. J. Schusterman, 1965,
    Echo-detection ability of the blind: size and distance factors.
    J. Exp. Psychol., 70, 246-251.

Riley, D. A. and M. Rosenzweig, 1957, Echolocation in rats. J.
    Comp. Physiol. Psychol. 50, 323-328.

Roberts, L. H., 1975, Confirmation of the echolocation pulse pro-
    duction mechanism of Rousettus. J. Mammal. 56, 218-220.

Rosenzweig, M. R., D. A. Riley, and D. Krech, 1955, Evidence
    for echolocation in the rat. Science 121, 600.

Supa, M., M. Cotzen and K. M. Dallenbach, 1944, "facial vision"
    The perception of obstacles by the blind. Am. J. Psychol.
    57, 133-183.

Schnitzler, H.-U., 1967, Discrimination of thin wires by flying
    horseshoe bats (Rhinolophidae). In "Animal Sonar Systems"
    (R.-G. Busnel, ed.) pp. 69-87. Lab. Physiol. Acoust.,
    Jouy-en-Josas, France.

Schnitzler, H.-U., 1968, Die Ultraschall-Ortunglaute der Hufeisen-
    Fled-ermause (Chiroptera-Rhinolophidae) in verschiedenen
    Orientierungssituationen. Z. vergl. Physiol. 57, 376-408.

Chapter II
Echolocation Signals and Echoes

Chairman : D.R. Griffin

co-chairmen

W.W.L. Au   and   D. Pye

- Underwater
  Functional and descriptive anatomy of the bottlenosed
  dolphin nasolaryngeal system with special reference
  to the musculature associated with sound production.
     R.F. Green, S.H. Ridgway and W.E. Evans

  Electromyographic and pressure events in the nasolaryngeal system of dolphins during sound production.
     S.H. Ridgway, D.A. Carder, R.F. Green,
     S.L.L. Gaunt, A.S. Gaunt,and W.E. Evans

  Echolocation signals of the atlantic bottlenose dolphin
  (Tursiops truncatus) in open waters.
        W.W.L. Au

  Acoustics and the behavior of sperm whales.
     W.A. Watkins

  Click sounds from animals at sea.
     W.A. Watkins

  Signal characteristics for target localization and
  discrimination.
     K.J. Diercks

- Airborne
  Echolocation signals and echoes in air.
     J.D. Pye

  Echolocation ontogeny in bats.
     P.E. Brown and A.D. Grinnell

FUNCTIONAL AND DESCRIPTIVE ANATOMY OF THE BOTTLENOSED DOLPHIN
NASOLARYNGEAL SYSTEM WITH SPECIAL REFERENCE TO THE MUSCULATURE
ASSOCIATED WITH SOUND PRODUCTION

R. F. Green, S. H. Ridgway, and W. E. Evans[†]

Naval Ocean Systems Center
San Diego, Ca. 92132 U.S.A.
† Hubbs-Sea World Research Institute
San Diego, Ca. 92109 U.S.A.

INTRODUCTION

The nasolaryngeal system of the bottlenosed dolphin (Tursiops
truncatus) is complex and highly specialized for respiration and
sound production. The exact source and mechanism of sound production
by dolphins is not understood. Electromyographic studies are
essential for understanding the functional anatomy and mechanisms
of sound production. Dissections and measurements have been made
for the purpose of identifying external landmarks and other details
necessary for inserting electrodes in the muscles of the nasolaryngeal area. Based upon these anatomic investigations, some
functions of this musculature have been suggested, and the first
electromyographic studies of dolphin acoustic mechanisms have been
initiated.

Previous Studies

The biological community generally accepts the production of
sound by the vast majority of living amniote vertebrates as beyond
question. Many assume that our understanding of how this vast array
of acoustic emissions are produced is at a high state of development.
This is far from true. With the exception of humans, our knowledge
of the mechanics of sound production in vertebrates, especially
cetaceans, is based on indirect evidence and speculation (Kelemen,
1963; Gaunt and Wells, 1973; Kinne, 1975).

The mechanisms of sound production in most vertebrates are
essentially modifications of the respiratory system and are in

reality an overlaid function. This factor more than any other has
contributed to compounding the problem of understanding the
mechanisms of sound production in marine mammals. We have not only
inherent complexities of the terrestrial mammalian nasolaryngeal
system but the added problems caused by modifications of the system
to permit a) breathing air while inhabiting an aquatic environment,
b) prolonged breath holding, c) diving to reasonably deep depth with
resulting pressure problems, and d) the ingestion of food without
taking water into the lungs. Despite years of discussions, special
seminars, experiments, morphological descriptions and several
studies of functional anatomy, uncertainty still exists, and the
details of production of the vast variety of sounds found in the
repertoire of cetaceans is still an unsolved mystery. Exact sites,
muscular or mechanical processes, aerodynamics (flow, pressure,etc)
and acoustical properties of the anatomic structures involved are
insufficiently known even in the bottlenosed dolphin (Tursiops
truncatus), the most intensively studied of all the species.

     In birds, and more recently in studies of rodents and
chiropterans(Hersh, 1966; Roberts, 1973, 1975a, 1975b; Gaunt et al.,
1976) additional insight has been gained by using inhalation of
light gases (HeO$_2$), measurement of air flow and pressures, and
electrical potentials of muscles thought to be involved in sound
production in birds(Gaunt and Gaunt, 1977). We are now in the process
of applying these tools to the study of sound production mechanisms
in Tursiops. The data from these studies combined with the
information already available from analytical measurements near
presumptive sound production sites (Evans, 1973; Diercks et al.,
1971; Romanenko, 1974), x-ray cinematography (Gurevich, 1973; Norris
et al., 1971) and production of sound in light air (HeO$_2$) and
atmosphere (Evans and Rieger,1977, in preparation), should provide
the base for an accurate description for the total system.

     There is no need for an additional detailed description of the
nasal sac system or the structure of the delphinid larynx; these
have been more than adequately provided by Lawrence and Schevill
(1956), Purves (1967), Mead (1972), and Schenkkan (1973). This
report is a documentation of the total nasolaryngeal system with
special reference to the musculature in relationship to external
features for use as a stereotaxic guide in electromyographic and
ultrasonic scanning studies.

     Interest in how dolphins and other cetaceans produce sound goes
back more than 100 years. Murie(1871), interpreted " a double
raised smooth membranous fold" in the lower part of the larynx in
Risso's dolphin (Grampus griseus), to represent the vocal cords and
therefore to be the site of sound production. Dubois (1886), did
not find vocal cords in any of the cetaceans he examined. He did
agree, however, with Watson and Young (1879) who worked on beluga

whales, and Turner (1872), who studied fin whales. These workers
assumed that the rear extensions of the arytenoid cartilages lie so
close together that they would vibrate as air passed over them thus
taking over the function of vocal cords. Lawrence and Schevill (1956)
suggested the sounds were laryngeal in origin. As recently as 1967
Purves concluded that the mechanism of sound production in Phocoena
phocoena is similar to sound production in man except that the
aryepiglottal folds are called into play instead of the thyroarytenoid (vocal) folds. Evans and Prescott (1962) suggested a dual
sound source; the larynx for whistles and the nasal plugs and
associated sacs for pulses. Norris (1964, 1969) favors the nasal
sac system as the site of sound generation. Diercks et al. (1971)
also suggested a sound production site in the region of the nasal
plugs. Norris et al. (1971), Mead (1972), Schenkkan (1973), Evans
and Maderson (1973), and Dormer (1974) further support the role of
the nasal sacs and particularly the edges of the nasal plugs as the
site of sound generation, especially for the pulses used in echolocation.

    Some of the above conclusions concerning the site of sound
production were derived by interpretation of anatomy while others
were supported by a variety of experimental evidence. Roberts
(1975a) has demonstrated that a comprehensive study of the anatomy
of the rodent larynx, although informative, did not add much insight
into a complete understanding of the function of this organ in the
production of either audible or ultrasonic sounds. In further
studies Roberts(1975b), using ablation of sections of the hypoglossal
nerve, demonstrated the involvement of the intrinsic musculature
of the larynx in the production of ultrasonic emissions. He also
demonstrated that the muscles supplied by the hypoglossal nerves
were not involved in the production of ultrasound. The use of this
technique on delphinids, necessitating sacrifice of the animals, is
impractical from an economic standpoint. Activity of specific
muscles suspected of contributing to sound production can be measured
using electromyographic (EMG) techniques. Studies of this nature also
present some definite advantages over the nerve ablation method in
that specific muscles can be studied. As far as can be determined
from the literature no one has approached the problem by EMG studies.
Since the muscles must ultimately produce most of the force to
activate the mechanisms by which sounds are produced, we view it as
essential to find out how they are functioning during the production
of sounds.

    Of several anatomic studies on the larynx and nasal sac systems,
little has been reported that would be useful for EMG studies. The
contents of this report developed out of a need for more detailed
anatomic information with reference to external landmarks. Our intent
is to supplement studies done by others so as to obtain the necessary
information to do EMG's.

MATERIALS AND METHODS

The bottlenosed dolphin, Tursiops truncatus (Figure 1), has been extensively investigated, and its echolocation capability is well documented (Evans, 1973; Altes, 1974). Since this is the species on which we plan to do EMG studies, all of the dissections, measurements, and illustrations were of this species.

Dolphins are valuable animals, and specimens for anatomic dissection are often hard to come by. The specimens for this study all came from animals that died of natural causes in the wild, in ocenariums or in our own laboratory.

The laryngeal dissections used as many as seven different dolphins. This explains why the various views of the laryngeal cartilages, muscles, etc. are not of the same size. The larynx was taken at postmortem examination and preserved in 10% formalin. The transverse, frontal and oblique sections are freehand sections and vary in thickness as indicated.

We were able to obtain the head of a Tursiops truncatus that had been perfused with 10% formalin shortly after death. The head was sectioned into more than 30 transverse sections 1 cm thick.

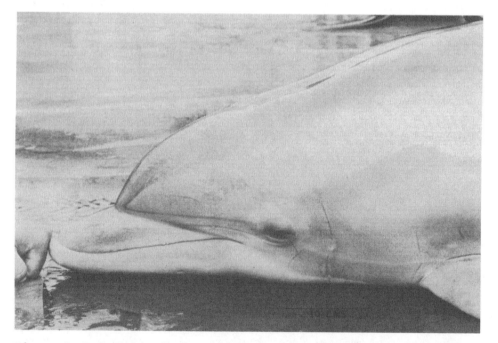

Figure 1. Left lateral view of the bottlenosed dolphin, Tursiops truncatus.

The sections ranged from 8-11 mm in average thickness and varied
no more than 2 mm from side to side (Figure 30). The tissues
exterior to the cranial cavity were found to be well fixed, but the
brain·was in such poor condition that the brain sections were dis-
carded and therefore not included in illustrations of the sections.
In some illustrations the details of cartilages, muscles and bone
have been outlined.

RESULTS AND DISCUSSION

The Larynx

The larynx (Figure 2 and 3) is so specialized it is considered
to be an organ useful in characterizing the cetacea. It is intranarial, projecting upward from the floor of the pharynx with its
anterior end held firmly within the internal nares by a sphincter
composed of palato-pharyngeal muscle (Negus, 1931). In odontocete
cetaceans the opening of the external nares (blowhole) has migrated
to a position on top of the head so that when the animal surfaces
the opening into the nasal passage is the first part of the head to
be exposed. The elongated intranarial larynx functions to keep the
nasal air passages continuous with the glottis while the porpoise
dashes through the water open mouthed after its prey (Huxley, 1888).

Laryngeal cartilages. With the exception of the so called
"aryteno-epiglottal beak", (Figure 2 and 3) the cetacean larynx is
similar in general structure to other mammalian larynges, being
composed of a cartilagenous framework held together by a number of
muscles.

The thyroid cartilage (Figures 4,6,12,15,16,17,18,19,20,21,22,
25,26,27,and 28) is made up of right and left laminae fused at their
ventral margins without forming even the slightest laryngeal
prominence. Each lamina has two well developed processes; the
anteriorly directed cranial cornua and the posteriorly directed
caudal cornua, the latter being the better developed. The thyroid
cartilage forms the main body of the larynx. Most of the extrensic
muscles, which function to move the complete larynx, are attached
to this cartilage.

The caudal cornua of the thyroid cartilage articulate with the
posterior-superior lateral margins of the cricoid cartilage by small
oval synovial joints. Anteriorly the thyroid cartilage also articulates with the epiglottal cartilage. This articulation is by a
fibrous joint that allows limited movement. Additional movement is
suggested, however, by a slight bending of the cartilage which is
especially thin just posterior to this joint.

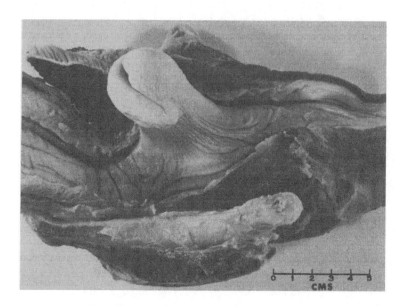

Figure 2. Left dorso-lateral view of the laryngeal beak of a
        bottlenosed dolphin.

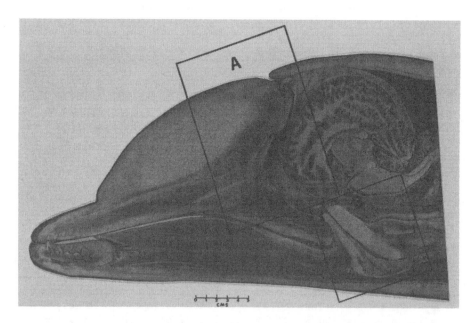

Figure 3. A midsagittal section through the head of a bottlenosed
    dolphin. A- the nasal area, B- the laryngeal area.

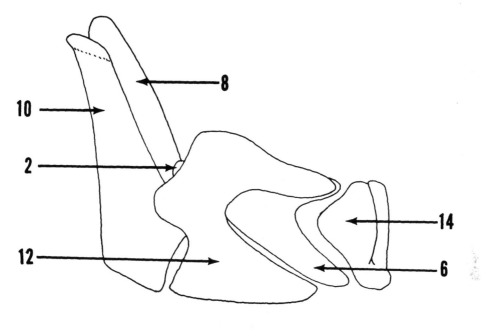

Figure 4. All laryngeal cartilages in approximate articulated position. 2- arytenoid cartilage, 6- cricoid cartilage, 8- cuneiform cartilage, 10- epiglottal cartilage, 12- thyroid cartilage, 14, 1st tracheal cartilage.

In Tursiops the cricoid cartilage (Figures 4,5,7,12,18,19,20, 21,22,24,25,26, and 27) is not a complete ring as is found in some cetaceans, but it is open ventrally, forming two rather long posteriorly projecting lateral cornua. These cornua almost fill the posterior lateral notch in each thyroid lamina.In addition to the joints formed between the cricoid and thyroid cartilages, there are much larger synovial joints along the anterior lateral margins of the body of the cricoid cartilage where the cricoid articulates with the arytenoid cartilages.

The delphinid epiglottal cartilage (Figures 4,5,8,12,13,14,15, 16,17,22,23,24,25,26,and 27) is much longer than the epiglottal cartilage in most mammals and has deepened medially to form a trough with thin lateral margins projecting posteriad to wrap around the anterior borders of the cuneiform and arytenoid cartilages, thereby forming the "laryngeal beak".

The arytenoid cartilages (Figures 4,5,9,12,17,18,19,22,24,25, and 26) articulate by well formed oval-shaped synovial joints with

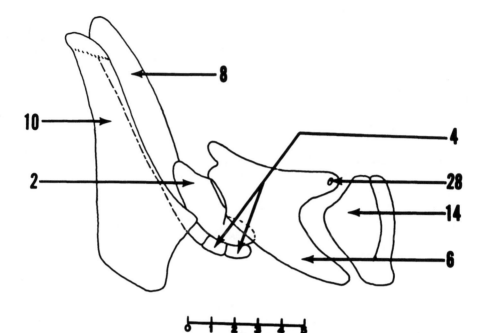

Figure 5. Laryngeal cartilages in approximate articulate position
except thyroid cartilage removed. 2- arytenoid cartilage,
4- corniculate cartilage, 6- cricoid cartilage, 8- cuneiform
cartilage, 10- epiglottal cartilage, 14- 1st tracheal cartilage,
28- articular facet to thyroid cartilage.

the anterior lateral margins of the cricoid cartilage. Each has a
well developed laterally projecting muscular process, but no vocal
processes are evident. The arytenoid cartilages also articulate
with the cuneiform cartilages.

The cuneiform cartilages (Figures 4,5,9,12,13,14,15,16,17,18,
19,20,22,23,24,25,26,and 27), form elongated blade-like processes
that face each other medially to lie in the trough formed by the
epiglottal cartilage. As indicated above, the cuneiform cartilages
articulate with the arytenoids. Two or more small cartilages,
probably representing the corniculate cartilages, attach to the
end of each cuneiform cartilage and also to the posterior two-fifths
of the ventral edge of the arytenoid.

The arytenoid, the cuneiform, and the corniculate cartilages
as described above have been referred to in various ways in the
literature on cetacean larynges. Benham (1901) called the structure
formed by these cartilages the arytenoid. D'Arcy-Thompson (1890)
called the blade-like cartilages the super-arytenoids. Howes (1879)

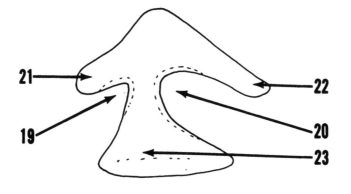

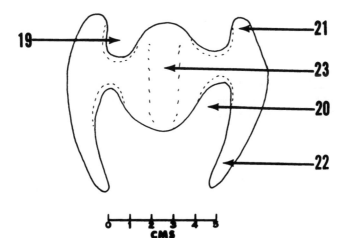

Figure 6. Lateral (top) and ventral (bottom) views of the thyroid cartilage. 19- anterior notch, 20- posterior notch, 21- cranial cornu, 22- caudal cornu, 23- lamina.

was the first to call the blades the cuneiform cartilages. Howes was supported by Cleland (1884), Purves (1967), and Dormer (1974).

Extrinsic laryngeal muscles. All the extrinsic muscles are paired except for the hyoepiglottal muscle which is single. To simplify description only the muscles on the left side are described.

The hyoepiglottal muscle (Figures 10,11,14,15,22,24,25,26, and 27) originates from the mid-dorsal basihyal segment of the hyoid bone, and inserts on the inferior one-half of the anterior edge of the epiglottal cartilage. This muscle probably functions to pull the epiglottal cartilage forward so as to enlarge the size of the anterior passageway in the laryngeal beak.

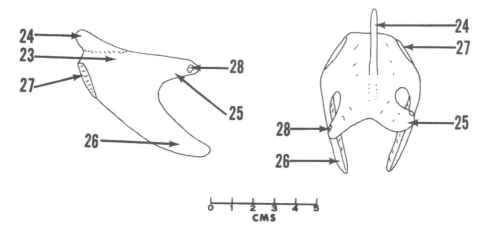

Figure 7. Left lateral (left) and dorsal (right) views of the cricoid cartilage. 23- lamina, 24- spine, 25- posterior dorsal cornu, 26- posterior ventral cornu, 27- articular surface to arytenoid cartilage, 28- articular facet to thyroid cartilage.

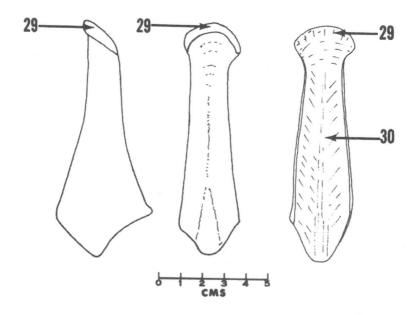

Figure 8. Lateral (left), anterior (middle), and posterior (right) views of the epiglottal cartilage. 29- lip, 30- trough.

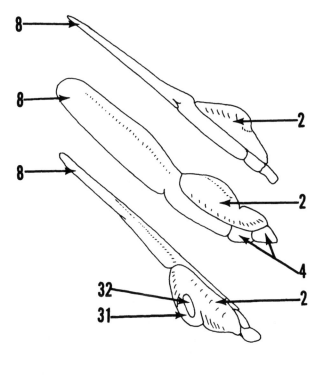

Figure 9. Ventral (top), lateral (middle), and dorsal (bottom)
views of the arytenoid, cuneiform and corniculate cartilages.
2- arytenoid cartilage, 4- corniculate cartilage, 8- cuneiform
cartilage, 31- muscular process, 32- articular surface to cricoid
cartilage.

The sternothyroid muscle (Figure 28) originates on the
anterior margin of the sternum, and inserts on the lateral surface
of the thyroid cartilage just superior to and anterior to the
insertion of the cricothyroid muscle. The sternothyroid probably
functions to pull the complete laryngeal apparatus posteriad.

The thyrohyoid muscle (Figure 28) originates on the lateral
surface of the thyroid cartilage just anterior and ventral to the
insertion of the sternohyoid muscle. Its insertion is at the
posterior margin of the basihyal and thyrohyal segments of the hyoid
bone. It probably functions to pull the laryngeal apparatus anteriad,
thus more firmly lodging the tip of the laryngeal beak in the
internal nares.

The larynx is suspended from the base of the cranium by a
highly complex series of muscles, the thyropalatine, the occipito-

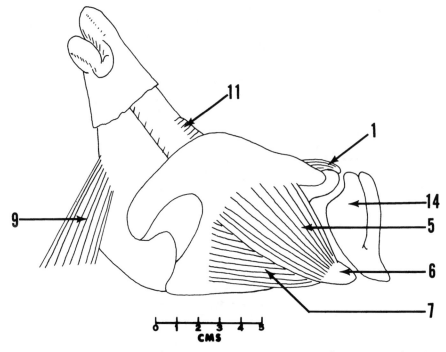

Figure 10.Left lateral view of the larynx. 1- Cricoarytenoid-dorsal
muscle, 5- cricothyroid-oblique muscle, 6- cricoid cartilage,
7- cricothyroid-straight muscle, 9- hyoepiglottal muscle, 11- interarytenoid muscle, 14- 1st tracheal cartilage.

thyroid and the thyropharyngeal. These muscles attach to the dorsal
margin and anteromedial surface of the thyroid cartilage, and then
pass dorsad and anteriad to attach to the floor of the cranium. The
most anterior of these muscles is the thyropalatine,(a posterior
part of the palatopharyngeal muscle), which attaches to the medial
surface of the thyroid lamina and the thyroid ligament. The middle
muscle, the occipitothyroid, attaches to the dorsal margin of the
cranial cornu of the thyroid cartilage. The most posterior muscle,
the thyropharyngeal, attaches to the dorsal margin of the posterior
cornu of the thyroid cartilage. The anterior fibers of the palato-
pharyngeal muscle together with the pterygopharyngeal muscle form
the sphincter that partially fills the lower part of the internal
nares and holds the laryngeal beak in place.

   Intrinsic laryngeal Muscles. The cricothyroid muscle, (Figures
10,12,18,19,20,21,22, and 27) has its origin to the lateral posterior
surface of the lateral cornu of the cricoid cartilage.Its insertion
is to the thyroid cartilage along the inferior margin of the caudal
cornu, to the posterior notch and to the lateral surface of the
lamina just below the posterior notch. This muscle has two distinct

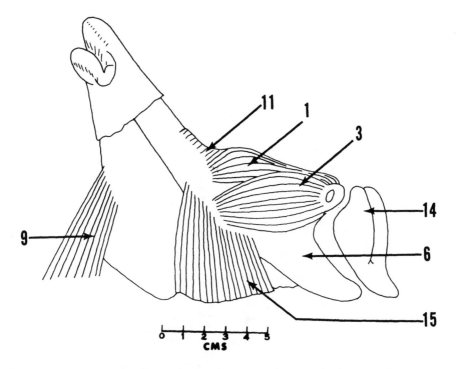

Figure 11. Lateral view of the larynx with thyroid cartilage and
cricothyroid muscles removed. 1- cricoarytenoid-dorsal muscle,
3- cricoarytenoid- lateral muscle, 6- cricoid cartilage, 9-hyo-
epiglottal muscle, 11- interarytenoid muscle, 14- 1st tracheal
cartilage, 15- thyroarytenoid muscle.

parts in the dolphin, a <u>dorsal</u> <u>oblique</u> <u>cricothyroid</u> and a <u>ventral</u>
<u>straight</u> <u>cricothyroid</u>.

     The <u>cricoarytenoid</u> <u>muscle</u>, (Figures 10,11,12,18,19,20,21,22,
24,25, and 26) has its origin over the dorsal surface of the cricoid
cartilage extending laterad to the medial surface of the caudal
cornu of the thyroid cartilage. The muscle is in two distinct parts,
the dorsal cricoarytenoid and the lateral cricoarytenoid. Because of
its attachment to the caudal cornu, Hosokawa (1950) calls this
muscle the ceratocricoarytenoideus. Both parts of the muscle insert
on the dorsal muscular process of the arytenoid cartilage. They
probably function to rotate the arytenoids lateral and posterio-
dorsad. This action increases inter-arytenoid space between the
edges of the arytenoids and thus the inter-cuneiform space between
the anterior edges of the cuneiform cartilages.

     The <u>thyroarytenoid</u> <u>muscle</u> (Figures 11,12,17,18,19,22,24,25,
26, and 27)lies on the medial surface of the thyroid cartilage and

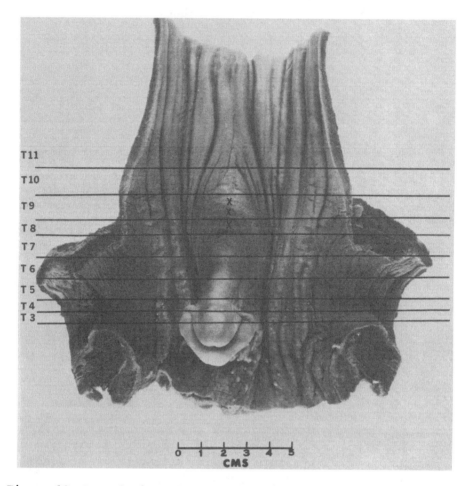

Figure 12. Dorsal view of laryngo-pharyngeal area showing location
of transverse sections T3-T11. XXX: indicates where the cricoid
crest can be palpated.

has its origin on the dorsal midline and medial surfaces. The
fibers pass dorsad to insert on the ventral surface of the muscular
process of the arytenoid cartilage. The anterior fibers pass laterad,
over the posterior angle of the epiglottal cartilage, but do not
appear to attach to the epiglottis in such a way as to act upon it.
The thyroarytenoid probably functions to pull the arytenoid and
cuneiform cartilages ventrad and mesad to decrease the angle between
their medial surfaces. If the arytenoid and cuneiform cartilages
are already approximated the thyroarytenoid muscles probably
function to pull the arytenoids and cuneiforms deeper into the
epiglottal trough.

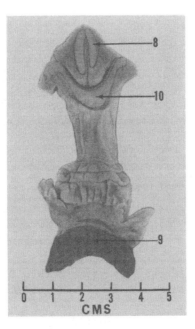

Figure 13- Anterior section of section T3. 8- cuneiform cartilage,
9- hyoepiglottal muscle, 10- epiglottal cartilage.

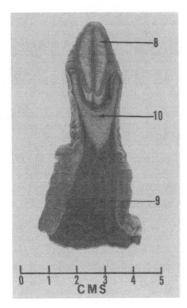

Figure 14- Anterior surface of section T4. 8- cuneiform cartilage,
9- hyoepiglottal muscle, 10- epiglottal cartilage.

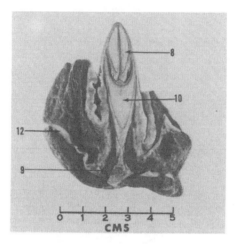

Figure 15. Anterior surface of section T5. 8- cuneiform cartilage, 9- hyoepiglottal muscle, 10- epiglottal cartilage, 11- thyroid cartilage.

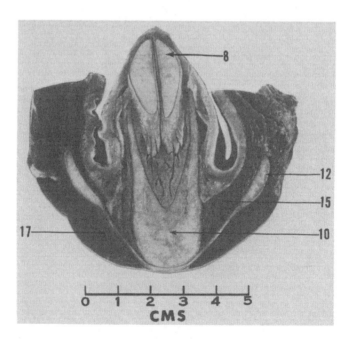

Figure 16- Anterior surface of section T6. 8- cuneiform cartilage, 9- epiglottal cartilage, 12- thyroid cartilage, 15- thyroarytenoid muscle, 17- thyrohyoid muscle.

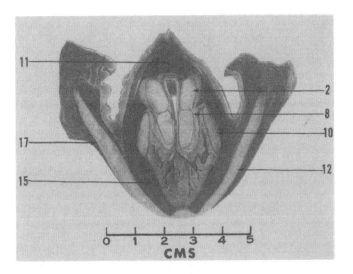

Figure 17. Anterior surface of section T7. 2- arytenoid cartilage, 8- cuneiform cartilage, 10- epiglottal cartilage, 11- interarytenoid muscle, 12- thyroid cartilage, 15- thyroarytenoid muscle, 17- thyrohyoid muscle.

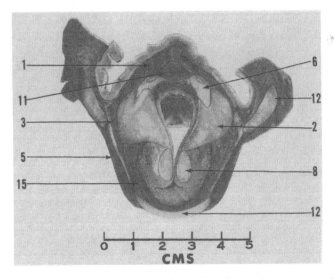

Figure 18. Anterior surface of section T8. 1- cricoarytenoid- dorsal muscle, 2- arytenoid cartilage, 3- cricoarytenoid- lateral muscle, 5- cricothyroid- oblique muscle, 6- cricoid cartilage, 8- cuneiform cartilage, 11- interarytenoid muscle, 12- thyroid cartilage, 15- thyroarytenoid muscle.

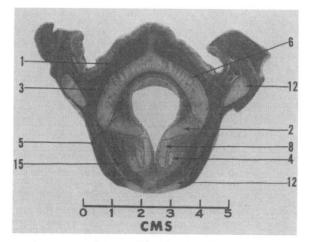

Figure 19. Anterior surface of section T9. 1- cricoarytenoid-
dorsal muscle, 2- arytenoid cartilage, 3- cricoarytenoid- lateral
muscle, 4- corniculate cartilage, 5- cricothyroid-oblique muscle,
6- cricoid cartilage, 8- cuneiform cartilage, 12- thyroid cartilage,
15- thyroarytenoid muscle.

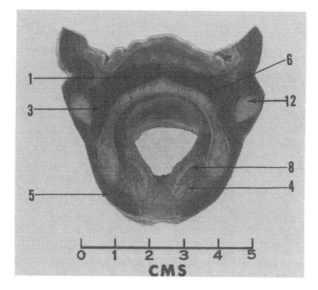

Figure 20. Anterior surface of section T10. 1- cricoarytenoid-
dorsal muscle, 3- cricoarytenoid- lateral muscle, 4- corniculate
cartilage, 5- cricothyroid-oblique muscle, 6- cricoid cartilage,
8- cuneiform cartilage, 12- thyroid cartilage.

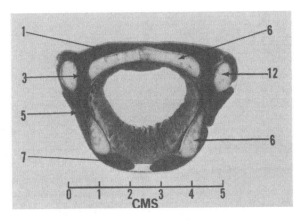

Figure 21. Anterior surface of section T11. 1- cricoarytenoid-dorsal muscle, 3- cricoarytenoid- lateral muscle, 5- cricothyroid-oblique muscle, 6- cricoid cartilage, 7- cricothyroid- straight muscle, 12- thyroid cartilage.

The interarytenoid muscle (Figures 10,11,12,17, and 18), joins the dorsal margins of the cuneiform and to a lesser degree the arytenoid cartilages. Some of the posterior fibers pass below and join into the anterior edge of the dorsal cricoarytenoid muscle. This muscle apparently functions to assist the cricoarytenoid muscles in increasing the interarytenoid and intercuneiform spaces.

EMG Studies On Laryngeal Muscles

Since none of the laryngeal muscles can be recorded by the use of surface electrodes because of the thick overlying blubber and subcutaneous muscle, electrodes must be inserted directly into the muscle to be studied. The bottlenosed dolphins available for these studies were young adults 125-150 kg in weight and 2.3-2.5 m in length. The measurements given below relate to animals in that size range.

There are two approaches that can be used to insert an electrode into the hyoepiglottal muscle. To begin the first approach (A in Figure 29), palpate the posterior edge of the hyoid bone along the midline. The electrode is inserted about 2.0 cm behind the posterior edge of the hyoid bone and directed anteriorly at about 50-55^0 so as to pass between the hyoid and the epiglottal cartilage. The muscle is between 4.5 and 5.0 cms deep and has a transverse width of about 1.5 cm.

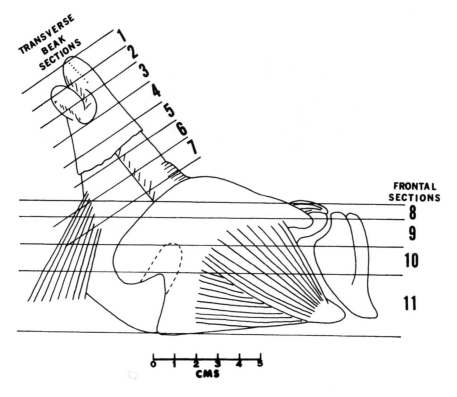

Figure 22- Left lateral view of the larynx indicating where
transverse beak (TB) sections 1-7 and frontal (F) sections 8-11
were made.

     The hyoepiglottal muscle can also be reached through the throat.
The electrode would then be directed as illustrated at B in
Figure 29, and the recording wires may be run out the subject's
mouth.

     To place an electrode into the sternothyroid muscle it was
oriented using the baseline as illustrated in Figure 28, moving
3.5 cm to the left (animals left) of the midline to insert into
the left sternothyroid muscle or 3.5 cm to the right of the mid-
ventral line to insert into the right sternothyroid muscle. The
electrode should be inserted at 90° to the body surface and placed
4.5 cm deep (A-Figure 28).

     The palato-pharyngeal muscle, the muscle that helps to hold
the larynx in place, and may also serve to differentially close
off the internal nares, can be recorded by inserting an electrode
as illustrated at C, Figure 29.

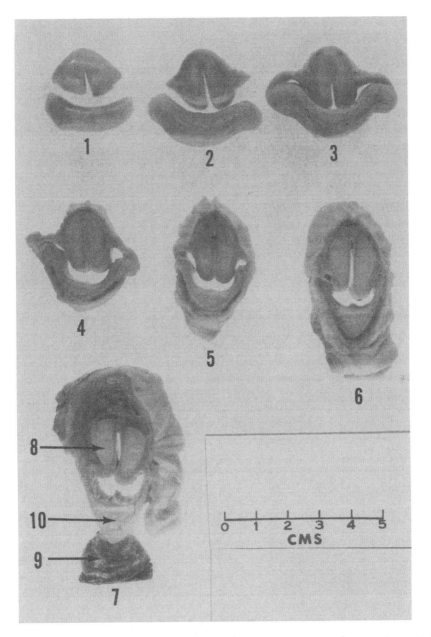

Figure 23. Dorsal surfaces of TB (transverse beak) sections 1-7. 8- cuneiform cartilage, 9- hyoepiglottal muscle, 10- epiglottal cartilage.

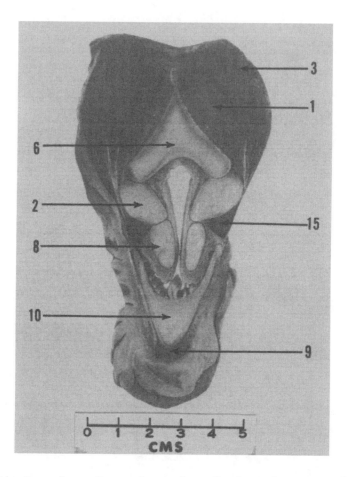

Figure 24. Dorsal surface of section F8. 1- cricoarytenoid-dorsal muscle, 2- arytenoid cartilage, 3- cricoarytenoid-lateral muscle, 6- cricoid cartilage, 8- cuneiform cartilage, 9- hyoepiglottal muscle, 10- epiglottal cartilage, 15- thyroarytenoid muscle.

     The electrode can be inserted into the cricothyroid muscle in much the same way as into the sternothyroid muscle, by using the line illustrated in Figure 28. Moving 2.5 cm lateral to the midventral line insertion can be made either to the left or right cricothyroid muscle. The electrode should be inserted at 90° to the body surface and placed 3.5-4 cm deep (C- Figure 28).

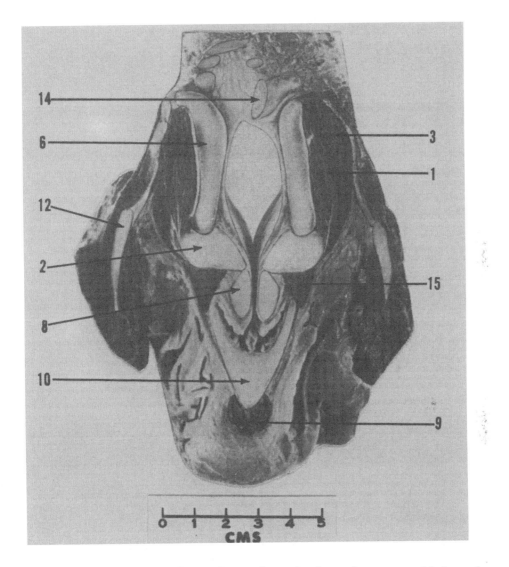

Figure 25. Dorsal surface of section F9. 1- cricoarytenoid-dorsal
muscle, 2- arytenoid cartilage, 3- cricoarytenoid-lateral muscle,
6- cricoid cartilage, 8- cuneiform cartilage, 9- hyoepiglottal
muscle, 10- epiglottal cartilage, 12- thyroid cartilage, 14- 1st
tracheal cartilage.

   To insert the electrode into the thyrohyoid muscle the
posterior hyoid is palpated on the midline. Then move posteriad
2-2.5 cm along the midline and then from 2-2.5 cm laterad. The
electrode should be inserted at 90° to the body surface. The

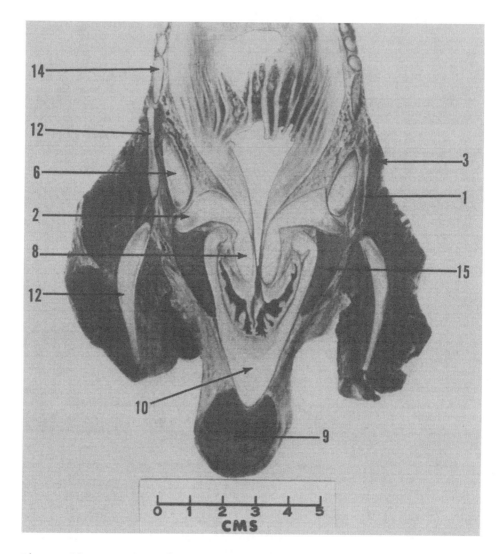

Figure 26. Dorsal surface of section F10. 1- cricoarytenoid-dorsal muscle, 2- arytenoid cartilage, 3- cricoarytenoid-lateral muscle, 6- cricoid cartilage, 8- cuneiform cartilage, 9- hyo-epiglottal muscle, 10- epiglottal cartilage, 12- thyroid cartilage, 14- 1st tracheal cartilage, 15- thyroarytenoid muscle.

muscle is 3.5-4.0 cm deep and is backed up by the thyroid cartilage and the base of the epiglottal cartilage (B- Figure 28).

The thyropalatine, the occipitothyroid and the thyropharyngeal muscles lie lateral to the pharyngeal and can probably be

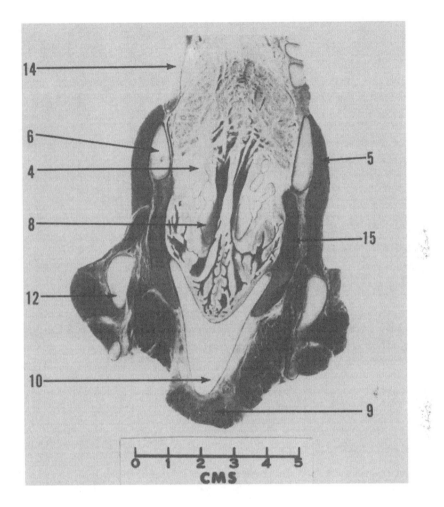

Figure 27. Dorsal surface of section F11. 4- corniculate cartilage, 5- cricothyroid-oblique muscle, 6- cricoid cartilage, 8- cuneiform cartilage, 9- hyoepiglottal muscle, 10- epiglottal cartilage, 12- thyroid cartilage, 14- 1st tracheal cartilage, 15- thyro-arytenoid muscle.

recorded by placing electrodes through the wall of the pharynx. If these muscles have an active role in sound production (other than suspending the larynx), we would speculate that they work together, therefore recordings from these muscles could be taken anywhere along the complete length of the larynx.

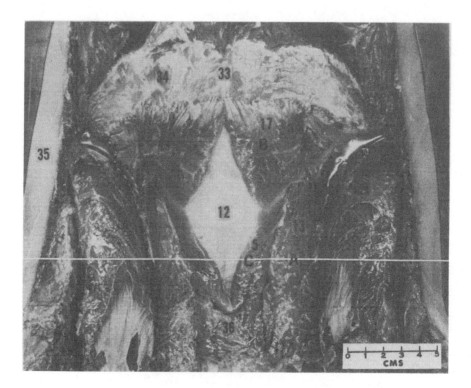

Figure 28. Ventral view of hyo-sternal area (the white line passes along the anterior edge of the flippers). 5- cricothyroid, 12- thyroid cartilage, 13- sternothyroid muscle, 17- thyrohyoid muscle, 18- panniculus carnosus muscle, 33- basihyal segment of hyoid bone, 34- thyrohyal segment of hyoid bone, 35-blubber, 36- thyroid gland. A- placement of sternothyroid muscle electrode, B- placement of thyrohyoid muscle electrode, C- placement of crico-thyroid muscle electrode.

The cricoarytenoid muscles are located lateral and posterior to the cricoid crest and can be approached by palpating the anterior margin of the crest just posterior to the base of the laryngeal beak. Electrodes can be inserted into either the left or right cricoarytenoid muscle 2 cm posterior to and 1 cm left or right of the crest. There is a sheet of esophageal muscle about 0.5 cm thick which passes between the pharyngeal wall and the crico-arytenoid muscle. The electrode must be inserted through this muscle layer to reach the cricoarytenoid.

The thyroarytenoid muscle is located approximately 2.3 cm lateral and ventral to the cricoid crest. The electrode should be

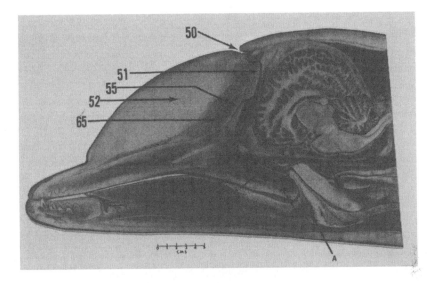

Figure 29. Mid-sagittal section of a dolphin head. 50- blowhole,
51- nasal plug, 52- melon, 55- premaxillary sacs, 65- nasal plug
muscle, A- placement of hyoepiglottal muscle electrode, B- alternate
placement of hyoepiglottal muscle electrode, C- placement of palato-
pharyngeal sphincter electrode, D- placement of interarytenoid
muscle elecrode.

inserted at a point 2 cm posterior to the posterior edge of the
epiglottal cartilage where the muscle is approximately 1 cm thick.

The interarytenoid muscle passes just anterior to the cricoid
crest. The muscle extends approximately 1.5 cm to the left and
1.5 cm to the right of the midline and is approximately 2.0 wide
from front to back) and 1.0 cm thick along the midline. The
electrode is directed at an angle parallel to the dorsal edge of
the cuneiform cartilages for insertion into this muscle (D-Figure29).

The Nasal System

The nasal system (Figure 3), includes the bony nasal passage;
the spiracular cavity; and the plugs, sacs, membranes and muscles
associated with the blowhole region, Due to their structure and
location, the internal nares with their plugs, sacs, and membranes
are considered to be the structures most likely involved in sound
production. Lawrence and Schevill (1956), Mead(1972), Schenkkan
(1973), and Dormer (1974) have given detailed descriptions and
illustrations of the nasal region in the bottlenosed dolphin.

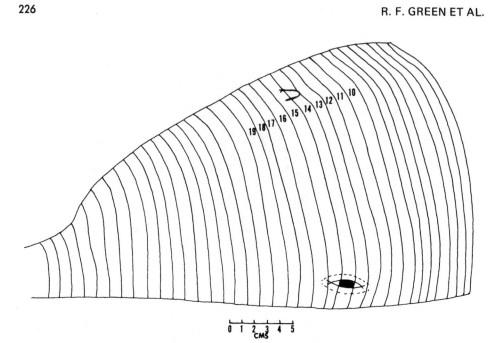

Figure 30. Left side of head showing the approximate location of the cross sections through the nasal region. The numbers lateral to the blowhole are section numbers referred to in the following figures.

We were interested in the anatomy of this area specifically for purposes of understanding dolphin sound production through ultrasonic scanning and EMG studies of function, therefore, we will present only a general description here.

The bony nasal passage passes upward through the skull in an anteriorly curved, almost vertical plane (Figure 29). Within the cranium the passage is separated by the bony and cartilagenous septum into two sub-oval tubes. The upper part of the passageway is a single chamber and is called the spiracular cavity (Von Baer, 1926). It passes upward a short distance to a single opening, the blowhole. The blowhole is more or less round when open, but is "C" shaped when closed.

The nasal sacs. The nasal sacs are extensively developed in the bottlenosed dolphin. Four pairs of sacs communicate either directly or indirectly with the spiracular cavity. The premaxillary sacs (Figures 29, 33, and 34) are the largest and lie on top of the premaxillary bones just anterior to the openings into the bony nasal passage. They open into the spiracular cavity. The vestibular sacs (Figures 32-37) are located just below the skin, projecting posteriolaterad from the spiracular cavity into which they open.

The nasofrontal sacs (Murie,1871) are "U" shaped tubular sacs that pass around the spiracular cavity below the vestibular sacs.

The accessory sacs (Schenkkan, 1971) project laterally from the inferior vestibules; passages that connect the posterior naso-frontal sacs to the posterior spiracular cavity.

The nasal plugs (Lawrence and Schevill, 1956) protrude from the anterior spiracular wall just dorsal to the premaxillary sacs. They fit tightly against the posterior wall of the spiracular cavity to form a seal. The distal margins of the plugs form into lips that fit into the openings between the inferior vestibules and the posterior spiracular cavity.

Nasal musculature. The several layers of muscles associated with the sacs, plugs, and lips are also complexly developed. Huber (1934) was of the opinion that the rostral muscles were derived from the pars labialis of the maxillonasolabialis muscle and the layers of muscle around the blowhole to be derived from the pars nasalis of the maxillonasolabialis. Lawrence and Schevill (1956) and Mead (1972) concur with Huber's interpretation of these origins. Mead, however, suggests the division should be made between medial and lateral layers of muscle rather than between anterior and posterior muscles. This would result in the lateral, rostral, and superficial nasal muscles being derived from the pars labialis, and the medial rostral and deep nasal musculature being derived from the pars nasalis of the maxillonasolabialis muscle.

The posteroexternus (Figure 40) is the most superficial of the nasal sac muscles. Its origin extends along the posterior 1/2 of the supraorbital process of the maxilla bone, the temporal crest of the frontal bone, and nuchal crest (occipito-frontal suture). The muscle is generally quite thin and tendinous, with anterodorsad directed fibers passing over the vestibular sac to attach to the dense connective tissue in the posterior and lateral wall of the nasal passage. These anterior fibers are loosely attached over the dorsal surface of the vestibular sacs. The posterior fibers are directed anteriomediad to insert just above the anterior fibers.

This description of the posteroexternus is much the same as that given by Lawrence and Schevill (1956). It does not agree with Mead (1972), who describes the insertion of this muscle as being to the vertex and contralateral muscle. Lawrence and Schevill also illustrate this muscle to show its origin over most of the nuchal crest thus completely covering the anteroexternus muscle below it. Mead, however, illustrates the posteroexternus with its origin over the lower nuchal crest, with anteroexternus muscle exposed posteriorly. We have found the muscle to be as illustrated by Lawrence and Schevill about 80 percent of the time. About 20 percent

of the specimens that we dissected conform to Mead's description.
On occasion one side of the specimen will appear as described by
Mead while the other side will be structured like that described
by Lawrence and Schevill.

This muscle probably functions to pull back on the posterior
wall and laterad on the lateral wall of the spiracular cavity. It
could also function as a compressor of underlying structures and to
restrict vertical expansion of the vestibular sac, as suggested by
Mead (1972). Such a compressor function could also help in the
conservation of water and heat that would ordinarily be lost with
expired air (Coulombe et al., 1965).

The intermedius muscle originates under the posteroexternus
along the posterior 1/2 of the supraorbital process as well as from
the fascia over the anteroexternus and posteroexternus muscles.
It is more easily identified at its insertion end where the fibers
pass loosely into the dense connective tissue mass dorsal to the
nasal plugs. Mead (1972) also found the more dorsal fibers to be
associated with the anterior dorsal surface of the vestibular sac
This we found to be true in about 50 percent of the animals we
dissected. It is also the case on one side of the head, but not on
the other. This muscle is sometimes missing from both sides of the
head. This muscle possibly functions to pull the nasal plugs down
against the internal nares, but does not seem to be of great importance, as indicated by the generally weak insertion and variability.

The anteroexternus muscle (Figures 32 and 33) is a wide expan-
sive muscle which takes its origin from the anterior supraorbital
process posterior to the temporal and nuchal crests. The anterior
margin of the muscle folds back over itself giving the appearance
of a separate muscle. The posterior dorsal margin occupies the angle
between the anterior nuchal crest and the lateral margin of the
vertex. The anterior fibers insert into the dense connective tissue
beneath the anterior lip of the blowhole and into the lateral
anterior wall of the spiracular cavity. The posterior fibers insert
on the posterior and lateral wall of the spiracular cavity as well
as on the dorsal edge of the blowhole ligament. This muscle passes
both below and above the vestibular sac, completely surrounding it.

Anteriorly, the anteroexternus is difficult to separate from
the anterointernus muscle. The anterior fibers lie just posterior
to the posterior extension of the lateral rostral muscle. Posteriorly, the muscle can be separated more easily from the postero-
internus muscle, as the fibers run in different directions.

This description of the anteroexternus muscle is much like the
description of the same muscle as redefined by Mead. Lawrence and
Schevill refer only to the anterior fibers as the anteroexternus.

We found these anterior fibers to be distinctly separate in a few
instances, but in most cases the anterior and posterior parts were
not separable.

The anterior fibers of the muscle probably function to pull
down on the nasal plug causing it to seal over the internal nares.
These anterior fibers may, as suggested by Lawrence and Schevill,
draw the vestibular sac forward and laterad. The posterior part of
the muscle, as described by Mead, probably functions to draw the
posterior lip backwards, opening the blowhole. Since the vestibular
sacs are completely surrounded by this muscle it very likely
functions to maintain the size and position of these sacs as well
as compressing the sacs.

The posteriorinternus muscle (Figure 40) lies deep to the posterior
three-quarters of the anteroexternus muscle as defined by Mead. Its
origin is to the frontal bone, above the eye; to the dorsal medial
margin of the supraorbital process, then dorsad to lie in the angle
between the nuchal crest and the vertex. Its fibers pass anteriad
and dorsad under the vestibular sacs to insert on the posterior and
lateral wall of the nasal passage as well as to the lateral margins
of the nasofrontal sacs. As Mead suggests, this muscle probably
functions to pull the posterior wall forward, putting pressure on
the blowhole ligament that in turn puts pressure on the lips of the
nasal plugs causing the plugs to more effectively close the external
bony passages.

The anterointernus muscle (Figures 32, 33, 35) is the deepest
and most massive of the nasal muscles. The anterior part of this
muscle as described by Mead includes what Lawrence and Schevill
called the profundus muscle. In our dissections, we most often
found the two parts of the muscle to be closely joined even though
there was occasionally a distinct boundry between them. Anteriorly
the fibers insert to the connective tissue mass anterior to the
spiracular cavity and over the nasal plugs. The median and posterior
fibers insert into the lateral wall of the nasal passage and to the
nasofrontal sac.

This muscle is probably the most powerful of the nasal muscles
and must therefore play a significant role in the function of the
blowhole region. By drawing the anterior wall of the spiracular sac
posteriad and by compressing the nasal plug mass it would play a
significant role in closing the nasal passage. Purves (1967)
suggested that the contraction of this muscle would compress the
premaxillary sacs and function in the recycling of air.

The rostral muscle (Figure 31) consists of an elongated mass
located on the lateral margins of the rostrum. Its origin is from
the dorsal surface of the maxilla. The muscle is divisible into

lateral and medial parts with the lateral fibers inserting into the
lip and connective tissue lateral to the melon. The medial fibers
project dorsally and medially to insert into the melon and lateral
margin of the nasal plug. The medial fibers probably function to
alter the shape of the melon and may function in sound production
by action on the nasal plug. The muscle may also function in concert
with sound production if the various theories of acoustic beam forming by the dolphin melon are correct (Wood, 1964).

The nasal plug muscle is an elongated mass of fibers having
their origin on the premaxillary bones anterior to the premaxillary
sacs. The fibers insert into the dense connective tissue and fat of
the nasal plugs. There appear to be more muscle fibers in the right
side than in the left side of this muscle. This is probably due in
part to the fact that the right plug is as much as 2 to 2-1/2 times
as large as the left plug. This muscle most likely functions to pull
the nasal plugs forward and may be involved in sound production.

The intrinsic muscles of the nasofrontal sacs almost completely
surround the nasofrontal sacs. The muscles originate from the dense
connective tissue on the anterior wall of the vertex. Some fibers
pass laterad below the lateral arc of the nasofrontal sacs to merge
with the nasal plug muscle. Other fibers are associated with the
inferior vestibule passages. There are an almost unlimited number of
actions these fibers might have upon the sacs and passages they are
associated with.

Mead (1972) has described the diagonal membrane muscle,
originating on the anterolateral vertex deep to the intrinsic muscle
and inserting in the attached margin of the diagonal membrane. Mead
prefaced his description by stating that due to the complexity of this
this region the muscle has escaped the attention of earlier workers.
We have been very careful in searching for this muscle during our
last 4 dissections of the Tursiops nasal region. We have been
successful in locating fibers which fit Meads description but the
muscle is very diffuse and not well developed.

EMG studies on nasal musculature. To place electrodes in the
various layers of nasal musculature the sagittal section illustrated
in Figure 29, and the cross sections illustrated in Figures 31-39
can be scaled up to fit an individual experimental animal. This is
most easily done by measuring the width of the animal's head at the
posterior canthus of the eyes. Measurements can then be taken directly
from adjusted illustrations so as to know where and how deep to place
the electrodes. The anterior surface of section 13 is used as a
baseline since it can be easily referenced from external landmarks.

Through dissection and morphometric evaluation we have determined external landmarks and stereotaxic coordinates necessary for

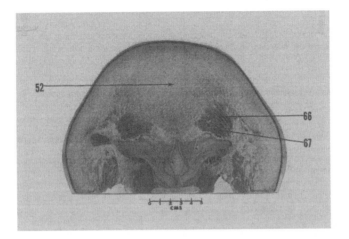

Figure 31. Anterior surface of section 19; 6 cm anterior to
baseline (section 13). 52- melon, 66- lateral rostral muscle,
67- medial rostral muscle.

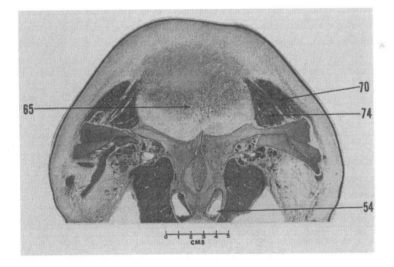

Figure 32. Anterior surface of section 16; 3 cm anterior to
baseline (section 13). 54- pterygoid sinus, 65- nasal plug muscle,
70- medial rostral muscle, 74- anterior externus muscle.

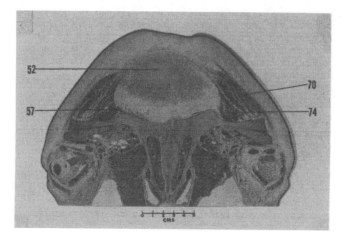

Figure 33. Anterior surface of section 15; 2 cm anterior to baseline (section 13). 52- melon, 57- premaxillary sac, 70- anterior externus muscle, 74- anterior internus muscle.

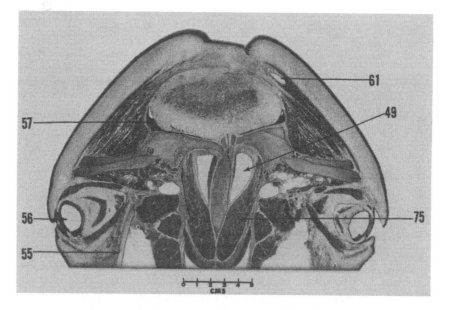

Figure 34. Anterior surface of section 14; 1 cm anterior to baseline (section 13). 49- nasal passage, 55- mandible, 56- eye, 57- premaxillary sac, 61- vestibular sac, 75- palato-pharyngeal sphincter.

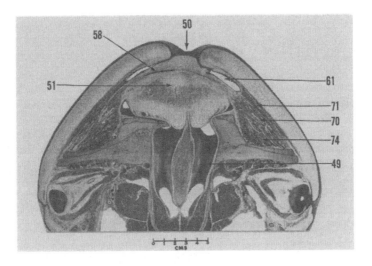

Figure 35. Anterior surface of baseline (section 13). This section passes 1/2 cm anterior to the back of the blowhole and 1/2 cm anterior to the posterior canthus of the eyes. 49- nasal passage, 50- blowhole, 51- nasal plug, 58- nasofrontal sac, 61- vestibular sac, 70- anterior externus muscle, 71- posterior externus muscle, 74- anterior internus muscle.

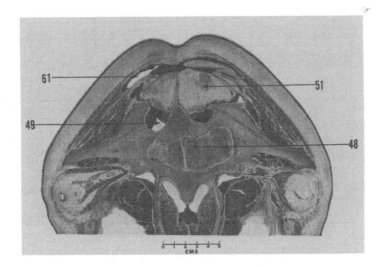

Figure 36- Posterior surface of section 13. This surface is 1 cm posterior to the anterior surface of section 13. 48- cranial cavity, 49- nasal passage, 51- nasal plug, 61- vestibular sac.

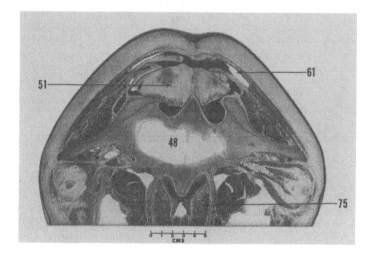

Figure 37. Anterior surface of section 12; 1 cm posterior to
baseline (section 13). 48- cranial cavity, 51- nasal plug,
61- vestibular sac, 75- palato-pharyngeal sphincter.

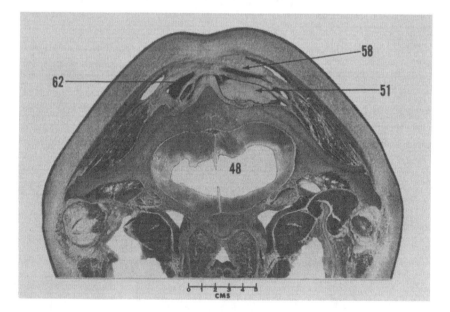

Figure 38. Posterior surface of section 12; 2 cm posterior to the
anterior surface of baseline (section 13). 48- cranial cavity,
51- nasal plug, 58- nasofrontal sac, 62-diagonal membrane.

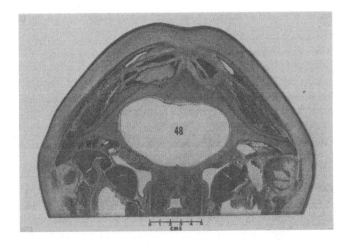

Figure 39. Anterior surface of section 11; 2 cm posterior to baseline (section 13). 48- cranial cavity.

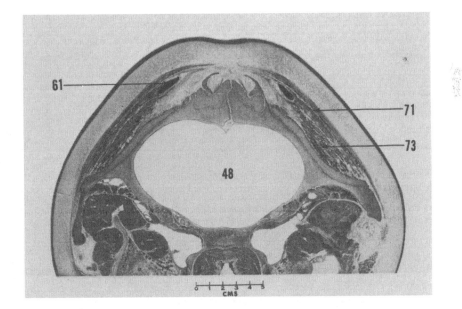

Figure 40. Anterior surface of section 10; 3 cm posterior to baseline (section 13). 48- cranial cavity, 61- vestibular sac, 71- posterior externus muscle, 73- posterior internus muscle.

placing electrodes in muscles suspected to be involved in sound production (see Ridgway, et al., this volume).

ACKNOWLEDGMENTS

We thank Dr. A. S. Gaunt for his many helpful suggestions and encouragement.

REFERENCES

Altes, R. A., 1974, Study of animal signals and neural processing with applications to advanced sonar systems, Rept. NUC TP 384, San Diego, Ca.

Benham, W. B., 1901, On the larynx of certain whales (Kogia, Balaenoptera and Ziphius, Proc. Zool. Soc., London, 286-300.

Cleland, P., 1884, On the viscera of the porpoise and white-beaked dolphin, J. Anat. Physiol., XVIII.

Coulombe, H. N., Ridgway, S. H., and Evans, W. E., Respiratory water exchange in two species of porpoise, Science, 149:86.

D'Arcy-Thompson, 1890, On the cetacean larynx, in: "Studies", Mus. Dundee.

Diercks, K. J., Trochta, R. T. Greenlaw, C. F., and Evans, W. E., 1971, Recording and analysis of dolphin echolocation signals, J. Acoust. Soc. Am., 49:1729.

Dormer, K. J., 1974, "The Mechanism of Sound Production and Measurement of Sound Processing in Delphinid Cetaceans", Ph. D. Diss., University of California, Los Angeles.

Dubois, E., 1886, Ueber den larynx (den cetacean), in: "Weber's Studien uber Saugetheire, 88-111.

Evans, W. E., and Prescott, J. H., 1962, Observations of the sound capabilities of the bottlenosed porpoise: a study of whistles and clicks, Zoologica, 47:121.

Evans, W. E., 1973, Echolocation by marine delphinids and one species of fresh-water dolphin, J. Acoust. Soc. Am., 54:191.

Evans, W. E., and Maderson, P. F. A., 1973, Mechanisms of sound production in delphinid cetaceans: a review and some anatomical considerations, Am. Zool., 13: 1205.

Evans, W. E., and Reiger, M. P., 1977, Production of sound in light air (HeO_2) atmosphere, in preparation.

Gaunt, A. S., and Wells, M. K., 1973, Models of syringeal mechanisms, Amer. Zool., 13:1227.

Gaunt, A. S., Gaunt S. L. L., and Hector, D. H., 1976, Mechanics of the syrinx in Gallus gallus. I. A comparison of pressure events in chickens to those in oscines, Condor, 78:208.

Gaunt, A. S., and Gaunt, S.L.L., 1977, Mechanics of the syrinx in Gallus gallus. II. Electromyographic studies of ad libitum vocalizations, J. Morphol., (in press).

Gurevich, V. S., 1973, A roentgenological study of the respiratory
    tract in Delphinus delphis, (Russ. Eng. Abstract), Zool. Zh.
    52:786.
Hersh, G., 1966, "Bird Voices and Resonant Tuning in Helium-Air
    Mixtures", Ph. D. Diss., University of California, Berkeley.
Hosakawa, H., 1950, On the cetacean larynx: with special remarks on
    the laryngeal sacs of the aryteno-epiglottal tube of the
    sperm whale, Whales Res. Inst., Science Report 3:23.
Howes, G. B., 1879, On some points in the anatomy of a porpoise,
    J. Anat. Physiol., 14:467.
Huber, E., 1934, Anatomical notes on Pinnipedia and Cetacea, Carn-
    egie Inst. Washington Publ. N°447:105.
Huxley, T. H., 1888, "A Manual of the Anatomy of Vertebrate Animals",
    D. Appleton and Co., 1,3,5 Bond St. New York.
Kelemen, G., 1963, Comparative anatomy and performance of the vocal
    organ in vertebrates, in: "Acoustic Behavior of Animals",
    R. G. Busnel, ed., Elsevier, Amsterdam.
Kinne, O., 1975, Orientation in space: animals-mammals, in: "Marine
    Ecology, Vol. 2", O. Kinne, ed., Wiley, London.
Lawrence, B., and Schevill, W. E., 1956, The functional anatomy of
    the delphinid nose, Mus. Comp. Zool. Bull., 114(4):103.
Mead, J. G., 1972, "On the Anatomy of the External Nasal Passages
    and Facial Complex in the Family Delphinidae of the Order
    Cetacea", Ph. D. Diss., University of Chicago.
Murie, J., 1871, On Risso's grampus, G. Rissoanus (Desm.), J. Anat.
    Physiol., 5:118.
Negus, V. E., 1931, "The Mechanism of the Larynx", C. V. Mosby Co.
Norris, K. S., 1964, Some problems of echolocation in cetacea ceta-
    ceans", in: "Marine Bioacoustics", W. N. Tavolga, ed., Perg-
    amon Press, New York.
Norris, K. S., 1969, The echolocation of marine mammals, in: "The
    Biology of Marine Mammals", H. T. Andersen, ed., Academic
    Press, New York.
Norris, K. S., Dormer, K. J., Pegg, J., and Liese, G. J., 1971,
    The mechanisms of sound production and air recycling in
    porpoises: a preliminary report," in:"Proc. 8th Annual
    Cong. Biol. Sonar and Diving Mammals".
Purves, P. E., 1967, Anatomical and experimental observations on the
    cetacean sonar system, in: "Animal Sonar Systems, Biology
    and Bionics", R. G. Busnel, ed., Imprimerie Louis Jean,
    GAP (Hautes Alpes), France.
Roberts, L. H., 1973, Cavity resonances in the production of orient-
    ation cries, Period. Biol., 75:27.
Roberts, L. H., 1975a, The functional anatomy of the rodent larynx
    in relation to audible and ultrasonic cry production, Zool.
    J. Linn. Soc., 56:255.
Roberts, L. H., 1975b, The rodent ultrasound production mechanism,
    in: "Ultrasonics".

Romanenko, Y. V., 1974, Physical fundamentals of bioacoustics, Fizi-
    cheskiye Osnovy Bioakustik, JPRS 63923, Moscow.
Schenkkan, E. J., 1971, The occurrence and position of the connecting
    sac in the nasal tract complex of small odontocetes, Mammalia
    (Cetacea), Beaufortia, 19:37.
Schenkkan, E. J., 1973, On the comparative anatomy and function of
    the nasal tract in odontocetes (Mammalia, Cetacea), Bijdr.
    Dierk., 43:127.
Turner, W., 1872, An account of the Great Finner Whale (Balaenoptera
    sibbaldii) stranded at Logneddery. Part I. The soft parts,
    Trans. Roy. Soc. Edinburgh, 26:197.
Watson, M., and Young, A. H. 1879, The anatomy of the Northern Beluga
    (Beluga catadon Gray Gray: Delphinapterus leucas Pallas)
    compared with that of other whales, Trans. Roy. Soc. Edin-
    burgh, 29:393.
Wood, F. G., 1964, "Marine Bioacoustics", W. N. Tavolga, ed., Perga-
    mon Press, New York.

ELECTROMYOGRAPHIC AND PRESSURE EVENTS IN THE NASOLARYNGEAL SYSTEM OF DOLPHINS DURING SOUND PRODUCTION

S. H. Ridgway, D. A. Carder, and R. F. Green

Naval Ocean Systems Center
San Diego, Ca. 92152 U.S.A.

A. S. Gaunt and S. L. L. Gaunt

Department of Zoology
Ohio State University
Columbus, Oh. 43210 U.S.A.

W. E. Evans

Hubbs-Sea World Research Institute
1720 South Shores Drive
San Diego, Ca. 92109 U.S.A.

INTRODUCTION

In electromyography fine wires are inserted into muscles to measure their activity periods. We have employed this technique to evaluate the activity of various muscles of the nasolaryngeal system of dolphins during the production of sounds. The anterior internus, the posterior internus, diagonal membrane muscle and the nasal plug muscle have been studied in greatest detail. These muscles fire just before and during the production of sounds whereas the hyoepiglottal muscle and the intercostal muscles do not. Simultaneous video recordings of blowhole movement and electromyogram (EMG) have been made to correlate such movement with EMG and sound production.

Catheters (5 to 8 Fr.) inserted into the nares, trachea and nasal sacs were used to measure pressure in various cavities during sound production. There were no pressure changes in the trachea.

This rules out the larynx as the primary sound source. Pressure
increased in the nares and premaxillary sacs prior to each sound and
dissipated after the sound. Catheters with inflated 15 ml balloons
were placed in the vestibular sacs, premaxillary sacs and nares without preventing sound production. Pressure increased in both nares
and both pre-maxillary sacs during sound production but not in the
vestibular sacs.

Sounds are produced in the nasal system by action of the muscular nasal plug. Pressure in the nares increases, but we do not
yet know exactly which membranes are vibrating or the direction of
air flow in the system.

Previous Studies

Dolphins emit numerous diverse sounds that are generally
described as whistles or pulses. A study of the mechanics of delphinid vocalizations is of interest from several points of view:
1) Delphinids appear to be remarkably effective in producing and
broadcasting sound at minimum energetic cost both at the surface and
to depths of at least a few hundred meters. 2) Some sounds (perhaps
most) are used in a very effective echolocation system which can be
better understood if the mechanism of sound generation is known.
3) If dolphins have an elaborate communications system based upon
sounds, a better understanding of how the sounds are made will be
useful in understanding their communications.

A number of anatomical studies have been done on the delphinid
nasolaryngeal system (see Green et al this volume) and various
theories have been proposed for how dolphins make sounds. Several
authors, including Lawrence and Schevill, (1956), Purves, (1967) and
Schenkken, (1973) have proposed that the sounds of cetaceans are
made in and by the larynx while Diercks et al, (1971), Gurevich, (1973),
Norris et al, (1971) and Dormer, (1979) have suggested sound production
in the nasal system. Evans and Prescott, (1962), have suggested a
combination system, whistles being produced by the larynx and pulses
by the nasal system - "anatomical and behavioral evidence, as well
as sound pressure measurements, indicate that echolocation clicks
are produced in the nasal sac system of porpoises. Within the sac
system the tubular sacs combined with the nasal plug nodes appear
to be the site of sound production." Lilly and Miller, (1961), were
the first to note that whistles and pulses could be produced simultaneously and Lilly (1963) reported that he "manually felt, over
the skull, the vestibular sacs participating in the vocal process as
these sacs filled and emptied during whistling. The mechanism of
production of this whistle as well as other whistles is still under
investigation. The number of the observed areas of production
(inside head with tightly closed blowhole, inside head with open
blowhole, and at blowhole slit) suggests that these animals can use

each of several parts of the nose for these whistles, rather than
just a single area" Lilly (1963).

Norris et al (1971), Hollien et al (1976) and Dormer (1979)
have used X-rays to observe tissue changes in living, phonating dolphins. All of these investigations noted movements in the nasal
system that correlated with sound production. There were no correlated movements in the larynx. Dormer (1979) says in summary that
"the mechanism of sound production in porpoises consists of a series
of muscular pumps, valves, and compliant sacs with the sound being
generated at the nasal plug."

Although the X-ray studies have given many helpful insights such
studies have not informed us as to how the sound is actually made.
Therefore, we set out to study the mechanism of sound production in
Tursiops and Delphinus using electromyography and pressure sensors
so that the events of muscular contraction, pressure changes and
sound production could be studied simultaneously.

MATERIALS AND METHODS

The experimental subjects were five Tursiops truncatus (three
males and two females).

For the experiments the dolphins were placed in a special
restraint device similar to the one figured elsewhere (Ridgway, this
volume). Water covered about two-thirds of the body. Only the
blowhole area and dorsum of the body was in air. An LC-10 hydrophone
in a suction cup was attached to the melon for the purpose of recording the animals sonic output. A similar hydrophone was dangled in
the water adjacent to the animal for a second channel of sound recording at 38 cm/sec on an FR-100 tape recorder. In addition, the
output of the hydrophones were integrated and displayed on a polygraph along with all the physiological measures. In some experiments sound was recorded on video tape simultaneous with the dolphin's blowhole movements and polygraph tracings of EMG and pressure
events.

Fine wires were placed into muscles (see Green et al, this
volume) through hypodermic needles according to the method of
Basmajian and Stecko (1962). The wires were connected by electrode
clips, amplified (Grass model 7P511B or 7P3B) and displayed by the
polygraph (Grass model 78D).

Catheters (5-8 Fr.) were inserted through the blowhole into the
nares, vestibular sacs or premaxillary sacs. Some catheters had
bulb tips that could be inflated with up to 15 ml of air or saline
solution. In two experiments a catheter was inserted through a
hypodermic needle into the trachea. The catheters were connected

to calibrated pressure transducers (Statham PM6TC + 5-350, PM6TC + 10-350, and/or PM131TC + 5-350) and the analog pressure was displayed on the polygraph.

RESULTS

EMG's from the following muscles have been studied during sound production: Posterior Internus (PI), Anterior Internus (AI), Nasal Plug Muscle (NPM), Hyoepiglottal Muscle (HM), Intercostal Muscles (IC), Palatopharyngeal Muscle (PPM) and Diagional Membrane (DM). PI, AI, NPM and DM were consistently associated with sound production. PPM was often active during sounds, but its firing was not as consistent as the former muscles during sound production. IC and HM were active during only one type of sound -- the "Bronx cheer" or "raspberry". This sound is made by the expulsion of large amounts of air through the fluttering lips of the blowhole.

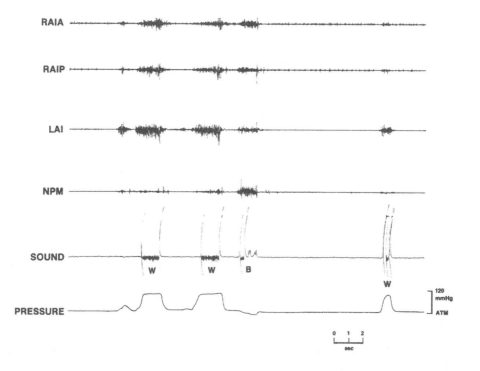

Figure 1. EMG, sound and pressure events in the nasal system of Tursiops -- Right anteriorinternus muscle, anterior part (RAIA), Right anteriorinternus muscle, posterior part (RAIP), Left anterior internus muscle (LAI), Nasal plug muscle NPM, Whistles (W) and Blows (B).

Muscles on both sides of the head were generally fired simultaneously during sounds and different parts of the same muscle had a similar EMG pattern (Figure 1). Occasionally less activity was seen in the right AI or PI than in the left and vice versa, but almost always there was some activity from both muscles of these two groups. In about ten percent of those sounds during which the DM was observed, it fired mainly on one side (Figures 2 and 3), but in most instances both sides were active (Figure 4).

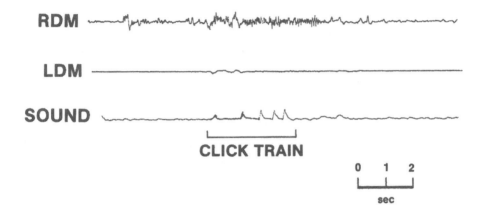

Figure 2. EMG of the right and left diagonal membrane muscles during a short click train.

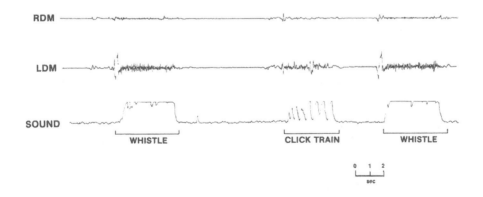

Figure 3. EMG of the right and left diagonal membrane muscles.

Sometimes the nasal plug (as viewed from above) moved toward the
right and sometimes toward the left. Although sound may be produced on one side or the other it seemed to require bilateral
muscle action in most cases.

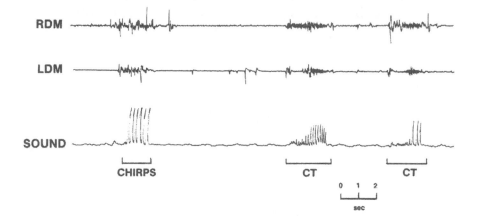

Figure 4. EMG of the right and left diagonal membrane muscles
during pulsed sounds described as chirps and click trains (CT).

Catheters were placed in both nares, in vestibular sacs on both
sides, in premaxillary sacs on both sides and in the trachea.
Dolphins produced both whistles and pulses in varied form and
intensity with catheters in place. Pressure in both nares increased
immediately following muscular contractions (AI, NPM, PI). The
sound occurs as pressure reaches its peak. Pressure subsides just
after the end of the sound as muscle firing ceases (Figure 5).

If we opened these intranarial catheters to atmospheric air,
sound production was greatly altered. If the catheter was small in
diameter, on the order of a millimeter or less, some short whistles
and short, very weak pulse trains were produced as air rushed from
the catheter. If the catheter was large in diameter, 3 or 4 mm,
the animal could not make sounds.

We were not able to maintain catheters in the premaxillary sac
for periods of more than a few minutes at a time -- the animal kept
ejecting them. The dolphin was able to produce whistles and pulses
with catheters in the premaxillary sacs. Pressure in these sacs
was identical with that in the nares. Inflated balloons (15 ml) at
the end of catheters in the premaxillary sacs or in the nares did
not prevent sound production.

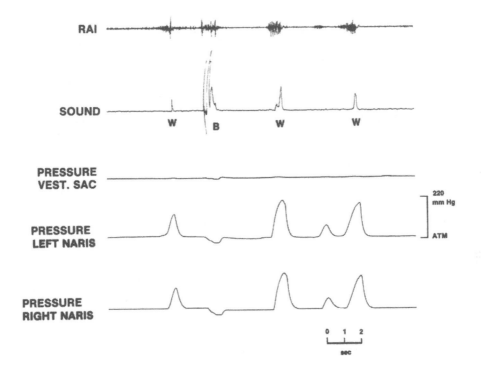

Figure 5. EMG and pressure events during three whistles (W) and a blow (B).

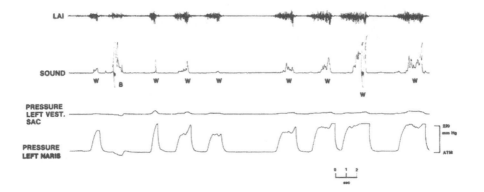

Figure 6. EMG and pressure events during several whistles (W) and a blow (B).

Pressure increases in the vestibular sacs often followed those in the nares, but were much smaller (Figure 6). Inflated balloon catheters placed in the vestibular sacs did not prevent sound production nor did the placement of open catheters, which allowed any pressure to escape, prevent sound production.

A positive pressure of about 25 mm Hg was maintained in the trachea between breaths. With each breath there was a negative pressure followed by a return to the positive pressure state. There were no changes in the intratracheal pressure during the production of whistles (Figure 7) or pulsed sounds.

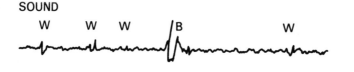

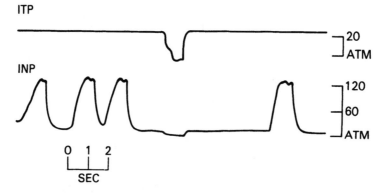

Figure 7. Pressure events in the trachea (ITP) and Nares (INP) during four whistles and a blow.

We have not yet been successful in placing a catheter in the very small nasofrontal sacs nor in placing electrodes in the muscles surrounding it.

DISCUSSION

Our studies clearly show that Tursiops produce sound in the nasal system and not in the larynx. Our most important evidence in support of this can be seen in Figure 7. Pressure increased in the nares before each whistle, but there was no change in intratracheal pressure except during the course of blows. In humans there is increased intratracheal pressure during phonation (Basmajian, 1978). Further support for this conclusion came from the EMG data. If vocalizations arose in the larynx, one would expect to see patterns like those in man. During human phonations the IC muscles are active. A preliminary burst of activity before each phrase is followed after an interval by increasing activity in the utterance of the phrase (Basmajian, 1978). In the dolphin IC was inactive during all of the common sounds. Only in the rarely produced "Bronx cheer" was this muscle seen to fire in concert with sound. Bronx cheers are apparently produced only when the dolphin's blowhole is out of the water. In addition, the EMG from the HM of the dolphin showed activity during the blow but not during sounds. In summary, the EMG and pressure events that are related to phonation in the human larynx were not observed during dolphin sounds, but EMG and pressure events in the nasal system are highly correlated with sound production.

The vestibular sacs are not essential for sound production since they can be isolated from the system without preventing sounds. Some records show small pressure increases in these sacs during sounds. We suspect that the vestibular sacs may be important as air reservoirs. Air may be blown into the sacs during sounds then recirculated, thus saving air for phonation during long periods of submergence. This function would be important at great depths where air carried by the animal is compressed into a relatively small volume.

Since dolphins must feed underwater, often at great depth, they are continually subjected to changes in pressure. Air in the thorax is compressed as the dolphin dives (Ridgway et al, 1969). If sound were produced in the larynx the resonant characteristics of the animal sounds would also be continuously changing. Since the volume of the nasal system is small, air can be leaked from the lungs to maintain volume as the lung is compressed. Because the air is equilibrated at any depth muscles will be able to operate the sacs and valves of the nasal system efficiently at depth and the acoustic resonance need not change with depth.

Our data on pressure events demonstrates that the larynx must be closed during sound production since large pressure changes occur in the nares anterior to it, but none in the trachea just posterior to and open to the larynx. Since sound is altered or prevented by an open tube in the nares, the nares must be in direct communication with structures that produced sound during the process. The open tube in the nares simply prevents the nares from acting as an efficient source of air pressure for the sound generating apparatus. The presence of an inflated balloon in the nares does not alter sound production since it does not prevent the space from pressurizing. If sounds were being produced within the air column of the nares the balloon would, however, be expected to alter the sound.

Based on data collected using an array of transducers attached to various locations on the gular, melon and rostral regions of a trained Tursiops, Diercks et al (1971) identified the site of the origin of sounds as an area 1.5 to 2.0cm beneath the blowhole margin. This appears to correspond to the region where the internal lip of the nasal plug abuts the diagonal membrane. Just lateral to this the nasal plug nodes insert into the opening of the naso-frontal sacs. Glandular structures in this area (Evans and Mader-son, 1973) might provide lubrication to tissue that might be subjected to a high velocity airstream associated with sound production. We do not agree with Evans and Maderson (1973) that the nasofrontal sacs can summarily be written off as an important component of the sound production mechanism. Their statement referring to the great individual variability observed in the size and form of these sacs that "the intra-specific variation which we have commented upon has not yet been quantified, it seems most unlikely that any significant variation would be permitted by natural selection if the nasofrontal sacs played an active major role in such a sophisticated function as delphinid phonation," is not equivocal. Just such variation could provide individual acous-tic differences which could be very adaptive for a highly social species.

It appears to us that during phonation air is compressed in the nares by depression of the nasal plug at the dorsal end and per-haps by action of the palatopharyngeal sphincter at the ventral end of the nares. Air must pass from the nares, past the diagonal mem-brane at their upper end, through the spiracular cavity along the posterior margin of the nasal plugs and perhaps past the nodes of the nasal plug to the nasofrontal sacs. Although we have not yet been able to measure pressure in the small nasofrontal sacs nor to record EMG's of the intrinsic muscles that surround them, we suspect that these sacs may be important in dolphin phonations. The sacs could serve as resonators. Mucus in the sacs and muscles around them could serve a damping function and help provide fine control over frequency.

The gross and microanatomical nature of the nasal plug nodes, diagonal membrane, and nasofrontal sacs, coupled with the acoustic and now electromyographic and pressure measurements reported in this paper strongly suggest that this total system is an excellent candidate as the source and mechanism of sound production in most delphinids.

REFERENCES

Basmajian, J. V. and Stecko, G. A., 1962, A new bipolar indwelling electrode for electromyography, J. Appl. Physiol., 17:849.

Basmajian, J. V., 1978, "Muscles Alive", The Williams and Wilkins Company, Baltimore.

Diercks, K. J., Trochta, R. T., Greenlaw, C. F., and Evans, W. E., 1971, Recording and analysis of dolphin echolocation signals, J. Acoust. Soc. Am., 49:1732.

Dormer, K. J., 1979, Mechanism of sound production and air recycling in delphinids: cineradiographic evidence, J. Acoust. Soc. Am., 65:229.

Evans, W. E., and Prescott, J. H., 1962, Observations of the sound capabilities of the bottlenosed porpoise: a study of whistles and clicks, Zoologica, 47:121.

Evans, W. E., and Maderson, P. F. A., 1973, Mechanisms of sound production in delphinid cetaceans: a review and some anatomical considerations, Am. Zool., 13:1205.

Green, R. F., Ridgway, S. H., and Evans, W. E., Functional and descriptive anatomy of the bottlenosed dolphin nasolaryngeal system with special reference to the musculature associated with sound production, this volume.

Gurevich, V. S., 1973, A roentgenological study of the respiratory tract in Delphinus delphis, (Russ. Eng. abstract), Zool. Zh., 52:786.

Hollien, H., Hollien, P., Caldwell, D. K., and Caldwell, M. C., 1976, Sound production by the Atlantic bottlenosed dolphin, Tursiops truncatus, Cetology, 26:1.

Lawrence, B., and Schevill, W. E., 1956, The functional anatomy of the delphinid nose, Mus. Comp. Zool. Bull., 114(4):103.

Lilly, J. C., and Miller, A. M., 1961, Sounds emitted by the bottlenosed dolphin, Science, 133:1689.

Lilly, J. C., 1963, Distress call of the bottlenosed dolphin: stimuli and evoked behavioral responses, Science, 139:116.

Norris, K. S., Dormer, K. J., Pegg, J., and Liese, G. J., 1971, The mechanisms of sound production and air recycling in porpoises: a preliminary report, in: "Proc. 8th Annual Cong. Biol. Sonar and Diving Mammals".

Purves, P. E., 1967, Anatomical and experimental observations on the cetacean sonar system, in: "Animal Sonar Systems, Biology and Bionics", R; G. Busnel, ed., Imprimerie Louis Jean, GAP (Hautes Alpes), France.

Ridgway, S. H., 1979, Electrophysiological experiments on hearing in Odontocetes, this volume.

Ridgway, S. H., Scronce, B. L., and Kanwisher, J., 1969, Respiration and deep diving in the bottlenosed porpoise, Science, 166:1651.

Schenkkan, E. J., 1973, On the comparative anatomy and function of the nasal tract in Odontocetes (Mammalia, Cetacea), Bijdr. Dierk., 43:127.

ECHOLOCATION SIGNALS OF THE ATLANTIC BOTTLENOSE DOLPHIN (<u>TURSIOPS</u>

<u>TRUNCATUS</u>) IN OPEN WATERS

Whitlow W.L. Au

Naval Ocean Systems Center

Kailua, Hawaii 96734

A wide variety of echolocation experiments performed with
the Atlantic bottlenose dolphin (<u>Tursiops truncatus</u>) have indicated that these dolphins possess a highly sophisticated and
adaptive sonar system. Results of discrimination experiments
have shown that <u>Tursiops</u> can detect a 10% difference in the
diameter of metallic spheres (Norris, Evans, Turner, 1967),
material composition and thickness differences as small as 0.1 cm
in metallic discs (Evans and Powell, 1967), differences between
plates shaped as circles, triangles and squares independent of
their cross sectional areas, as well as a 6% change in the
diameter of the circles (Barta and Evans, 1970; Fish, Johnson,
and Ljungblad, 1976), and a 0.8 dB difference in the target
strength of corprene cylinders (Evans, 1973). Murchison and
Penner (1975) have demonstrated detection ranges of 72.2 m for a
2.54-cm diameter solid steel sphere and 76.8 m for a 7.6-cm diameter water-filled sphere. Range resolution capabilities of 0.9,
1.5 and 2.8 cm for target ranges of 1,3, and 7 m have been
reported by Murchison (1976). Nachtigall, Murchison and Au (1978)
reported on the ease with which a <u>Tursiops</u> could discriminate
between foam cubes and cylinders of different sizes. Hammer and
Au (1978) found that an animal could recognize differences in the
wall thickness of aluminum cylinders that were as small as 0.16
cm. They also found that the animal could detect differences
between free-flooded aluminum, bronze and steel cylinders of
equal dimensions.

The ability of <u>Tursiops</u> to perform these difficult target
detection, recognition and discrimination tasks should be dependent on the kinds of echolocation signals that are being
emitted, on the way the signals are beamed out and echoes

received by the animal, and on the signal processing schemes
being utilized. A considerable amount of information can be
obtained by an animal scanning across a target and examining the
target from different spatial aspects. Target information can
also be contained entirely within the echoes, independent of any
spatial clues. In this review, the echolocation signals used by
a number of T. truncatus in the open waters of Kaneohe Bay, Oahu,
Hawaii will be discussed. Since the natural habitat of T.
truncatus consists primarily of bays, harbors, inlets and other
littoral areas, having the capability of performing echolocation
experiments in the open waters of a bay is important from the
standpoint of working in an environment very similar to the
natural environment of these animals.

   Measurements of the echolocation signals emitted by two
Tursiops performing a target-detection experiment in Kaneohe Bay
led Au et al. (1974), to discover certain previously unknown
properties of the animals' sonar signals. They found that the
signals had peak energies at frequencies (peak frequencies)
between 120 and 130 kHz, which was over an octave higher than
previously reported peak frequencies between 30 and 60 kHz
(Evans, 1973). They also measured an average peak-to-peak click
source level of 220 dB re 1μPa at 1 yard, which represented a
level on the order of 30 dB higher than previously measured for
Tursiops. The use of high-frequency echolocation clicks was
attributed to the characteristics of the high ambient noise
environment of Kaneohe Bay. The effects of the ambient noise
on the animals will be discussed in Section II-B entitled
Spectrum Adaptation. The high source levels were attributed to
the absence of boundaries which could cause high reverberation
and to the long distances between the animals and the target.

   An example of a typical echolocation click train for an
animal performing a target detection task is shown in Fig. 1,
with the frequency spectrum plotted as a function of time on the
left and the individual click waveform displayed on the rights.
Five parameters are written beside each click waveform. The top
notations are the peak-to-peak source level in dB re 1μPa at 1
meter, followed by the peak frequency and the 3-dB bandwidth
which are separated by a slash (/). The parameters listed below
the signals are the cross-correlation coefficient between the
present signal and the previous one, followed by the time of
occurrence of the signal relative to the start of the click
train. The amount of information present in Fig. 1 is considerable and can be unwieldy especially if many tens of trials and
sessions are considered. In order to simplify the examination of
these signals, a process of averaging both the time domain and
frequency domain waveforms will be used. The averaging is
performed in software and is similar to the kinds of averaging

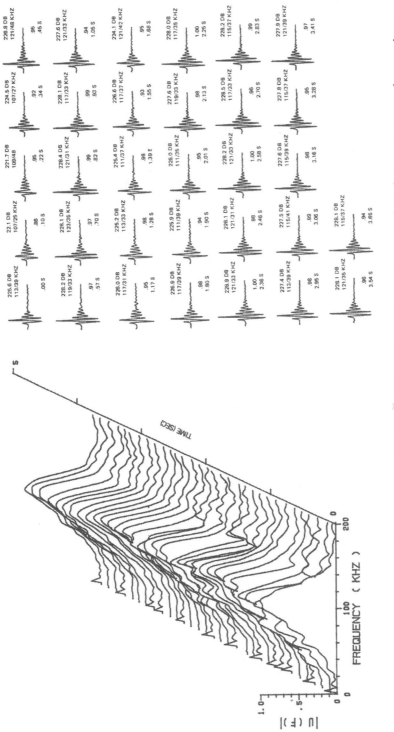

Figure 1. An echolocation click train for a Tursiops truncatus performing a target detection task in Kaneohe Bay.

performed with real-time spectrum analyzers, and other digital
equipments. The average waveform and spectrum of the click
train displayed in Fig. 1 is shown in Fig. 2. Note the similarity between the average waveform and the individual signals of
Fig. 1. The higher level signals in most click trains are highly
correlated, whereas fluctuations in the waveshape are generally
associated with the lower level signals. Most of the waveforms
presented in this review will be in terms of their average waveforms. Some of the results that will be discussed have been
reported by Au et al. (1974); and Au, Floyd and Haun (1978),
although the bulk of the results have not yet been published.

I. ACOUSTIC FIELD OF AN ECHOLOCATING DOLPHIN

In order to facilitate the discussion of the acoustic field
of a dolphin, it is helpful to establish a frame of reference
with respect to the site at which the echolocation clicks are
produced. Although the exact location of click production and
the mechanisms involved are not yet known, indirect evidence
from anatomical investigations (Evans and Maderson, 1973),

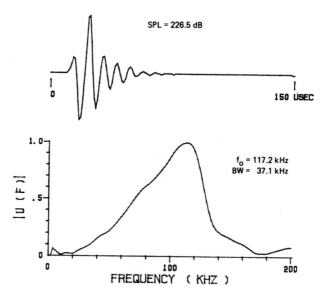

Figure 2.   The averaged waveform and spectrum for the click
train of Fig. 1.

acoustic measurements (Diercks, et al., 1971), ambient pressure measurements within the nares (Ridgway et al., 1974), and x-ray measurements (Norris et al., 1971; Hollien et al., 1976) all suggest that click signals are produced in the vicinity of the external nasal passage. Diercks et al., (1971), using multiple contact hydrophone data, places the source in the vicinity of the nasal plugs in the nares. Therefore, the region of the nasal passage approximately 3 cm below the blowhole will be assumed to be the site of click production in Tursiops.

A. Contact Hydrophone Measurements

The acoustic field in the immediate vicinity of a dolphin's head should be extremely complex because of the presence of air sacs, cranial bones and tissue having differing sound velocities, that could cause reflection and refraction of the acoustic pulses generated within the animal's head. The signal that is emitted into the water is formed by the superposition of the direct component of the generated pulse and the pulses that are modified by the various structures in the animal's head. Romenenko (1975) gave an example of an emitted pulse formed by the superposition of the direct pulse and a pulse reflected off the premaxillary sac. Dubrovskiy and Zaslavskiy (1975), using an optical technique with 5 mm diameter light bulbs inserted in the region of the plugs that shut the aditus to the tubular sacs, found the intermaxillary bone to be a principal contributor to the formation of the emission field. Bel'kovich and Dubrovskiy (1977), give an exellent summary of the Russian studies on the role of air sacs and cranial bones on the formation of the sound field of dolphins.

Simultaneous measurements of the signals in the near and far fields of a dolphin were made by Au et al., (1978) using a contact hydrophone placed either on the animal's melon or rostrum and a hydrophone located one meter from the animal's rostrum. Examples of the average signals of two echolocation click trains measured with the contact hydrophone placed on the melon and two click trains with the contact hydrophone placed on the animal's rostrum are shown in Fig. 3. Also included in Fig. 3 are the average signals measured by a hydrophone located one meter in front of the animal's rostrum. A comparison between the average signal measured at 1 m with the signal measured at the melon seems to indicate that there was a considerable amount of interference to the direct pulse by internally reflected and/or refracted pulses. Each cusp in the time-domain signal measured at the melon may indicate the arrival of a reflected or refracted component of the generated pulse. The peak frequencies for the signals measured at the melon were 8 and 6 kHz higher respectively, than the signals at 1 m. The 3-dB bandwidths were also smaller for the

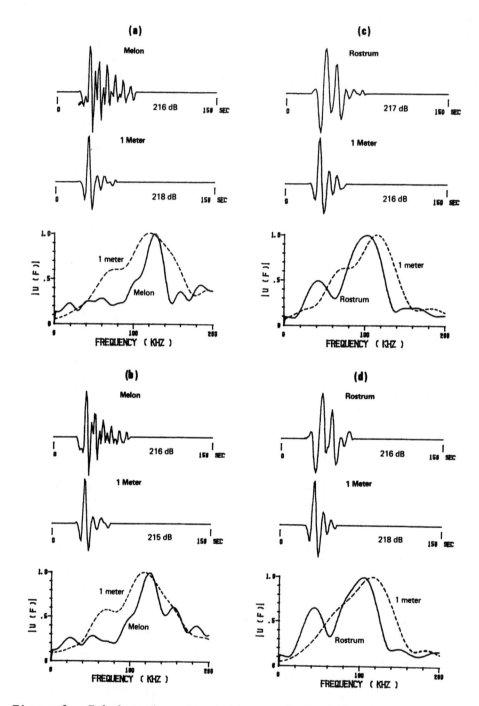

Figure 3.   Echolocation signals measured simultaneously at the
             melon or rostrum and at one meter.

signals at the melon. The higher peak frequencies and lower
bandwidths were probably due to the many cusps associated with
the signals measured at the melon.

    The signals measured at the rostrum were not as complex as
those at the melon. The waveforms for the rostrum signals seem
to suggest that they were formed by the superposition of a direct
and a single internally reflected or refracted pulse. The peak
frequency of each rostrum signal was 11.7 kHz lower than the
signal at 1 m. This is not surprising since the signals that
were detected by the hydrophone on the rostrum traveled away from
the major axis of the projected beam.

    Although the signals measured at 1 m in all cases were
similar in shape both in the time and frequency domain, the
corresponding signals measured at both the rostrum and melon were
considerably different. The melon signals seemed to be formed
by a large number of internally reflected and refracted pulses
that arrived very close together and soon after the arrival of
the direct pulse. This type of propagation is not surprising
since the melon region is close to the air sacs and areas of the
skull which could reflect acoustic energy into melon region.

    The amplitude of the signals measured by contact hydrophones
placed on the melon and rostrum of an animal in relationship to
the signals measured in the far field was also studied by Au,
Floyd and Haun (1978). An animal was trained to station in a chin
cup that was supported by surgical rubber tubing. A vertical
array of three hydrophones was placed 1 m from the animal's
rostrum and a hydrophone was placed at 85 m which was 5 m beyond
the target. Two separate cases were considered, the first with
the contact hydrophone placed on the rostrum and the second
with the hydrophone on the melon. Since, the mean value of the
sound pressure level (SPL) measured at 1 m were within 0.9 dB in
both cases, the results of the two conditions were combined. The
results in terms of the peak-to-peak SPL are shown in Fig. 4.
with the SPL at 1 m obtained from the hydrophone in the vertical
array which detected the largest signal.

    The dash curve represents the SPL at 1 m referenced to any
range r by considering the spherical spreading and absorption
losses. The absorption coefficient used was 0.44 dB/m which
corresponds to a typical Kaneohe Bay water temperature of 24°C
and salinity of 35 ppt and a signal frequency of 120 kHz. There
is good agreement between the SPLs at 1 and 85 m after accounting
for the transmission loss. The mean SPLs at 1 m and at the
rostrum and the melon were 217.7, 217.2, and 215.8 dB re 1 μPa,
respectively. Evans (1973), using an array of contact hydrophones,
found that the amplitude of a signal measured at the melon was
approximately equal to that measured at the rostrum. The 1.4 dB

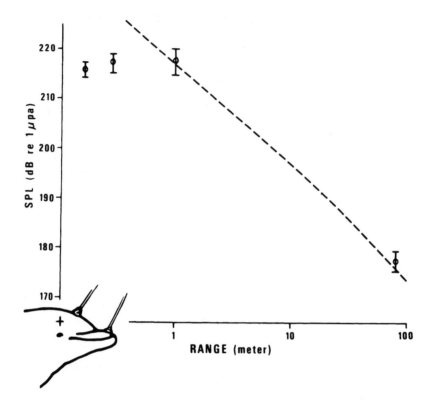

Figure 4.   Peak-to-Peak Sound Pressure level of echolocation
            signals as a function of range

difference between the melon and rostrum in fig. 4 is in general
agreement with the results of Evans (1973), considering the
variability in the data.   The reason for the amplitude of the
outgoing echolocation signal at the surface of the animal's head
being approximately the same as at a distance of 1 m may be
explained by the fact that the contact hydrophone was placed in
the near field of the animal's sound projection system.   Within
the near field, the SPL should vary considerably with distance
along the major axis of the beam, forming a number of peaks and
nulls.

B.   Near-to-Far Field Transition

A series of measurements was made by Au, Floyd, and Haun
(1978), to determine the distance from the tip of an animal's

rostrum where the amplitude of the acoustic pressure of the echolocation signals began to satisfy the far-field condition by decreasing as a function of 1/r. An array of hydrophones was placed at different distances from the tip of a rigid chin-cup station aligned in the direction of the target to measure the variation in the SPL as a function of distance away from the animal. Initially all the hydrophones were submerged to the same depth, with the acoustic center of each hydrophone at a depth corresponding to the top of the animal's rostrum. It was expected that this arrangement would measure the signal close to the major axis of the vertical beam. However, the vertical beam-pattern measurement to be discussed in Sec. IC showed that the major axis was directed at an elevation angle of 20° above the rostrum. Therefore a second series of measurements was made with the hydrophones aligned along the major axis of the sound beam.

The results of both series of measurements are shown in Fig. 5. Both sets of results agree fairly closely, with the measured SPL difference departing from the 1/r function between 0.375 and 0.500 m in the first case and 0.500 and 0.625 m in the second. Since the second series of measurements was performed with the hydrophones aligned along the major axis of the vertical beam, it should more accurately represent the propagation of the

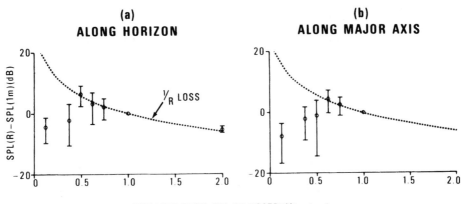

DISTANCE FROM TIP OF ROSTRUM(meters)

Figure 5. Relative sound pressure levels of transmitted signals across the transitional region between the acoustic near and far fields. The dash curve in each figure represents the spherical spreading loss with the reference point at 1 m.

signals close to the animal. Therefore, the region in which the
acoustic near field makes the transition to the far field exists
between 0.500 and 0.625 m from the tip of the animal's rostrum
for the high-frequency signals emitted by the animals in Kaneohe
Bay.· The location of the transition region for animals emitting
signals at peak frequencies different from the 110 to 130 kHz
typically used in Kaneohe Bay (see Section II-B) should vary
linearly with frequency.

C. Directional Pattern of Emitted Echolocation Signals

The echolocation signals of dolphins and other odontocetes
have been shown by a number of investigators to be directional.
Norris et al, (1961), observed that a blindfolded Tursiops could
not detect targets below its jaws and at elevation angles
greater than 90° above the rostrum. Evans, Sutherland, and Beil
(1964), used a Stenella longirostris cadaver and a Tursiops skull
in order to study the directional properties of echolocation
signals. With a cw sound source placed in the region of the
nasal sacs, they found that a definite beam was formed, and it
was directed 15° above the rostrum in the vertical plane and
forward in the horizontal plane. The beam was highly dependent
on frequency becoming narrower as the frequency increased.
Norris and Evans (1966), measured the directionality of Steno
bredanensis echolocation clicks by systematically moving a
single hydrophone placed at a fixed depth to different azimuths
as the animal echolocated on a target. The sound was found to be
directed forward in a narrow beam and was highly frequency
dependent. Schevill and Watkins (1966), studied the directionality of the clicks produced by an Orcinus orca and found that
the frequency content and amplitude of the clicks varied with
the orientation of the animal. As the whale turned away from
the hydrophone, the higher frequency components of the clicks
would progressively diminish in amplitude. Evans, (1973), used
an array of hydrophones spaced 5° apart in the horizontal plane
to measure the horizontal beam pattern of a restrained Tursiops.
Although several elevation angles were used, there is a distinct
possibility that none of Evan's, (1973), measurements were made
along the major axis of the beam in the vertical plane.

The vertical and horizontal beam patterns of a Tursiops
named Sven were measured by Au, Floyd and Haun (1978), with the
animal trained to station in a chin cup. The composite broadband
vertical beam pattern from nine trials is shown in Fig. 6a, and
the composite horizontal beam pattern from twelve trials shown
in Fig. 6b. These two plots show that the echolocations signals
were projected 2 to 3° to the left of center and at an elevation
angle of 20°. The 3-dB beamwidths were approximately 10° in the
vertical plane and 9.8° in the horizontal plane. The horizontal

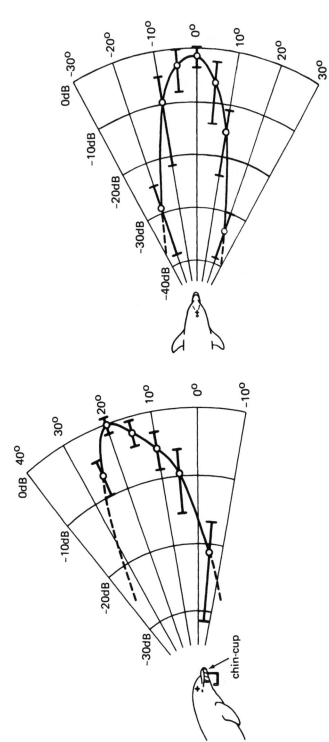

Figure 6.   Composite broadband beam patterns for the echolocation signals projected by a *Tursiops truncatus* (a) in the vertical plane and (b) in the horizontal plane.

beam was measured along the axis of the vertical beam. The 20°
elevation of the vertical beam is in general agreement with the
15° angle measured by Evans, Sutherland, and Beil (1964), using a
Tursiops skull. The elevation of the beam axis would allow the
animal to direct its beam forward, since in a free swimming
posture, the rostrum of the animal frequently appears to be
pointed downwards at a 15° to 20° angle.

The responses of the LC-10 hydrophones used by Au, Floyd and
Haun (1978), to transient-like signals such as porpoise clicks
were not matched. The hydrophones indicated differences as much
as 10% in the peak frequency of the simulated porpoise clicks used
to calibrate the arrays. Therefore, a follow-on effort was made
using a BM-101 and four B&K-8103 hydrophones in order to obtain
better data on the waveform and frequency spectrum of the
projected signals as a function of angle in both the vertical
and horizontal planes. In response to simulated porpoise clicks,
each of the hydrophones indicated approximately the same peak
frequency, bandwidth and wave shape.

The follow-on series of array measurements was conducted
with an animal named Ekahi. Instead of a chin cup, a 40-cm rubber
hoop was used as the stationing device, which allowed the animal
to tilt its head downwards while echolocating. The target which
was located 6 meters from the hoop was submerged to the same
depth (76.2 cm) as the center of the hoop. The vertical array
was placed 1.5 m and the horizontal array 2.5 m from the hoop.
Ekahi was trained to enter the hoop and remain silent until an
audio tone was turned on cueing the animal to begin echolocating.
This technique of conditioned echolocation is discussed by
Schusterman, Kersting and Au (1979).

The composite broadband beam patterns for ten trials in each
plane are displayed in Fig. 7. The two sets of composite beam
patterns of Figs. 6 and 7 are similar, with Ekahi's beam being
slightly broader than Sven's. The 3-dB beamwidths were approximately 11.7° in the vertical plane and 10.7° in the horizontal
plane. Some of the differences between the two experiments may
have been due to different angular separations of the hydrophones
near the beam axis.

Examples of the average waveform and the corresponding
frequency spectrum as detected by the five hydrophones in the
vertical array of the second series of measurements are displayed
in Fig. 8. As the angle departed from the beam axis, the signals
in the time domain became progressively distorted, with respect
to the signal at 0°. In the frequency domain, the peak frequencies
decreased as the hydrophone angle departed from 0°. The presence
of multipaths is apparent for the signals measured above the

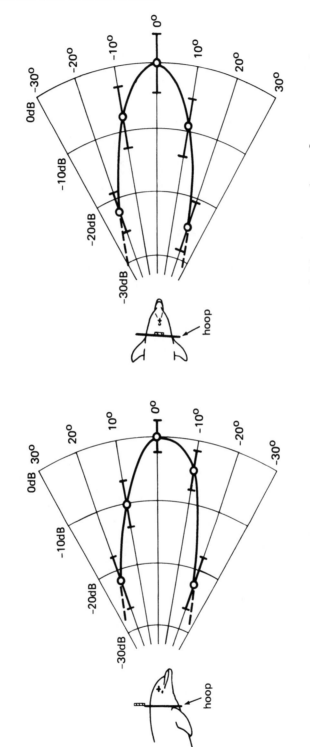

Figure 7. Composite broadband beam patterns obtained in the follow-on series of measurements using closely matched sets of hydrophones, (a) in the vertical plane and (b) in the horizontal plane.

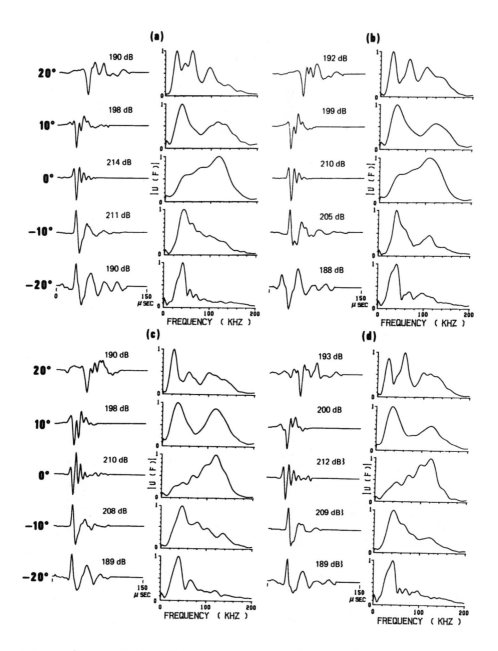

Figure 8.   Examples of the average waveforms and frequency spectra
            as a function of the elevation angle in the vertical
            plane.

beam axis at +10° and +20°. The multipaths may be due to internal reflections and refraction of the signals within the animal's
head, to the signals being radiated from different portions of a
finite source region, or to combinations of these possibilities.
For the signals measured above the major axis there are still
considerable amounts of energy at frequencies close to the peak
frequencies of the 0° signals, causing the spectra to have
multiple humps or peaks. This is not true for the signals
measured below the major axis. There seems to be a general shift
to lower peak frequencies with relatively little energy at the
frequencies close to the peak frequencies of the signals
measured along the major axis. The signals in the time domain
appear to be stretched with little multi-path effects.

The average waveforms and their corresponding frequency
spectra as detected by the five hydrophones in horizontal array
are displayed in Fig. 9 for three trials. As the angle departed
from the beam axis, the signals became progressively more distorted with respect to the signals measured at 0°. The signals
measured at -10° were the least distorted, having peak frequencies
close to those along the beam axis. As the angle departed from
the beam 0°, the peak frequencies shifted to lower values, with
the signals measured at -20° experiencing the greatest amount of
shift. The signals were not symmetrical about the beam axis
which is expected since the structure of the skull is not
symmetrical about the mid-line of the animal.

Besides the nasal sacs and the cranial bones, the melon may
also play a large role in the spatial directivity of the emitted
beam. In 1957, F. Wood (1964) suggested that the melon of a
porpoise might be a sound transducer helping to couple internally
generated sounds to sea water. He also speculated that the melon
may act as an acoustic lens to focus the sound into a narrow
beam. The beam patterns measured by Evans, Sutherland, and Beil
(1964), using a Tursiops skull were considerably broader than
the beam patterns of Figs. 6 and 7. This suggests that the melon
may play a role in focusing the outgoing acoustic energy, causing
a narrower beam to be radiated than would otherwise exist.
Romenenko (1975), studied the effects of a dolphin's skull, with
and without soft tissue present, on the horizontal beam pattern.
He found that for frequencies of 80 kHz and higher, the beams
measured with the whole head were approximately half as wide as
the beams obtained with the skull alone. Norris and Harvey (1974)
examined the sound velocity of the lipids in the melon and found
a definite structure that could cause the focusing of the outgoing
acoustic energy. By slicing the melon of a dead animal in a regular fashion and measuring the sound velocity of various sections
within each slice, they found a low-velocity core extending from
just below the anterior surface towards the right nasal plug and

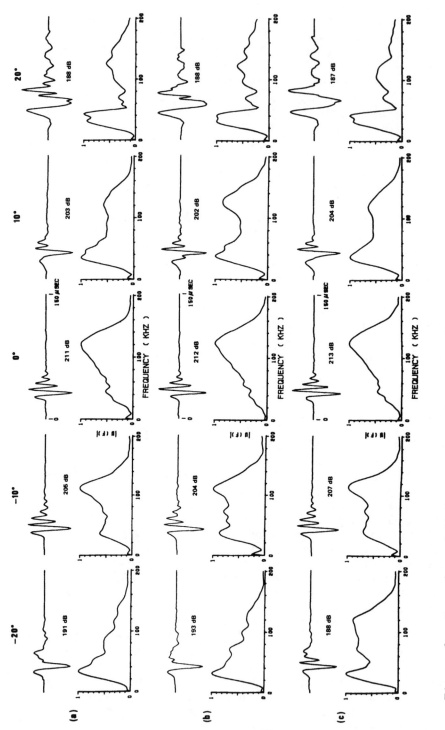

Figure 9.   Examples of the average waveforms and frequency spectra as a function of angle in the horizontal plane.

a graded outer shell of higher velocity tissue. Such a velocity gradient would cause the signal to be focused both in the vertical and horizontal planes. The amount of refraction and the subsequent effect on the overall beam patterns have not been determined yet.

II. CHARACTERISTICS OF ECHOLOCATION SIGNALS

A. Click Intervals

The click intervals associated with two animals echolocating on a target located at ranges between 54.9 and 73.2 m was examined by Au et al. (1974). Two examples showing the variations in the click interval on a pulse-to-pulse basis, one with the target present and the other with the target absent, are shown in Fig. 10. One interesting feature shown in Fig. 10 is the variation in the interval from pulse to pulse. The variation seems to be fairly regular, with the click interval increasing, then decreasing, and once again increasing, forming a cyclic pattern. The variations depicted in Fig. 10 were typical of all the click trains examined. The click intervals when the target was present were all greater than the two-way transit time for an acoustic signal to travel to the target and back, indicating that the animals emitted each successive pulse after the arrival of the echo signal from the preceding pulse. This fact was also observed by Morozov et al. (1972), in their study. It is interesting to note that the click intervals for the target-absent case were also greater than the two-way transit time had a target been present. This implies that the animals had a general expectation of hearing the target echo at a specific time and may have been performing a "mental time-gating" of the echoes.

Accurate measurements of the time difference between the click intervals and the two-way transit time can only be made if the animal's position with respect to the target is well known. Several experiments have been conducted at the Naval Ocean Systems Center (NOSC) with an animal either stationing in a hoop or in a chin cup so that the distance between animal and target was fixed and could be accurately measured to within several cm. The average click interval per trial for a target distance of 6 m and 1 m is displayed in Fig. 11. In the 1 m case, the animal wore a blindfold. It is clear from the figure that the click intervals were longer than the two-way transit times. The average lag time between the reception of the echo and the transmission of the next pulse was 21.6 ± 10.6 msec at 6 m and 18.6 ± 8.3 msec at 1 m. Morozov, et al., (1972), reported a mean difference of 20 msec, which was fairly constant over a distance of 40 to 4 m as the animal closed in on a target. The click intervals for Tursiops, measured by Evans and Powell (1967) with a simultaneous

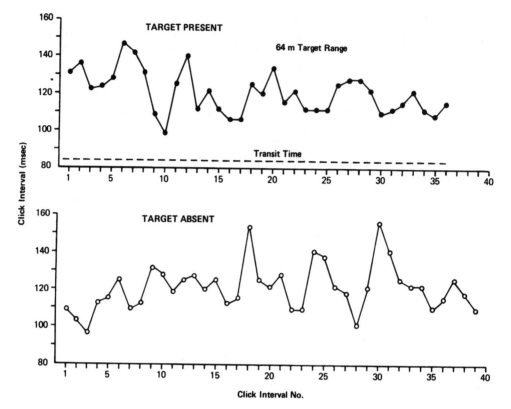

Figure 10.   Examples of the variations in the click interval
during two echolocation trials.

audio-video monitor system, showed an almost constant mean lag
time of 15.4 msec for target ranges from 1.4 m to 0.4 m.   As the
animal closed within 0.4 to 0.03 m the mean lag time decreased
to a minimum of 2.5 msec.   Therefore, the mean lag time for
target ranges greater than 0.4 m is approximately 15-22 msec.
This implies that the maximum mean repetition rate should be
between 45 and 67 pulses per second, for targets at ranges
greater than 0.4 m, and that it may take an animal between 15 and
22 msec to process an echo.   At the very close ranges (less than
0.2 m), the animal may possibly process several echoes at a time
instead of single echoes, as seems to be the case involving
targets at ranges greater than 0.4 m.

B.   Peak Frequency and Bandwidth

Histograms of the peak frequencies and the 3-dB bandwidths
of the signals used by four animals performing different echolocation tasks in Kaneohe Bay are presented in Fig. 12.

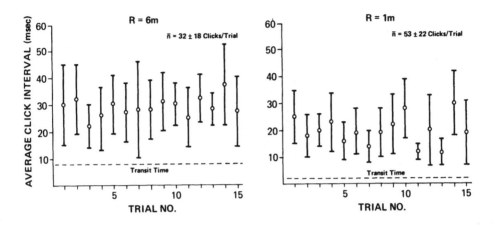

Figure 11. Average click intervals per trial for two ranges.

The average peak frequency and bandwidth along with the time bandwidth product of the signals used to determine the histograms are shown in Table 1. The dolphins, Ehiku and Heptuna, were involved in a target detection task at ranges greater than 50 m with a 7.6 cm diameter water-filled sphere as the target. The signals used by Ehiku and Heptuna were measured with a hydrophone located at 73 m. Ekahi was involved in a shape discrimination study at a range of 6 m with foam spheres and cylinders as targets. The cylinders had lengths that were less than 5 cm and diameters that were less than 3.81 cm. Sven was involved in a target recognition and discrimination task at a range of 6 m with different diameter and material composition cylinders that were 17.8 cm in length. The same hydrophone was used in the measurements of the signals emitted by these four animals.

The frequency histograms indicate that with the exception of Heptuna, at least 70% of the clicks had peak frequencies between 110 and 130 kHz. Only 60% of Heptuna's clicks had peak frequencies between 110 and 130 kHz. The peak frequencies most favored by Ehiku, Ekahi and Heptuna fell in the interval between 110 and 120 kHz. The average peak frequencies for these three animals in Table 1 are similar, with values close to 115 kHz. Sven showed a preference for the frequency interval between 120 and 130 kHz, although its average value of 121.1 kHz is close to the lower limit of the 120–130 kHz interval.

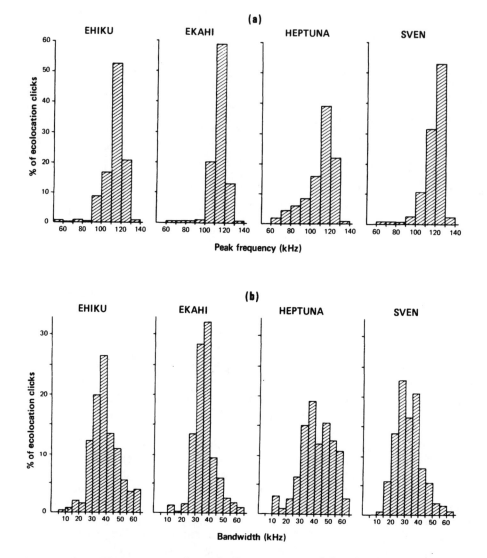

Figure 12. Histograms of peak frequency and bandwidth for four
Tursiops truncatus performing echolocation tasks in
Kaneohe Bay.

The half-power bandwidth of the animal's echolocation signal
is a parameter that has not received much attention in the past.
However, this parameter is an important one since the information
content of a signal is directly related to its bandwidth
(Berkowitz, 1965). The bandwidth histograms indicate that at
least 75% of the signals had bandwidths greater than 25 kHz with
the frequency interval between 30 and 40 kHz being the most common.

Signals that are representative of the average signal used by each animal are shown in Fig. 13. These signals have peak frequencies and bandwidths that were closest to the average peak frequency and bandwidth displayed in Table 1. Included in Fig. 13 is the envelope of the autocorrelation function $\rho(\tau)$ of the signals. The time resolution constant $\Delta\tau$ and the time-bandwidth product $\beta\tau$ included in Fig. 13 and Table 1, are signal parameters commonly used in radar and sonar signal analysis. The time resolution constant is directly related to the autocorrelation function of the signal and was originally defined by Woodward (1953) as

$$\Delta\tau = \frac{\int_{-\infty}^{\infty} |\rho(\tau)|^2 \, d\tau}{|\rho(0)|^2} = \frac{\int_{-\infty}^{\infty} |U(f)|^4 \, df}{[\int_{-\infty}^{\infty} |U(f)|^2 \, df]^2} \tag{1}$$

The rms bandwidth β and duration τ about the energy centroid of the signal are generally used to compute the time-bandwidth product. The parameters β and τ are defined as (Burdic, 1968)

$$\beta^2 = \frac{\int_{-\infty}^{\infty} (f - f_0)^2 \, |U(f)|^2 \, df}{\int_{-\infty}^{\infty} |U(f)|^2 \, df} \qquad \tau^2 = \frac{\int_{-\infty}^{\infty} (t - t_0)^2 \, |\mu(t)|^2 \, dt}{\int_{-\infty}^{\infty} |\mu(t)|^2 \, dt} \tag{2}$$

where the centroids f_0 and t_0 are defined as

$$f_0 = \langle f \rangle = \frac{\int_{-\infty}^{\infty} f \, |U(f)|^2 \, df}{\int_{-\infty}^{\infty} |U(f)|^2 \, df} \qquad t_0 = \langle t \rangle = \frac{\int_{-\infty}^{\infty} t \, |\mu(t)|^2 \, dt}{\int_{-\infty}^{\infty} |\mu(t)|^2 \, dt} \tag{3}$$

$U(f)$ is the transform of the signal envelope $\mu(t)$. The time-bandwidth product using the expressions for β and τ given in equation 2 will have a lower limit of $1/(4\pi)$.

Table 1. The average signal parameters for 4 Tursiops truncatus

animal	f_p (kHz)	BW (kHz)	$\beta\tau$	$\Delta\tau$ (μsec)
Ekahi	114.9±5.0	41.5±12.4	0.12	15.3
Ehiku	115.0±3.9	40.8± 3.6	0.12	11.8
Heptuna	116.2±5.4	45.6± 6.9	0.15	14.5
Sven	121.1±3.2	38.0±12.2	0.12	13.6

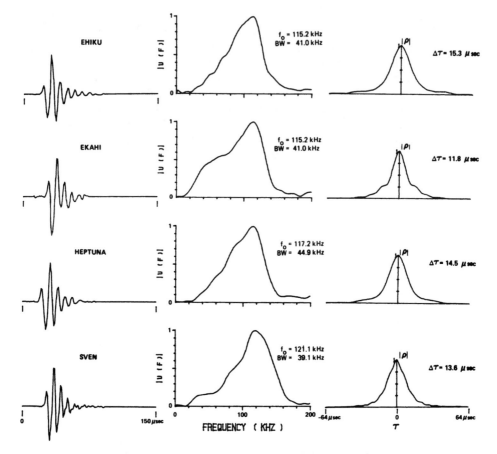

Figure 13.   Representative signals having approximately the same
peak frequency and bandwidth as the average values of Table 1.

C.   Spectral Adaptations

        Spectral adaptation of the echolocation signals used by the
animals in Kaneohe Bay due to the ambient noise condition has been
reported by Au, Floyd and Haun (1978).   The peak frequencies in
Table 1 indicate the animals typically used signals with peak
frequencies between 100 and 130 kHz.   These peak frequencies are
over an octave higher than previously reported values between 30
and 65 kHz, (Evans, 1973), for Tursiops.   Au, et. al., (1974),
noted that the ambient noise in Kaneohe Bay, is dominated by the
presence of snapping shrimp, and has a typical spectrum that is
relatively flat between 40 and 100 kHz, dropping off rapidly for
frequencies above 100 kHz.   They estimated that the animals should

be masked by the noise up to a frequency of approximately 137 kHz, so that in the reception of the target echoes, the received signal-to-noise ratio becomes the most critical parameter. Substantial improvement to the signal-to-noise (S/N) ratio could be achieved by operating at peak frequencies above 100 kHz. The average peak frequencies listed in Table 1 are very similar, yet the tasks the animals were involved in were completely different. This suggests the dominant role of the ambient noise on the selection of the proper peak frequency by the animals. It will be shown in Section IID that Ekahi and Sven enjoyed at least a 20 dB greater signal-to-noise ratio than Heptuna and Ehiku, and therefore did not need to operate at frequencies beyond 100 kHz. However, less acoustic energy was required to achieve a given or desired S/N ratio by emitting clicks with frequencies greater than 100 kHz.

The question of whether a dolphin adaptively changes the spectrum and shape of its echolocation signals when examining different targets has generated a considerable amount of specula-tion and a number of hypotheses with very little data from meticu-lously conducted experiments. The mean peak frequencies and the standard deviations of the echolocation signals shown in Fig. 14 can be used to examine this question. Ekahi was initially involved in an experiment to determine his ability to discriminate between foam spheres and cylinders. In any given trial, one target out of a set of three foam spheres and five foam cylinders would be pre-sented six meters from the animal stationed in a hoop. The six targets were designed to have similar target strengths to eliminate target strength differences as clues. Ekahi was later used in a discrimination experiment with metallic cylinders as targets. The metallic cylinders were at least 3.5 times longer than the foam cylinders of the initial experiment. Sven was involved in an experiment to test his ability to recognize two specific aluminum cylinders from cylinders made of the different materials listed in Fig. 14. Although the metallic cylinders used with Ekahi were completely different from the foam spheres and cylinders, there was no significant shift in the peak frequency of the signals used by Ekahi. The same could be said for the data collected with Sven. It is important to realize that the data of Thompson and Herman (1975), indicate that for frequencies between 100 and 130 kHz, a bottlenose dolphin cannot perceive pitch differences of less than 8 kHz. Therefore, the small frequency differences shown in Fig. 14 should not be perceived by the animals. The data of Fig. 14 seems to indicate that for the conditions existing in Kaneohe Bay, the animals do not purposefully adjust the peak frequency of the emitted signals to "match" the targets being investigated. Bel'kovich and Dubrovskiy (1977), arrived at a similar conclusion for dolphins kept in tanks at their facility.

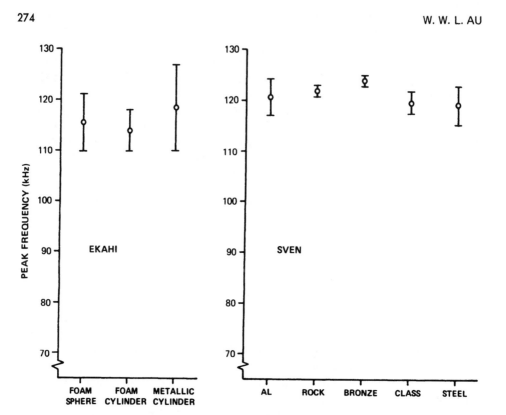

Figure 14. Average peak frequencies for the signals used by Ekahi and Sven while performing in different discrimination experiments.

Changes in the spectral composition of the signals in a pulse train do occur. However, Bel'kovich and Dubrovskiy (1977), argued that any attachment of significance to these changes are most likely due to incorrect interpretation of the results, or are based on insufficient statistics. To illustrate the need to examine dolphin signals from a statistical basis, consider the spectral-time plots for six echolocation trials by Ekahi on the same target in Fig. 15. The signals were measured with Ekahi performing in the shape discrimination experiment, and the spectral-time plots pertain to trials in which a 10.2 cm foam sphere was presented to the animal. In cases (a) through (c), there are obvious changes in the spectral composition of the signals, and observations based on a single or a few displays may lead to the conclusion that the animal was adaptively adjusting his pulses to maximize the information available in the echoes. However, when

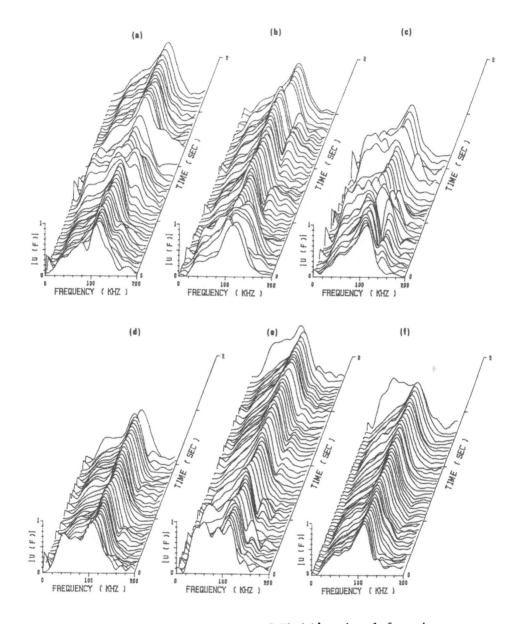

Figure 15. Spectral-time plots of Ekahi's signal for six different trials with the same target present.

comparing plots a through c, there does not seem to be any consistent pattern in which the signals were being altered.
Furthermore, the animal also emitted signals that were fairly
consistent and steady, as can be seen in plots d through f. Yet
all of these spectral-time plots were for the same target. The
spectral changes may be due to the animal turning momentarily
away from the target, internal steering of the beam even when
its rostrum was directed to the target, fluctuations due to its
click mechanism, or lapses in the attention given by the animal
during a trial.

It is possible that spectral adaptations were not observed
for these animals because of the dominant influence of the
ambient noise condition. Since the animals preferred to use
signals with peak frequencies between 110 and 130 kHz, any changes
or frequency adjustment within this range may have insignificant
effects on the echoes. In an environment where an animal is not
masked by the noise, and can achieve approximately the same S/N
ratio per unit of effort at any frequency between 30 and 130 kHz,
spectral adaptations may possibly be present.

D. Click Source Levels

The source level of the echolocation signals used by the
animals should depend on the S/N ratio of the received echoes
that the animals require or desire to have in order to perform a
given task. The SPL of the echoes from a target can be expressed
by the equation

$$EL(dB) = SL(dB) - 40 \log r - 2\alpha(f_p)r + TS(dB) \qquad (2)$$

Where: EL = echo level in dB

SL = source level in dB

r = target range

$\alpha(f_p)$ = absorption coefficient at the
peak frequency of the signal

TS = target strength of the target
in dB

Equation 2 simply states that the amplitude of the return echo is
equal to the amplitude of the outgoing signal minus the propagation and reflection losses. The peak-to-peak source level as
a function of the total loss due to propagation and reflection
for five different animals under different circumstances is
depicted in Fig. 16. The data show that there is a general

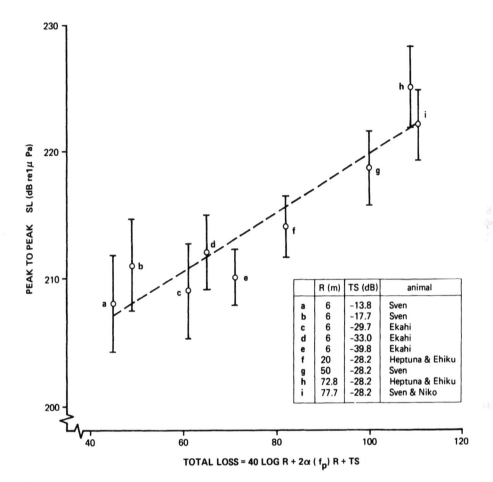

Figure 16. Peak-to-peak source level as a function of the total
acoustic energy loss for five different animals
performing different echolocation tasks.

increase in the source level as the total losses increase. However, the source level does not increase at the same rate as the
total loss. This indicates that the animals seem to prefer to
operate at a high S/N ratio. The decrease in the loss in moving
from case i to case a is over 62 dB, and the corresponding
decrease in the source level was only 12 dB.

The results of Fig. 16 also seem to indicate that range has a more powerful influence on the source level than target strength. In cases a and e, with both targets at 6 m, there is a 26 dB difference in the target strength but only a 4 dB difference in the source level. In cases c and e involving the same animal, the difference in target strength is 10.1 dB, whereas the difference in the source level was only 1 dB. These small changes in source level with target strength suggest that the animals operated at a high S/N ratio when the targets were at 6 m and changes in the target strength were not large enough to cause appreciable changes in the source level. There is hardly any difference in the source level for cases a through e with the targets at 6 m. However, when the target range increased causing large changes in the total loss, the source level also increased considerably.

The maximum average peak-to-peak source level in a trial was recorded at 227.6 dB for one of the trials in case h of Fig. 16. The largest single click measured was 230 dB, emitted by Heptuna in case h. These source levels may seem inordinately high, especially when comparing with conventional man-made sonars, and considering the amount of energy required to project high level signals into the water. However, one must realize that peak-to-peak levels are being considered for the animals, the signals are very short, and the beam is relatively narrow. The rms pressure for these dolphin signals is typically 15 to 20 dB lower than the peak-to-peak amplitudes. For comparing transient signals similar to dolphin signals, it is more meaningful to refer to the energy flux density of the acoustic wave than to the sound pressure level (Urick, 1967). The energy flux density is defined as the time integral of the instantaneous intensity. For signals reported in case h of fig. 16, the average energy flux density was 167.4 ± 1.3 dB re $(1 \; \mu Pa)^2$ s. The rms pressure of a cw pulse tone that would have an equivalent energy flux density as the average echolocation signal of case h can be expressed as

$$SPL(dB) = 167.4 - 10 \log (t) \tag{3}$$

For a 10 msec long pulse, an rms source level of 187.4 dB re 1 μPa would have the same energy flux density as the dolphin signals of case h of Fig. 16.

III. DISCUSSION

There are many difficulties inherent in the area of measurement and analysis of echolocation signals used by dolphins, and therefore there are large gaps in our understanding of these signals and how they are used. Most of the results presented in this review are averaged kinds of results. Within the range of average characteristics, there are often many variations that are not explainable or easily characterized. One of the major problems in studying dolphin signals involves the sheer mass of data that is collected in just one recording session. We typically record approximately 60 to 70 trials during a recording session, with each trial resulting in anywhere from tens of clicks to several hundred. If these numbers are compounded by several days of recording, the total amount of data recorded per animal per experiment can be overwhelming. It generally requires several orders of magnitude more time and effort to digitize the analog tapes, analyze the digital information, and finally display the results. Imagine having 60 to 70 spectral history and waveshape plots like those shown in Fig. 1, for each recording session. Added to these plots could be the frequency distribution plots and graphs of the variations of certain parameters associated with the signals on a click-to-click basis.

The approach taken in the measurement and analysis of dolphin signals is one in which the animals were allowed to achieve a solution or solutions to a specific sonar problem before the signals were recorded. Diercks (1972), suggested that a more meaningful approach would be to vary the animal's sonar problem and monitor the development of its solution, which would enable us to understand the signal design utilized by the animal in order to achieve a solution. This approach is universally recognized as having considerable merit, but has not been used in the United States because of the cost that would be involved. It is simpler to record a short sequence after stabilization in the animal's performance is achieved. The alternative would be to perform many recordings over an indeterminate period during the development of the solution. A feature that would be desirable in the second approach is to have the capability of real-time or near real-time analysis of the results. However, before the second approach is even considered, effective ways of handling the mass of data that would be collected and require analysis must be devised.

For the present, we are confined to results that at best can only approximate the acoustic field of these animals. A dolphin is not like a standard transducer, which can be positioned precisely in a given location and excited in a desired manner by an electrical signal. Instead, we are dealing with a relatively large aquatic animal often weighing in excess of 130 kg, and operating in an environment which is not conducive for tightly controlled acoustic experiments. Even if good physical control could be exercised over a dolphin, other factors such as the possibility of internal beam steering, behavioral control, motivation, attention spans, and other psychological influences must be considered. If an animal has the freedom of exciting its source region in a spatially non-uniform and irregular manner, it could cause considerable variations in the signal within a click train on a click-to-click basis. The attachment of significance to these variations or any variations within any click train is difficult since it is not possible to know if fluctuations exist because the animal decided to momentarily "look" at something else besides the target, or vary the degree of attention paid to each echo. Fluctuations in the signals may also be present due to changes in the degree of effort expended by the animal to produce the pulses in the train. Our experience has shown that these animals tend to optimize their behavior to accomplish a task using the least amount of energy as possible. In an echolocation experiment involving noise masking, and using two separate animals, when the masking level is high making the task difficult or impossible, the animals would often go through the motions of echolocating and responding without emitting a single click, or emit only a small fraction (less than 20%) of the number of clicks they would otherwise use. All of these difficulties are inherent when dealing with a complex creature such as a dolphin, and can be overcome only by continual refinements in methodology, behavioral control, and acoustic measurements and analysis.

REFERENCES

Au, W. W. L., Floyd, R. W., Penner, R. H., and Murchison, A. E., 1974, Measurement of echolocation signals of the Atlantic bottlenose dolphin, Tursiops truncatus Montagu, in open waters, J. Acoust. Soc. Am., 56:1280.
Au, W. W. L., Floyd, R. W., and Haun, J. E., 1978, Propagation of Atlantic bottlenose dolphin echolocation signals, J. Acoust. Soc. Am., 64:411.
Barta, R. E., and Evans, W. E., 1970, Private communications in Fish, Johnson, and Ljungblad (1976).
Bel'kovich, V. M., and Dubrovskiy, N. A., 1977, Sensory bases of Cetacean orientation, U.S. Joint Publication Research Service JPRSL/7157, May 27.
Burdic, W. S., 1968, "Radar Signal Analysis", Prentice-Hall, Inc., Englewood Cliff, N. J.
Diercks, K. J., Trochta, R. T., Greenlaw, C. F., Evans, W. E., 1971, Recording and analysis of dolphin echolocation signals, J. Acoust. Soc. Am., 49:1729.
Diercks, K. J., 1972, Biological sonar systems: a bionics survey, Applied Research Laboratories Technical Report N° 72-34 (ARL-TR-72-34), 190.

Dubrovskiy, N. A., and Zaslavskiy, G. L., 1975, Skull bone, dolphin sounding pulses,"Dolphin Echolocation" (Joint U.S. publication Research Service, JPRS 65777, Sept. 79).

Evans, W. E., Sutherland, W. W., and Beil, R. G., 1964, The directional characteristics of delphinid sounds, in: "Marine Bio-Acoustics,"W. N. Tavolge, ed., Pergamon Press, New York.

Evans, W. E., and Powell, B. A., 1967, Discrimination of different metallic plates by an echolocating delphinid, in: "Proc. Symp. Animal Sonar Systems: Biology and Bionics", R. G. Busnel, ed., Laboratoire de Physiologie Acoustique, Jouy-en-Josas, France.

Evans, W. E., 1973, Echolocation by marine delphinids and one species of fresh-water dolphin, J. Acoust. Soc. Am., 54:191.

Evans, W. E., and Maderson, P. F. A., 1973, Mechanisms of sound production in delphinid cetaceans: a review and some anatomical considerations, Am. Zool., 13:1205.

Fish, J. F., Johnson, C. S., and Ljungblad, K. K., 1976, Sonar target discrimination by instrumented human divers, J. Acoust. Soc. Am., 59:602.

Hammer, C., and Au, W. W. L;, 1978, Target recognition via echolocation by an Atlantic bottlenose porpoise (Tursiops truncatus), J. Acoust. Soc. Am., 64, Suppl. N° 1:587.

Hollien, H., Hollien, P., Caldwell, D. K., and Caldwell, M. C., 1976, Sound production by the Atlantic bottlenose dolphin Tursiops truncatus, Cetology, 26:1.

Morozov, B. P., Akapiam, A. E., Burdin, B. I., Zaitseva, K. A., Sokovykh, Y. A., 1972, Tracking frequency of the location signals of dolphins as a function of distance to the target, Biofizika, 17:139.

Murchison, A. E., and Penner, R. H., 1975,Open water echolocation in the bottlenose dolphin (Tursiops truncatus): metallic sphere detection threshold as a function of distance, in: "Proceedings of Conference on the Biology and Conservation of Marine Mammals," University of California, Santa Cruz, Ca.

Murchison, A. E., 1976, Range resolution by an echolocating dolphin (Tursiops truncatus), J. Acoust. Soc. Am., 60:S5.

Nachtigall, P. E., Murchison, A. E., and Au, W. W. L., 1978, Discrimination of solid cylinders and cubes by a blindfolded echolocating bottlenose dolphin (Tursiops truncatus), J. Acoust. Soc. Am., 64, Suppl. 1:587.

Norris, K. S., Prescott, J. H., Asa-Dorian, P. V., and Perkin, P., 1961, Experimental demonstration of echolocation behavior in the porpoise Tursiops truncatus (Montagu), Biol. Bull., 120:163.

Norris, K. S., and Evans, W. E., 1966, Directionality of echolocation clicks in the rough-tooth porpoise, Steno bredanensis (Lesson) in: "Marine Bioacoustics", W. N. Tavolga, ed., Pergamon Press, New York.

Norris, K. S., Evans, W. E., and Turner, R. N., 1967, Echolocation in an Atlantic bottlenose porpoise during discrimination,

    in: "Proc. Symp. Animal Sonar Systems: Biology and Bionics",
    R. G. Busnel, ed., Laboratoire de Physiologie Acoustique,
    Jouy-en-Josas, France.

Norris, K. S., Dormer, K. J., Pegg, J., and Liese, G. J., 1971, The
    mechanisms of sound production and air recycling in porpoises:
    a preliminary report, in: "Proc. 8th Annual Cong. Biol. Sonar
    and Diving Mammals".

Norris, K. S., and Harvey, G. W., 1974, Sound transmission in the
    porpoise head, J. Acoust. Soc. Am., 56:659.

Romanenko, Y. V., 1973, Investigating generation of echolocation
    pulses in dolphins, Zool. Zh., 11:1698, JPRS 61553.

Romanenko, Y. V., 1974, Physical fundamentals of bioacoustics,
    Fizicheskiye Osnovy Bioakustik, JPRS 63923, Moscow.

Schevill, W. E., and Watkins, W. A., 1966, Sound structure and dir-
    ectionality in Orcinus (Killer Whale), Zoologica, 51:71.

Schusterman, R. J., Kersting, D. A., and Au, W. W. L., 1979, Stim-
    ulus control of echolocation pulses in Tursiops truncatus,
    this volume.

Thompson, R. K. R., and Herman, L. M., 1975, Underwater frequency
    discrimination in the bottlenose dolphin (1-140 kHz) and
    human (1-8 kHz), J. Acoust. Soc. Am., 57:943.

Urick, R., 1967, "Principle of Underwater Sound", McGraw-Hill, New
    York.

Wood, F. G., 1964, "Marine Bioacoustics", W. N. Tavolga, ed., Perg-
    amon Press, New York.

Woodward, P. M., 1953, "Probability and Information Theory with
    Applications to Radar", Pergamon Press, New York.

ACOUSTICS AND THE BEHAVIOR OF SPERM WHALES

William A. Watkins

Woods Hole Oceanographic Institution

Woods Hole, Massachusetts 02543

The first good recordings of underwater sperm whale sounds that we have were made 27 years ago (22 April 1952, R/V Caryn at 38°29'N, 69°29'W) and five years later the sounds were identified definitely and described (Worthington and Schevill 1957). At every opportunity since then, we have stopped with these whales and listened underwater (Watkins 1977). Repeated observations have gradually sorted out many of the sounds. The overlapping clatter of click series that are usually heard in the presence of these whales have given way with time to acoustic sequences that apparently have characteristics identifiable with individuals. The unorganized welter of sound can be simplified to series of pulses that are related and that can sometimes be traced for hours. Whales underwater have been located and tracked by their own sounds. Yet with all this, the sounds are still not well enough correlated with the whales' actions to provide anything but glimpses into the role that acoustics plays in the behavior of sperm whales.

The literature dealing with the underwater sounds from sperm whales shows the sequence of our perception of that behavior: The sounds were described first by Worthington and Schevill (1957). A sample recording and analysis of sperm whale sounds was published by Schevill and Watkins (1962). A short description was given by Schevill, Backus, and Hersey (1962), and Backus and Schevill (1966) presented a more detailed analysis of these sounds. A variety of sounds were reported in the presence of sperm whales by Perkins, Fish, and Mowbray (1966). A description of sounds produced by these whales during a whale hunt was given by Busnel and Dziedzic (1967). Sonobuoy measurements of the sound characteristics were made by Dunn (1969) and Levenson (1974).

Norris and Harvey (1972) suggested mechanisms for sound production. Arrays of hydrophones (Watkins and Schevill 1972,
Watkins 1976) were used with sperm whales to study acoustic responses and behaviors including their reaction to pinger sounds
and sub-surface swimming and diving (Watkins and Schevill 1975),
spatial distribution underwater (Watkins and Schevill 1977a),
and exchanges of stereotyped pulse sequences (Watkins and
Schevill 1977b). Diving behavior of whales being chased during
marking operations was assessed by means of sonar equipment and
reported by Lockyer (1977). The study of acoustic behavior of
sperm whales was reviewed by Watkins (1977).

The characteristics of sperm whale sounds, the comparisons
with equivalent sounds of other odontocetes and the behavioral
associations that are discussed here summarize our understanding
to date, both from published and as yet unpublished observations
at sea.

METHODS

Whenever we have located sperm whales at sea, either acoustically or by sight, we have tried to move upwind, stop engines,
and listen underwater. Several hundred hours of sperm whale
sounds have been recorded, from both single hydrophones and
three-dimensional arrays of hydrophones. We have attempted to
correlate the sounds with any behaviors that were observed,
usually short glimpses of parts of whales occasionally seen at
the surface.

The acoustic recordings have been analyzed for spectra of
their sounds and for relationship to other sounds. Sound sequences from groups of whales and from individuals have been
traced over several hours of listening. Array data has been
analyzed for location of nearby sound sources and for direction
to more distant sources. Intensities and frequency relationships
have been compared. A wide spectrum of listening gear has been
utilized, including narrowband sonar systems, medium bandwidth
(30 kHz) gear, and calibrated wide bandwidth (to 200 kHz) instrumentation equipment. Analysis systems were designed to permit detailed scrutiny of particular sounds or sequences of sound,
and included display by means of oscillographs for time series,
sound spectrographs for frequency comparisons, and computer
analyzers for location analyses. The published studies have
specified equipment and techniques used.

CHARACTER OF SPERM WHALE SOUNDS

The sounds we have heard from sperm whales have always been
impulsive—clicks at a variety of repetition rates. These were
sharp onset, broadband pulses with energy as high as 30 kHz or

more, when the whales were within 20 m. From such nearby whales, major emphases in frequency were often found in the 10 to 16 kHz range. With distance, of course, the higher frequencies were attenuated by the environment so that at 2 km most of the audible energy was in frequencies below 5 or 6 kHz, with apparent emphases at 2 to 4 kHz. After listening for a time to a particular sperm whale, it was possible to judge its relative distance from the hydrophone by an assessment of the amount of high-frequency energy audible in the click sounds. The lower frequency content of sperm whale clicks was variable but usually extended well below 100 Hz. Secondary pulse structure in the clicks often provided strong low-frequency energy. We have not found much directionality in the clicks of sperm whales and we have found no differences that could be attributed to depth.

The level of sperm whale sounds was found to be highly variable. As much as 26 dB difference was observed between sounds a few seconds apart in the same series, with approximately equivalent spectra from apparently the same whale. Calculations of maximum sound levels for sperm whale clicks also were variable on different occasions from 65 to about 80 dB, relative 1 dyne/cm^2. Thus, with substantial energy below 1 kHz, and favorable local oceanographic conditions, sperm whales could be heard at distances that often were found to be greater than 10 km. Under the same conditions, it was possible also to barely hear whales clicking at distances of only a few hundred meters. These differences were estimated to be as much as 40 dB. The whales varied the level of their sounds apparently at will.

The duration of individual clicks from sperm whales varied from 2 to 30 msec or more, depending on the structure of secondary pulses present in many of the clicks (see Backus and Schevill 1966). The longer sequence of sound in individual clicks enhanced their low frequency content so that the longer clicks sometimes blended into the background reverberation. Multiple sound paths and surface reflections often made detailed analyses of the sounds difficult to sort out. Clicks from individual whales (equivalent high frequency spectra, same direction to source, and without change in repetition sequence) sometimes switched from longer reverberant pulses to short clicks with little internal structure.

The sperm whale clicks were produced occasionally in very slow sequence, almost singly, one in 5 or 10 sec for sequences of 2 to 10 or more clicks. Usually, however, the clicks were heard at repetition rates of 1.5 to 3 per sec, and characteristically repeated at very regular intervals (see Backus and Schevill 1966). Click series without interruption or much change in rate have been measured for sequences lasting 20 min or more.

Click series of only 1 to 10 sec were also commonly heard. Longer series were usually produced at a slower and more regular rate,
shorter series were sometimes at fast rates, to 60 or more per
sec. Sometimes the terminal portion of long series and more
often the end of short series were variable and became irregular
in click rate (Watkins and Schevill 1977b).

Click Patterns

The irregular click patterns that sometimes formed the
terminal portions of some of the click series were occasionally
repeated at the end of other click series by the same whales.
These end patterns or codas were repeated sometimes also as
separate stereotyped click sequences. The codas were formed of
3 to 40 or more clicks in about one second and produced in
specific temporal patterns of click repetition. Each coda pattern appeared to be unique for specific whales within acoustic
range over at least a few hours. Though codas were not often
heard, sometimes they were repeated almost exactly, up to 66
times in a 15 min period. Codas were produced in particular
situations and were occasionally heard as exchanges between whales
(Watkins and Schevill 1977b).

Because sperm whales tended to maintain relatively regular
clicking rates during their longer click sequences, it was often
possible to follow the sounds of individuals. During these
sequences, it was possible also to note patterns of reverberation
and reflection as well as relative spectra and click structure
that served to help identify succeeding sound sequences from the
same whales. In addition, analyses of multiple hydrophone array
recordings often provided direction (in the vertical as well as
the horizontal plane) to the sound source. And occasionally,
this could be refined to give a location for the source. Thus,
in spite of irregularity in clicking patterns and silent periods,
we were able sometimes to trace much of the acoustic output of
individual whales over as much as four hours. This is also the
maximum that we have been sure we were still with the same sperm
whales.

Individual whales did not produce sounds continuously.
They often were silent at the surface, then as a dive began,
they would click for a short time after submergence. Of a
group of whales, there was usually one or two that were dominant
acoustically, and clicked more often, more continuously than the
others. But even these whales were silent underwater for periods
of a few seconds to 10 min at a time. Whales sometimes apparently took turns clicking. Sometimes they all clicked at
once, producing as many overlapping click series as there were
whales in the group. And sometimes all were silent for short

periods. Usually, at least one whale in a group (the acoustically dominant whale) could be heard to start clicking again
within one to two minutes. Some whales produced only one or
two identifiable click sequences within a dive cycle of 20 min
or more. Usually, however, when several whales were together
or within a few km of each other, at least a few click series
were audible from each whale.

ACOUSTIC BEHAVIOR

Lone sperm whales out of acoustic range of other whales
seldom produced any audible sounds. But, if sounds could be
heard underwater from distant sperm whales, the whale that was
apparently by itself produced occasional click sequences. A
lone whale, that joined other whales became as loquacious as
the others increasing its acoustic activity as it approached
the group. As whales separated, less sounds were produced.

At the surface, whales in a group together were relatively
silent, and they usually dove within a short time of each other.
They often began their dives at steep vertical angles with flukes
raised above the water as they started down. From our four-
hydrophone array data, we found that the whales often began
clicking at 5 to 25 m depth, their vertical dive angles were now
only 10° to 15°, and their swimming speeds were 2 to 4 km per
hour. Underwater, the whales invariably spread out so that received angles to the sound sources indicated a widening distribution. The whales separated as they dove and apparently occupied a large volume of water, perhaps foraging as they maintained
separations of several hundred meters (Watkins and Schevill
1977a). Maximum dive depths were estimated at 500 to 800 m.
Whales from each of the different source directions periodically
clicked for a few minutes usually at slow (2 to 3 per second),
regular rates, often overlapping with the click sequences of
others. Though widely separated underwater, as the whales came
toward the surface, the sound source directions also came together and merged. Then the whales reappeared at the surface
within a short distance of each other, usually less than 50 m.

Sperm whales that passed close to our drifting ship or near
the buoyed hydrophone cables have often produced long series of
relatively rapid clicks (25 to 50 per sec) usually at a very
regular rate. The endings of such series have sometimes changed
to irregular patterns of click repetition with variable rates and
missed beats. The last few clicks have sometimes been repeated
at the same repetition patterns as separate codas or as the same
coda at the end of another click sequence. During these close
approaches by sperm whales, click repetition rates usually have
not varied relative to distances from obstacles or other whales.

One whale came toward the ship and passed a few meters below the
keel on its way down for a dive. Its click series was rapid and
very regular with little obvious variation, starting at the surface and continuing till the whale was well away from the ship
and our cables.

Why Do Sperm Whales Click?

Our own bias has assumed echolocation for sperm whales to
conform with the demonstrated capabilities of many other odontocetes, and it would seem convenient for them if they did. The
spectra and general form of sperm whale clicks has seemed to
support this notion. The frequency range is suited for echolocation of obstacles, and prey. Other features of their sounds,
however, do not so easily fit echolocation: (1) Sperm whale
clicks do not appear to be highly directional. (2) Click repetition rates are generally very regular and have not varied with
changing distances to an approaching obstacle. (3) Sperm whales
are silent for long periods, especially when they are alone.
(4) The level of their clicks appears to be generally greater
than that required for echolocating prey or obstacles. The
sounds may be heard over such long ranges that environmental
distortion of the sound paths doubtless would be limiting. (5)
Individual clicks are usually too long for convenient echolocation (a 30 msec click would obscure echo information from
the first 22 m), and the duration of clicks usually remains
constant throughout the click series.

Other odontocetes encountered at sea and monitored with the
same equipment have shown different characteristics from sperm
whales. These have included *Stenella plagiodon, Stenella
coeruleoalba, Delphinus delphis, Lagenorhynchus acutus,
Lagenorhynchus albirostris, Grampus griseus, Globicephala
melaena,* etc. All of these are considered to be echolocators
(though most have not yet been tested). (1)Their clicks have been
highly directional, so much so that hydrophones at less than
100 m to the side or rear of the animal have often failed to
pick up the sounds. (2) Click repetition rates have varied with
the distance of the animal to the target. (3) Clicks are produced nearly continuously with only short silent periods. (4)
The levels of their clicks seem to match the requirements of
the echolocation tasks, and the energies are efficiently distributed. (5) Individual clicks are usually short and become even
shorter at higher repetition rates.

Though their sounds are not arranged in ways that would seem
convenient for echolocation, nevertheless, sperm whales may
utilize the echo information. Echoes from prey and obstacles at
some ranges may be heard between steady outgoing pulses, and the

rate of clicking may be modified enough to accommodate some changing distances. Increasing click repetition rate with shortening distance to a target (echolocation run) is commonly heard from other echolocating odontocetes, but we have only heard a similar sequence once from sperm whales--a recording by Leanne Hinton and Kenneth Bloom in 1968 off northern Chile (Norris and Harvey 1972). Other echo information such as distances from the surface or bottom probably could also be noted, as well as general details of sound scattering layers nearby.

The acoustic behavior of sperm whales, though not apparently convenient for echolocation, does seem to fit a context of communication, perhaps to keep in contact with other sperm whales underwater. A number of our observations make this plausible: the exchange of codas, the relative silence of whales at the surface together, the increase in clicking underwater from widely separated whales, their return to the surface together, the silence of lone whales, the usual composition of the sounds that makes click sequences audible and distinctive even at a distance.

Probably, both echo information and communication are a part of their acoustic behavior. Our observations seem to relate the sounds more clearly to social interaction, but we may also be dealing with a different bio-sonar system for sperm whales.

Acknowledgements

This work has been supported by the Oceanic Biology Programs of the Office of Naval Research, N00014-74-C0262 NR083-004. The analyses and observations of sperm whales at sea have been shared by many, particularly by William E. Schevill, whose work in 1949 began the scientific investigation of whale sounds. He participated directly in most of the sperm whale observations and has reviewed this manuscript. I am grateful for his helpful encouragement. The manuscript has been prepared by Karen E. Moore and Shirley Waskilewicz. This is contribution No. 4252 from the Woods Hole Oceanographic Institution.

REFERENCES

Backus, R. H., and Schevill, W. E., 1966, Pyseter clicks, in: "Whales, Dolphins, and Porpoises", K. S. Norris, ed., Univ. of California Press, Berkeley.

Busnel, R. G., and Dziedzic, A., 1967, Observations sur le comportement et les émissions acoustiques du cachalot lors de la chasse, Bocagiana, 14:1.

Dunn, J. L., 1969, Airborne measurements of the acoustic characteristics of a sperm whale, J. Acoust. Soc. Am., 46:1052.

Levenson, C., 1974, Source level and bistatic target strength of the sperm whale (Physeter catodon) measured from an oceanographic aircraft, J. Acoust. Soc. Am., 55:1100.

Lockyer, C., 1977, Observations on diving behaviour of the sperm whale Physeter catodon, in: "A Voyage of Discovery", M. Angel, ed., Pergamon Press, Oxford.

Norris, K. S., and Harvey, G. W., 1972, A theory for the function of the spermaceti organ of the sperm whale (Physeter catodon L.) NASA Special Publication 262:397.

Perkins, P. J., Fish, M. P., and Mowbray, W. H., 1966, Underwater communication sounds of the sperm whale, Physeter catodon, Norsk Hvalfangst-Tidende, 55(12):225.

Schevill, W. E., Backus, R. H., and Hersey, J. B., 1962, Sound production by marine animals, in: "The Sea, Vol. 1", M. N. Hill, ed., J. Wiley and Sons, New York.

Schevill, W. E., and Watkins, W. A., 1962, "Whale and Porpoises Voices. A Phonograph Record", Woods Hole Oceanographic Institution, Woods Hole, Mass., 24 pages and phonograph record.

Watkins, W. A., and Schevill, W. E., 1972, Sound source location by arrival-times on a nonrigid three-dimensional hydrophone array, Deep-Sea Research, 19:691.

Watkins, W. A., and Schevill, W. E., 1975, Sperm whales (Physeter catodon) react to pingers, Deep-Sea Research, 22:123.

Watkins, W. A., 1976, Biological sound-source locations by computer analysis of underwater array data, Deep-Sea Research, 23:175.

Watkins, W. A., 1977, Acoustic behavior of sperm whales, Oceanus, 20:50.

Watkins, W. A., and Schevill, 1977a, Spatial distribution of Physeter catodon (sperm whales) underwater, Deep-Sea Research, 24:693.

Watkins, W. A., and Schevill, W. E., 1977b, Sperm whale codas, J. Acoust. Soc. Am., 62:1485.

Worthington, L. V., and W. E. Schevill, 1957, Underwater sounds heard from sperm whales, Nature, 180:291.

CLICK SOUNDS FROM ANIMALS AT SEA

William A. Watkins

Woods Hole Oceanographic Institution

Woods Hole, Massachusetts   02543

Echolocation signals from passing cetaceans (*Delphinidae, Phocoenidae,* etc.) at sea generally exhibit somewhat different characteristics from those of the same species in captivity that are involved in discrimination experiments in echolocation. Presumably, these differences are partly because the echolocation task at sea would have been primarily for orientation (as suggested by Norris, Evans, and Turner 1967; Norris 1969). Generally, the clicks from groups of cetaceans passing within a few meters of a hydrophone at sea are (1) relatively low frequency (rarely energy above 50 kHz from any species), (2) mostly low level, and (3) often at slow, variable repetition rates. Under these circumstances, the clicks may be used primarily in social contexts.

Such click sequences from individual animals usually are very difficult to follow for more than a few seconds. It is seldom possible to know which animal produced the sounds that were heard, how deep it was, how far away it was, or what its orientation to the hydrophone might have been. Overlapping sounds from several sources often occur as other animals are apparently stimulated to sound production by the sounds they hear. Overriding louder squeals or bursts of clicks from nearby animals often overload and distort the recording of the lower level clicks.

The highly directional character of echolocation clicks, so noticeable for captive animals (Norris et al. 1961; Schevill and Watkins 1966; Norris and Evans 1967; and Au, Floyd, and Haun 1978) also is very evident in listening to animals at sea. As groups of delphinids pass and individuals approach a hydrophone within 20 m or less, the click sounds often are rich in frequencies to 30 or

50 kHz and the levels increase predictably with shortened distances. Then, as an animal changes orientation relative to the hydrophone, the high frequencies decline sharply and the levels of even the lower frequency click components may drop off rapidly even though distances to the animal may continue to shorten as it comes abreast. When these cetaceans investigate hydrophones and cables at close range, their echolocation clicks dramatically increase in apparent intensity and bandwidth and often overload linear recording circuits. Frequencies to 150 kHz may be heard at these times.

These differences in sound level and frequency spectra that depend on the orientation of the animals also make it difficult to locate or to track animals by their own sounds. With multiple hydrophones separated by only 30 m, it is the lower frequency click components only that are picked up by the hydrophones to the side or below or to the rear of the animal. These transient lower frequency components of clicks are easily masked by ambient noise, and their level often is not high enough to be heard well on all hydrophones in an array.

It is seldom possible, therefore, to describe accurately the echolocation clicks of cetaceans at sea. Their general character-istics and similarities are noticeable, but specific sound structures are described best from more controlled situations. For example, we were aware of the click sounds of *Phocoena phocoena* (harbor porpoise) from encounters with them at sea, but it was not until a sequence of four increasingly more controlled experiments with captive animals that additional details of their click sounds were available (Busnel, Dziedzic, and Andersen 1963; Schevill, Watkins, and Ray 1969; Dubrovskii, Krasnov, and Titov 1970; Møhl and Andersen 1973). The older acoustic measurements with limited bandwidth, lower frequency equipment did not detect high frequency emphases and consequently assessments of source level were often too low, though lower frequency components were accurately repre-sented (Watkins 1974). Some of the differences in the characteris-tics of echolocation clicks that have been reported for the same species of cetaceans can be attributed to differences in equip-ment bandwidth, as well as differences in behavior, and in echolocation tasks.

Comparisons of cetacean sounds at sea also must account for their voluntary control of sound production. Frequency, spectral emphasis, pulse rate and sound level can be varied by the animal at will (demonstrated particularly through hydrophone array analyses, as in Watkins and Schevill, 1974). Depending on behavioral activity and level of interest, the click sounds from the same group of animals can vary widely. The longer duration tonal (squeal, whistle) and pulsed (squawk) sounds from cetaceans often are species-specific and appear to be louder at sea than their click

sounds. These longer sounds are probably not used for echolocation, and they are not considered here.

The assessments of click characteristics from animals at sea
outlined below are mostly from our unpublished observations, based
in large part on wide bandwidth (200 kHz) recording using three-
dimensional arrays with 30-m or more separation between hydrophones (Watkins and Schevill 1972). Typically, our ship is
stopped and quieted upwind or in the path of the animals so that
they may pass close to the hydrophones. In most instances the
animals are in groups of three to several hundred, depending on
species and behavior, they are apparently undisturbed, and usually
they ignore the ship and its hydrophone array. The animal sounds
appear to be representative of behaviors that do not require much
echolocation effort; therefore, these click sounds may be primarily
for social contact. Some of the distinctions in click structure
stand out and appear to be species specific, while other click
variations are more readily explained by circumstantial or behavioral differences. Care in obtaining undistorted recordings (particularly difficult in high frequency and transient recording) and
in utilizing equivalent listening systems for each encounter vali-
dates such comparisons.

Clicks heard at sea from many of the smaller cetaceans are
not easily separable by species, particularly true for most delphinids (4.5 m or less), even though their squeals (whistles) and
pulsed sounds often are species-specific. *Delphinus delphis,
Grampus griseus, Lagenorhynchus acutus, L. albirostris, L. obscurus,
Stenella coeruleoalba, S. longirostris, S. plagiodon, Tursiops
aduncus,* and *T. truncatus* that have passed by our hydrophones at
sea have produced clicks that were superficially very much alike.
Differences usually were attributable to variable orientations of
the animals. Anomalous click spectra (such as those from *Lagenorhynchus australis* in an exceptionally quiet background, Schevill
and Watkins 1971) may represent responses to behavioral or environmental situations. The clicks heard at sea from delphinids passing
by our hydrophones were short, broadband, rapid rise-time pulses
with maximum frequencies at 30 to 150 kHz, and variable frequency
emphases. Click terminations were often lost in reverberation and
surface reflections, and though the higher frequency components
were produced in less than 0.5 msec, there usually was a longer
total pulse envelope with considerable energy in frequencies well
below 500 Hz. Levels varied widely with orientation and behavior,
but maximum source levels under these circumstances were about
60 dB at 1 m (re 1 dyne/cm^2), about half the dB level measured in
discrimination echolocation by Au, Floyd, and Haun (1978). The
differences between echolocating captives and clicks from the same
species at sea have been noted also by other investigators (Busnel
and Dziedzic 1966, for example), and for other species (*Steno*

bredanensis, Norris and Evans, 1966).

Under the same "at sea" conditions as for the smaller species listed above, delphinid species of 5 m or larger produced clicks that often had more low frequency energy and less obvious high frequency emphasis, slower average click rates, and somewhat highter maximum click levels at the lower frequencies. The stronger lower frequency content of the clicks (relative to clicks of the smaller delphinids) was particularly noticeable when two species of two sizes (*Globicephala melaena* and *Stenella coeruleoalba,* for example) passed by the hydrophone within a few seconds of each other. Apparently, a species size gradient (*Tursiops, Pseudorca, Globicephala, Orcinus*) roughly matches the increasing low frequency emphasis or perhaps decreasing high frequency emphasis in the clicks of animals recorded at sea. Differences in the clicks of echolocating captives and those of *Globicephala* at sea were noted by Evans (1973). Smaller animals of larger species also fit the general pattern for their species (Schevill and Watkins, 1966). In at least the largest of these, *Orcinus,* there also is a matching reduction in the high frequency hearing capability (Hall and Johnson, 1972). With the animals' behavior and orientation to the hydrophone, clicks from these larger species also varied widely in repetition rate, in level, and in frequency content. Sometimes the clicks had emphases at particular frequencies, and sometimes they had broader bandwidth characteristics that may have reflected individual preference.

Phocoena phocoena is very difficult to hear at sea. Their click sounds are relatively low level (maximum about 30 dB at 1 m re: 1 dyne/cm^2 when facing a hydrophone) and they have characteristic low frequency emphases often near 2 kHz (Schevill, Watkins, and Ray, 1969); Møhl and Andersen, 1973; Watkins, 1974). High frequency emphases above 100 kHz (Dubrovskii, Krasnov, and Titov, 1970; Møhl and Andersen, 1973) and also noted in *Phocoena* clicks during echolocation experiments. These higher frequency components have not been recorded at sea perhaps because of rapid attenuation and their low level, their high directionality, or like the delphinids, *Phocoena* may not emphasize the high frequencies unless they need to do so. We have recorded somewhat similar sounds from *Phocoenoides dalli* (see also Ridgway, 1966), and from four species of *Cephalorhynchus* (Watkins, Schevill, and Best, 1977). These species have many of the same characteristics in the low frequencies at sea, suggesting that if echolocation were tested, other higher frequency click components also would be found.

Clicks with tonal emphases at specific frequencies from 2 to 20 kHz are produced by *Monodon monoceros* , narwhal (Watkins, Schevill and Ray, 1971; Ford and Fisher, 1978) and in two of the three observations reported, the animals appeared to be feeding and perhaps echolocating. If so, as in *Phocoena,* the clicks probably also have higher frequency emphases (not recorded by the 24 kHz equipment). In contrast, another species of the family Monodontidae, *Delphinapterus Leucas*, the beluga, produced short clicks with broadband characteristics. Our recordings of beluga clicks at sea were with equipment bandwidths of 4 kHz, 30 kHz, and 200 kHz, and all indicate short (often less than 0.1 msec), highly directional clicks with little low frequency energy. The click sounds from passing beluga at sea were not readily detected unless the animal was close to a hydrophone or turned toward a hydrophone (often suddenly overloading the equipment as it turned).

The differences between the click sounds heard from cetaceans passing in the open ocean and the clicks produced by the same species during echolocation experiments appear to be related to differences in (1) behavior - social vs. task motivation, (2) types of echolocation - relative position vs. target discrimination, (3) orientation - seldom facing hydrophone vs. careful positioning, and (4) competing ambient sound. The sounds heard from social groups of cetaceans at sea, therefore, do not demonstrate their acoustic capabilities but are indicative of their usual patterns of sound production. Similarly, analyses of clicks produced during echolocation experiments show the animal's careful adjustment of click formation to suit difficult tasks and meet training requirements, but may not give the acoustic parameters that can be expected during other behaviors.

Acknowledgements

Support for these studies has been from the Oceanic Biology Program of the Office of Naval Research, N00014-74-C0262 NR083-004. Wm. E. Schevill shared in these observations and read the manuscript. We thank F. G. Wood for his review. Karen E. Moore and Elaine M. Ellis have prepared the manuscript. This is contribution N° 4300 from the Woods Hole Oceanographic Institution.

REFERENCES

Au, W. W. L., Floyd, R. W., and Haun, J. E., 1978, Propagation of
    Atlantic bottlenose dolphin echolocation signals, J. Acoust.
    Soc. Am., 64:411.
Busnel, R. G., Dziedzic, A., and Andersen, S., 1963, Acoustique
    physiologie - sur certaines caracteristiques des signaux
    acoustiques du marsouin Phocoena phocoena, L. C. R. Acad.
    Sc. Paris, 257:2545.
Busnel, R. G., Dziedzic, A., 1966, Acoustic signals of the pilot
    whale Globicephala melaena and of the porpoise Delphinus
    delphis and Phocoena phocoena, in: "Whales, Dolphins, and
    Porpoises", K. S. Norris, ed., Univ. California Press, Berk-
    eley.
Dubrovskiy, N. A., Krasnov, P. S., and Titov, A. A., 1971, On the
    emission of echolocation signals by the Azov sea harbor
    porpoise, Akust. Zh., 16:521; Sov. Phys. Acoust., 16:444.
Evans, W. E., 1973, Echolocation by marine delphinids and one species
    of fresh-water dolphin, J. Acoust. Soc. Am., 54:191.
Ford, J. K. B., and Fisher, H. D., 1978, Underwater acoustic signals
    of the narwhal (Monodon monoceros), Can. J. Zool., 56:552.
Hall, J. D., and Johnson, C. S., 1972, Auditory thresholds of a
    killer whale Orcinus orca Linnaeus, J. Acoust. Soc. Am., 51:
    515.
Møhl, B., and Andersen, S., 1973, Echolocation: high-frequency
    component in the click of the harbour porpoise Phocoena ph. L.,
    J. Acoust. Soc. Am., 54:1368.
Norris, K. S., Prescott, J. H., Asa-Dorian, P. V., and Perkin, P.,
    1961, Experimental demonstration of echolocation behavior in
    the porpoise Tursiops truncatus (Montagu), Biol. Bull., 120:
    163.
Norris, K. S., and Evans, W. E., 1966, Directionality of echoloca-
    tion clicks in the rough-tooth porpoise, Steno bredanensis,
    (Lesson), in: "Marine Bioacoustics", W. N. Tavolga, ed.,
    Pergamon Press, New York.
Norris, K. S., Evans, W. E., and Turner, R. N., 1967, Echolocation
    in an Atlantic bottlenose porpoise during discrimination,
    in:"Proc. Symp. Animal Sonar Systems: Biology and Bionics,"
    R. G. Busnel, ed., Laboratoire de Physiologie Acoustique,
    Jouy-en-Josas, France.
Norris, K. S., 1969, The echolocation of marine mammals, in: "The
    Biology of Marine Mammals", H. T. Anderson, ed., Academic
    Press.
Ridgway, S. H., 1966, Dall porpoise, Phocoenoides dalli (True):
    observations in captivity and at sea, Norsk Hvalfangst-
    Tidende, 5:97
Schevill, W. E., and Watkins, W. A., 1966, Sound structure and
    directionality in Orcinus (Killer Whale), Zoologica, 51:71.
Schevill, W. E., and Watkins, W. A., 1971, Pulsed sounds of the
    porpoise Lagenorhynchus australis, Breviora, 366.

Schevill, W. E., Watkins, W. E., and Ray, C., 1969, Click struc-
    ture in the porpoise, Phocoena phocoena, J. Mamm., 50:721.
Watkins, W. A., Schevill, W. E., and Ray, C., 1971, Underwater
    sounds of Monodon (narwhal), J. Acoust. Soc. Am., 49:595.
Watkins, W. A., and Schevill, W. E., 1972, Sound source location
    by arrival-times on a nonrigid three-dimensional hydrophone
    array, Deep-Sea Research, 19:691.
Watkins, W. A., 1974, Bandwidth limitations and analysis of ceta-
    cean sounds, with comments on delphinid sonar measurements
    and analysis, (K. J. Diercks, R. T., Trochta, and W. E.
    Evans, J. Acoust. Soc. Am., 54:200 (1973)). J. Acoust. Soc.
    Am., 55:849.
Watkins, W. A., and Schevill, W. E., 1974, Listening to Hawaiian
    spinner porpoises, Stenella cf. longirostris, with a three-
    dimensional hydrophone array, J. Mamm., 55:319.
Watkins, W. A., Schevill, W.E., and Best, P. B., 1977,Underwater
    sounds of Cephalorhynchus heavisidii (Mammalia: Cetacea),
    J. Mamm., 58:316.

SIGNAL CHARACTERISTICS FOR TARGET LOCALIZATION AND DISCRIMINATION

K. Jerome Diercks

Applied Research Laboratories

University of Texas, Austin, Texas 78712

Included in this review are hypotheses regarding exploitable clue structure in target echoes, analyses of echolocation signal forms and how the characteristics of signals might influence echolocation behavior, and explicit, although heuristic, models of specific echolocation phenomena.

EXPLOITABLE ECHO CLUE STRUCTURE

A target in water may present a complex idiosyncratic response to broadband insonification, a phenomenon of the geometrical and physical (elastic) properties of the target. Targets differing in one or more properties will present different responses (echoes) to the same insonification (Hickling, 1962, 1964; Metsaveer, 1971; Ryan, 1978). A target that is insonified by a signal that is long in duration (relative to the time extent of the target) and swept in frequency will yield an echo whose amplitude fluctuates as the frequency changes (describing the "frequency response" of the target over the swept band), a result of interference between waves reflected from different boundaries or traversing different paths within the scatterer. Locally intense amplitude peaks spanning narrow frequency bands within the swept band of the echo may occur, and it is speculated that these may be detected and exploited by the animal to effect discrimination between differing targets. Also, the amplitude fluctuations within the echo may be periodic with one or more periods idiosyncratic of the target, and differences in periodicities or period patterns may be exploited for target differentiation.

In a purely theoretical work Koshovyy and Mykhaylovs'kyy (1972) assess the utility of both the local peak and periodicity clues for differentiating simple geometric forms (cylinders) that differ in both size and material. They calculate the frequency response for a specific metallic cylinder (Duralumin) over the nondimensional frequency range $0 < ka < 25$ ($ka = 2\pi a/\lambda$, where a = target radius and λ = wavelength in the surrounding medium). This response displays a large amplitude peak localized around $ka = 2$, with lesser peaks (or well delineated nulls) occurring at intervals of $\Delta ka \simeq 3$. It is assumed by the authors that cylinders of other materials will exhibit analogous peaks (at other frequencies) in their responses over the same frequency band--or, by implication, the band that is relevant to the echolocator--which differentiate the materials. The occurrence of such peaks may or may not occur (Hickling, 1962, 1964; Ryan, 1978) rendering this clue unreliable.

However, these authors further show a linear relationship between the velocity of a compressional wave in the target material and the principal (reviewer's emphasis) periodicities of the target's responses, which does provide for a reliable (and psychophysically manageable) discriminatory clue. Targets that differ slightly in the periodicities of their frequency responses would be more difficult to discriminate than those with larger differences, a prediction borne out by experiment (Babkin et al, 1971; Dubrovskii and Titov, 1975). Dubrovskii and Titov, with a bottlenose dolphin T. truncatus, showed a nominal 0.5 ka difference between frequency response periodicities was required for reliable discrimination between two sets of different metal spheres.

Diercks et al. (1968, 1975), using recorded swept frequency (FM) echoes from different metallic spheres, demonstrated the utilization of these clues--local frequency peaks, and response periodicities--by human listeners. Their results also indicated a 0.4-0.5 ka difference in response periodicities required for reliable discrimination. They further showed that although discrimination could be effected by the amplitude variations of the echo envelope (AM-only echo), the task was noticeably easier when each amplitude peak was uniquely associated with a different localized frequency band (FM echo).

While this "model" of echo (target) discrimination has received some experimental support, its validity hinges on the assumed signal form, a relatively long duration swept frequency pulse, which is not characteristic of delphinid echolocators (at least in captivity). It is generally accepted that the delphinid echolocation signal form is impulsive in nature; specific parameter values are species dependent (Diercks, 1972; Evans, 1973).

A target that is insonified by an impulsive signal will return a train of image, or at least similar pulses to the receiver,
depending upon the structural and acoustical complexity of the
scatterer (Diercks and Hickling, 1967; Shirley and Diercks, 1970;
Ryan, 1978). This echo form is akin to the impulse response of
the target, and the Fourier spectrum of the echo is the frequency
response of the target (over the bandwidth of the signal).

It is possible, although unlikely, that the animal performs
a time-amplitude analysis of echoes (impulse trains) from
differing targets, detects differences in the times of occurrence
and relative amplitudes of the pulses making up the echoes from
each, and discriminates on this clue structure. It is more likely
that the animal performs a time-frequency transformation of the
echo yielding a neural equivalent of the target's frequency
response which is then assessed in a manner analogous to that
postulated above for swept frequency signals, i.e., detection of
differences in (transformed) echo "periodicities or period
patterns" and of pitch differences between localized amplitude
peaks.

Thus, the model advanced by Koshovyy and Mykhaylovs'kyy is
applicable to both signal forms, depending on the assumptions
made about the transmitter or receiver.

ANALYSES OF SIGNAL FORMS

While the Russian literature on delphinid echolocation
abundantly displays diverse signal forms recorded during various
echolocation tasks (Ayrapet'yants et al., 1969; Dubrovskiy et al.,
1971; Zaslavskii, 1972; Romanenko, 1973, 1974; Ayrapet'yants and
Konstantinov, 1974; Golubkov et al., 1975), analyses and charac-
terizations of these have appeared simplistic or do not exist.
In general, their results have impacted speculation on the anatomical nature and location of the signal generator. Other results,
discussed in context below, have caused or influenced the development of conceptual models of specific echolocation phenomena.

Dziedzic, Escudié, and their colleagues in France have examined signal forms of various species of echolocating dolphins,
including the low frequency emissions of Phoecena phoecena
(Escudié et al., 1971; Dziedzic et al., 1974; Dziedzic et al.,
1975; Dziedzic and Alcuri, 1977; Dziedzic et al., 1977). Using
the ambiguity function

$$\chi_s(\tau,\eta) = \int_T S(t)S(\eta t-\tau)dt \quad ,$$

where $S(t)$ = echolocation signal,

     τ = echo signal delay, and

     η = Doppler coefficient,

these investigators have shown that the generic forms of signals
employed by echolocating dolphins--impulsive "clicks" and short
duration, frequency modulated "chirps"--are highly tolerant to
relative motion between target and emitter, i.e., they are
Doppler tolerant. For example, for the impulsive signals of
Delphinus delphis, for relative velocities spanning the range −75
to 75 m/sec, much larger than is likely ever to occur in reality,
the Doppler coefficient η varies from 0.95 to 1.05, and over this
range $\chi_s(\tau,\eta) \simeq \chi_s(0,1)$. Similar results were obtained for a
short duration chirp recorded from this species, however, with
increased ambiguity in range resolution (Dziedzic et al., 1974).

Analysis of low frequency chirp and click signals from
Phoecena phoecena revealed similar Doppler tolerances (Dziedzic
et al., 1977). These signals displayed increasing frequency modulation, i.e., signal period decreasing with duration, in contrast
to comparably Doppler tolerant signals with decreasing modulation
examined by other investigators (Altes, 1974; Altes and Skinner,
1977).

The import of these results is that with a broad bandwidth,
e.g., 3 to 4 octaves, phase perturbations of the signal, regardless
of its form are ineffectual.

From the study of the low frequency signals of Ph. ph. it was
concluded that the FM chirp, in addition to its Doppler tolerance,
is characteristically like certain theoretically optimal signals
for use in a noisy and/or reverberant environment, implying its
use in these conditions, with employment of a relatively much
higher frequency band emission when required for increased resolution (Dziedzic et al., 1977).

An acoustical simulation by Dziedzic et al. (1975) of a
dolphin-like echolocation click showed effects of spatiotemporal
filtering on the signal parameters: the spectrum was narrowed
and displaced toward a lower frequency band. The ambiguity surfaces recorded for the signal recorded on the axis of transmission
and at an angle of 12° off this axis both revealed a strong tolerance to Doppler, i.e., insensitivity to relative velocities
vectored along off-transmission axes.

A study of form recognition by Tursiops truncatus revealed
pronounced changes in signal characteristics occurring at a

"threshold" distance from the targets (4 m distance from 20 cm diam polygonal plates) (Dziedzic and Alcuri, 1977). A typical signal for animal-target distances greater than 4 m was like a damped sinusoid, or an impulse followed by ringing, with a spectral peak at 50-55 kHz and range resolution (first zero of autocorrelation) of nominal 55 cm. At distances shorter than 4 m the typical signal was like a single sine wave, at much lower frequency, 9-30 kHz, with 3 to 5 times less range resolution, 130-260 cm.

The conclusion reached in this study was that target differentiation occurred during the initial phase of approach, at distances greater than 4 m, using the higher frequency broader bandwidth signal. That the animal significantly degraded its resolution during the final 4 m of approach to the targets is difficult to rationalize. It is perhaps worth noting that the "character" of these results supports the observation by Au et al., (1974) of use of a significantly higher frequency emission by Tursiops at long ranges in open water than that reported by others for this species at relatively shorter distances in pools (Diercks et al., 1971, 1973; however, see also, Golubkov et al., 1975).

MODELS OF ECHOLOCATION PHENOMENA

Livshits (1975) postulates a pattern recognition model for target detection, identification (or differentiation), and ranging that employs a dual channel correlator with a pattern memory in one channel. Echoes from targets at different ranges are crosscorrelated with a delayed stored replica of the emitted signal. Spatiotemporal filtering, Doppler, etc. will distort the echo waveform, leading to correlations less than maximal. At the same time these same echoes are being compared, by crosscorrelation, with stored (remembered) echo patterns in the "identification" channel of the processor. The stored pattern that correlates best with each signal is then crosscorrelated with the delayed stored replica of the emission in the detection channel.

Only echoes whose crosscorrelations exceeded an initial threshold value during the first correlation are strobed for the second correlation, which, if this exceeds a second, higher threshold, constitutes target recognition (through memory). It is possible that the initial correlation of an echo signal with the stored emission could exceed the higher (second) threshold; memory scanning and re-correlation would still need be performed to accomplish recognition.

A corollary procedure in this model would be for the animal to emit one or more of the stored signal patterns identified during the first memory search in hopes of achieving an enhanced, suprathreshold crosscorrelation of the echo following this second emission, i.e., a process of "signal design." If the latter correlation exceeded the higher threshold value, target recognition would again have been accomplished. However, it would not be unlikely that the correlation with the modified stored emission might be degraded relative to the first time, leaving the echolocator uncertain of how to proceed.

Echolocation signal modifications that are apparently target related have been reported, but the data are equivocal. Although signal design is an appealing notion, incontestable evidence of its occurrence in echolocating animals is lacking.

It is possible that this model could be tested by accumulating and examining large numbers of signals obtained using acoustically known targets to (1) yield a reliable evaluation of signal variation, and (2) to compare reliable signal variations with target acoustical characteristics to yield an assessment of the apparent occurrence and nature of signal "design."

Livshits (1974) has also postulated a model for rapid "beam scanning" by dolphins based on the assumptions of a swept frequency, or frequency modulated transmission and a dispersive medium between the signal generator and the radiating surface of the emitter. Although anatomy is not specifically addressed, it is implicit that the signal generator is thought to lie near the nasal plugs or sacs and that the dispersive medium is the melon.

In its simplest expression, the model assumes that the transmission path lengths from the generator to the radiating surface are unequal, as shown below.

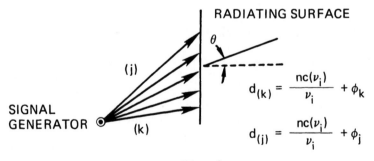

Fig. 1.

This causes at frequency ν_i a "phase gradient" across the aperture of the radiator leading to propagation along an axis oblique to the radiating surface. A change to frequency $\nu_i + \Delta\nu$ will cause a different phase gradient or pattern across the aperture leading to propagation along an axis directed at a slightly different angle from the surface. It is seemingly unnecessary to invoke a frequency dependence of the transmission velocity in the medium (melon), i.e., dispersion, except to bound the phase gradient across the aperture, e.g., $-\pi < \phi \leq \pi$.

Assuming that a narrow beam is formed at each discrete frequency ν_i and radiated in a direction Θ_i, then sweeping frequency from ν_i to ν_k will cause the radiated beam to scan from Θ_i to Θ_k. The scan rate will depend on the rate of frequency change. It follows that only a fractional part of the duration of the swept frequency signal is radiated in any direction Θ--akin to a rotating directed transmission (RDT)--thereby increasing range resolution.

An echo, which is a fractional part of the generated waveform, is crosscorrelated with a delayed stored replica of the of the latter to yield a narrow correlation peak at the locus of ν corresponding to Θ thereby providing a target bearing measure (assuming synchronous transmission), and through pulse compression by correlation, further enhanced range resolution.

While there seems to be evidence for the occurrence of rapid scanning (Bel'kovich and Reznikov, 1971) and of lateral beam steering (Ayrapet'yants and Konstantinov, 1974), the many shortcomings of the model should be evident. The formation of a narrow beam at ν_i directed along Θ_i assumes a steady state across the radiating aperture. The nonsteady state inherent in a swept frequency transmission does not preclude beam sweeping but necessitates an elaborate frequency dependent velocity function for the transmission medium to provide the well behaved beam forming and scanning process described (which very well _may_ exist). The occurrence and characteristics of radiated sidelobes and the ambiguities they might cause are circumvented by assuming that they do not exist (and if they do, phase shading _may_ be employed to suppress them).

The processing scheme seems particularly naive. Crosscorrelation of the expected short duration echo--perhaps only one or two cycles of the swept frequency signal--with the swept frequency signal would result in correlation sidelobes of amplitude nearly equal to the correlation "peak" and, thus, would cause appreciable ambiguity in target bearing resolution. Also, such an echo would have a small time-bandwidth product $\simeq 1$, providing little processing gain and negligible enhancement of range resolution.

If it is accepted that a narrow beam is formed and rapidly scanned to present effectively only a short duration transmission, like a transient, that differs in frequency with direction, which result has been observed in animal studi-s (Diercks, et al., 1971; Romanenko, 1973) and in syntheses (Diercks, et al., 1973; Dziedzic et al., 1975), then the specification of a long duration FM signal is not damaging. Indeed, the model could readily manage either the stability (Diercks et al., 1971) or instability (Romanenko, 1973) of the signals recorded from animals.

The appearance of direction dependent frequency differences is not sufficient to "demonstrate" the model; these would also appear with a broad bandwidth radiation directed along the axis of the radiator (i.e., rostrally in the animal). The model provides impetus for careful observation of apparent and obvious scanning during animal echolocation. Reliable observation of apparent scanning not correlated with related head or body motions would indicate that the transmission model described could be operating and would thus impact many extant concepts and hypotheses regarding the anatomical structures used for and the mechanics of signal generation.

REFERENCES

Altes, R. A., 1974, Study of animal signals and neural processing with applications to advanced sonar systems, Rept. NUC TP 384, San Diego, Ca. 92132.

Altes, R. A., and Skinner, D. P., 1977, Sonar velocity resolution with a linear-period-modulated pulse, J. Acoust. Soc. Am., 61:1019.

Au, W. W. L., Floyd, R. W., Penner, R. H., and Murchison, A. E., 1974, Measurement of echolocation signals of the Atlantic bottlenose dolphin. Tursiops truncatus, Montagu, in open waters, J. Acoust. Soc. Am., 56;1280.

Ayrapet'yants, E. S., Golubkov, A. G., Ershova, I. V., Zhezherin, A. R., Zvorykin, V. N., Korolev, V. I., 1969, Echolocational differentiation and description of radiated pulsations in dolphins, Doklady Akad. Nauk SSSR, 188:1197.

Ayrapet'yants, E. S., and Konstantinov, A. E., 1974, "Ekholokatsiya V Prirode", JPRS 63328-1, 2 , Leningrad.

Babkin, V. P., Dubrovskiy, N. A., Krasnov, P. S., Titov, A. A., 1971, Discrimination of the materials of spherical targets by the bottlenose dolphin, Tr. Akust. Inst., 17:80.

Bel'kovich, V. M., and Reznikov, A. Y., 1971, What's new in dolphin sonar?, Priroda, 11:84.

Diercks, K. J., 1972, "Biological Sonar Systems: a Bionics Survey," Applied Research Laboratories Tech. Rpt. N° 72-34 (ARL-TR-72-34), 190 (5 September, 1972).

Diercks, K. J., 1975, Listener discrimination of broadband FM echoes from simple elastic targets, J. Acoust. Soc. Am., 58, Suppl. 1, S103.

Diercks, K. J., and Hickling, R., 1967, Echoes from hollow aluminum spheres in water, J. Acoust. Soc. Am., 41:380.

Diercks, K. J., Ryan, W. W., Mikeska, E. E., and Weisser, F. L., 1968, (see Diercks, 1975), Applied Research Laboratories Tech. Memo., N° 68-22 (ARL-TM-68-22), 46, (25 November, 1968).

Diercks, K. J., Trochta, R. T., and Evans, W. E., 1973, Delphinid sonar: measurement and analysis, J. Acoust. Soc. Am., 54: 200.

Diercks, K. J., Trochta, R. T., Greenlaw, C. F., and Evans, W. E., 1971, Recording and analysis of dolphin echolocation signals, J. Acoust. Soc. Am., 49:1729.

Dubrovskiy, N. A., Krasnov, P. S., and Titov, A. A., 1971, On the emission of echolocation signals by the Azov Sea harbor porpoise, Akust. Zh., 16:521; Sov. Phys. Acoust., 16:444.

Dziedzic, A., Escudié, B., Guillard, P., Héllion, A., 1974, Evidence of tolerance to the effect of Doppler in the sonar emissions of Delphinus delphis, C. R. Acad. Sci. Paris, 279(Ser. D):1313.

Dziedzic, A., Escudié, B., and Héllion, A., 1975, Realistic methods for analysis of biological sonar signals, Ann. Telecomm., 30,270.

Dziedzic, A., and Alcuri, G., 1977, Acoustic recognition of forms and characteristics of the sonar signals of Tursiops truncatus, C. R. Acad. Sci. Paris, 285(Ser. D):981.

Dziedzic, A., Chiollaz, M., Escudié, B., and Héllion, A., 1977, Some properties of low frequency sonar signals of the dolphin Phocoena phocoena, Acustica, 37:258.

Escudié, B., Héllion, A., and Dziedzic, A., 1971, Results from studies of air and marine sonars through signal processing and spectral analysis, in: "Proc. Troisième Colloque sur le Traitement du Signal et ses Applications", Nice, 1-5 June, 1971:533.

Evans, W. E., 1973, Echolocation by marine delphinids and one species of fresh-water dolphin, J. Acoust. Soc. Am., 54:191.

Golubkov, A. G., Korolev, V. I., Antonov, V. A., and Ignat'eva, E. A., 1975, Comparison of dolphin echolocation signals with results of optimal signal calculations, Doklady Akad. Nauk SSSR, 223:1251.

Hickling, R., 1962, Analysis of echoes from a solid elastic sphere in water, J. Acoust. Soc. Am., 34:1582.

Hickling, R., 1964, Analysis of echoes from a hollow metallic sphere in water, J. Acoust. Soc. Am., 36:1124.

Koshovyy, V. V., and Mykhaylovs'kyy, V. M., 1972, The operating principle of the echolocating system of marine animals, Dopovidi Akad. Nauk Ukrayins'kayi RSR, Ser. A, Fizyko-techhn. ta matemat. Nauky, Ukranian, 12:1097, JPRS 58344.

Livshits, M. S., 1974, Some properties of dolphin hydrolocator from the viewpoint of correlation hypothesis, Biofizika, 19: 916, JPRS 64329 (17 March, 1975).

Livshits, M. S., 1975, Correlation model of the recognition of objects by echolocating animals, Biofizika, 20:920.

Metsaveer, J., 1971, Algorithm for calculation of echo-pulses from elastic spherical shells in fluid by summation of wave groups, Inst. of Cybernetics, Acad. Sci. Eston. SSR Tallinn, Preprint 3, 40.

Romanenko, Y. V., 1973, Investigating generation of echolocation pulses in dolphins, Zool. Zh., Moscow, 11:1698, JPRS 61553.

Romanenko, Y. V., 1974, Physical fundamentals of bioacoustics, Fizicheskiye Osnovy Bioakustik, Moscow, JPRS 63923.

Ryan, W. W., Jr., 1978, Acoustical reflections from aluminum cylindrical shells immersed in water, J. Acoust. Soc. Am., 64: 1159.

Shirley, D. J. and Diercks, K. J., 1970, Analysis of the frequency response of simple geometric targets, J. Acoust. Soc. Am., 48:1275.

Zaslavskii, G. L., 1972, Investigation of the location signals of the bottlenosed dolphin using a two-channel system of recording, Biofizika, 4:717, Biophysics, 17:753.

ECHOLOCATION SIGNALS AND ECHOES IN AIR

J. David Pye

Department of Zoology and Comparative Physiology,

Queen Mary College, Mile End Road, London E1 4NS

1. INTRODUCTION : A PERSONAL VIEW

Since the Frascati meeting in 1966, there has been considerable expansion in our understanding of the acoustic structures
used by bats and birds for echolocation in air. In part this has
been due to the development of instrumental techniques which were
already undergoing something of a revolution at the time. High-
speed tape-recorders suitable for bat recording had only become
available a few years previously and their use in conjunction with
Sonagraph sound spectrum analysers was then becoming widespread.
This combination was rapidly replacing the analysis of waveforms
by photographic sampling in real time from an oscilloscope screen
which, due to complex phase changes, is open to ambiguous interpretation of harmonics or frequency changes.

The sonagraph has sometimes been accused of generating
spurious harmonics, but if used with oscilloscope level monitors
which can cope with the peak levels of brief signals in a way that
V-U meters cannot, it is capable of displaying the harmonic
structure with considerable clarity and fidelity. It is true that
phase information is lost and the resolution of rapid transient
structures is poor, but these features can be depicted in
oscilloscope waveform displays which then become complementary
to the spectrum x time display of the sonagram.

Now the fashion is again changing and analyses are increasingly
being undertaken by digital computers instead of the preprogrammed
analogue sonagraph. In the case of spectrum x time displays, the
advantages are somewhat debatable since the 'mountain range' display

may conceal smaller 'foothills' lying 'behind', whereas the
Z-modulation of a sonagram reveals all, especially if used with
contouring.   But the advantage of the digital computer lies in
its flexibility of programming and in the variety of other derived
data that it can present.   Thus the analysis of echolocation
signals is again moving into a new era of greater subtlety, where
description is giving way to statements of real functional
significance.

This development was foreshadowed 13 years ago at Frascati by
the pioneering work of D. A. Cahlander. He not only presented
the first application of radar theory to the echolocation signals
of bats but also derived formulations of the theory which are
applicable to the 'wideband' nature of many such signals.
Understanding of the need for such an approach, and of the need to
revise the assumptions that are made with relatively narrowband
radar signals, is now rapidly increasing.

Other developments during the period under review include the
discovery of some new kinds of pulse structure and a much wider
taxonomic sample of species studied: it is now possible to say
something about pulses detected from about 200 species from 16
families of bats.   That is nearly one third of all species, the
major taxonomic omissions being three tiny families comprising
only four species in all.

On the other hand, the miniaturisation of recording equipment,
so that it can easily be used from battery power in the field, has
made it increasingly clear that many species are flexible in their
operation and cannot be specified by a single 'definitive' pulse
structure.   Several species that are highly stereotyped in their
behaviour indoors are found to vary from apparently pure Doppler
operation to high-resolution, Doppler-tolerant range measuring
when observed under varying and more natural conditions.   This
should increase our caution when speaking of 'the pulses' of a
given species, since these may only be used in the situation
observed and really represent part of a larger repertoire.   Moreover the data available at present do contain some taxonomic
surprises and raise puzzling problems about the evolution of
echolocation signals in bats.

Finally there has been some progress in understanding the
generation of the signals, their propagation in the atmosphere and
the properties of certain important targets as echo reflectors.
An increasing number of nocturnal insects are now known to detect
bats and to take evasive action;   three more families have been
added in the last 13 years and the details of this behaviour are
now better understood.

These are the matters that form the content of this review and
they will now be set out formally and as objectively as possible.
It is hoped that the above introduction will help to put them into
context historically, functionally and with regard to the
reviewer's brief for this meeting.

2. FREQUENCY PATTERNS

The signals emitted by, and assumed to be used by, different
bats show a great variety of structure which makes them difficult
to classify in a single, systematic continuum. To these must be
added the sounds emitted by other animals, especially the birds.
Frequency, frequency pattern, intensity, duration and harmonic
content all show a range of characteristics, many of them assorted
independently in the total population. Simmons (in press) has
discussed the development of these signals mainly by reference to
frequency patterns, duration and harmonic content, while Novick
(1971) discussed pulse design almost exclusively by reference to
duration.

It is, however, now clear from radar/sonar theory that
frequency pattern plays a major role in determining the accuracy
and resolution of different signals. Frequency pattern also
correlates closely with the characteristics of the auditory system
and the general behaviour of each species. It is frequency pattern,
therefore, that probably forms the best basis for classification and
comparison, although other factors must not be overlooked.   The
sounds that have been recorded from bats then fall into three
general categories: broadband impulse sounds, frequency modulated
(f.m.) sweep sounds and constant frequency (c.f.) sounds.   Each
of these will be discussed separately but, since the distinctions
are not always sharp, a more general treatment must be given first.

Broadband impulse sounds are those in which there is no
clearly defined, coherent 'carrier' frequency, no evidence of
frequency modulation and an amplitude pattern that is rapid and
transient - ie. 'clicks'. A number of species come close to this
description, notably Rousettus and the birds. It is debatable,
however, whether the 15-20 ms multi-click bursts of Aerodramus
maximus and Steatornis show a coherent wave structure or merely
consist of independent transient impulses. Even the very brief
clicks of Aerodramus vanikorensis and other swiftlets can be
considered as simply a packet of a very few cycles (Figure 3).

F.m. sweep and c.f. signals would seem to be self-explanatory
terms and radar theory suggests that they are used to measure
target range and radial velocity respectively. However, purely
c.f. pulses are rare and much more commonly a long c.f. portion

is associated with an initial sweep (upwards or downwards) and/or
a terminal downsweep.   The variety of such intermediate patterns
is shown diagrammatically in Figure 1, while some actual analyses
are given in Figure 2.

Furthermore, some pulses have such brief periods of constant
frequency that it is doubtful whether they confer any appreciable
degree of velocity sensitivity.   Simmons et al (1975) proposed
two categories of c.f./f.m. pulses: 'short' c.f./f.m. with
durations less than 5 ms, and 'long' c.f./f.m. with durations
greater than 6 ms.   This scheme has recently been challenged by
Gustafson and Schnitzler (in press) on the grounds that some
species emit pulses in each category (it is not clear whether the
authors wish in this way to categorise individual pulses or the
species that emit them; the subject of flexibility of acoustic
behaviour by a single species is discussed in Section 7).

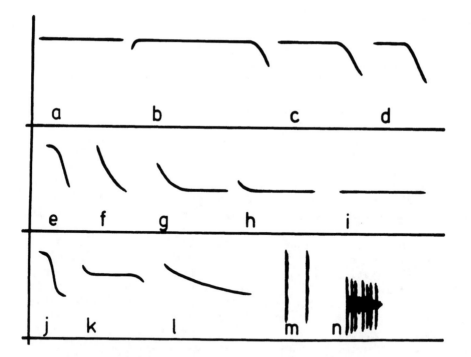

Figure 1. Some frequency patterns of echolocation sounds used in
air; no harmonic structure is indicated. (a) to (i) form a
sequence from high c.f. through c.f./f.m., pure f.m. and f.m./c.f.
to low c.f.   The lowest row shows two forms of 'sigmoid' pulse
(j & k), shallow f.m. (l), a double broadband impulse sound (m)
and a multi-click burst (n).

Pye (1978) has calculated the velocity sensitivity of a range
of c.f. signals by comparison with the value of 0.1 m.s^{-1} measured
in <u>Rhinolophus ferrumequinum</u> by Simmons (1974); the theoretical
limits are better than this by a factor of just over two.   Even
c.f. durations of only 5 ms should permit a sensitivity better
than 1 m.s^{-1} at 35 kHz, reducing to 0.23 m.s^{-1} at 150 kHz.
Although Simmons et al (1975) appear to disagree, Pye argued that
a sensitivity little better than flight velocity, say 5-10 m.s^{-1},
would be useful at least for target detection.   On the basis of
this argument, and for convenience, no distinction will be made
here between different durations of constant frequency;   all
signals containing any perceptible c.f. component will be classed
as c.f. pulses, whether or not this is associated with f.m. sweeps.

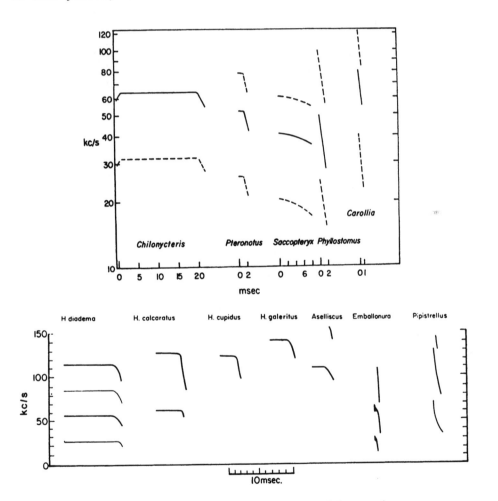

Figure 2.   Diagrams of frequency patterns and harmonic structures
in pulses of bats from the New World (Grinnell, 1970) and the Old
World (Grinnell and Hagiwara, 1972).

Figure 1 also shows a 'sigmoid' f.m. pattern (j) and the pattern which Griffin called a 'pothook' (e). The frequency is not quite constant in either of these cases and they are perhaps best classed as f.m. pulses for want of information about their functional significance.

Descriptions of bat signals owe much to the pioneering comparative studies of Griffin, Novick, Möhres and Kulzer in 1955-63 although more recent techniques have led to a revision of some details. In the following three sections material from papers published since Frascati will be reviewed together with a general picture of our present understanding. Not all papers will be cited individually since the overall account owes much to previous work. Further details have been culled from: Bradbury (1970), Griffin and Simmons (1974), Grinnell (1970), Grinnell and Hagiwara (1972a, b), Konstantinov and Sokolov (1967), Möhres and Neuweiler (1966), Peff and Simmons (1971), Pye (1968a, 1972, 1973, 1978), Pye and Roberts (1970), Roberts (1972a, b; 1973), Schnitzler (1968, 1970a, b; 1971, 1973), Simmons (1969, 1970, 1971, 1973), Simmons et al (1979), Suthers (1967), Suthers et al (1972) and Suthers and Fattu (1973). Larger reviews for the period include: Airapet'yants and Konstantinov (1970/1973), Sales and Pye (1974) and Novick (1977).

3. BROADBAND IMPULSE SOUNDS

Sounds of this general category are used in air by the two known groups of echolocating birds: the South American oil bird, Steatornis caripensis, and some of the South-East Asian cave swiftlets, Aerodramus (formerly Collocalia). They are also emitted by megachiropteran fruit bats of the genus Rousettus and probably by the closely related genus Stenonycteris.

All these sounds share one feature: they are composed of brief, click-like waveforms that are not coherent, or only barely so, within each pulse; the rapid, transient-like nature of their amplitude patterns leads to considerable band-spreading. There is no clear carrier frequency or indication of frequency modulation. The sounds fall into two groups as recently pointed out by Medway and Pye (1977): double clicks with an internal interval of 15-40 ms as in Rousettus and some Aerodramus spp, and 'single' clicks as in Steatornis and Aerodramus maximus. The 'single' clicks, however, are individually complex, each consisting of a rapid burst of sound impulses lasting for up to 25 ms. The spectral range is largely ultrasonic in Rousettus (though also clearly audible to man) while in all the birds there is little or no energy above 10-15 kHz.

No new work on Steatornis is known to this reviewer since 1966. However, Sales and Pye (1974) and Medway and Pye (1977) have analysed an original recording (made in Trinidad in 1963) and

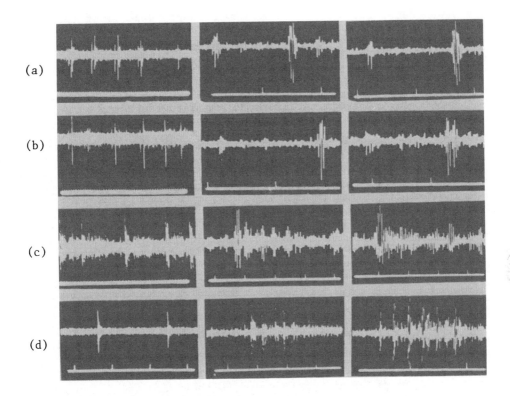

Figure 3. Oscillograms of some broadband impulse sounds.
(a) Aerodramus fuciphagus: a train of 5 double clicks and 2
clicks in detail; (b) Aerodramus vanikorensis: a train of 6
double clicks and 2 clicks in detail; (c) Aerodramus maximus:
a train of 5 multi-click bursts and 2 bursts in detail; (d) a
double click of Rousettus aegyptiacus (left) and two multi-click
bursts of Steatornis caripensis in detail. All time markers =
10 ms. From Medway and Pye (1977).

have generally confirmed the earlier (1954) descriptions by
Griffin.   Energy was found from 1 kHz to 12 kHz with the peak at
2-4 kHz and a lesser one at 5-7 kHz (Griffin reported 7-8 kHz).
Each click consisted of up to seven impulses lasting up to 20 ms
in all (Figure 3d); click rate varied from 4-7 s^{-1}, when flying
round a large chamber of the cave, to 11-12 s^{-1} when landing on
the nest.   Observation of birds feeding at a nut tree suggested
that no signals are emitted outside the cave although the night
was dark.

Medway and Pye (1977) have reviewed the recent literature on
Aerodramus, including papers by Fenton (1975), Griffin and Suthers
(1970), Medway (1967, 1969), Medway and Wells (1969), Pecotich

(1974) and Penny (1975). There is now evidence that 11 of the 15 species can echolocate by emitting clicks, while 2 appear not to and the status of 2 is unknown. Four species have been studied in sufficient detail for descriptions of their sounds to be compared:

Table 1. Echolocation sounds in Aerodramus spp.

Species	Click	kHz	Source
A.h.hirundinaceus	Double	2-15+	Fenton (1975)
A.vanikorensis granti	Double	2-16	Griffin & Suthers (1970)
A.v.salangana	Double	2-9+	Medway & Pye (1977)
A.v.natunae	Double	2-10	Medway & Pye (1977)
A.f.fuciphagus	Double	2-9+	Medway & Pye (1977)
A.maximum lowi	'Single'	1.5-7+	Medway (1959, Medway & Pye (1977).

Double clicks have a constant internal interval of 15-16 ms, but Griffin and Suthers (1970) reported that in A.v.granti the clicks were occasionally treble. Although the spectrum is much lower and the first click is of lower amplitude than the second, the double clicks strongly resemble those of Rousettus (Figure 3, a, b, d). The frequency ranges given above are revised from earlier recordings. This conflicts with the argument of Sales and Pye (1974) that a difference in wavelength could account for the superior obstacle avoidance ability of A.vanikorensis (avoids 6 mm wires with 4.5-7.5 kHz; Griffin and Suthers, 1970) when compared with A.fuciphagus (hits wooden rods of 1 cm using 1.5-4.5 kHz; Medway, 1967, 1969). It seems that further work with consistent techniques is necessary, but Fenton (1975) found that both sounds and the ability to avoid thin rods were similar in A.hirundinaceus to those reported for A.vanikorensis by Griffin and Suthers (1970).

Aerodramus maximus clicks appear to be complex and 'messy' when recorded indoors, as if superimposed on multiple echoes. In a few of the clearest cases, however, it is apparent that each click consists of a burst of about six impulses lasting for a little over 20 ms (Figure 3c) and is thus very similar to the clicks of Steatornis (Medway and Pye, 1977). The clicks of A.unicolor (then Collocalia brevirostris unicolor), described briefly by Novick (1959), may also be of this kind.

There has been little new work on the echolocation sounds of rousettine fruit bats. Sales and Pye (1974) examined new recordings of Rousettus aegyptiacus from Uganda and of R.amplexicaudatus from Malaysia. No differences could be distinguished in the click structures or spectra (Figure 4) although the interclick interval was about 50% greater in R.amplexicaudatus. A computer interval

histogram run on a recording of hundreds of clicks from several
R.aegyptiacus flying in a large outdoor aviary, showed a peak at
20 ms and negligible scatter outside 18-22 ms. This is considerably
less than observed by Kulzer (1958, 1960) in this species from
Egypt (20-44 ms) and may represent a geographical variation in
behaviour.   Grinnell and Hagiwara (1972b) found that clicks of
R.amplexicaudatus from New Guinea had a spectrum of at least
10-90 kHz (peak 20-60 kHz) and an interclick interval that was
consistently about 30 ms. (Woolf, 1974, briefly reported pairing
of f.m. pulses in young Eptesicus fuscus, with an interpulse
interval of 20-40 ms).

     Finally, Kingdon (1974) has reported that the closely related
Stenonycteris lanosus lives in dark caves like Rousettus and is
able to echolocate by emitting clearly audible clicks.   No
recordings or analyses were made but the sounds are said to be
"chinks, that seem .... quieter and more 'musical' than the clicks
of Rousettus". If there were "only the faintest glimmer" of
light, they flew silently.   This is in contrast to the less closely
related Lissonycteris angolensis which Lawrence and Novick (1963)
showed to have no echolocation.

     Echolocation by shrews was first reported for three species
of Sorex and Blarina brevicauda by Gould et al in 1964.   The
sounds they recorded showed a rapid series of closely spaced
clicks of wide bandwidth.   Grunwald (1969) was unable to confirm
echolocation in two species of Crocidura although ultrasounds were
recorded while sniffing. Sales and Pye (1974) reported clicks with
energy from 10 kHz to 70-110 kHz from C.russula (a species used by
Grunwald) as they investigated a microphone;   these clicks were
emitted singly or in rapid bursts of 8-25 with interclick
intervals as short as 1 ms. Buchler (1976) has produced fresh

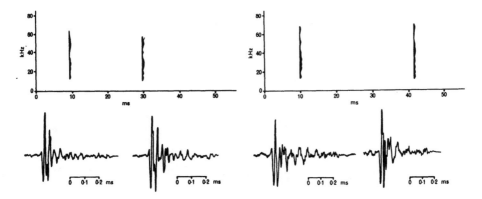

Figure 4. Sonagrams and oscillograms of double clicks from
Rousettus aegyptiacus (left) and R.amplexicaudatus (right).
From Sales and Pye (1974).

evidence of echolocation in <u>Sorex vagrans</u>.   Recordings from this
species showed double clicks with a spectrum of 18-60 kHz (peak at
20-40 kHz) and an interclick interval averaging 8.2 ms (range
5.3-11.7 ms).   The first click was slightly "shorter" than the
second.   The resemblance between these sounds and those of birds
and fruit bats is striking.

Sounds produced by other terrestrial mammals, including tenrecs
and rats are reviewed in posters at this conference by Buchler and
Mitz and by Chase respectively.

4.   F.M. SWEEP PULSES

The short, harmonically rather pure, deep frequency sweep
pulse was originally discovered and intensively investigated by
Griffin and his coworkers (see eg. Griffin, 1958) in <u>Myotis
lucifugus</u>.   It was for some time thought to be typical of all
Vespertilionidae and still seems to be so for many genera under all
conditions observed.   In cruising flight the duration is 1-4 ms
during which the frequency falls by about an octave.   Amplitude
increases to a maximum about the middle of the pulse and then
decays again so that precise duration and frequency limits of the
sweep are difficult to define - values are extended by better
signal-to-noise ratios.

The swept carrier wave is generally the fundamental, for a
weak second harmonic component often appears towards the end of
the pulse.   The sweep has often been referred to as linear f.m.
and sometimes as a logarithmic sweep or a hyperbolic sweep.
However, Cahlander (1967) at Frascati presented analyses made with
an 'uncorrected' period meter (one giving period instead of its
reciprocal, instantaneous frequency) and showed that the period
(1/f) is linear.   Whether this holds more generally for f.m.
sweeps is not yet clear.   But Altes and Titlebaum (1970) showed
theoretically that, because of the broadband nature (ie. high
bandwidth to centre-frequency ratio, often approaching unity), a
linear period sweep is optimally Doppler tolerant (see Section 8).
It is thus ideal for range measurement even if the range is
changing, although a Doppler-coupled ranging error is then introduced.   This kind of pulse thus appears to lie at one end of the
velocity/range measuring continuum and to be highly specialised.

During the interception of prey, negotiation of an obstacle
or landing, the pulse repetition rate increases from 10-15 s^{-1} to
100-200 s^{-1}, the pulse duration falls to 0.25 ms or so, the centre
frequency and sweep width are decreased;   there is often a
pronounced increase in the second harmonic intensity though the
total intensity falls due to weakening of the fundamental.   This
pattern is shown in Figure 5.   The high repetition rate produces

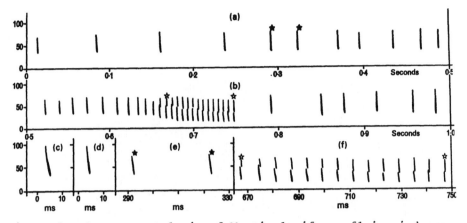

Figure 5. Sonagram analysis of <u>Myotis</u> <u>lucifugus</u> flying indoors.
(a) and (b) are a continuous record including an avoidance 'buzz';
(c) and (d) are loud pulses from another recording; (e) and (f)
show the starred pulses from (a) and (b) in detail. From Sales
and Pye (1974).

a brief buzzing note from an ultrasound detector in real time and
the term 'interception buzz' has come to be applied to the
sequence itself. Cahlander (1967) showed at Frascati that range
discrimination is improved in interception pulses and Doppler
coupling is virtually eliminated from the ambiguity diagram of a
single pulse.

The pure f.m. sweep is not restricted to the Vespertilionidae,
but some other families and even many vespertilionids show a much
greater harmonic content. Here the depth of the fundamental sweep
is reduced but bandwidth is maintained by harmonics up to the fourth
or fifth. A strong second harmonic is seen for instance in the
cruising pulses of <u>Plecotus</u>, <u>Nyctalus</u> and <u>Pipistrellus</u> (Figures 6
and 12). Such pulses are also found in some Emballonuridae,
Natalidae and Molossidae. Higher order harmonics are seen in
orally emitted pulses of some Emballonuridae, Rhinopomatidae,
Mormoopidae and Mystacinidae, and in nasally emitted pulses of
Nycteridae, Megadermatidae, Phyllostomidae and Desmodontidae.

The early work of Griffin and others was achieved by photographic sampling of oscilloscope waveforms. Since some of the
waveforms are complex in short, multiple harmonic sweeps, they are
easily misinterpreted unless supported by spectral analysis as
is now provided by tape recordings. Thus the relative intensities
and order of harmonics may be obscured and it is now known that,
especially in nasally emitting bats, the fundamental is often weak
or indetectable. Also frequency modulation is difficult to detect

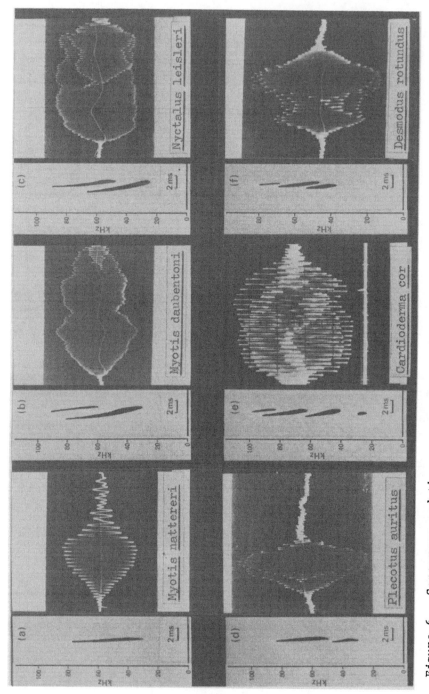

Figure 6. Sonagrams and the corresponding oscillograms of six f.m. pulses with different harmonic complexity. From Sales and Pye (1974).

in short pulses and this is exacerbated by fixed formant filtering
(see Section 9) which accentuates different harmonics in turn at
the same frequency. Thus pulses of Megadermatidae and Nycteridae
were initially described (Novick, 1958a; Mðhres and Neuweiler,
1966) as c.f. pulses with no frequency sweep. Novick (1977) and
others persist in this interpretation although it can be clearly
shown (Sales and Pye, 1974, and elsewhere) that a sweep is present
as in most pulses of the Phyllostomidae (Figure 6). The reviewer
can now confirm this general picture for Lavia frons, Cardioderma
cor, Megaderma lyra, M.spasma, Nycteris thebaica, N.hispida, N.arge,
N.macrotis and N.aethiopica, although the waveforms closely
resemble those published by other authors.

Another variant of the f.m. sweep pulse is shown by Molossidae
and some Vespertilionidae such as Nyctalus (when hunting in the
open). This is a long, shallow sweep (Figure 1, 1). Here there
appears to be little harmonic content although the propagation
distance may suppress higher harmonics. The bandwidth is therefore often quite small and the sweep may become so shallow that
the pulses become effectively constant frequency. Frequency
itself is generally low in these pulses, ranging from 15 kHz to
35 kHz or so. Occasionally, hunting molossids (which show a
great variety of signal structures) produce pulses of a deeper
'sigmoid f.m.' shape as in Figure 1, j.

Other variations of f.m. are associated with periods of
constant frequency and with flexible behaviour as discussed in
Sections 5 and 7.

5. CONSTANT FREQUENCY PULSES

The first c.f. pulses were discovered by Mðhres in 1953 in
Rhinolophus ferrumequinum. Because of their extreme length (c.f.
commonly over 70 ms) the signals of Rhinolophidae might be regarded
as the extreme form of velocity sensing systems. But such signals
always contain a downward frequency sweep at the end and commonly
a smaller upward one at the beginning (Figure 1, b). Pure c.f.
signals with no sweeps are uncommon but are found for instance in
Noctilio or in the outdoor cruising pulses of molossids and some
vespertilionids. These signals may be of 'high' frequency (ie. at
the top of sweeps seen in other pulses) as in Noctilio (Figure 1, a)
or 'low' frequency (at the bottom) as in molossids and vespertilionids (Figure 1, i).

Since Frascati, Schnitzler and his associates have clearly
shown (1968-in press) that Rhinolophus detects upward Doppler
shifts in c.f. echoes and adjusts the emitted frequency so as to
keep wanted echoes at a fixed frequency. This Doppler compensation thus results in a fine variation of emitted c.f. values from

a 'resting' frequency to as much as 8 kHz below this (in
R.ferrumequinum). This behaviour is correlated with striking
auditory adaptations shown by electrophysiology (Neuweiler and his
associates), by behavioural tests (Long) and by aural anatomy
(A. Pye and Bruns). Simmons (1974) has measured velocity sensitivity in R.ferrumequinum by behavioural methods, showing a
response to 0.1 m.s^{-1}, corresponding to the theoretical limit for
a pulse of this frequency and a duration of 20 ms.

Schnitzler has also demonstrated Doppler compensation in
Pteronotus parnellii of the Mormoopidae (Schnitzler 1970a,b, 1971)
and in Asellia tridens of the Hipposideridae (Gustafson and
Schnitzler, in press), both of which emit pronounced c.f. portions
with terminal sweeps. Actual durations of c.f. vary between
families and with behaviour: typical values are 8-70 ms in
Rhinolophidae, 5-25 ms in Hipposideridae and 5-27 ms in P.parnellii.
In all cases the longest pulses are emitted singly at low repetition
rates in cruising flight but in interception or when negotiating an
obstacle the c.f. part shortens and repetition rates rise as high as
80-100 s^{-1} (Figure 7).

C.f. pulses with various degrees of terminal sweep have also
been found in some Emballonuridae such as Rhynchonycteris and
there is circumstantial evidence that in crowded caves they are
emitted by some Phyllostomidae such as Lonchophylla, Carollia or
Glossophaga (see Section 7). Thus c.f. pulses may occur in all
four Superfamilies of the Microchiroptera; it remains to be seen
whether they are always associated with velocity sensing as theory
would suggest.

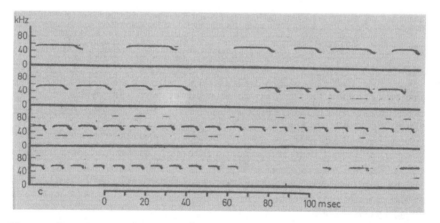

Figure 7. Sonagram analysis of Pteronotus parnellii flying
indoors. Long c.f./f.m. pulses, emitted singly or in groups in
cruising flight, give way to shorter c.f. in the extended avoidance sequence. Predominant second harmonic. From Schnitzler
(1970a).

It is a striking feature of the 'obligate' c.f. signals that
the second harmonic is generally the strongest component. This was
first noticed by Novick (1958a) in some African rhinolophids and
now seems to be the general rule. Even though the fundamental and
third harmonic may be of negligibly low intensity when the bat is
in good condition, they can appear quite strongly in bats that are
ailing or if the vocal tract is off-tuned by light bases (see
Section 9). This is especially true of the nasally emitting bats
(Rhinolophidae and Hipposideridae); for the orally emitting c.f.
species (Emballonuridae and Mormoopidae) there are appreciable
levels of fundamental, third harmonic and even traces of the fourth
harmonic. In Noctilio and the c.f. signals of Molossidae and
Vespertilionidae ('facultative' c.f. forms) the fundamental is
predominant.

Very short periods of c.f. before an f.m. sweep (Figure 1, d,
e) have been discussed in Section 2. They have been recorded from
all the Mormoopidae except Pteronotus parnellii. This poses an
evolutionary problem, for P.parnellii shows extreme specialisations
that are largely convergent on the Rhinolophidae and Hipposideridae;
yet P.personata especially is very similar in form but acoustically very different. This also shows in the auditory systems of
these bats as shown physiologically (Grinnell, 1970) and anatomically (A. Pye, poster, this conference).

A further feature to emerge since Frascati is the use of two
constant frequencies, or a bimodal spectrum. Some pulses have a
higher c.f. frequency than others, with the intermediate band
occupied only by down-sweeps from the higher pulses. This can take
two forms: bimodal populations and bimodal emission by single
individuals. The first was seen in Hipposideros commersoni (Pye,
1972) from Kenya (Figure 8). It seems likely that this was a
mixed roost of H.c.gigas and H.c.marungensis, but Triaenops
persicus (=afer) and to a lesser extent Hipposideros caffer from
the same cave showed a tendency to be bimodal in the presence of
conspecifics though not when isolated. A functional interpretation
of the 'silent band' in terms of Doppler spreading was tentatively
proposed.

Individual bimodality has been found in three species of
Emballonuridae and in all of them the pulse pattern tends to be
sigmoidal c.f. (Figure 1, k). In both Saccopteryx leptura and
S.bilineata the individual bats emit pulses of higher and lower
frequency alternately (Pye, 1973, 1978). This occurs whether the
bats are flying singly or in company, indoors or hunting freely.
The frequency interval between the two kinds of pulses is 2-2.5 kHz,
corresponding to the Doppler shift produced by the apparent flight
speed. In outdoor recordings the lower frequency of an approaching
bat is commonly the same as the upper frequency when it recedes

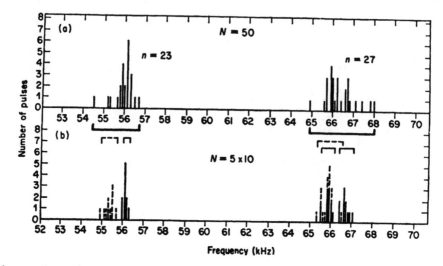

Figure 8. Histograms of c.f. frequency in <u>Hipposideros</u> <u>commersoni</u>
from Kenya. (a) 50 pulses recorded in the roost cave; (b) 10
pulses from each of 5 bats recorded separately. From Pye (1972).

(Figure 9).   But considerable variation in the frequency interval
is found at times so that one or both frequencies appear to change
at the source.   Interception uses only the lower frequency and
non-flying bats use only the upper frequency.   Again an interpretation was attemped on the hypothesis that the lower frequency
is pre-adapted for Doppler compensation.

     In the third form, <u>Coleura</u> <u>afra</u>, two frequencies are only
found in bats hunting freely (mentioned briefly by Pye, 1978).
When flying indoors or even when emerging from the roost cave, only
the upper frequency was detected.   The possibility of Doppler
compensation in these species and in other bats that use c.f. only
outdoors (such as <u>Nyctalus</u> and <u>P.pipistrellus</u>) may be resolved by
the use of Doppler radar observation (Pye, 1978; Halls, 1978).

6.   AMPLITUDE AND AMPLITUDE PATTERNS

     A number of factors make it difficult to obtain reliable and
comparable measurements of the amplitude or intensity of ultrasonic
signals.   Suitable calibrated microphones and other equipment
certainly exist and it is easy to produce measures of given pulses
<u>as detected</u>.   The difficulty with defining absolute values lies
with the animal, since bats are often quite directional in their
emissions (see Section 10) and actively turn their heads unless
intensively trained.   Also there appear to be considerable pulse-
to-pulse variations and even hand-held or perching bats are
difficult to keep at the correct distance from the microphone,

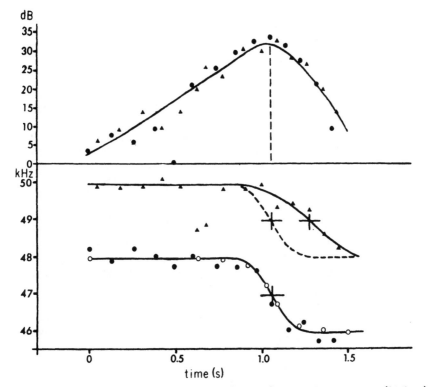

Figure 9. Analysis of amplitude (above) and frequency (below) of
Saccopteryx bilineata flying past the microphone in the open. Open
circles represent an imaginary source of 47 kHz moving at 7 m.s^{-1}
within 77 cm of the microphone.   From Pye (1978).

which, at the often used value of 10 cm, is quite critical.
Really useful values would have to be obtained from flying bats and
this is clearly impossible at present. Detailed statistical analysis might be useful but does not appear to have been attempted.

Most statements about the 'loudness' of bat signals are
therefore rather subjective and relative, relying on the observer's
experience of how much amplifier gain is necessary in making
recordings.   Novick has recently (1977) presented similar conclusions and has withdrawn the maximum values of over 1000 µbar
(100 Pa) at 10 cm reported earlier. It seems that the loudest bats
may reach 100-200 µbar at 10 cm while the quieter, 'whispering'
bats only give 2-5 µbar - a total range of about 40 dB.

Most purely insectivorous bats, or at least those that hunt
for flying insects, are loud, almost independently of frequency
pattern.   The most intense include Molossidae, Vespertilionidae,
Emballonuridae, Noctilionidae, Rhinolophidae, Hipposideridae and
Mormoopidae.   Insectivorous bats that feed on perching insects,

such as Nycteridae and Plecotus of the Vespertilionidae, tend to be soft; nevertheless, with a strategically placed microphone, this reviewer has recorded Nycteris arge hawking for insects in the open with very faint pulses. Megadermatidae, Phyllostomidae and Desmodontidae are also very soft although again some phyllostomids and some megadermatids may sometimes hawk for insects. Rhinopomatidae and Mystacinidae are also of fairly low intensity.

It seems that despite the lack of hard data, intensity is the one characteristic for which a correlation with feeding habits can be attempted although the above exceptions suggest caution. It is also the only measure which is reasonably consistent within each family. Neither of these points is at all true for frequency patterns since short f.m., multiple harmonic sweeps are used for every type of feeding (except fishing?), while insects are caught by pulses of every type of frequency pattern. Taxonomic contrasts of frequency pattern are numerous: eg. Pteronotus parnellii versus the other Mormoopidae, or Rhynchonycteris and Saccopteryx versus Taphozous and Peropteryx in the Emballonuridae.

Amplitude patterns within each pulse are very interesting and give useful clues to pulse generation (see Section 9). Early observations by Griffin and Novick showed that short f.m., multiple harmonic sweeps of some phyllostomids may have an envelope with two peaks. Pye (1967, 1968a) suggested that this is due to a fixed formant filter accentuating first one harmonic and later the one above during the f.m. sweep. Thus both amplitude peaks have the same frequency as seen in the periodicity of the waveform.

The same argument applies to the terminal sweep of c.f./f.m. pulses of Rhinolophidae, Hipposideridae and Pteronotus parnellii. The amplitude of the sweep may decay rapidly from that of the intense c.f. part (sweep-decay pulses) or it may rise to a new peak from a weaker c.f. part before decaying again (sweep-peak pulses). This variation has been confirmed in Rhinolophus by Schnitzler (1967) who showed that sweep-decay is used by a resting or cruising bat while sweep-peaks are used at higher repetition rates when landing (see Section 9).

One of the most striking cases is that of Asellia tridens in which the c.f. amplitude may fall so low that it disappears into the baseline on the oscilloscope (-20 dB = 1/10 amplitude). This led Möhres and Kulzer (1955) to suppose that the c.f. part can be eliminated and for practical purposes perhaps it is (as intensity if proportional to amplitude squared). However, Pye and Roberts (1970) showed that the duration of the c.f. part is unaffected and a sonagraph display with a 40 dB dynamic range (100:1 in amplitude) shows much the same frequency pattern (Figure 10).

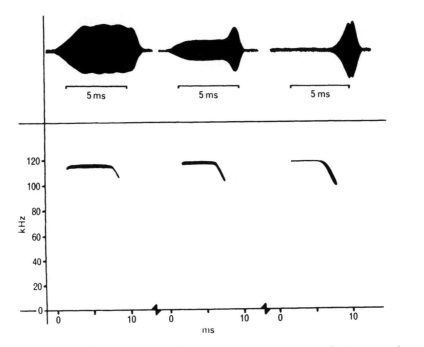

Figure 10. Oscillogram envelopes and sonagrams of three pulses from Asellia tridens. Left: a sweep-decay pulse; centre: a sweep-peak pulse; right: a pulse in which the c.f. part is present but is so attenuated that it is masked by noise in the linear-amplitude oscilloscope picture and may be effectively f.m. only. From Sales and Pye (1974).

7. FLEXIBILITY OF PULSE STRUCTURE

At one time it seems to have been tacitly assumed that each species used a characteristic pulse structure with only limited changes such as the cruising and interception pulses. The advent of field recording since Frascati has considerably changed this picture. Although it is often difficult to identify bats in the wild, this can be done often enough to show that some species behave very differently under different conditions (reviewed by Pye, 1973, 1978; Sales and Pye, 1974). This is not always the case, for Griffin and Simmons (1974) found the outdoor hunting signals of Rhinolophus ferrumequinum to be just like those of bats avoiding obstacles indoors. They found interception 'buzzes', although Airapet'yants and Konstantinov (1970/73) found that Rhinolophus falls silent just before interception, when recorded in a flight cage.

But striking changes are seen in some vespertilionids when recorded in the wild. Scotophilus nigrita then introduces a low-c.f. component while cruising (Figure 11) while Pipistrellus

pipistrellus and Nyctalus noctula commonly eliminate the f.m.
sweep altogether (Figure 12) as first reported by Pye (1967).
There appears to be a close correlation here with the altitude of
the bat.   Nyctalus commonly flies high and eliminates f.m.;
P.pipistrellus only eliminates f.m. when flying above about 5 m.
At lower altitudes, cruising pulses are like those of Scotophilus,
while close to the ground or indoors the c.f. part is rare.   Even
Myotis, which previously seemed to be quite stereotyped, may also
have flexible species: Fenton and Bell (in press) have recently
described a variety of cruising pulses from species in the western
U.S.A., including f.m./c.f. with a low c.f. of up to 5 ms in
M.volans.

Oscillograph records by Griffin (1958) of Eptesicus fuscus
flying in the open appeared to show similar behaviour with very
long pulses.   Simmons, Lavender and Lavender (1978) recently
rexamined these records and confirmed this conclusion. Furthermore
they were able to reproduce the behaviour in the laboratory by
subjecting trained E.fuscus to broadband random noise. At high
noise levels the bats lengthened their pulses to include a terminal
c.f. part and also reduced the relative levels of harmonics. It
seems that a longer pulse (of greater energy) and a narrower

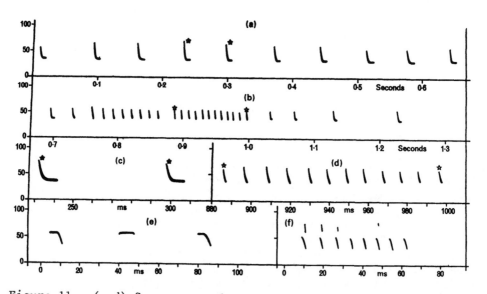

Figure 11. (a–d) Sonagram analyses of Scotophilus nigrita hunting
in the open. (a) and (b) form a continuous sequence; (c) shows
the starred cruising pulses and (d) the starred buzz sequence in
more detail. (e) and (f) show a cruising triplet and a short buzz
from Noctilio leporinus flying in a large outdoor cage.
From Sales and Pye (1974).

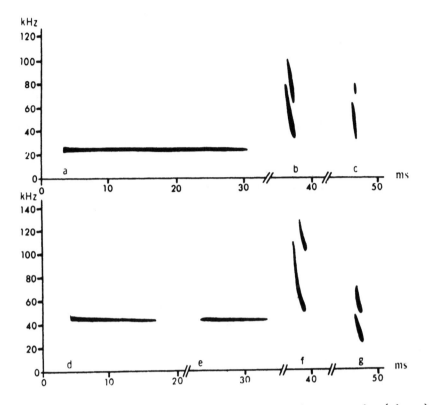

Figure 12.   Sonagrams of pulses from Nyctalus noctula (above) and
Pipistrellus pipistrellus (below).   (a, d, e): outdoor cruising
pulses; (b, f): indoor cruising pulses; (c, g): interception
buzz pulses. From Pye (1978).

bandwidth would be ideal adaptations to echo detection in noise.
But since the precision of range measurement is thereby reduced,
the bats may have then to rely on more velocity (Doppler) information.   A. Pye (1978) has shown that an African serotine, E.rendalli,
has a cochlear structure with many features of c.f. bats and has
suggested that this species may use c.f. extensively although it did
not when recorded in a quiet laboratory.

     The most extreme example in the Vespertilionidae has been
described by Simmons and O'Farrell (1977) in Plecotus phyllotis.
When hunting outdoors, this bat emits a pure c.f. of 27 kHz for
20-200 ms followed after a few ms silence by a sweep from 24 kHz
to 12 kHz with some harmonics.   The second harmonic sweep of 40-22
kHz shows an amplitude reduction at 28-25 kHz, the c.f. frequency.
Sweeps can also be emitted alone, without the c.f. part.   Indoors
the bats emit short sweeps with two components, equal to the second
and fourth harmonics of the outdoor pulses, or with one component

equal to the fifth harmonic.   This repertoire offers formidable
problems to a formant interpretation.

Further changes have been reported for unidentified, high-
flying molossids (probably Tadarida spp in Africa and Molossus spp
in Panama) by Sales and Pye (1974).   They found considerable
variability in duration and frequency pattern.   Simmons, Lavender,
Lavender et al (1978) have described similar diversity in Tadarida
brasiliensis and T.macrotis which were highly adaptive to acoustic
conditions: from pure c.f. through f.m./c.f. to pure f.m. sweeps.
Noise interference indoors produced an increase in intensity and
duration and (surprisingly) there was some increase in bandwidth
in T.brasiliensis.

Another kind of variation was discussed in Section 6:   the
pulses of Asellia, described by Möhres and Kulzer (1955) as
'kombinierter-typ' pulses.   Not only do they combine c.f. and f.m.
but either can be accentuated to the virtual exclusion of the
other (Figure 10).

Variability has also been described in both species of Noctilio
by Suthers (1965, 1967;   Suthers and Fattu, 1973), whose findings
were confirmed for N.leporinus by Sales and Pye (1974).   Cruising
pulses are c.f./f.m. but when fishing two such pulses commonly occur
together with a c.f. pulse (lasting up to 11.6 ms with little or no
perceptible sweep) to form a cruising triplet (Figure 11, e).
Suthers has also described c.f./f.m. pulses with a slow sweep lasting up to 10 ms or so.   Bats approaching each other on collision
courses emit 'honks' consisting of a brief c.f. part followed by an
extended deep sweep of as much as two octaves.   Interception buzzes
eliminate the c.f. part while the f.m. sweeps shorten and fall in
frequency as in Myotis (Figure 11, f).

Bradbury (1970) has reported striking flexibility in the
phyllostomid, Vampyrum spectrum.   Indoor discrimination flights
used short pulses with a double sweep, shown to be the third harmonic plus a subsequent but overlapping fourth harmonic.   Duration
shortened from 1.5 ms to 0.4 ms as the target was approached.
This pattern suggests a very sharply tuned formant filter.   In a
large outdoor flight cage there were also long pulses up to 15 ms
in duration with a very shallow sweep.   These had a predominant
fundamental and decreasing harmonics up to the seventh and thus
showed little sign of formant filtering.   Figure 13 shows one
pulse of each type.

Pye (1973) produced circumstantial evidence that in a cave in
Panama other phyllostomids, principally Lonchophylla robusta but
also Carollia perspicillata and/or Glossophaga soricina, may emit
long c.f./f.m. pulses similar to those of Pteronotus parnellii.
When recorded in the laboratory, all these gave only short f.m.,

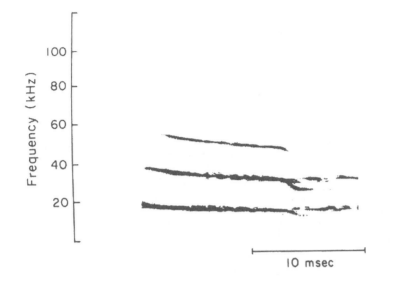

Figure 13. Two consecutive pulses (with some echoes) of a
Vampyrum spectrum flying in a large outdoor cage. The shorter
pulse is typical of pulses used in discrimination flights indoors.
From Bradbury (1970).

multiple harmonic pulses 'typical' of their family.

From the above it is now clear that, if observations are made
under a variety of conditions, cases of pronounced flexibility can
be found in all four Superfamilies of the Microchiroptera. Further
species are sure to be found as observations and techniques are
extended. It must also be assumed that the changes seen are
adaptive. Simmons et al (1975; Simmons, in press) discussed the
concept of various strategies in the detection of echoes from
different kinds of targets under varying conditions such as clutter
and ambient noise; it seems that at least some bats can vary
their stratagems in order to optimise their performance.

It is also clear that if pulse structures cannot be 'assigned'
to each species as characteristic, they cannot be used with any
confidence as a taxonomic criterion. Recently Simmons (in press)
has attempted a discussion of the 'phylogeny and evolution' of
echolocation in bats but without implying a relationship to the
phylogeny of the bats themselves. Despite the stimulatory nature
of this synthesis, three fixed points seem fairly safe. The
primitive type of pulse is probably the multiple harmonic type with
shallow or no f.m.; the pure fundamental, deep f.m. sweep
(apparently used almost invariably by Myotis) is the most advanced
form of range-measuring signal; and the very long c.f./f.m. signal
(apparently used invariably by Rhinolophus) is the most advanced

form of velocity-measuring signal.

8. RADAR SIGNAL THEORY

At Frascati Cahlander (1967) presented a pioneering study of
the application of radar theory to the echolocation signals of
bats. After a theoretical consideration of wideband signals, he
computed ambiguity diagrams for six pulses of Myotis lucifugus
and an outdoor cruising pulse of Lasiurus borealis.

The need for reworking the theory arises from the invalidity
of certain assumptions made in radar theory. In radar the bandwidth
is so low compared with the centre frequency that Doppler shifts
may be regarded (eg. Bird, 1974) as an incremental offset, thus
simplifying the calculations without loss of accuracy. But with
wideband signals, which in bats may exceed an octave, the true
Doppler shift must be treated as a factor, compressing or expanding
the time scale, which Cahlander called $\propto$ (see also Kelly and
Wishner, 1965).

Similar treatments have been applied to wideband sonar signals,
for example by Russo and Bartberger (1964) and by Kramer (1967).
Further differences due to propagation in air and to the nature of
bats' targets are outlined by Pye, J.D. (this conference, poster).

The ambiguity diagram may be prepared either by assembling a
series of cross-correlations between the signal and Doppler-shifted
versions of the signal or by assembling Fourier spectra of products
between the signal and time-shifted versions of the signal (Halls,
this conference, poster). Bird (1974) has shown that these two
methods correspond to two different approaches to receiver design:
the correlation or matched filter receiver and the Fourier transform
or beatnote receiver. Either system may correspond to an 'ideal'
receiver and both have been suggested as models of echo processing
in bats.

It is often assumed that a sharply peaked ambiguity surface
represents both high accuracy and good resolution. But Bird (1974)
has shown that these are related but distinct concepts. A sharp
peak gives high accuracy but its 'foothills' may mask weaker echoes
from other targets. Improved resolution may therefore be gained
by degrading the accuracy, ie. by deliberately mismatching the
filter in correlation receivers.

It is clear that ambiguity diagrams offer detailed information
about the potential capability of bat signals, either as single
pulses or as groups or trains of pulses. It seems reasonable to
assume that the animals will exploit such capability due to

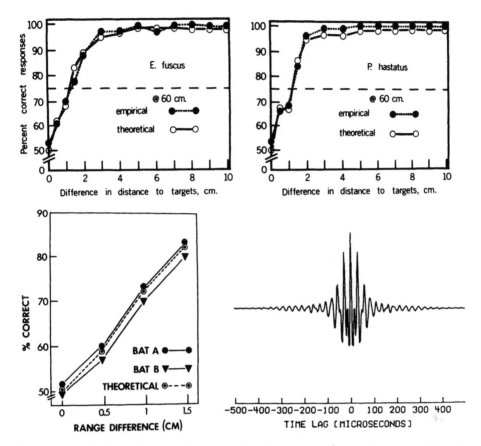

Figure 14. Range discrimination in three species of bats. Theoretical performance is compared with actual achievement in distinguishing the nearer of two targets. Top left: <u>Eptesicus fuscus;</u> top right: <u>Phyllostomus hastatus;</u> lower left: <u>Tadarida brasiliensis;</u> lower right: the autocorrelation function of a <u>T.brasiliensis</u> pulse. From Simmons (1970); Simmons, Lavender, Lavender et al (1978).

adaptiveness under survival pressure. This has actually been demonstrated for some species by Simmons (1970, 1971, 1973; Simmons, Lavender, Lavender et al, 1974, 1978). He prepared autocorrelation functions (without Doppler shifts) of the pulses of three species and from these calculated the theoretical limits of range discrimination. The performance of trained, non-flying bats in distinguishing between two targets at different ranges was then measured and the combined results are shown in Figure 14.

Altes and Titlebaum (1970) developed theoretical aspects of Cahlander's work and showed that for wide-band f.m., optimum Doppler tolerance is provided by a linear period sweep. The close

approximation of many bats' sweeps to this form was described in
Section 4. Glaser (1971, 1974) has also tackled the application of
theory to bat signals; he assumes linear f.m. sweeps and some
other postulates that are frankly puzzling. Escudié et al (1976
and poster, this conference) have also been active in the theoretical consideration of bat signals in relation to echolocation
performance; they assume hyperbolic f.m. in Myotis and other
vespertilionids.

     Despite the insights provided by theoretical work, this must
be supported by computation of ambiguity diagrams for actual signals.
This second aspect of Cahlander's work has now been taken up by
three groups: Beuter (in press and this conference), Pye and Halls
(this conference, posters) and Tupinier et al (in press). All have
chosen to effect the computation by cross-correlation and this has
been achieved digitally (Beuter; Escudié; Tupinier et al, in
press), by analogue methods (Cahlander) or by a 'hybrid' system
using digitally operated CCD delay lines and analogue correlation
(Halls). At first sight analogue computation by the Fourier
spectra method looks attractive but it requires single sideband
mixers of exceptional bandwidth.

     The results of Beuter form a separate review of this
conference.

                    9.   SOUND GENERATION

     Before Frascati the main work on sound production in bats was
that of Kulzer (1958, 1960) on Rousettus and Novick and Griffin
(1961) on five families of the Microchiroptera. Since then the
literature has increased considerably.

     The click sounds of Rousettus are, as Kulzer showed, produced
by clicking the tongue. Pye (1967) suggested that resonance
of the buccal cavity would then influence both the spectrum and the
envelope of emitted pulses. Roberts (1975) tested this by recording R.aegyptiacus and R.amplexicaudatus in air and in 80% He:
20% O_2 gas mixtures (He-O_2). Velocity of sound was monitored by
the resonant frequency of a spark-excited tube in order to allow for
air leakage. The spectra were complex but all identifiable points
were raised in He-O_2: in one case between 1.56 and 1.58 times for
a measured increase in sound velocity of 1.59 times.

     Little is known about the production of the somewhat similar
sounds by echolocating birds. The slender evidence available for
Aerodramus has been discussed by Medway and Pye (1977) who favoured
some form of oral emission rather than a direct linkage from wing
movements in flight as postulated by Harrison (1966).

In the Microchiroptera it appears that ultrasound is vocally generated by the vibration of laryngeal membranes driven by respiratory air-flow (Novick and Griffin, 1961). Several aspects of this mechanism have been confirmed and developed by more recent work. Laryngeal structure has been examined in 13 species of 7 families by Wassif and Madkour (1968/9). They described skeletal structure, drawing attention to the pronounced tracheal and ventricular pouches that may act as resonators in Rhinolophus, Asellia and Nycteris.

Correlation of pulse production with respiratory air-flow has been achieved with a hot-wire anemometer by Schnitzler (1968) in Myotis myotis, Rhinolophus ferrumequinum and R.euryale, and by Roberts (1972b) in R.luctus, Eptesicus serotinus and Plecotus auritus. Suthers et al (1972; Suthers and Fattu, 1973) used a thermistor anemometer on Phyllostomus hastatus and Eptesicus fuscus. All authors agree that pulse production occurs primarily during exhalation although it can occur during inhalation when interruptions of air-flow(mini-exhalations) could be seen. Schnitzler (1971) also correlated pulse production with wing-beat in Myotis lucifugus and Carollia perspicillata, while Suthers et al (1972) observed respiratory air-flow, wing-beat and pulse production simultaneously in flying Phyllostomus hastatus. Inhalation occurs on the down-beat of the wing, exhalation and the majority of pulses on the up-beat.

Laryngeal action has been further studied in Eptesicus fuscus by Suthers and Fattu (1973) who measured sub-glottic pressures. The values during phonation were high, up to 60-70 cm H_2O. They suggested that the vocal folds form a powerful glottal stop while the thin, paired vocal and ventricular membranes are ultrasonic generators during the pulse. This is consistent with their own histological and myophysiological observations and with the denervation experiments of Novick and Griffin (1961). Tidal volume as measured with a body plethysmograph varied from 0.3 ml for single pulses to 0.6 ml for rapid pulse trains.

Schuller and Suga (1976) denervated laryngeal muscles, recorded from active cricothyroid muscle and electrically stimulated the cricothyroid in Rhinolophus. Unilateral denervation of the superior laryngeal nerve reduced the c.f. frequency by 4-6 kHz while bilateral denervation reduced it by as much as 30 kHz with the introduction of prominent harmonics. Both spike recording and stimulation suggested that the cricothyroid muscle is responsible for tuning the c.f. and is thus the mediator of Doppler compensation behaviour (See Section 5).

At Frascati Pye (1967) suggested that vocal tract resonance should act as a formant in filtering the sounds actually emitted, thus accounting for the relative amplitudes of harmonics, progressive phase changes during f.m. and some aspects of envelope shape.   An electrical model of such a system was described later (Pye, 1968a).   Light gases should not affect the glottal funda- mental or harmonic frequencies but should strongly influence their relative intensities   as   in human vowel sounds (this argument appears to have been misunderstood by Novick, 1977).   Preliminary results with hydrogen-air mixtures were described for Artibeus jamaicensis (1967) and Rhinolophus hipposideros (1968a).   In Rhinolophus even slight mis-tuning caused the fundamental to appear strongly, suggesting that this is normally suppressed by a tuned- reject resonance.   It is therefore striking that Schuller and Suga (1976) found that harmonics emitted after bilateral superior laryngeal nerve section in R.ferrumequinum were "much weaker" if they fell close to the normal fundamental c.f. frequency.

More careful and sophisticated experiments have now been performed using He-O_2 by Schnitzler (1970a, b, 1973) on Rhinolophus ferrumequinum and Pteronotus parnellii, and by Roberts (1973) on R.ferrumequinum, R.luctus, Hipposideros diadema, Desmodus rotundus, Eptesicus serotinus, Pipistrellus pipistrellus and Plecotus auritus. All c.f. bats behaved as predicted, with no change in frequency but a progressive shift of energy from the second to the third harmonic and strengthening of the fundamental.   Roberts also found this last effect in Desmodus and suggested that the nasal cavity suppresses the fundamental in other nose-leaf bats.   He was, however, unable to find any influence of He-O_2 in vespertilionids and concluded that here the purity and amplitude pattern of the fundamental sweep are determined at the laryngeal level.

Roberts (1972a) also investigated the suggestion that sweep- peaks and sweep-decay patterns in c.f. bats could be selected by a change in glottal frequency.   High-resolution analysis by Lissajou's figures of c.f. frequency in pulses of each type from six species of Rhinolophidae, five species of Hipposideridae and in Pteronotus parnellii failed to show any correlation (such a correlation would of course be inconsistent with present knowledge of Doppler compensation).   He concluded that the changes may be brought about by retuning the formant frequency or by 'source generated' amplitude changes.

Recently Griffiths (1978) has examined the cricothyroid muscle in Mormoopidae and suggested that its posterior branch may be responsible for varying the vocal tract resonance and thus the harmonic content of the pulse.   Suthers and Durrant (poster, this conference) recorded from and denervated both branches of this muscle in Pteronotus parnellii.   Results show that both the anterior and posterior branches contribute to the frequency of the c.f. pattern of the pulse - the former branch being particularly important at the start of the pulse, the latter affecting the

terminal f.m. sweep.   Of special interest is the fact that the
posterior, but not the anterior branch also plays an important role
in controlling the harmonic content of the emitted sound.   During
vocalization in 20% oxygen and 80% helium, the harmonic emphasis
is switched from the second to the first and third harmonics.

An important development in studies of sound production was
the demonstration by Suga and Schlegel (1972, 1973;   Suga, Schlegel
et al 1973) that normal orientation pulses can be elicited from
awake bats by electrical stimulation of the midbrain.   This was
followed up by Suga, Schlegel et al (1973) in seven species of
three families (Noctilio leporinus, Pteronotus parnellii,
P.suapurensis, Eptesicus fuscus and three species of Myotis),
representing a wide range of pulse types.   The emitted sounds were
typical of those used in echolocation and various pulses of each
known repertoire could be elicited by exciting different regions of
the brain.   Pulses produced by this technique have been used in
the study of laryngeal mechanisms (Suthers and Fattu, 1973) and of
directional radiation patterns (Shimozawa et al, 1974;   Schnitzler
and Grinnell, 1977;   see Section 10).

Jen and Suga (1976) have also shown that motor control of
laryngeal and middle-ear muscles is coordinated, potentials appearing in the former about 3 ms before the latter.   Both sets of
muscles also respond to acoustic stimuli, with the laryngeal
muscles contracting about 3 ms after the middle-ear muscles.   Jen
and Ostwald (1977) used this cricothyroid muscle response as a
test of hearing: response threholds with downward sweeping f.m.
stimuli were lower than with either upward sweeps or tone pulses at
the best frequency.

10. SOUND RADIATION PATTERNS

Griffin and Hollander (1973) discussed the difficulties of
measuring the sound field around active bats, but considerable
success has been achieved with animals under various degrees of
constraint.

Simmons (1969) measured directionality in Pteronotus parnellii
at 56 kHz, its c.f. frequency, and in Eptesicus fuscus at its
greatest amplitude(about 30 kHz depending on direction).   He used
four microphone positions in the horizontal plane: 0° (straight
ahead), 22.5°, 45° and 90° to one side.   Both species showed half
the pressure amplitude (-6 dB) at 22.5°;   at greater angles
Pteronotus was more directional(-20 dB at 45°, -26 dB at 90°)
than Eptesicus (-9 dB at 45°, -14 dB at 90°).   These results were
compared with Möhres figures (1967) for Rhinolophus ferrumequinum
(quoted as -6 dB at 22.5°) and Megaderma lyra(-6 dB at 35°)

although Möhres (1967) and Möhres and Neuweiler (1966) actually showed "intensity" plots with -6 dB at 55° for Megaderma.

Schnitzler (1968) referred to "beamwidth" in Rhinolophus ferrumequinum and R.euryale as 20° both horizontally and vertically. Thus it appears that, at least near the axis, a mouth-horn can be as directional as a nose-leaf. Strother and Mogus (1970) calculated theoretical patterns for a number of species, predicting -6 dB at 28° for Pteronotus parnellii and at 43° for R.ferrumequinum.

Griffin and Hollander (1973) compared 0° and 90° amplitudes in Myotis lucifugus and Eptesicus fuscus. Higher frequencies at the top of the fundamental sweep were more directional (25 dB) than lower ones at the bottom (10 dB). In E.fuscus the second and third harmonics were also examined and despite their higher frequency gave only about 10 dB for 0°/90°. But the end of the second harmonic sweep (frequency equal to the start of the fundamental sweep) rose to a value of 23 dB. This suggests that the polar diagram has side-lobes at harmonic frequencies.

Shimozawa et al (1974) measured detailed patterns around Myotis grisescens and M.lucifugus. The bats' heads were fixed and stereotyped pulses were elicited by brain stimulation. Readings were taken every 10° at 55 kHz, 75 kHz and 95 kHz, and two detailed plots are shown in Figure 15. In the horizontal plane -6 dB points are at 40° for 55 kHz, 33° for 75 kHz and 30° for 95 kHz; in the median plane the main beam axis was 5-10° below the horizontal.

A similar examination of the c.f. pattern in Rhinolophus ferrumequinum was made by Schnitzler and Grinnell (1977) and the results are shown in Figure 15; -6 dB points are at 25° laterally and above, but at 70° below (similar values were given by Airapet'yants and Konstantinov, 1970/73). Interference with the upper part of the nose-leaf produces distortion of the pattern, especially in the vertical plane (Figure 15), while blocking one nostril disrupts it severely. These results and those of Shimozawa et al have been combined with directional hearing patterns to indicate the total pattern of echolocation sensitivity (Grinnell and Schnitzler, 1977; Shimozawa et al, 1974).

Peff and Simmons (1971) measured horizontal angular discrimination by behavioural methods. E.fuscus could distinguish 6-8° and Phyllostomus hastatus (-6 dB beamwidth 40-50°) gave 4-6°. Clearly directionality is much sharpened by the receiver, and transmitter directionality serves mainly to enhance signal energy at the target.

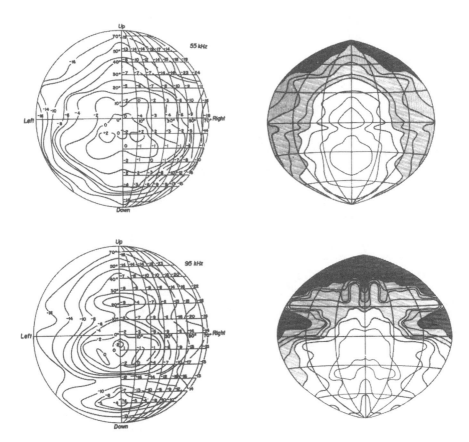

Figure 15. Directional emission patterns for two bats.
Left: patterns at 2 dB intervals for Myotis grisescens at two
frequencies; from Shimozawa et al (1974); right: patterns at
3 dB intervals for Rhinolophus ferrumequinum, above, the normal
bat, below after covering the upper part of the nose-leaf with
vaseline; from Schnitzler and Grinnell (1977).

11. PROPAGATION

At Frascati Pye (1967) expressed some concern about the
possibility of velocity dispersion in the propagation of ultrasound
in air.   It now seems that this is unlikely to be a significant
effect over the biological spectrum, at least in reasonably normal
concentrations of carbon dioxide.   No progressive phase shift can
be seen between a fundamental and its harmonics for a range of
frequencies over at least several metres.   So it seems likely
that Pierce's original observations (1925) were perturbed by 'tube

effects' in his interferometer. Further confirmation of this point is still required, however.

Another propagation effect, dispersive absorption in air, has been reviewed in detail by Griffin (1971). The effect may be profound at higher frequencies and is strongly dependent on the water vapour content of the air: 150 kHz at 3% water vapour suffers a loss of about 15 dB.m^{-1}. Water vapour content can be derived from temperature and humidity. For bats, most conditions will be bounded by 3% water vapour (say 100% relative humidity at 24°C) and 0.5% water vapour (50% R.H. at 8°C or 25% R.H. at 20°C). Griffin presented the absorption data as a function of % water vapour, with frequency as the parameter, but for some purposes it may be more convenient to read against frequency, with % water vapour as the parameter. Some of Griffin's data have therefore been replotted in this way in Figure 16. Recent figures from an unpublished Boeing lab report (Evans, L.B., 1974) and a National Physical Laboratory report (Bazley, 1976) suggest that all these values should be approximately halved. Evans also states that temperature effects are small for a given humidity. Delany (1977) has discussed the historical appreciation of this factor and has reviewed other aspects of ultrasound propagation in air.

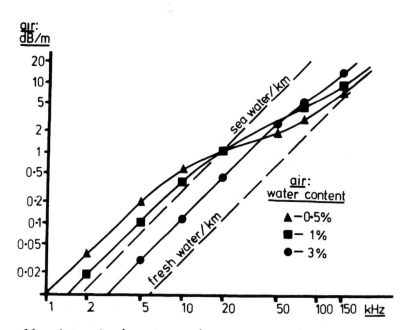

Figure 16. Atmospheric attenuation as a function of frequency at three values of water vapour content. The dashed lines for under-water attenuation are in dB.km^{-1}. Atmospheric data in dB.m^{-1} replotted from Griffin (1971); (Pye, in press).

Discussing the strong attenuation, Griffin (1971) pointed out that because of its 'low-pass filter' properties, air is a "very murky medium" especially if it is damp and the frequency high. Thus bats needing to detect echoes from long distances would be obliged to use low frequencies and indeed high flying species (molossids and some vespertilionids such as Nyctalus) use lower frequencies than others.   It may be that a ground echo is necessary for orientation and maximum altitude could thus be frequency dependent.

Möhres (1953) suggested that vectorial addition of sounds from the two nostrils of Rhinolophus, half a wavelength apart, could partially overcome the inverse square law in the axial direction. He plotted actual measurements of sound pressure amplitude against distance.   But it is intensity that decreases as the inverse square, so that pressure (proportional to the square root of intensity) decreases as the inverse distance.   Möhres' graphs are, indeed, a very good fit for an inverse distance relation.   His vectoral argument, when calculated quantitatively, shows that, because the nostrils are so close together, there is negligible interaction on the axial direction except within a few mm of the nose-leaf.   For 80 kHz and nostril spacing ($\lambda/2$) 2.1 mm, the vector sum is 99.9% effective (less than 0.01 dB deficit) at only 2.37 cm.   At all distances of importance, the 'far field', the axial propagation behaves as if there were a single source.

A further influence on propagation could be the presence of water droplets in the air (Pye, 1971).   Lord Rayleigh (1896) calculated that spherical liquid drops in air show a conformational

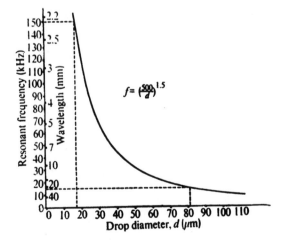

Figure 17.   Resonant frequencies for spherical water droplets in air from a formula given by Rayleigh.   From Pye (1971).

vibration at a frequency related to their diameter. Extrapolation suggests that droplets of 100 μm resonate at 11 kHz and 18 μm at 150 kHz (Figure 17). Stable fogs contain droplets from 100 μm to 1 μm and such suspended droplets should be very nearly spherical, giving sharp tuning (high Q). Thus fog should absorb biological ultrasound and reradiate it slowly over a considerable time. To echolocation, fog would appear 'inky black' with perhaps a faint after-glow. Artificial fogs in the laboratory certainly produce extremely high attenuation at all ultrasonic frequencies and field observations suggest that bats avoid discrete fog banks; indeed their sounds cannot be detected by a bat detector from inside the fog wall.

12. ECHOES

The echo return from natural objects with broadband signals is complex. Much radar theory assumes a point target so that the echo is identical to the transmitted signal except for a delay due to round-trip propagation, a frequency shift due to radial velocity and an attenuation due to propagation losses and target cross-section. But real targets present a more complicated problem and many, such as insect prey, will have dispersive cross-sections as well as being aspect sensitive and temporally variable. The reflective properties of many targets, even simple geometrical ones, are too complex for theoretical analysis and are probably best tested empirically as in Simmons' experiments with trained bats (eg. Simmons and Vernon, 1971).

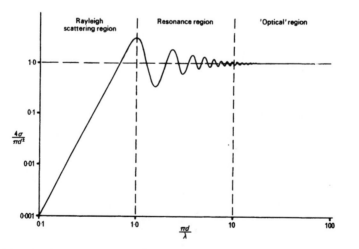

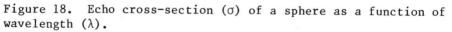

Figure 18. Echo cross-section (σ) of a sphere as a function of wavelength (λ).

The simplest, aspect insensitive, case is a sphere whose back-scattering cross-section as a function of wavelength is shown in Figure 18. At long wavelengths (greater than the target circumference, πd) the echo intensity is proportional to wavelength to the minus four (classical Rayleigh scattering). Only when wavelength is less than one tenth of the circumference is echo strength effectively independent of wavelength. Thus small targets are high-pass filters in respect of the echoes they produce. Small insects are neither spherical nor aspect-insensitive and their beating wings must also lead to temporal fluctuations; nevertheless they too must act as high-pass reflectors.

In some insects, such as moths, the wings are large enough to produce strong, specular echoes at normal incidence. The reflective properties of actual moths in tethered flight have been measured by Roeder (1963, 1967c); he showed very clearly that the echo strength is very aspect-sensitive and increases enormously when the wings are at certain critical positions. Johnson et al (1974) briefly drew attention to the Doppler shifts to be expected from moving insect wings. Recently Schnitzler (1978; and in press) has shown that flight movements of insect wings produce an asymmetrical echo spectrum due to interaction between amplitude modulation and Doppler shift; he has discussed in some detail the implications of this complex echo for the echolocation of c.f. bats.

Many insect prey species are not merely passive reflectors but detect the ultrasound emitted by bats up to 30-40 m away. Responses range from turning away from distant bats to erratic, evasive manoeuvres or power diving to the ground from nearer ones. Members of the Arctiidae and Ctenuchidae may reply by emitting a train of ultrasonic clicks, sometimes at very high repetition rates (recent references Fenton and Roeder, 1974; Fullard and Fenton, 1977; Fullard, 1979). These moth sounds may mask the echo of the bat's cry or warn of distastefulness or even mimic the warning signals of other distasteful species (Dunning, 1968).

Much has been learned of these various echolocation counter-measures in the last 13 years (partly reviewed by Sales and Pye, 1974) and their study may give useful insight to the operation of echolocation systems. Noctuid moths have been studied in great detail by Roeder (1966 a-d, 1967 a-c, 1969 a-c, 1970, 1971, 1972, 1974 a,b, 1976; Roeder and Fenton, 1973; Roeder and Payne, 1966) and by Adams (1971, 1972), Agee (1967, 1969 a-c, 1971), Ghiradella (1971), Goldman and Henson (1977), Lechtenberg (1971), Paul (1973, 1974), and Payne, Roeder and Wallman (1966). The abdominal tympanic organ of geometrids has been examined by Roeder (1974a) and a strikingly different, 'non-tympanic' ear has been discovered in the mouthparts of certain sphingids (Roeder, Treat and Vandeberg, 1968, 1970; Roeder and Treat, 1970; Roeder, 1972,

1976).   An ultrasonic ear that can initiate evasive behaviour
has also been discovered on the forewing of certain neuroptera
(Miller and MacLeod, 1966) and Miller has studied this system and
its associated behaviour in some detail (1970, 1971, 1974, 1975,
in press and this conference).

Thus no less than five different ultrasonic hearing organs
are now known to mediate the evasion of bats in at least seven
distinct groups of nocturnal flying insects.   The evolutionary
significance of this variety, in view of the relatively recent
predation pressure of insectivorous bats, has been indicated by
Pye (1968b).

Larger, distributed targets such as foliage and rough ground
must give complex reflections although flat surfaces, such as some
single leaves, act as specular reflectors.   Smooth ground or water
surfaces would ideally give an 'image' of the bat at a range equal
to twice the bat's altitude.   Ripples on water, however, give
strong reflections over wide angles.   A charming and thought-
provoking picture of the world as seen (or not seen) by a bat
using only ultrasonic echoes has been given by Lewin (1978).

References

Adams, W.B., 1971. Intensity characteristics of the noctuid acoustic
    receptor, J.gen.Physiol, 58: 562-579.
Adams, W.B., 1972. Mechanical tuning of the acoustic receptor of
    Prodenia eridania (Cramer) (Noctuidae), J.exp.Biol., 57: 297-308
Agee, H.R., 1967. Response of acoustic sense cell of the bollworm
    and tobacco budworm to ultrasound, J.econ.Entomol., 60:
    366-369.
Agee, H.R., 1969a, Response of flying bollworm moths and other
    tympanate moths to pulsed ultrasound, Anns.ent.Soc.Amer.,
    62: 801-807.
Agee, H.R., 1969b, Response of Heliothis spp. (Lepidoptera:
    Noctuidae) to ultrasound when resting, feeding, courting,
    mating or ovipositing, Anns.ent.Soc.Amer., 62: 1122-1128.
Agee, H.R., 1969c, Acoustic sensitivity of the European corn
    borer moth, Ostrinia nubilalis,Anns.ent.Soc.Amer., 62:
    1364-1367.
Agee, H.R., 1971, Ultrasound produced by wings of adults of
    Heliothis zea, J.Insect Physiol., 17: 1267-1273.
Airapet'yants, E.Sh. and Konstantinov, A.I., 1970/73,
    "Echolocation in Animals", Nauka, Leningrad, published 1970
    (English translation 1973, Israel Program of Scientific
    Translations).
Altes, R.A. and Titlebaum, E.L., 1970, Bat signals as optimally
    doppler tolerant waveforms, J.acoust.Soc.Amer., 48: 1014-1020.
Bazley, E.N., 1976, Sound absorption in air at frequencies up to
    100 kHz, National Physical Laboratory Acoustics Report No.
    Ac 74.
Bird, G.J.A., 1974, Radar Precision and Resolution, Pentech Press,
    London.
Bradbury, J.W., 1970, Target discrimination by the echolocating
    bat, Vampyrum spectrum, J.exp. Zool, 173: 23-46.
Buchler, E.R., 1976, The use of echolocation by the wandering
    shrew (Sorex vagrans), Anim.Behav., 24: 858-873.
Cahlander, D.A., 1967, in "Animal Sonar Systems: biology and
    Bionics", ed. Busnel, R.-G., Jouy-en-Josas, pp. 1052-1081.
Delany, M.E., 1977, Sound propagation in the atmosphere: a
    historical review, Acustica, 38: 201-223.
Dunning, D.C., 1968, Warning sounds of moths, Z.Tierpsychol., 25:
    129-138.
Escudié, B., Hellion, A., Munier, J. and Simmons, J.A., 1976,
    Etude theoretique des performances des signaux sonar
    diversité de certaines chauves-souris à l'aide du traitement
    du signal, Rev.d'Acoust., 38: 216-229.
Evans, L.B., 1974, Atmospheric absorption of sound: temperature
    dependence, Report No. D3-9190 (code no. 81205)., Boeing
    Co., Wichita, Kansas.

Fenton, M.B., 1975, Acuity of echolocation in Collocalia hirundinacea
    (Aves: Apodidae), with comments on the distributions of
    echolocating swiftlets and molossid bats, Biotropica, 7: 1-7.

Fenton, M.B. and Bell, G.P. (in press), Echolocation and feeding
    behaviour in four species of Myotis (Chiroptera), Can.J.Zool.

Fenton, M.B. and Roeder, K.D., 1974, The microtymbals of some
    Arctiidae, J.Lepid.Soc., 28: 205-211.

Fullard, J.H., 1979, Behavioral analyses of auditory sensitivity
    in Cycnia tenera Hübner (Lepidoptera: Arctiidae), J.comp.
    Physiol., 129: 79-83.

Fullard, J.H. and Fenton, M.B., 1977, Acoustic and behavioural
    analyses of the sounds produced by some species of Nearctic
    arctiidae (Lepidoptera), Can.J.Zool., 55: 1213-1224.

Ghiradella, H., 1971, Fine structure of the noctuid moth ear.
    I. The transducer area and connections to the tympanic
    membrane in Feltia subgothica, Haworth, J.Morph., 134: 21-46.

Glaser, W., 1971, Zur Fledermausortung aus dem Gesichtspunkt der
    Theorie gestörter Systeme, Zool.Jb.Physiol., 76: 209-229.

Glaser, W., 1974, Zur Hypothese des Optimalempfangs bei der
    Fledermausortung, J.comp.Physiol., 94: 227-248.

Goldman, L.J. and Henson, O.W., 1977, Prey recognition and selection
    by the constant frequency bat, Pteronotus parnellii, Behav.
    Ecol.Sociobiol., 2: 411-419.

Gould, E., Negus, N.C. and Novick, A., 1964, Evidence for
    echolocation in shrews, J.exp.Zool., 156: 19-38.

Griffin, D.R., 1954, Acoustic orientation in the oil bird,
    Steatornis, Proc.nat.Acad.Sci., 39: 884-893.

Griffin, D.R., 1958, "Listening in the Dark", Yale Univ. Press.

Griffin, D.R., 1962, Comparative studies of the orientation sounds
    of bats, Symp.zool.Soc.Lond., 7: 61-72.

Griffin, D.R., 1971, The importance of atmospheric attenuation
    for the echolocation of bats (Chiroptera), Anim.Behav., 19:
    55-61.

Griffin, D.R. and Hollander, P., 1973, Directional patterns of
    bats' orientation sounds, Period.Biol., 75: 3-6.

Griffin and Novick, 1975, Acoustic orientation of neotropical
    bats, J.exp.Zool., 130: 251-300,

Griffin, D.R. and Simmons, J.A., 1974, Echolocation of insects by
    horseshoe bats, Nature, Lond, 250: 731-732.

Griffin, D.R. and Suthers, R.A., 1970, Sensitivity of echo-
    location in cave swiftlets, Biol.Bull., 139: 495-501.

Griffiths, T.A., 1978, Modification of M.Cricothyroideus and the
    larynx in the Mormoopidae, with reference to amplification
    of high-frequency pulses, J.Mammal., 59: 724-730.

Grinnell, A.D., 1970, Comparative auditory neurophysiology of
    neotropical bats employing different echolocation signals,
    Z.vergl.Physiol., 68: 117-153.

Grinnell, A.D. and Hagiwara, S., 1972a, Adaptations of the auditory
    nervous system for echolocation: studies of New Guinea bats,
    Z.vergl.Physiol., 76: 41-81.

Grinnell, A.D. and Hagiwara, S., 1972b, Studies of auditory neurophysiology in non-echolocating bats and adaptations for echolocation in one genus, Rousettus, Z.vergl.Physiol., 76: 82-96.

Grinnell, A.D. and Schnitzler, H.U., 1977, Directional sensitivity of echolocation in the horseshoe bat, Rhinolophus ferrumequinum; II Behavioral directionality of hearing, J.comp.Physiol., 116: 63-76.

Grünwald, A., 1969, Untersuchungen zur Orientierung der Weisszahnspitzmäuse (Soricidae-Crocidurinae), Z.vergl.Physiol., 65: 191-217.

Gustafson, Y. and Schnitzler, H.U., in press, Echolocation and obstacle avoidance in the hipposiderid bat, Asellia tridens, J.comp.Physiol.

Halls, J.A.T., 1978, Radar studies of bat sonar, Proc.Fourth Int. Bat Res.Conf., eds. Olembo, R.J., Castelino, J.B. and Mutere, F.A., Kenya Lit. Bureau, Nairobi, 137-143.

Harrison, T., 1966, Onset of echo-location clicking in Collocalia swiftlets, Nature, Lond., 212: 530-531.

Jen, P.H.S. and Ostwald, J., 1977, Response of cricothyroid muscles to frequency-modulated sounds in FM bats, Myotis lucifugus, Nature, Lond., 265: 77-78.

Jen, P.H.S. and Suga, N., 1976, Coordinated activities of middle-ear and laryngeal muscles in echolocating bats, Science, 191: 950-952.

Johnson, R.A., Henson, O.W. and Goldman, L., 1974, Detection of insect wingbeats by the bat, Pteronotus parnellii, J.acoust. Soc.Amer., 55: S53 (Abstract).

Kelly, E.J. and Wishner, R.P., 1965, Matched-filter theory for high-velocity accelerating targets, I.E.E.E. Trans.mil., Electr., Mil-9: 56-69.

Kingdon, J., 1974, East African Mammals: an Atlas of Evolution in Africa, Vol. IIA, Academic Press, London & New York.

Konstantinov, A.I. and Sokolov, B.V., 1967, Use of active and passive location by bats in catching insects, 3rd Conf.Ecol. Physiol.Biochem.Morphol., Novosibirsk (in Russian).

Kramer, S.A., 1967, Doppler and Acceleration Tolerances of High-gain Wideband Linear F.M. Correlation Sonars, Proc. I.E.E.E., 55: 627-636.

Kulzer, E., 1958, Untersuchungen über die Biologie von Flughunden der Gattung Rousettus Gray, Z.Morph.und Ökol der Tiere, 47: 374-402.

Kulzer, E., 1960, Physiologische und Morphologische Untersuchungen über die Erzeugung der Orientierungslaute von Flughunden der Gattung Rousettus, Z.vergl.Physiol., 43: 231-268.

Lawrence, B. and Novick, A., 1963, Behavior as a taxonomic clue; relationships of Lissonycteris (Chiroptera), Breviora, 184: 1-16.

Lechtenberg, R., 1971, Acoustic response of the B cell in noctuid moths, J.Insect Physiol., 17: 2395-2408.

Lewin, R., 1978, Bats-eye view, Trends in Neurosciences, 1: 38.

Medway, Lord, 1959, Echo-location among Collocalia, Nature, Lond., 184: 1352-1353.

Medway, Lord, 1967, The function of echo-navigation among swiftlets, Anim.Behav., 15: 416-420,

Medway, Lord, 1969, Studies of the biology of the edible-nest swiftlets of South East Asia, Malayan Nature Jour., 22: 57-63.

Medway, Lord and Pye, J.D., 1977, Echolocation and the systematics of swiftlets, in: Evolutionary Ecology, eds. Stonehouse, B. and Perrins, C., Macmillan, London, 225-238.

Medway, Lord and Wells, D.R., 1969, Dark orientation by the giant swiftlet, Collocalia gigas, Ibis, 111: 609-611.

Miller, L.A., 1970, Structure of the green lacewing tympanal organ (Chrysopa carnea, Neuroptera), J.Morph., 131: 359-382.

Miller, L.A., 1971, Physiological responses of green lacewings (Chrysopa, Neuroptera) to ultrasound, J.Insect Physiol., 17: 491-506.

Miller, L.A., 1974, (Abstract) The behavioral response of green lacewings to ultrasound, Amer.Zool., 13: 1258.

Miller, L.A., 1975, The behavior of flying green lacewings, Chrysopa carnea, in the presence of ultrasound. J.Insect Physiol., 21: 205-219.

Miller, L.A., in press, Interactions between bats and green lacewings (Chrysopa, Insecta) in free flight, Proc. 5th Int. Bat Res.Conf., Texas Tech.Univ.Press.

Miller, L.A. and MacLeod, E.G., 1966, Ultrasonic sensitivity: a tympanal receptor in the green lacewing, Chrysopa carnea, Science, 154: 891-893.

Möhres, F.P., 1953, Über die Ultraschallorientierung der Hufeisennasen (Chiroptera: Rhinolophinae), Z.vergl.Physiol., 34:547-588.

Möhres, F.P., 1967, Ultrasonic orientation in megadermatid bats, Animal Sonar Systems: biology and bionics, ed. Busnel, R.-G., Jouy-en-Josas, 115-127.

Möhres, F.P. and Kulzer, E., 1955, Ein neuer, kombinierter Typ der Ultraschallorientierung bei Fledermäusen, Naturwissenschaften, 42: 131-132.

Möhres, F.P. and Kulzer, E., 1956a, Über die Orientierung der Flughunde (Chiroptera, Pteropodidae), Z.vergl.Physiol., 38: 1-29.

Möhres, F.P. and Kulzer, E., 1956b, Untersuchungen über die Ultraschallorientierung von vier afrikanischen Fledermaus-familien, Verh.der Deutsch.zool.Ges. in Erlangen, 1955, 59-65.

Möhres, F.P. and Neuweiler, G., 1966, Die Ultraschallorientierung der Grossblat-Fledermäuse (Chiroptera, Megadermatidae), Z.vergl.Physiol., 53: 195-227.

Novick, A., 1958a, Orientation in paleotropical bats: I. Microchiroptera, J.exp.Zool, 138: 81-154.

Novick, A., 1958b, Orientation in paleotropical bats: II.
    Megachiroptera, J.exp.Zool., 137: 443-462.
Novick, A., 1959, Acoustic orientation in the cave swiftlet,
    .Biol.Bull., 117: 497-503.
Novick, A., 1962, Orientation in neotropical bats: I. Natalidae
    and Emballonuridae, J.Mammal., 43: 449-455.
Novick, A., 1963, Orientation in neotropical bats: II.
    Phyllostomidae and Desmodontidae, J.Mammal., 44: 44-56.
Novick, A., 1971, Echolocation in bats: some aspects of pulse
    design, Amer.Scient., 59: 198-209.
Novick, A., 1977, Acoustic orientation, in "Biology of Bats",
    ed. Wimsatt, W.A., 3: 73-287.
Novick, A. and Griffin, D.R., 1961, Laryngeal mechanisms in bats
    for the production of orientation sounds, J.exp.Zool., 148:
    125-146.
Paul, D.H., 1973, Central projections of tympanic fibres in
    noctuid moths, J.Insect Physiol., 19: 1785-1792.
Paul, D.H., 1974, Responses to acoustic stimulation of thoracic
    interneurons in noctuid moths, J.Insect Physiol., 20: 2205-2218.
Payne, R.S., Roeder, K.D. and Wallman, J., 1966, Directional
    sensitivity of the ears of noctuid moths, J.exp.Biol., 44:
    17-31.
Pecotich, L., 1974, Grey swiftlets in the Tully River Gorge and
    Chillagoe caves, Sunbird, 5: 16-21.
Peff, T.C. and Simmons, J.A., 1971, Horizontal-angle resolution by
    echolocating bats, J.acoust.Soc.Amer., 51: 2063-2065.
Penny, M., 1975, The Birds of Seychelles and the Outlying Islands,
    Collins, London.
Pierce, G.W., 1925, Piezoelectric crystal oscillators applied to
    the precision measurement of the velocity of sound in air and
    carbon dioxide at high frequencies, Proc.Amer.Acad.Arts Sci.,
    60: 271-302.
Pye, A., 1978, Aspects of cochlear structure and function in bats,
    Proc.Fourth Int.Bat Res.Conf., eds. Olembo, R.J., Castelino,
    J.B. and Mutere, F.A., Kenya Lit.Bureau, Nairobi, 73-83.
Pye, J.D., 1967, Synthesizing the waveforms of bats' pulses,
    Animal Sonar Systems: biology and bionics, ed. Busnel, R.-G.,
    Jouy-en-Josas, 43-65.
Pye, J.D., 1968a, Animal sonar in air, Ultrasonics, 6: 32-38.
Pye, J.D., 1968b, How insects hear, Nature, Lond., 218: 797.
Pye, J.D., 1971, Bats and fog, Nature, Lond., 229: 572-574.
Pye, J.D., 1972, Bimodal distribution of constant frequencies in
    some hipposiderid bats (Mammalia: Hipposideridae), J.Zool.,
    166: 323-335.
Pye, J.D., 1973, Echolocation by constant frequency in bats,
    Period.biol., 75: 21-26.
Pye, J.D., 1978, Some preliminary observations on flexible echo-
    location systems, Proc.Fourth Int.Bat Res.Conf., eds. Olembo,
    R.J., Castelino, J.B. and Mutere, F.A., Kenya Lit.Bureau,
    Nairobi, 127-136.

Pye, J.D., in press, Why ultrasound?, Endeavour.

Pye, J.D. and Roberts, L.H., 1970, Ear movements in a hipposiderid bat, Nature, Lond., 225: 285-286.

Rayleigh, Lord, 1896, The Theory of Sound (2nd edn), Macmillan, London.

Roberts, L.H., 1972a, Variable resonance in constant frequency bats, J.Zool., 166: 337-348.

Roberts, L.H., 1972b, Correlation of respiration and ultrasound production in rodents and bats, J.Zool., 168: 439-449.

Roberts, L.H., 1973, Cavity resonances in the production of orientation cries, Period.biol., 75: 27-32.

Roberts, L.H., 1975, Confirmation of the echolocation pulse production mechanism of Rousettus, J.Mammal., 56: 218-220.

Roeder, K.D., 1963, Echoes of ultrasonic pulses from flying moths, Biol.Bull., 124: 200-210.

Roeder, K.D., 1966a, Auditory system of noctuid moths, Science, 154: 1515-1521.

Roeder, K.D., 1966b, Acoustic sensitivity of the noctuid tympanic organ and its range for the cries of bats, J.Insect Physiol., 12: 843-859.

Roeder, K.D., 1966c, A differential anemometer for measuring the turning tendency of insects in stationary flight, Science, 153: 1634-1636.

Roeder, K.D., 1966d, Interneurones of the thoracic nerve cord activated by tympanic nerve fibres in noctuid moths, J.Insect Physiol., 12: 1227-1244.

Roeder, K.D., 1967a, Prey and predator, Bull.ent.Soc.Amer., 13: 6-9.

Roeder, K.D., 1967b, Turning tendency of moths exposed to ultrasound while in stationary flight, J. Insect Physiol., 13: 873-888.

Roeder, K.D., 1967c, Nerve cells and insect behaviour, Harvard Univ. Press, Cambridge, Mass., revd. edn.

Roeder, K.D., 1969a, Acoustic interneurons in the brain of noctuid moths, J.Insect Physiol., 15: 825-838.

Roeder, K.D., 1969b, Brain interneurons in noctuid moths: differential suppression by high sound intensities, J.Insect Physiol., 15: 1713-1718.

Roeder, K.D., 1969c, Acoustic interneurons in the brain of noctuid moths, J.exp.Zool., 134: 127-157.

Roeder, K.D., 1970, Episodes in insect brains, Amer.Scient., 58: 378-389.

Roeder, K.D., 1971, Acoustic alerting mechanisms in insects, in "Orientation: Sensory Basis", ed. Adler, H.E., Ann.N.Y. Acad.Sci., 188: 63-79.

Roeder, K.D., 1972, Acoustic and mechanical sensitivity of the distal lobe of the pilifer in choerocampine hawkmoths, J.Insect Physiol., 18: 1249-1264.

Roeder, K.D., 1974a, Responses of the less sensitive acoustic sense cells in the tympanic organs of some noctuid and geometrid moths, J.Insect Physiol., 20: 55-66.

Roeder, K.D., 1974b, Acoustic sensory responses and possible bat
    evasion tactics of certain moths, Can.Soc.Zool.ann.Meet.Proc.,
    Univ.New Brunswick, 71-78.
Roeder, K.D., 1976, Joys and Frustrations of doing research,
    Perspectives in Biol.Med., winter, 231-245.
Roeder, K.D. and Fenton, M.B., 1973, Acoustic responsiveness of
    Scoliopteryx libatrix L. (Lepidoptera: Noctuidae), a moth that
    shares hibernacula with some insectivorous bats, Can.J.Zool.,
    51: 681-685.
Roeder, K.D. and Payne, R.S., 1966, Acoustic orientation of a moth
    in flight by means of two sense cells, Symp.Soc.exp.Biol., 20:
    251-272.
Roeder, K.D. and Treat, A.E., 1970, An acoustic sense in some
    hawkmoths (Choerocampinae), J. Insect Physiol., 16: 1069-1086.
Roeder, K.D., Treat, A.E. and Vandeberg, J.S., 1968, Auditory
    sense in certain sphingid moths, Science, 159: 331-333.
Roeder, K.D., Treat, A.E. and Vandeberg, J.S., 1970, Distal lobe
    of the pilifer: an ultrasonic receptor in choerocampine
    hawkmoths, Science, 170: 1098-1099.
Russo, D.M. and Bartberger, C.L., 1964, Ambiguity diagram for
    linear F.M. sonar, J.acoust.Soc.Amer., 38: 183-190.
Sales, G.D. and Pye, J.D., 1974, "Ultrasonic Communication by
    Animals", Chapman and Hall, London.
Schnitzler, H.U., 1967, Discrimination of thin wires by flying
    horseshoe bats (Rhinolophidae), Animal Sonar Systems: biology
    and bionics, ed. Busnel, R.-G., Jouy-en-Josas, 69-87.
Schnitzler, H.U., 1968, Die Ultraschall-Ortungslaute der Hufeisen-
    Fledermäuse (Chiroptera-Rhinolophidae) in verschiedenen-
    Orientierungssituationen, Z.vergl.Physiol., 57: 376-408.
Schnitzler, H.U., 1970a, Echoortung bei der Fledermaus, Chilonycteris
    rubiginosa, Z.vergl.Physiol., 68: 25-38.
Schnitzler, H.U., 1970b, Comparison of the echolocation behaviour
    in Rhinolophus ferrumequinum and Chilonycteris rubiginosa,
    Bijd.tot de Dierk, 40: 77-80.
Schnitzler, H.U., 1971, Fledermäuse im Windkanal, Z.vergl.Physiol.,
    73: 209-221.
Schnitzler, H.U., 1973, Die Echoortung der Fledermäuse und ihre
    hörphysiologischen Grundlagen, Fortschritte der Zool., 21:
    136-189.
Schnitzler, H.U., 1978, Die Detektion von Bewegungen durch
    Echoortung bei Fledermäusen, Vehr.Dtsch.Zool.Ges., 1978: 16-33.
Schnitzler, H.U., in press, Detection of the fluttering movements
    of insects by constant frequency bats, Proc. 5th Int.Bat Res.
    Conf., Texas Tech.Univ.Press.
Schnitzler, H.U. and Grinnell, A.D., 1977, Directional sensitivity
    of echolocation in the horseshoe bat, Rhinolophus ferrumequinum;
    I. Directionality of sound emission, J.comp.Physiol., 116:
    51-61.

Schuller, G. and Suga, N., 1976, Laryngeal mechanisms for the emission of CF-FM sounds in the Doppler-shift compensating bat, Rhinolophus ferrumequinum, J.comp.Physiol., 107: 253-262.

Shimozawa, T., Suga, N., Hendler, P. and Schuetze, S., 1974, Directional sensitivity of echolocation system in bats producing frequency-modulated signals, J.exp.Biol., 60: 53-69.

Simmons, J.A., 1969, Acoustic radiation patterns for the echolocating bats, Chilonycteris rubiginosa and Eptesicus fuscus, J.acoust.Soc.Amer., 46: 1054-1056.

Simmons, J.A., 1970, Distance perception by echolocation: the nature of echo signal-processing in the bat, Bijd.tot de Dierk, 40: 87-90.

Simmons, J.A., 1971, Echolocation in bats: signal processing of echoes for target range, Science, 171: 925-928.

Simmons, J.A., 1973, The resolution of target range by echolocating bats, J.acoust.Soc.Amer., 54: 157-173.

Simmons, J.A., 1974, Response of the Doppler echolocation system in the bat, Rhinolophus ferrumequinum, J.acoust.Soc.Amer., 56: 672-682.

Simmons, J.A., in press, Phylogenetic adaptations and the evolution of echolocation in bats (Chiroptera), Proc.5th Int.Bat Res. Conf., Texas Tech.Univ.Press.

Simmons, J.A., Fenton, M.B. and O'Farrell, M.J., 1979, Echolocation and pursuit of prey by bats, Science, 203: 16-21.

Simmons, J.A., Howell, D.J. and Suga, N., 1975, Information content of bat sonar echoes, Amer.Scient., 63: 204-215.

Simmons, J.A., Lavender, W.A. and Lavender, B.A., 1978, Adaptation of echolocation to environmental noise by the bat, Eptesicus fuscus, Proc.Fourth Int.Bat Res.Conf., eds. Olembo, R.J. Castelino, J.B. and Mutere, F.A., Kenya Lit.Bureau, Nairobi, 97-104.

Simmons, J.A., Lavender, W.A., Lavender, B.A., Childs, J.E., Hulebak, K., Rigden, M.R., Sherman, J., Woolman, B and O'Farrell, M.J., 1978, Echolocation by free-tailed bats (Tadarida), J.comp.Physiol., 125: 291-299.

Simmons, J.A., Lavender, W.A., Lavender, B.A., Doroshow, C.A., Kiefer, S.W., Livingston, R., Scallet, A.C. and Crawley, D.E., 1974, Target structure and echo spectral discrimination by echolocating bats, Science, 186: 1130-1132.

Simmons, J.A. and O'Farrell, M.J., 1977, Echolocation by the long-eared bat, Plecotus phyllotis, J.comp.Physiol., 122: 201-214.

Simmons, J.A. and Verton, J.A., 1971, Echolocation: discrimination of targets by the bat, Eptesicus fuscus, J.exp.Zool., 176: 315-328.

Strother, G.K. and Mogus, M., 1970, Acoustical beam patterns for bats: some theoretical considerations, J.acoust.Soc.Amer., 48: 1430-1432.

Suga, N. and Schlegel, P., 1972, Neural attenuation of responses
    to emitted sounds in echolocating bats, Science, 177: 82-84.

Suga, N. and Schlegel, P., 1973, Coding and processing in the
    auditory system of FM-signal-producing bats, J.acoust.Soc.
    Amer., 54: 174-190.

Suga, N., Schlegel, P., Shimozawa, T. and Simmons, J., 1973,
    Orientation sounds evoked from echolocating bats by electrical
    stimulation of the brain, J.acoust.Soc.Amer., 54: 793-797.

Suthers, R.A., 1965, Acoustic orientation by fish-catching bats,
    J.Zool., 158: 319-348.

Suthers, R.A., 1967, Comparative echolocation by fishing bats,
    J.Mammal., 48, 79-87.

Suthers, R.A. and Fattu, J.M., 1973, Fishing behaviour and acoustic
    orientation by the bat (Noctilio labialis), Anim.Behav., 21:
    61-66.

Suthers, R.A., Thomas, S.P. and Suthers, B.J., 1972, Respiration,
    wing-beat and ultrasonic pulse emission in an echolocating
    bat, J.exp.Biol., 56: 37-48.

Tupinier, Y., Biraud, Y., Chiollaz, M. and Escudie, B., in press,
    Analysis of vespertilionid sonar signals during cruise,
    pursuit and prey capture, Proc.Fifth Int.Bat Res.Conf., Texas
    Tech.Univ.Press.

Wassif, K. and Madkour, G., 1968-9, The structure of the hyoid bone,
    larynx and upper part of trachea in some Egyptian bats,
    Bull.Zool.Soc.Egypt, 22: 15-26.

Woolf, N.K., 1974, Ontogeny of bat sonar: paired FM signalling,
    J.acoust.Soc.Amer., 55: S 53 (abstract).

ECHOLOCATION ONTOGENY IN BATS

Patricia E. Brown and Alan D. Grinnell

Departments of Biology and Physiology, and Ahmanson
Laboratory of Neurobiology, Brain Research Institute;
University of California, Los Angeles; Los Angeles,
California 90024

ECHOLOCATION ONTOGENY IN BATS

The ontogeny of echolocation in bats has received relatively
little attention. Some bats do not hear at birth and are incapable
of emitting the high frequency sounds necessary for accurate echolocation. No bat can fly at birth. Even bats born in a relatively
advanced state, with the ability to hear and emit ultrasounds, still
must learn to fly and correlate their outgoing pulses and returning
echoes in a three-dimensional world before they are weaned. A study
of echolocation ontogeny must integrate physiology, morphology, and
behavior. The emission of short, high frequency sounds depends on
the development of laryngeal musculature, but these ultrasonic pulses
are of little value if the auditory system cannot process them or if
the animal cannot maintain the high body temperatures necessary for
ultrasonic hearing. All these systems develop simultaneously, and
we have yet to assess the role of learning versus maturation in echolocation ontogeny. Our understanding of echolocation in adult bats
can be enhanced through knowledge of how this complex sensory system
develops.

Development of Vocalizations

All bats are born in a relatively helpless state and are totally
dependent on their mothers for a period of weeks or months postnatally
(Orr, 1970). Gould (1975) has classified baby bats of eight genera
as either precocial (eyes and ears open at birth, body covered with
fur, emitting more than one type of vocalization), altricial (eyes
and ears closed, skin naked, emitting only one type of vocalization),

355

or as intermediate between these two extremes. The degree of morphological development at birth is not necessarily related to phylogeny.
An example would be the vespertilionids Antrozous pallidus (Brown,
1973, 1976; Gould, 1975) and Myotis (see refs. below). Antrozous
is truly altricial, with eyes and ears sealed for at least four days
postnatally (Fig. 1), while Myotis velifer (Brown, unpublished data)
and M. lucifugus (Gould, 1971, 1975) are alert and mobile at birth.
Smaller bats generally produce relatively large young compared to
the adult (Orr, 1970). Bats of the family Phyllostomatidae are typically born in an advanced state (body covered with fur and eyes open)
and develop quite rapidly.

FM Bats

In most instances, there is a correlation between the degree of
morphological development at birth and the types of sounds emitted
by the infant. For example, newborn Antrozous emit only continuous
loud isolation calls when separated from their mothers (Brown, 1973,
1976). These calls which are longer, louder, and lower in frequency
than echolocation pulses, serve to promote contact between the mother
and young. Isolation calls appear to be ubiquitous among baby bats,
and whereas they differ somewhat in structure between species, they
all serve a common function (Gould, 1970, 1971, 1975a, 1975b, 1977,
1979; Gould, Woolf, and Turner, 1973; Brown, 1973, 1976; Matsumura,
1979; Schmidt, 1972; Porter, 1977; Orr, 1954; Davis et al., 1968; Jones,
1967; Douglas, 1967; Pearson et al., 1952; Kulzer, 1962; Kay and
Pickvance, 1963; Kunz, 1973). A double-note (a pair of long-short
or short-long pulses) is also present at birth in some species, but
appears at a later age in others (Gould et al., 1973; Gould, 1975a;
Woolf, 1974). Short FM sounds, the apparent precursors of later FM
echolocation pulses, are emitted at birth by some precocial species
(i.e. Macrotus and Carollia) but generally do not appear for several
days (Eptesicus) to a week (Antrozous) postnatally. Woolf (1974) rec-
ognizes isolation calls, double notes, and FM pulses as an ontogenetic
sequence, and has grouped them into a "sonar family". Whereas these
categories of sounds may be structurally similar, isolation calls are
not necessarily the precursors of adult echolocation pulses, although
in some species this may be the case. In only four genera of bats
have the sound emissions in known-age bats been analyzed throughout
their postnatal development. These include Myotis (M. lucifugus,
Gould, 1971, 1975a; M. velifer, Brown, unpublished observations; M.
oxygnathus, Konstantinov, 1973), Eptesicus fuscus (Gould, 1971, 1975b;
Woolf, 1974), Antrozous pallidus (Brown, 1973, 1976), and Rhinolophus
ferrumequinum (Konstantinov, 1973; Matsumura, 1979; Glenis
Long, personal communication). At present it is not possible to postulate a general plan for bat sonar ontogeny, and each species must
be examined individually. The vespertilionids Antrozous, Eptesicus,
and Myotis all differ in the time course of their vocal development.
In Antrozous, it appears that echolocation pulses and communication
sounds have different origins. As previously stated, Antrozous emits
only FM isolation calls at birth (Fig. 2). As the bat matures, the
isolation call shortens and increases in frequency, until by 20 days

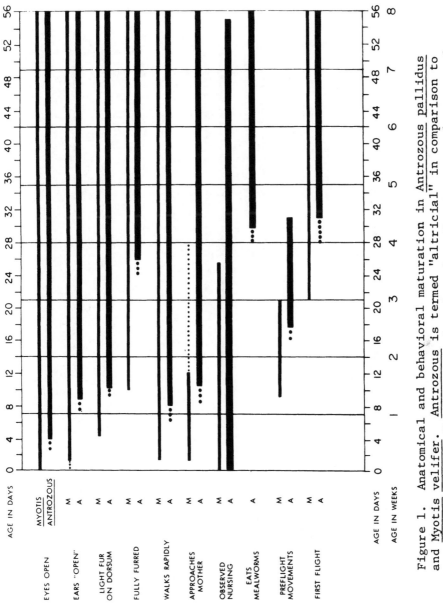
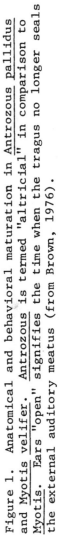

Figure 1. Anatomical and behavioral maturation in Antrozous pallidus and Myotis velifer. Antrozous is termed "altricial" in comparison to Myotis. Ears "open" signifies the time when the tragus no longer seals the external auditory meatus (from Brown, 1976).

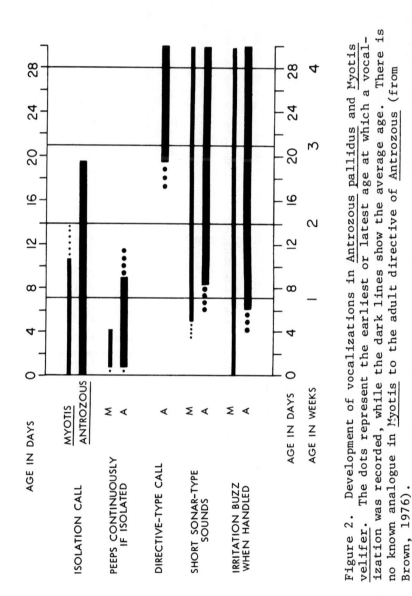

Figure 2. Development of vocalizations in Antrozous pallidus and Myotis velifer. The dots represent the earliest or latest age at which a vocalization was recorded, while the dark lines show the average age. There is no known analogue in Myotis to the adult directive of Antrozous (from Brown, 1976).

postnatally it resembles the adult "directive," a communication sound.
Double note isolation calls are emitted by some baby Antrozous and
not others. These may function as a vocal signature for individual
recognition (Brown, 1976). On about the fifth day postnatally, a
shorter 10-20 msec FM pulse is emitted in addition to the longer (60
msec) isolation call. This is the apparent precursor of the short
orientation sound first emitted between seven and nine days after
birth (Fig. 3a). This short ultrasonic pulse is emitted as the now-
mobile baby crawls around exploring its environment. It gradually
shortens and increases in frequency and repetition rate (Table 1),
evolving into the adult orientation pulse before the first flight at
four to five weeks (Fig. 3b,c). The early sonar pulses are rich in
harmonics. In some instances, it appears that the fundamental is
dropped during ontogeny, with subsequent sound energy being funneled
into the second harmonic. In the majority, however, there is a gradual increase in fundamental frequency and a subsequent loss of harmonics.

Myotis velifer (Brown, unpublished data) follows the same sequence as Antrozous except that the time course is compressed (Fig.
1,2), and this more advanced bat crawls actively at two days after
birth and emits short FM pulses in addition to longer isolation calls
and double notes. Since all of these categories of sounds are present
simultaneously, it does not appear that the isolation call evolves
into the double note and sonar pulse, nor that they represent a continuum. It is possible that even though isolation calls are the
acoustic precursors of echolocation pulses (i.e. they preceed them
developmentally), the isolation calls still remain to serve a different function. No analogue to the adult "directive" of Antrozous
exists for Myotis. The isolation call is emitted less frequently as
the bat matures, disappearing entirely by fourteen days postnatally.
Infant Myotis lucifugus were found to emit isolation calls, double
notes, and short FM pulses (Gould, 1971, 1975a). These babies interspersed sonar signals and isolation calls depending on their state
of excitation. If a day old Myotis lucifugus was held by the nape
of the neck and moved up and down, it emitted sonar calls. In contrast, Konstantinov (1973) reported that the larger European Myotis
oxygnathus emitted only loud, audible isolation calls for the first
ten days after birth. These isolation calls apparently evolved into
adult-like echolocation signals by 25 days postnatally, although the
higher frequencies utilized by adults were not recorded until 45 days,
an age when insect pursuit is normally initiated in this species.
This observation supports the hypothesis proposed by Gould (1970,
1971, 1977) and Woolf (1974) that sonar calls are derived from isolation calls.

Eptesicus is intermediate between Myotis and Antrozous both in
its degree of morphological development and its vocalizations (Gould,
1971). For the first three days postnatally, only an isolation call
is emitted in response to stressful situations. Although infant

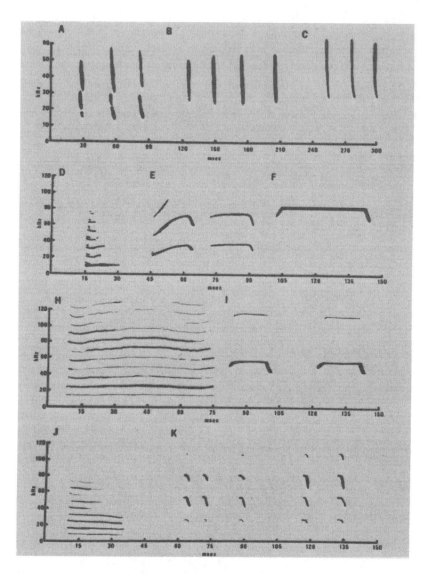

Figure 3. A-C: Tracings of sound spectrographs of orientation pulses
                 for <u>Antrozous</u> <u>pallidus</u>: A - 9 days; B - 16 days;
                 C - adult.
           D-F: <u>Rhinolophus</u> <u>ferrumequinum</u>: D - newborn isolation
                 call; E - 8 days double note; F - 14 days orientation
                 pulse. The CF portion of adult pulses is 3 kHz
                 higher. These tracings are from recordings made by
                 Long of a wild colony.
           H,I: <u>Pteronotus</u> <u>parnellii</u>: H - newborn isolation call;
                 I - adult orientation pulse.
           J,K: <u>Pteronotus</u> <u>suapurensis</u>: J - newborn isolation call;
                 K - adult orientation pulse.

Table 1. Average characteristics of <u>Antrozous pallidus</u> echolocation
pulses during ontogeny. Repetition rate and interpulse interval vary
with activity. Before 28 days when they can fly, measurements of
repetition rates are from crawling bats, while rates for the 46-day
and adult are from bats flying in a room without obstacles.

AGE days	FREQUENCY KHZ	PULSE DURATION msec.	INTERPULSE INTERVAL msec.	REPETITION RATE pulses/sec.
6	30→10	20	158	8
17	40→17	4	30	24
26	60→23	3	22	28
46	70→20	2.5	17	35
82 (adult)	80→27	2	12	45

<u>Eptesicus</u> continue to emit isolation calls under certain situations;
they are replaced at three to five days by double notes, either in
long-short or short-long combinations (Gould <u>et al</u>., 1973; Gould,
1975b; Woolf, 1974). At approximately five days of age, short (3
msec) sonar sounds are produced often as paired signals. They differ
from double notes in being of equal duration. Double notes continue
to function in communication,while the short FM sounds evolve into
echolocation pulses by day 21. <u>Eptesicus</u> begins to fly at approximately 25 days postnatally. Woolf (1974) contends that double notes
are the precursors of sonar sounds.

     Woolf (1974) demonstrated that the development of the FM pulses
was not affected by deafening due to bilateral malleus removal in
infant <u>Eptesicus</u>. In order to alter the normal acoustic environment,
Gould (1975b) hand-raised some infant <u>Eptesicus</u> while others were
raised by mothers whose superior laryngeal nerves had been cauterized.
These infants emitted normal sonar pulses when recorded as adults,
suggesting a strong genetic component in the ontogeny of the sonar
signal.

CF Bats

The three genera discussed in detail up to this point all utilize FM pulses for echolocation as adults, and it is not surprising that the earliest sounds emitted by these bats are long FM isolation calls that, though longer and lower in frequency, are recognizable as members of the same "sonar family" (Woolf, 1974). It is therefore not surprising that neonatal vocalizations of bats employing CF or CF/FM pulses as adults should contain primarily constant frequencies. Several recent studies have documented constant frequency vocalizations in juvenile bats of the Old World families Hipposideridae (Gould, 1979; Grinnell and Hagiwara, 1972) and, Rhinolophidae (Kay and Pickvance, 1963; Matsumura, 1979; and Long, personal communication). Whereas the adult bats emit all sonar pulses via the nostrils, neonates appear to emit high intensity isolation calls, double notes, and sonar calls primarily via the mouth. This same pattern has also been observed in phyllostomatids (Gould, 1977), where adults emit relatively low intensity sounds nasally, but neonates produce higher intensities orally. The nature of this ontogenetic shift from oral to nasal vocalizations in Rhinolophus was first reported by Konstantinov (1973), and has been subsequently documented by Sumiko Matsumura (1979 and personal communication) under laboratory conditions, and by Glenis Long (personal communication) in a natural roost. Bats less than one week old emit predominantly "noisy" or broad band isolation calls through the mouth (Fig. 3d), although some pure tone pulses are emitted nasally. Often intermediate combinations of nasal and oral sounds occur in the same pulse. These infant pure tone pulses differ from those of the adults. Infant pulses contain slight FM descending components both at the beginning and end of the pulse (Fig. 3e), while the adults begin each pulse with a long CF portion and end with an FM tail (Fig. 3f) (Schnitzler, 1968). The infant sounds are rich in harmonics, these being supressed and filtered in the adult sonar pulses with most sound energy present in the second harmonic (Novick, 1958; Pye, 1968). Halls (in prep.) has found that within roosts, adults do emit "modified" shortened CF pulses, with more FM present and harmonics, presumably for communication. As Rhinolophus matures, nasal signals replace oral ones, so that at three weeks the oral mode is found only in the first syllable of the "attractive" (= isolation) call. By four weeks when this bat begins to fly, all oral calls have disappeared with the exception of communicative distress calls. In addition to the shift from oral to nasal, and "noisy" to pure tones, the character of the nasal tones changes with age. The fundamental frequency increases from 12 kHz in neonates to 36 kHz characteristic of the adults. During development, the fundamental and third harmonic are suppressed, so that only the second harmonic remains. The anatomical basis for the shift from oral to nasal sound emission arises from the closing of the laryngo-nasal junction which is incomplete in the week-old bat. This allows air flow either through the nose or mouth, and accounts for the high degree of variability in infant vocalizations (Matsumura, 1979).

Long (personal communication) found a similar acoustic sequence in the European horseshoe bat (Rhinolophus ferrumequinum ferrumequinum) while recording vocalizations in a natural maternity roost in the attic of a house. However the time frame was compressed as this subspecies begins to fly at 14-16 days postnatally. The CF portion of the adult echolocation pulse is 83 kHz compared to 72 kHz for R. ferrumequinum nippon. In both cases this is the second harmonic. Long found that in infant vocalizations the greatest sound energy was present in the fundamental and third harmonic. She also noted that, in contrast to FM bats, the duration of the pulse (CF component) increased with age and showed less variability than in infant bats.

Matsumura (in preparation) has made an observation in the context of mother-young communication that might have important implications in the development of echolocation behavior in young Rhinolophus. During reunions, before the infant can fly, antiphonal calling by mother and infants ensues. In the final stages of this "duetting", the infant appears to copy the pure tone emission pattern of the mother (albeit 3 kHz lower). At the beginning of a vocal exchange the mother seems to actively synchronize her "lead signals" (pure tone pulses rich in harmonics analogous to "directives" in Antrozous) with the infant's two to three syllable pure tone pulse (nasal sound). Later the infant begins to emit single syllable pulses, and actively synchronizesthese so that they regularly overlap with the mother's emissions, in effect copying her pattern of vocalizations.

Adult Hipposideros emit echolocation pulses containing an initial CF component followed by an FM tail. The duration and frequency of the CF portion varies with the species, as does the sweep of the FM tail. Among five species of Hipposideros recorded by Gould (1979) in Malaysia and four by Grinnell and Hagiwara (1972) in New Guinea, H. diadema emits a long CF/short FM pulse, structurally similar to Rhinolophus ferrumequinum but shorter in duration, while H. ridleyi produces a short CF/long FM. The sonar calls of the other species are intermediate between these. Due to problems of maintaining healthy nursing bats in captivity, Gould was unable to trace sonar development in any individual, but rather obtained cross-sectional data by recording sounds from different size/age classes. Infant Hipposideros, like Rhinolophus, emit via the mouth high intensity isolation calls, double notes, and CF pulses that are rich in harmonics. In addition to isolation calls, one day old H. diadema produced a CF/FM signal similar to that of the adult with predominant energy in the second harmonic ($\sim$40 kHz) whereas only the third harmonic is presented in the adult sonar pulse ($\sim$60 kHz). In H. diadema from New Guinea (Grinnell and Hagiwara, 1972) four harmonics are present in the adult pulse with the second (58 kHz) and fourth (116 kHz) being dominant. The long low CF isolation calls recorded in infant H. bicolor and cineraceus are similar to those emitted by newborn Rhinolophus (Fig. 3d) and Pteronotus (Fig. 3h,j). In Hippo-

sideros the double notes and precursors of sonar pulses contain ascending and descending FM components (Gould, 1979; Grinnell and Hagiwara, 1972) which also resemble those of week-old Rhinolophus (Fig.
3e).

In Panama, an attempt was made to trace the ontogeny of CF/FM
sonar pulses in Pteronotus parnellii and suapurensis, members of the
New World family Mormoopidae (Brown and Grinnell, unpublished data).
Mother bats refused to care for their babies in captivity, and hand-
rearing was not possible. Recordings were made of newborn vocalizations which were limited to long, low frequency CF pulses rich in
harmonics (Fig. 3h,j), similar to those recorded in Rhinolophus (Fig.
3d) and Hipposideros. Although the adult sonar calls are quite
different in P. parnellii and suapurensis, the infant isolation calls
are structurally quite similar. Unfortunately it was not possible
to determine whether these early vocalizations were precursors of
the adult orientation pulse. Just as Woolf (1974) has postulated an
FM "sonar family", a similar "family" grouping appears to exist for
CF bats of diverse phylogenies.

The Development of Hearing

When new-born pallid bats (Antrozous) failed to behaviorally
respond to their mothers' directives (communication sounds), one
logical conclusion was that they could not hear. This hypothesis was
confirmed by neurophysiological recordings of single units and evoked
potentials from the posterior colliculus in anesthetized bats in
response to auditory stimuli (Brown, Grinnell, and Harrison, 1978).
In fact there was no auditory response in bats until six days postnatally, and then only to loud, low frequency sounds such as contained in the mother's directives. By this age, the infant's ears
are open and erect, the tragus no longer sealing the external meatus.
Absolute sensitivity and frequency range increase rapidly until by
24 days after birth, the evoked potential audiogram resembles that
of the adult (Fig. 4). Short ultrasonic pulses are not emitted until
day nine, and parallel the ability to hear frequencies in that range.
This corresponds to the age at which the baby bat is able to crawl
rapidly and explore its environment. Except for the first week of
its life, the infant bat emits only sounds which it can actually
hear (Fig. 5).

Besides the high frequency sensitivity lacking in infant Antrozous, another important adaptation for echolocation is a high degree
of temporal resolution or fast auditory recovery time. Forty msec
or longer are required for the auditory system to respond to the
second of two identical stimuli (i.e. the outgoing pulse and returning
echo) compared with less than four msec in the adult (Fig. 6). This
also develops rapidly in the juvenile bat, so that fast temporal
resolution has developed by the time of first flight at one month.

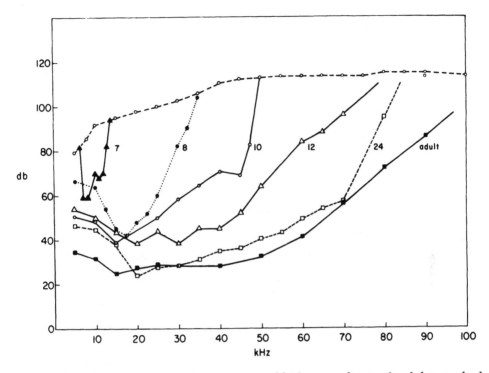

Figure 4. Audiograms of <u>Antrozous</u> <u>pallidus</u> as determined by evoked potentials recorded from the posterior colliculus. The age in days is to the right of each curve. No evoked responses were observed in bats younger than 6 days. By 24 days, both frequency range and sensitivity resembled that of adults. Top dotted line represents maximum speaker output in this and subsequent graphs of evoked potentials. (From Brown <u>et</u> <u>al</u>., 1978.)

Even after they begin to answer their mothers' vocalizations, infant <u>Antrozous</u> appear to have difficulty in localizing the sound source. This behavioral observation is substantiated by collicular responses to sounds presented at different angles. While evoked potentials show little directionality at any age, single unit recordings in bats older than three weeks (but not younger than this) show a high degree of directionality suggesting that binaural interactions develop relatively late.

<u>Myotis</u> <u>velifer</u> are born in a more advanced state than <u>Antrozous</u> (Fig. 1), and at two days after birth are capable of emitting short sonar sounds while crawling. Evoked potentials were first recorded in two-day old bats (Brown and Grinnell, unpublished data). Increased sensitivity to high frequency sounds developed more rapidly than in <u>Antrozous</u>, with the evoked potential audiogram in a two-week old

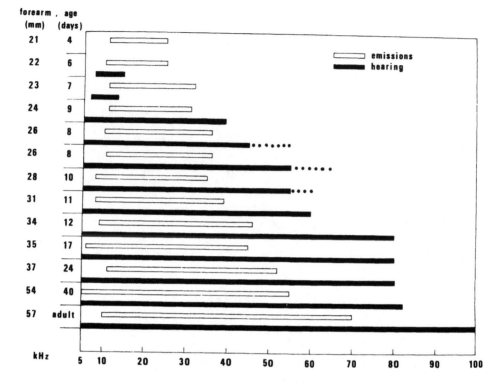

Figure 5. Range of frequencies emitted (first harmonic) compared to frequencies actually heard by <u>Antrozous</u> <u>pallidus</u> at different ages. Bats are arranged according to forearm length, which is a better index of maturation than absolute age. Range of hearing was determined by evoked potentials recorded with gross tungsten electrodes. Dotted lines indicate responses to higher frequencies in single units. Except during the first week postnatally, pallid bats emit sounds which they can hear. (From Brown <u>et</u> <u>al</u>., 1978.)

<u>Myotis</u> resembling that of an adult (Fig. 7). As in <u>Antrozous</u> the ability to hear high frequency sounds parallels the bat's ability to emit them (Fig. 8).

In contrast to <u>Myotis</u> <u>velifer</u>, young <u>Myotis</u> <u>oxygnathus</u> develop more slowly, with no behavioral or electrophysiological responses to sound before ten days of age, with adult sensitivity attained by 25 days. During this period temporal resolution improved from a partial recovery time of 20 msec in ten day old <u>M</u>. <u>oxygnathus</u> to 4 msec in adults. There was also a dramatic improvement in the ability of the auditory system to follow a train of stimuli (Konstantinov, 1973; Konstantinov and Stosman, 1972).

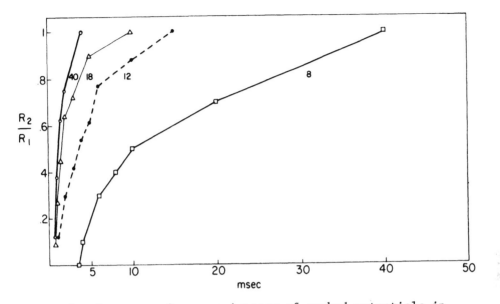

Figure 6. Recovery of responsiveness of evoked potentials in
Antrozous pallidus of different ages (age in days next to each
curve). R_2/R_1 represents the ratio of amplitudes of the second
response to the first when two identical stimuli are presented.
A value of 1 signifies full response recovery. The abscissa shows
the interval in msec separating the two stimuli. (From Brown et al.,
1978.)

Development of the auditory system is more rapid in the CF/FM
bat Rhinolophus ferrumequinum. Stosman and Konstantinov (1972) first
recorded evoked potentials to auditory stimuli in week old bats, but
both frequency range and sensitivity rapidly increased until by 18-
21 days the audiograms resembled those of the adult. However the
extremely sharp decrease in auditory threshold in the region (∿70 kHz)
corresponding to the CF component of the adult echolocation signal
was not noted until 25 days postnatally. Recovery of full responsiveness to the second of two identical stimuli within 10 msec was
first recorded in 12-16 day old bats. Adult levels of temporal resolution were attained by 18-20 days, just before Rhinolophus begins
to fly.

Grinnell and Hagiwara (1972) studied the electrophysiological
responses of a juvenile Hipposideros galeritus and an immature, non-
volant H. calcaratus. Although the exact age of these bats was not
known, the evoked potentials were tuned to lower frequencies than in
the adults, corresponding to the lower emitted pulse frequencies of
the juvenile bats. They postulated that "the range of frequencies
emitted may be regulated to correspond to the gradual changes in

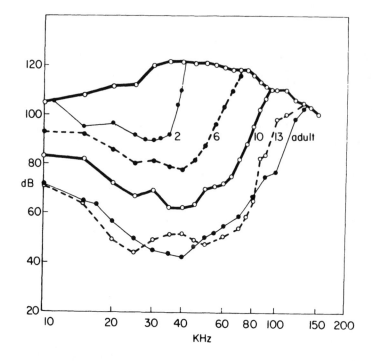

Figure 7. Audiograms of <u>Myotis velifer</u> as determined by evoked
potentials recorded from the posterior colliculus. The age in days
is to the right of each curve. Responses were first recorded in
two-day old bats, and by 13 days the audiogram resembled that of
an adult.

auditory capabilities as the inner ear matures" (Grinnell and Hagiwara, 1972).

     Baby <u>Pteronotus parnellii</u> and <u>suapurensis</u> are morphologically
immature and helpless at birth, and relatively immobile with their
eyes closed and skin naked. They emit a continuous isolation call,
which unfortunately is ignored by their mothers in captivity, resulting in surplus of one-day old subjects and none of later ages.
They would certainly be classified as altricial (Gould, 1975a), yet
their auditory responsiveness is the most developed of any bat studied
to date. Day old <u>P. suapurensis</u> have a tuning curve which closely
parallels that of the adult although slightly less sensitive (Fig. 9).
Day-old <u>Pteronotus parnellii</u> exhibit almost the full range of frequency response found in the adult, but lack the sharp tuning present
near the frequency of the CF portion of the sonar pulse (Brown and
Grinnell, unpublished data). They do exhibit the greatest sensitivity
at 20 kHz, a frequency represented in neonatal sound emissions (Fig.
10).

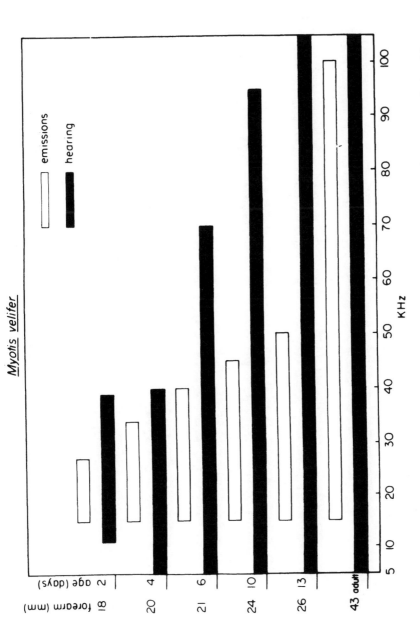

Figure 8. The range of frequencies emitted (first harmonic) compared to those heard by *Myotis velifer* at different ages and forearm lengths. After the age of two days, this species emits sounds containing frequencies that they can hear.

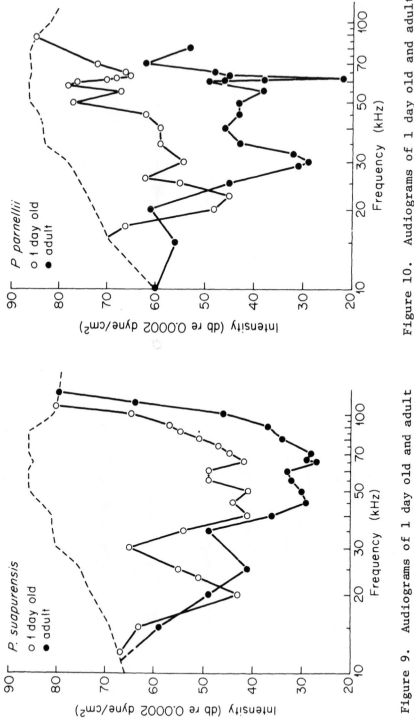

Figure 9. Audiograms of 1 day old and adult _Pteronotus suapurensis_ as determined by evoked potential recordings from the posterior colliculus.

Figure 10. Audiograms of 1 day old and adult _Pteronotus parnellii_ as determined by evoked potential recordings from the posterior colliculus.

Unfortunately, the development of the auditory system has not been followed in most bats, even those in which the adult capabilities have been well-documented. In those species that have been studied in detail, we find dramatic changes in the auditory system, observable from day to day, throughout ontogeny. An increase in absolute sensitivity is coupled to a shift in tuning toward higher frequencies correlated with an increase in the frequencies contained in vocalizations. There is a rapid acceleration in the rate of recovery of responsiveness, following initial exposure to a sound, probably one of the most impressive adaptations of the auditory system for echolocation (Grinnell, 1963b; Grinnell and Hagiwara, 1972b). Unfortunately the accompanying anatomical changes necessary to elucidate the observed neurophysiological maturation have not been described in baby bats, but probably reflect postnatal development both in the cochlea and neural circuitry (Brown, Grinnell and Harrison, 1978). Studies on other altricial mammals have shown that the cochlea and interneuronal connections in the brain continue to develop after birth. For example, the mouse inner ear is not fully differentiated until ten days postnatally (Weibel, 1957; Kikuchi and Hilding, 1965; Mikaelian and Ruben, 1965; Sher, 1971). In rats, major developments. of neural circuits in the forebrain continues for a month after birth (Caley, 1971), while brain stem connections proliferate for at least three weeks postnatally (Endroczi and Hartmann, 1968). It is conceivable that the long latencies and lack of synchrony observed in the first auditory evoked potentials recorded in baby bats arise from neural information being carried by a few, small, non-myelinated fibers with non-uniform diameters. The change in frequency range and sensitivity in bats can probably be attributed to the maturation of the cochlea. Studying tonotopic organization of best frequencies of units of Antrozous has shown that in an eight-day old bat they are uniformly distributed throughout the colliculus with best response at 15-20 kHz (Brown, Grinnell and Harrison, 1978). With an increase in age the surface units remain unchanged in frequency preference, but the tuning in deeper units is to higher frequencies (20-40 kHz in 10-day old bats, and 50-80 kHz in adults). Since even the earliest units detected appear to have complex response areas with narrowly tuned and closed curves, suggesting convergence and lateral inhibition, it is postulated that hair cells and primary afferents change their tuning to account for the frequency shifts. All of these hypotheses await actual confirmation by detailed ontogenetic neuroanatomical studies of the bat's auditory system.

Development of Echolocation Behavior

It is difficult to assess the roles of learning versus maturation in echolocation ontogeny since perfection of flight and echolocation skills must both occur simultaneously. Must bats learn to associate objects with returning echoes? How important is early experience to later mastery of echolocation skill? What is innate and what is learned?

A few observations have been made on the early flight behavior of bats, but no comprehensive studies have been published on early echolocation behavior either in natural or laboratory situations. From our developmental studies of hearing and vocalizations, we are now aware that these systems mature before the age of first flight. However, juvenile bats may continue to nurse for as long as a month after they actually begin to fly. Schmidt and Manske (1973) reported nine-month old vampire bats nursing in a captive colony. This overlap between first flight and weaning allows time for perfection of echolocation behavior before a bat has to forage on its own. Probably a certain period is necessary in all species. Brown (unpublished data) has noted a low survival rate (as judged by band recoveries) among captive-born Antrozous, later released as adults into a wild colony. Their chance of survival increased if they were released with their mothers, but was never as great as if they were born and reared in the wild colony. Captive-born bats may be still searching for the dish of mealworms, never having developed an accurate auditory image of their natural food (i.e. ground-dwelling arthropods for Antrozous). The basic problem with field studies of early echolocation behavior, is the inability to record from known-age bats. While this is rectified in captive-born bats, the natural context and relevant stimuli may be absent. We have studied echolocation ontogeny in Antrozous in captivity for ten years and recognize the natural sequence of sonar pulse development. This past spring and summer (1978), Brown monitored a natural maternity colony of pallid bats on Santa Cruz Island, California. The problem of individual identification was overcome by collaring mothers and pre-volant young (age determined by forearm length) with FM radio transmitters tuned to narrow frequency bands. This enabled her to follow the movements and tape record the vocalizations of individual bats on their first flights. On Santa Cruz Island, pallid bats roost in two-story brick barns, utilizing spaces within the thick brick walls as diurnal retreats. Adults night roost from the attic rafters. Infant bats spend the first weeks of life between the bricks, where their opportunity for acoustic feedback is limited. Between three and four weeks postnatally they crawl around on top of the bricks after the mother departs for nightly foraging periods. Within the next week they begin to make short flights between rafters in the barn. The FM orientation pulses emitted at this time contain lower frequencies than those of the adults and are interspersed with communication sounds (directives). The repetition rate is relatively high (20-30/sec) and fairly constant, with no apparent increase before take-off or landing. Around five weeks postnatally the young bats venture out of the roost for the first time, often accompanying their mothers. These initial flights are usually directly to another building within a 200 meter radius. Sometimes they spend most of the night flying back and forth between buildings. During this time the juveniles' echolocation pulses are distinguishable from those of the adults by their slightly lower frequency content. Between five and six weeks the young bat will spend periods of five to ten minutes away from the roost, and during this time they are often detected circling low over adjacent fields. The echolocation pulses of adults and juveniles are now indistinguishable. By seven weeks, the bats' nocturnal activity pattern is similar to the adults. They are away from the roost for approximately two hours after dark, gather with others in the night roost for most of the night, and often engage in a predawn hunting foray. In the night roost, juveniles less than eight weeks old were observed nursing. Therefore, even if unsuccessful in their early hunting attempts, they were assured of a meal. Since our recording equipment is not portable enough to allow us to follow a flying bat in the field, we are now developing a crystal-controlled transmitter with a condenser microphone to remotely monitor sonar signals.

Buchler (in preparation) has also conducted field observations on early echolocation behavior in _Myotis_ _lucifugus_. His colony roosted under the shingles of a building and did not have access to the interior of the building in which to practice initial flights. Although _Myotis_ develops more rapidly, the basic pattern was the same as in _Antrozous_. He attached radioactive tantalum tags to six infant bats in order to follow their movements. On some bats capable of flight he attached temporary chemoluminescent spheres. At about fifteen days of age, young bats were observed to leave the roost after the adults had departed. They would fly slowly near the roost, landing on large stationary objects such as walls and tree trunks. During this time they emitted pairs of echolocation pulses, a phenomenon not observed in older foraging bats or adults. Buchler notes that paired ultrasonic pulses are also produced by wandering shrews, tenrecs and swiftlets, animals that use their pulses presumably for obstacle avoidance. They may also serve this function for naive bats. Adult _M_. _lucifugus_ do not emit the double note while hunting. At this early age the juvenile bat may be forming an "auditory" picture of its immediate environment. Within a few nights the young bats venture into the surrounding woods and attempt their first insect captures. Some bats employ a "flycathcatcher" technique, flying out from a tree trunk to ambush an insect, then returning to it. At this intermediate stage, the juvenile bat remains stationary and the target moves, while in the final and most difficult step, a moving bat is confronted with moving prey. Older bats gradually increase their foraging range, flight duration and stamina, but do not yet perform the elaborate aerial pursuit maneuvers observed in adults. They do begin to emerge at the same time as the adults and fly with them along traditional routes to feeding areas. It is difficult to assess how much information is transferred from adult to juvenile bats about foraging areas and techniques.

In the laboratory we can study echolocation development under controlled conditions although it is possible that some aspects of behavior may not develop at natural rates. Wire avoidance tests have been a standard method for assessing bat echolocation ability. Normally, flying bats will increase the repetition rate of echolocation pulses as they approach and detect an object (Griffin, 1958). By measuring the point at which this increase occurs, the distance of initial detection can be determined.

The initial flights of five week old pallid bats through vertical strings were recorded in our laboratory. Often only a slight increase in repetition rate or none at all occurred on string approach, and more likely than not a collision resulted. By the second or third pass an increase in repetition rate was noted prior to negotiating the obstacle. Many flights were usually necessary before their accuracy at avoidance equalled that of adults. Many did not possess the flight skills to accomplish this maneuver. In some instances, a week of practice in a large flight cage free of obstacles was necessary before they could avoid the strings. This may be analogous to the practice period for young wild bats in the barn attic or between buildings.

Pallid bats were raised by their mothers under a variety of conditions in an attempt to alter their early acoustic environment and hence auditory experience. The control situation was a large flight room. The most natural of these altered environments was raising the bats in small containers approximating a crevice. When removed from their homes at five weeks of age, these bats had no difficulty in echolocating or flying, yet did not possess the stamina of those raised in the large flight room. In an attempt to prevent them from hearing their own echoes, some bats were reared in foam-lined, sound-proof, "anechoic" boxes, and others in chambers in which they were constantly bombarded with loud "white noise". Neither condition affected their future ability to echolocate. Deficits were apparent, however, in bats that had been fitted with monaural or binaural ear plugs. If the plugs remained intact between seven and twenty-one days postnatally, echolocation ability as an adult was impaired. Although there were a few exceptions, the majority of these bats were unable to avoid string obstacles less than 50% of the time compared with 70% for controls. All bats of any group that could not fly well were eliminated from the testing procedure. The monaural group showed consistently poorer performance than the binaurals. This deficit still exists more than two years later. We could not block the hearing of self-emitted sounds, but possibly the perception of echoes is important during a critical developmental period.

Gould (in preparation) also attempted to alter the early acoustic environment of _Eptesicus_ _fuscus_ by raising them in a helium-oxygen atmosphere, where sound travels approximately 1.5 times faster than in air. The performance of these bats on initial flights in air through wire obstacles was tested and compared to that of control, captive-raised young, and wold bats with the same forearm length. Surprisingly, no difference was observed in bats raised in the He-O_2 chamber. All bats on their initial flights increased their pulse repetition rate upon obstacle approach. For young bats the distance at which this change in repetition rate occurred was half that of adults. Even on initial flights, many naive bats could avoid the obstacles. Pulse patterns changed with age and experience. There was an increase in pulse interval during the search phase and a decrease in pulse interval during the terminal phase of target approach, coupled with a decrease in pulse duration.

These preliminary observations suggest that some aspects of echolocation ontogeny such as the production of sonar pulses and auditory tuning are genetically predetermined and refractory to experimental manipulation. Others, such as the development of echolocation behavior, may be influenced by experience and environment. Current research is now aimed at clarifying these relationships.

Acknowledgements

This research was supported by National Science Foundation grants. We would like to thank Drs. Jim Simmons, Brock Fenton, and Jean Harrison for the use of their equipment; Mr. R. Roverud, Ms. C. Brown and Dr. Timothy Brown for assistance in the field; and Ms. N. Meyers for the development of the bat radio transmitter. Special appreciation is due Dr. T. Brown for comments on clarity and style of this manuscript.

References

Brown, P.E., Vocal communication and the development of hearing in the pallid bat, Antrozous pallidus. Ph.D. Thesis, UCLA, 144pp., 1973.

Brown, P.E., Vocal communication in the pallid bat, Antrozous pallidus. Zeit Tier Psychol., 41, 34-54, 1976.

Brown, P.E., Grinnell, A.D. and Harrison, J.E., The development of hearing in the pallid bat, Antrozous pallidus. J. Comp. Physiol., 126, 169-182, 1978.

Caley, D.W., Differentiation of the neural elements of the cerebral cortex in the rat. In: Cellular aspects of neural growth and differentiation. Ed., D. Pease, University of California Press, Los Angeles, pp. 73-102, 1971.

Davis, W.H., Barbour, R.W. and Hassell, M.D., Colony behavior of Eptesicus fuscus. J. Mammal, 50, 729-736, 1968.

Douglas, A.M., The natural history of the ghost bat, Macroderma gigas in Western Australia. W. Austral. Nat., 10, 125-137, 1967.

Gould, E., Echolocation and communication in bats, In: About Bats. Eds. B. Slaughter and D. Walton, Southern Methodist Univ. Press, Dallas, pp. 144-161, 1970.

Gould, E., Studies of maternal-infant communication and development of vocalizations in the bats Myotis and Eptesicus, Communications in Behavioral Biology. Part A, 5, No. 5, pp. 263-313, March, 1971.

Gould, E., Neonatal vocalizations in bats of eight genera. J. of Mammalogy, 56, no. 1, 15-29, February, 1975a.

Gould, E., Experimental studies of the ontogeny of ultrasonic vocalizations in bats. Dev. Psychobiol., 8, 333-346, 1975b.

Gould, E., "Echolocation and Communication" in Biology of Bats of the New World Family Phyllostomatidae, Part II, Ed. Baker, R.J., J.K. Jones, Jr., and D.C. Carter, Special Publications the Museum Texas Tech. University, No. 13, pp. 247-279, 1977.

Gould, E., Infant calls of 10 species of Malaysian bats (Micro and Megachiroptera). Poster presented at Animal Sonar Systems Jersey Symposium, 1979.

Gould, E., Woolf, N.K. and Turner, D.C., Double-note communication calls in bats: occurrence in three families, J. of Mammal., Vol. 54, No. 4, 998-1001, 1973.

Grinnell, A.D. and Hagiwara, S., Adaptations of the auditory nervous system for echolocation. Studies of New Guinea bats, Z. vergl. physiologie, 76, 41-81, 1972a.

Grinnell, A.D. and Hagiwara, S., Studies of auditory neurophysiology in nonecholocating bats, and adaptations for echolocation in one genus, Rousettus. Zeit. Vergl. Physiol., 76, 82-96, 1972b.

Grinnell, A.D., The neurophysiology of audition in bats: Temporal parameters. J. Physiol.(Lond)., 167, 67-96, 1963.

Jones, C., Growth, development, and wing loading in the evening bat Nycticeius humeralis (Rafinesque). J. Mammal, 48, 1-19, 1967.

Kay, L. and Pickvance, T.J., Ultrasonic emissions of the lesser horseshoe bat, Rhinolophus hippersideros (Bech), Proceed. Zool. Soc., Lond., 141, 163-171, 1963.

Kikuchi, K. and Hilding, D.A., The development of the organ of Corti in the mouse. Acta Oto-Laryngal, 60, 207-222, 1965.

Konstantinov, A.I., Development of echolocation in bats in postnatal ontogenesis. Period Biol., 75, 13-19, 1973.

Konstantinov, A.I. and Stosman, I.M., Electrical activity in the inferior colliculus under the effect of ultrasound stimuli in ontogeny in bats of the genus Myotis oxygnathus.Zh. Evol. Biokhim. Fiziol., 8, 182-188, 1972.

Kulzer, E., Über die jugendentwicklung der Angola-Bulldog fleder maus Tadarida (Mops) condylura, (A. Smith, 1933) (Molossidae). Säugeteirkundl. Mett., 10, 116-124, 1962.

Kunz, T.H., Population studies of the cave bat (Myotis velifer): reproduction, growth, and development. Occ. papers, Mus. Nat. Hist., Univ. of Kansas, 15, 1-43, 1973.

Matsumura, S., Mother-infant communication in a horseshoe bat (Rhinolophus ferrumequinum nippon), I. Development of vocalization. Submitted to J. Mamm., 1979.

Mikaelian, D. and Ruben, R.J., Development of hearing in the normal CBA-J mouse. Acta Oto-Laryngal, 59, 451-461, 1965.

Orr, R.T., Natural history of the pallid bat, Antrozous pallidus. Proc. Calif. Acad. Sci., 28, 165-248, 1954.

Orr, Robert T., "Development: Prenatal and Postnatal" in Biology of Bats. Vol. 1, Chapter 6, Ed. by W.A. Wimsatt, Academic Press, New York, pp. 217-231, 1970.

Pearson, O.P., Koford, M.R. and Pearson, A.K., Reproduction of the lump-nosed bat (Corynorhinus rafinesquii) in California. J. Mammal., 33, 273-320, 1952.

Porter, F.L., Social behavior and acoustic communication in the bat, Carollia perspicillata. Ph.D. dissertation, Washington University, 169pp., 1977.

Schmidt, V., Social calls of juvenile vampire bats (Desmodus rotundus) and their mothers. Bonn. Zool.. Beitr., 4, 310-316, 1972.

Schmidt, V. and Manske, V., Die jugendentwicklung der vampirfleder- mäuse (Desmodus rotundus). Z. Saugetierkunde, 38, 14-33, 1973.

Schnitzler, H.U., Die ultraschull-ortungslaute der Hufeisen-Fleder- mäuse (Chiroptera-Rhinolophidae) in verschiedenen Direntierung- ssituationen. Z. Vergl. Physiol., 89, 275-286, 1968.

Sher, A.E., The embryonic and postnatal development of the inner ear of the mouse. Acta Oto-Laryngal (Suppl), 285, 1-77, 1971.

Stosman, I.M. and Konstantinov, A., Characteristics of evoked poten- tials of inferior colliculus of the bat Rhinolophus ferrum- equinum. Zh. Evol. Biokhim i Fiziol., 8, 612-616, 1972.

Weibel, E.R., Zur kenntnis der differenzieurungs-vörgange im epithel des ductus cochlearis. Acta Anatomica, 29, 53-90, 1957.

Woolf, N., The ontogeny of bat sonar sounds: with special emphasis on sensory deprivation. Unpublished doctoral thesis, Johns Hopkins University, Baltimore, Maryland, 140pp., 1974.

Chapter III
Adaptiveness of Echolocation

Chairman : G. Neuweiler

co-chairmen

F.G. Wood   and   M.B. Fenton

- <u>Underwater</u>
  Adaptiveness and ecology of echolocation in toothed
  whales.
  >F.G. Wood and W.E. Evans

- <u>Airborne</u>
  Adaptiveness and ecology of echolocation in terrestrial
  (aerial) systems.
  >M.B. Fenton

ADAPTIVENESS AND ECOLOGY OF ECHOLOCATION IN TOOTHED WHALES

F. G. Wood* and W. E. Evans**

*Naval Ocean Systems Center, San Diego, CA 92152

**Hubbs-Sea World Research Institute, San Diego, CA 92109

Of all aquatic animals only odontocete cetaceans are known to have a well developed sonar system. While certain other marine vertebrates will be discussed briefly in the latter part of this review, we are primarily concerned with the toothed whales.

EARLY BEHAVIORAL OBSERVATIONS

Before discussing the ecological and adaptive aspects of echolocation in odontocetes, it may be of value to briefly review the 30 year history of observations of this phenomenon.

In 1947, Arthur F. McBride, curator of Marine Studios (later Marineland of Florida), recorded some observations of dolphin (*Tursiops truncatus*) behavior that suggested to him the existence of "some highly specialized mechanism enabling the porpoise to learn a great deal about his environment through sound." The observations were made in the course of capture operations undertaken in very turbid water and usually at night. McBride noted that the animals: (a) would avoid a fine-mesh net but readily charge a large (10-inch square) mesh net, and (b) would immediately go over the net at the point where the cork line was momentarily pulled below the surface by others striking the net from below. The possibility of bioluminescent clues was considered, but no bioluminescence was visible to a human eye above the surface. McBride was reminded of the "sonic sending and receiving apparatus which enables the bat to avoid obstacles in the dark, "and, further, saw possible significance in the enormous development of the dolphin's cerebral cortex and the obvious importance of the acoustic

sense to these animals. This was the first direct inference
backed by good evidence that dolphins use sound for navigation.

Following McBride's untimely death, his notes remained in
the files of the Marineland of Florida research laboratory until
one of us (F.G.W.) happened on them in the early 1950's. During a
visit to Marineland by W. E. Schevill in 1955, the notes were
called to his attention. By then there was evidence that the
bottlenose dolphin could both emit and hear sounds in the ultrasonic frequency range (> 20 kHz). This led Winthrop Kellogg to
speculate increasingly on the existence of dolphin sonar (Kellogg
and Kohler, 1952; Kellogg, Kohler and Morris, 1953; Kellogg, 1953).
Schevill thought that McBride's priority in this matter should be
recognized, and offered to submit the notes for publication. Thus,
McBride's observations, with an introductory note by Schevill,
became a historically important item in the literature (McBride,
1956).

Well before McBride's observations became known, a number of
workers had harbored the suspicion that cetaceans could echolocate.
The first rather crude experiments to determine the hearing range
of the Atlantic bottlenose dolphin revealed that these animals
would respond to ultrasonic frequencies (Kellogg and Kohler, 1952;
Kellogg, Kohler and Morris, 1953; Schevill and Lawrence, 1953).
The turbid coastal waters inhabited by *Tursiops truncatus* suggested
that this species, and logically others, might well use echolocation for food finding and navigation. However, Schevill and
Lawrence (1953), in their auditory response experiment with a
dolphin in the near-opaque waters of a fenced off lagoon at
Florida's Marineland, listened carefully for echolocation signals
but heard none. To them the question remained open.

When one of us (F.G.W.) came to the Florida Marineland as
curator in 1951, he found that one of his predecessors had
acquired a U.S. Navy underwater sound projector (type CFF 78187).
Using this as a hydrophone, in conjunction with an amplifier, tape
recorder, and speaker or headset, he began to record and listen to
the sounds produced by the dolphins, *Tursiops* and, usually, one or
two spotted dolphins, *Stenella plagiodon*, that were maintained in
a large oceanarium tank. The hydrophone was suspended in front of
a porthole so that it was possible, while listening, to also record
notes on the behavior of the animals. From this work (Wood, 1952,
1953) two observations are relevant here. Whenever the transducer
was hung in the tank, the dolphins exhibited their normal curiosity
of any strange object or new introduction. As they swam by a few
feet from the transducer, they would peer at it and at the same
time emit rasping and grating sounds. This suggested that they
were "echo-investigating" the hydrophone. The same response could
be elicited by other objects, such as a bucket or length of pipe.

In retrospect, it seems apparent that although the animals could see the object clearly, they were also relying on the different kind of information derived from the echoes of their sound emissions.

The other observation of interest in the present context had to do with the animals during feeding. When a diver descended into the tank and began his circuits of the bottom, handing out fish, a constant subdued combination of rasping and mewing sounds were heard. When the last fish was handed out, the diver would raise the empty basket above his head, and the dolphins, which had been clustering about him, would immediately leave and display no further interest. At this time the undercurrent of sound would cease or diminish quite abruptly. Again, in light of later knowledge it seems clear that the mewing and rasping sounds were echolocation pulse trains which sounded subdued because the animals were echolocating directly on the fish being handed out some distance from the hydrophone. When they echolocated on the hydrophone they often overloaded the equipment. It could be inferred from this that echolocation is very closely tied to feeding activity, even occurring when the food is clearly visible.

The later experiments of Schevill and Lawrence (1956) and Kellogg (1958) effectively removed any doubt that the Atlantic bottlenose dolphin does indeed have a sonar capability, although in none of this work was vision completely excluded. Incontrovertible evidence was provided by Norris et al. (1961) who successfully blindfolded a dolphin by covering its eyes with latex suction cups. Their trained Atlantic bottlenose swam normally when blindfolded and avoided obstacles, including pipes suspended vertically to form a maze. The dolphin oriented to fragments of fish tossed in the water and took them as they drifted downward. Norris noted that when a bit of food drifted below the level of the melon, the animal did not respond to it. From this he speculated that the sounds were directional and being projected from the melon.

During the same period in the latter half of 1959 when Norris and his colleagues were attempting to devise a blindfold for dolphins, one of us (F.G.W.) was similarly engaged, unsuccessfully, at the Florida Marineland. Hearing of Norris' success using suction cups to occlude a dolphin's vision, Wood used the same technique to blindfold a recently caught *Tursiops*. The animal was untrained and had to be caught in order for the suction cups to be applied. Once they were in place, however, the dolphin swam about the tank with no indication that it lacked vision, and went directly to fish tossed in the water, taking them as they drifted down.

This Atlantic *Tursiops* had been taken in turbid inshore waters. Norris (1969) described three later observations of dolphins taken from clear oceanic waters which became disoriented when unable to use vision. A *Delphinus delphis*, which emits typical click trains at sea, when introduced into a tank filled with dirty water "may hit the walls frantically, force its head out of water, and sometimes may drown in its frenzy." A trained Pacific whitesided dolphin, *Lagenorhynchus obliquidens*, when blindfolded promptly ran into the side of its tank, and only after a period of days learned to navigate without vision. In the third instance a *Tursiops* captured in Hawaiian waters was taught to push a paddle on command. After being blindfolded the animal failed utterly to find the paddle despite the fact that it was 20 cm in diameter and covered with acoustically reflective foam Neoprene.

From these observations Norris concluded that there is very likely a large learned component in the echolocation behavior of marine mammals, and that animals living in clear seas may use their systems in quite different ways from those that inhabit turbid inshore waters.

The early speculation regarding a sonar capability in dolphins was based on behavioral observations. It was reinforced by the assumption that such a capability would have ecological adaptiveness for animals that navigate and catch their actively swimming prey in turbid water or in darkness, as at night or at depths.

THE INVESTIGATION OF ODONTOCETE ECHORANGING AND CONSIDERATION OF ITS ADAPTIVENESS

Observations and experiments in the 1950's confirmed the existence of odontocete echolocation and provided some insight into its adaptive features. The work that followed in the 60's concentrated on refining knowledge of the effectiveness and complexity of this remarkable sensory specialization.

Norris et al. (1967) and Evans and Powell (1967) experimentally demonstrated that blindfolded Atlantic bottlenose dolphins not only detect, avoid, or find objects, but can also discriminate finite differences in "targets" that affect their acoustic reflectivity, e.g., size, shape, and mass (density). During this same time period Busnel and colleagues in France and Denmark had demonstrated that the harbor porpoise, *Phocoena phocoena*, is also an echolocator (Busnel and Dziedzic, 1967).

The results of these studies were presented, along with correlative data on the hearing of the species concerned, at the international conference on biosonar held in Frascati, Italy, in 1966. The work received comment and critique from biologists who had been studying the sonar system of bats and from sonar engineers who also participated in the conference (Busnel, 1967). This critical look at echolocation in dolphins and porpoises raised significant questions which were to shape some of the research that followed. The demonstrated capabilities were impressive, but how important was this sophisticated sensory capability in navigation or food finding? What acoustic information was being used by the animals studied? The bottlenose dolphin and harbor porpoise inhabit inshore waters and even rivers where the water is murky and sometimes full of suspended debris that could represent false targets and navigational hazards. Did open-sea species such as the common dolphin and pilot whales echolocate? Although short-duration clicks had been heard from all odontocetes recorded, evidence of production of click trains was not evidence of an echolocation capability, although many authors referred to these acoustic emissions as echolocation clicks.

In 1965 Soviet scientists began to replicate some of the work of Western investigators using those species they had available in the Black Sea: the harbor porpoise, *Phocoena phocoena*, the common dolphin, *Delphinus delphis*, and a subspecies of *Tursiops*. By 1973 the literature had increased immensely. The number of species for which an echolocation capability had been demonstrated experimentally was growing, the source levels and frequency content of the pulses used were being measured with increasing accuracy, and signal processing techniques were being employed to gain a greater understanding of the total delphinid sonar system (Evans, 1973).

In 1967 the odontocete species that were known to use echolocation were *Phocoena phocoena* and *Tursiops*. By 1973 several additional species had been added to this list. Penner and Murchison's work (1970) on *Inia* indicated that this species could discriminate the difference in diameter of thin wires presented as pairs behind a visually opaque, acoustically transparent screen. The common dolphin, *Delphinus delphis*, could detect differences in complex geometric figures as well as *Tursiops* could (Belkovich et al., 1969). A preliminary study was done with a Pacific pilot whale, *Globicephala*, trained to wear latex eye cups and retrieve rings thrown into the water (W. E. Evans and L. L. Clarke, 1972, unpub.). As mentioned earlier, a Pacific whitesided dolphin, *Lagenorhynchus obliquidens*, learned to avoid obstacles without the aid of vision (Norris, 1969). This same species was tested with the task used by Evans and Powell (1967) to demonstrate echolocation discrimination in *Tursiops* and performed as well as the latter species (Evans, 1973). During a study designed to determine hearing thresholds, a captive killer whale, *Orcinus orca*, was also trained to retrieve a ring while blindfolded (Hall and Johnson, 1971).

The instrumentation now being used was considerably more sophisticated than that used during the 1950-1960's. The total recording system of hydrophones, associated amplifiers, and tape recorders had much broader responses — 100 Hz - 120 kHz typically, compared to 50 Hz - 20 kHz. Information on frequency and waveform characteristics of the pulse increased significantly, but the complexity of interpreting the data also increased.

Most of the experiments being done under laboratory conditions in the 1970's were still asking questions as to just how good the echolocation system is, without much reference to what this all means to the animal in the real world. In contrast, much of the information on bats was based on field observations and recordings of sounds produced by various species in free flight and during feeding.

Questions as to the ecological and adaptive nature of echoranging in various species of small toothed whales were raised in 1969 by Norris, and were again briefly addressed by Evans (1973). In Evans' discussion of the echolocation pulse waveforms used by various species, he divided the animals by type of habitat. The small odontocetes inhabit very diverse aquatic environments. Some species are found exclusively in rivers and lakes (e.g., *Inia*, *Platanista*). Some prefer tidal flats, muddy bays, and estuaries (e.g., *Phocoena*, *Tursiops*). Some species occur almost exclusively in the open sea (e.g., *Delphinus*), while others are generally found in the open sea but venture from time to time into the shallower waters around islands (e.g., *Globicephala* and *Lagenorhynchus*). A few primarily inhabit arctic waters, close to the ice pack or up rivers and fjords (e.g., *Delphinapterus* and *Monodon*). Two odontocetes, *Orcinus* and *Tursiops*, appear to have a greater zoogeographic range and occupy a wider variety of aquatic environments than any others. The echolocation pulses produced by various species were placed by Evans in specific categories: 1) narrow-band, low-frequency sonic pulses, 400 Hz to 20 kHz, with peak energy at 2 to 4 kHz; 2) broadband sonic to low-frequency ultrasonic, 100 Hz to 30 kHz, with peak energy at 16 to 20 kHz; 3) broadband, high-frequency ultrasonic, 16 to 150 kHz, with peak energy at 60 to 80 kHz; and 4) broadband sonic to ultrasonic, 200 Hz to 150 kHz, with peak energy occurring at 30 to 60 kHz.

The first of these categories (narrow-band, low-frequency sonic) was originally believed to be descriptive of *Phocoena*, *Inia*, and *Delphinapterus*, but improvements in instrumentation and measuring techniques revealed that assignment to this category was more a function of limitations in the recording systems than it was the nature of the signals used by the animals. However, this category still has some validity, since later studies have shown that these species may operate in more than one mode.

An additional error can result from the directional characteristics of echolocation signals, the frequency content of recorded
pulses depending on the angle of the animal to the hydrophone
(Norris and Evans, 1967; Au et al., 1978; Watkins, this volume).
The high frequency content of the pulse is directed essentially to
the front of the animal. At 10^0 to 15^0 to the right or left the
frequency decreases rapidly (Norris and Evans, 1967).

Although the harbor porpoise, Amazon River dolphin, and
beluga were originally thought to produce pulses with predominantly
or only low-frequency content, this was not consistent with the
observed echolocation detection and discrimination capabilities,
nor, in the case of *Phocoena*, with the finding that its hearing
extended beyond 120 kHz (Mohl and Andersen, 1970). The later
studies using broadband instrumentation demonstrated that while
all of these species produced pulses with low-frequency components,
when echolocating under conditions of poor or no visibility their
pulses contained considerable energy in the ultrasonic range
(Dubrovsky et al., 1971; Herald et al., 1969; Penner and Murchison,
1970; Gurevich and Evans, 1977; Mohl and Andersen, 1973).

It is of interest to note that these species, as well as the
Indus River dolphin, *Platanista indi*, which also falls in this
category (Herald et al., 1969), share certain environmental and
ecological similarities. They inhabit turbid waters, feed on very
small prey, and must frequently have to distinguish their prey from
drifting debris.

While the only other member of the Monodontidae, the narwhal,
differs from the beluga ecologically in some respects, being a deep
water species (Sergeant, 1978), teuthophagous, and apparently
diving to considerable depths when feeding (Tomilin, 1957), its
pulses, like those of the beluga, are relatively narrow band and
quite structured. The narwhal has a low band from .5 to 5 kHz
and a high band from 12 to 24 kHz (Ford and Fisher, 1978; see also
Watkins and Schevill, 1971).

In contrast to the species discussed above, all of which
occupy rather narrow ecological niches, *Tursiops* emits broadband
pulses. As mentioned earlier, populations of *Tursiops* occur in a
wide range of habitats, from the open sea to bays and rivers.
From measurements made in tanks, most of the energy in *Tursiops*
pulses was found to peak between 35 and 60 kHz (Evans, 1973). However, in a target detection experiment conducted with two Atlantic
bottlenose dolphins in the open waters of a Hawaiian bay, Au et al.,
(1974) recorded pulses with peak energies between 120 and 130 kHz.
The conditions at the site--shallow water with high ambient noise
conditions caused primarily by snapping shrimp--were very similar
to environments in which *T. truncatus* is often found. By shifting

the peak frequencies of their pulses to well above 100 kHz the
animals were presumably optimizing the signal-to-noise ratio, and
in so doing revealed an adaptive capability not previously observed
under the less realistic conditions of a laboratory tank.

In Figure 1 representative pulse waveforms for several of the
species discussed are shown. Detailed acoustic and ecological data
for *Tursiops*, *Delphinus*, *Orcinus*, *Inia*, *Phocoena*, and *Delphinapterus*
are presented in the Appendix.

In April 1969, in the course of conducting a series of echolocation discrimination tests (Diercks et al., 1971) with a female
Atlantic bottlenose dolphin named Scylla, it was decided to conduct
an experiment that had been suggested many times: testing the
efficiency of a blindfolded *Tursiops* in catching a live free-
swimming fish. The dolphin, as in the discrimination tests, was
instrumented with seven broadband hydrophones fabricated at the
Applied Research Laboratories, University of Texas. Five of the
hydrophones were directed inward to monitor the transmitted echolocation signals, and two, attached to the blindfolding eyecups,
were directed forward to detect target echo signals (Fig. 2). The
seven hydrophones were connected through an umbilical to poolside
preamplifiers and magnetic tape recorders. In addition, two
Atlantic Research (now Celesco) Type LC-10 broadband hydrophones
were positioned in front of the two targets and a third hydrophone,
USRD/NRL Type BM101A, was located behind the target plane, all
three connected via amplifiers to tape recorders. The movements
of the dolphin were monitored by two video cameras, one above the
pool and one underwater. The configuration of the test tank and
equipment is shown in Fig. 3.

Echolocation signals were detected by the monitoring hydrophones during all discrimination trials prior to the live fish
experiment. In the last trial before the fish was introduced the
targets were a metal sphere and a dead fish suspended by its tail.
Following this, a metal sphere was suspended in the left target
location to maintain semblence of another routine echolocation
trial. As Scylla approached the target area, emitting echolocation
clicks, a live fish (a 20 cm sargo, *Anisotremus davidsoni*) something she had not encountered during her several years in captivity,
was dropped into the water at the right hand target location.

The dolphin, without emitting any detectable echolocation
signals, positioned herself adjacent to the fish as indicated in
Fig. 4. She maintained this position with respect to the fish as
it swam around the perimeter of the pool. No echolocation signals
were detected by any of the ten hydrophones during this circuit.
When the fish reached the nylon net suspended in the passage
connecting the two pools it darted through. Scylla rolled and

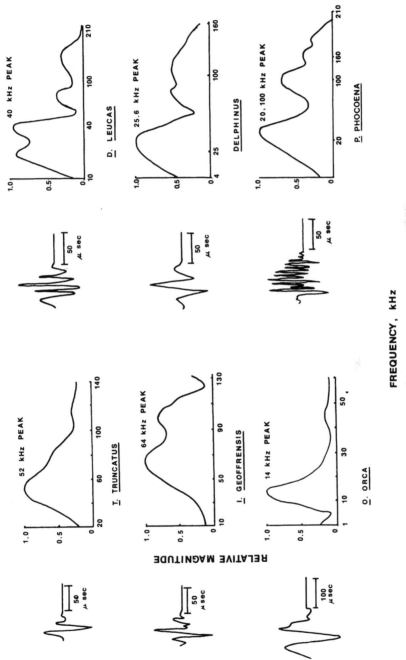

FREQUENCY, kHz

FIGURE 1. Representative echolocation signal, pulse waveforms, and associated spectra for six Odontocete cetaceans.

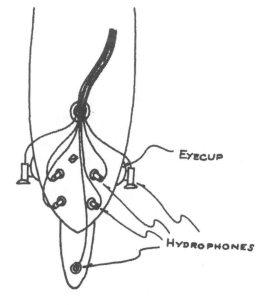

FIGURE 2.   Transducer locations of the head of the bottlenose dolphin
used in live fish experiment.

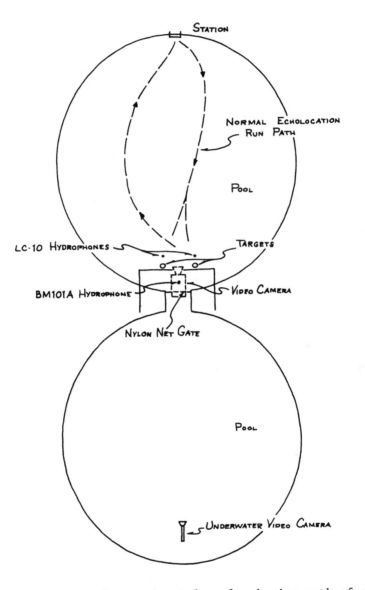

FIGURE 3. Schematic of experimental pools showing path of animal
during 'normal' echolocation run on non-moving targets.

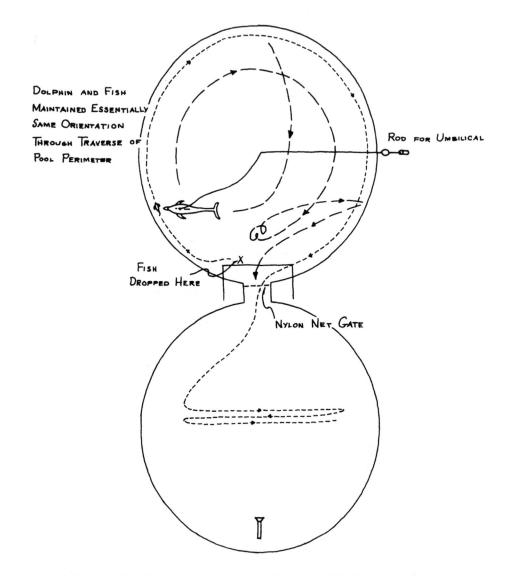

FIGURE 4.   Path of instrumented bottlenose dolphin and fish during
            the first trial.

became entangled in the hydrophone umbilical. She was then held at the side of the pool while all of the attached hydrophones and the two eyecups were removed. When released she returned to the nylon net where she could now see the fish in the adjacent pool. At this time she began to emit clicks at a rate that rose as high as 1500 per second. The repetition rate reached a level that she was apparently unable to maintain; her head would jerk, she would stop emitting, then begin again to the next head jerk, and so on. Subsequently, the fish was netted and tossed back into the test pool where Scylla, again blindfolded, now caught it and returned it to her trainer, all without emitting a detectable sound. This was repeated four more times, with the fish returned dead after the fifth retrieval. It was then hand-fed to her, and she readily ate it.

This experiment is described in some detail because in its various circumstances it is probably unique, and it appears to be significant. For the first part of the test we can only surmise that passively received acoustic information enabled the blind-folded dolphin to track the fish. Since she made no attempt to catch it during a complete circuit of the tank, and later retrieved it five times, we can also surmise that she didn't con-sider the live fish a food object. Scylla's performance indicated that these animals are able to detect, track, and catch fish with-out using echolocation or vision. This mode of acoustic behavior may be more commonly used in nature than we tend to suppose, and perhaps is attributable to the "mixed blessings" of echolocation discussed by Fenton (this volume).

Scylla's vocal behavior and head jerks when she could see the fish in the other pool but not get at it because of the net is probably best explained as an emotional outburst. Long-term obser-vations of several *Delphinus*, recorded on magnetic and video tape at the Hubbs-Sea World Research Institute, strongly suggest that click train emissions often occur in social contexts and represent emotional and perhaps communicative behavior.

We can study the odontocete echolocation capability as a relatively well-defined system with three primary functional compo-nents: 1) sound production, 2) echoreception, and 3) acoustic echo processing. In the context of ecological adaptations, however, as this last experiment demonstrated, echolocation cannot be viewed as distinct from another of the animals' acoustic attributes: that having to do with reception and interpretation of sounds occurring in their environment. This latter capability poses difficult prob-lems for the investigator. Our perception of the world is inherently limited by our particular ensemble of sensory equipment. We cannot know, or even imagine, the perceptual world of other animals that differ markedly from us in sensory capabilities. We can experimentally "ask questions" of them using behavioral or electrophysiological techniques, and obtain certain information regarding their sensory threshold ranges and apparent limitations. But we cannot know the complete gestalt of their perception and how that perception is adapted to the environment in which they live.

Still other aspects of sound production in odontocetes that apparently cannot be viewed as completely distinct from echolocation are those having to do with emotional state (as seemed to be demonstrated in the live fish experiment) and communication. It has commonly been assumed that clicks are echolocation signals, while whistles and other phonations are the ones that have communicative or emotional significance (see, for example, McBride and Kritzler, 1951; Wood, 1953; Schevill and Lawrence, 1956; Kellogg, 1961). However, there is growing evidence that clicks may serve other purposes. Lilly and Miller (1961) assigned a communicative function to slow click trains of *Tursiops truncatus*. Schevill and Watkins (1962) and Watkins (this volume), while not precluding sperm whale clicks for echo information, provide evidence that clicks are used as communication signals. In a well controlled experiment designed to test whether one bottlenose dolphin could tell another, separated by a visual barrier, to push one or the other of two paddles, depending on whether the first animal saw a flashing or steady light, the two performed with a very high percentage of correct responses. Although both whistles and clicks were emitted by the first dolphin, analysis of the tapes revealed that performance success depended directly on her emission of click trains (Bastian, 1967; see also Evans and Bastian, 1969, p. 433, for the conclusion that the animals had, in effect, trained themselves during the long period of preparation for the tests, and that intentional communication was not demonstrated).

One other possible function of clicks has been suggested. In consideration of the problem sperm whales have in catching squid in the darkness of great depths, the fact that otherwise healthy sperm whales with badly deformed lower jaws have been taken by whalers, and other factors, Soviet investigators have speculated that a sperm whale may search the water in front of it with lower-frequency echolocation. When a squid or fish is discovered and approached, the whale alters its click structure to higher frequencies, which narrows and concentrates the beam so that the prey is stunned and can be readily seized (Berzin, 1971).

ODONTOCETE ACOUSTICS AND ANATOMICAL MODIFICATIONS

The importance of echolocation to toothed whales is indicated by modifications of the skull, ear bones, and soft structures of

the head. Some of these features are apparent in early fossil
remains from the Lower Miocene. We consider it possible that echolocation was at one time more important to the success of these
animals than it is now, and that other acoustic attributes and
capabilities have been acquired (perhaps because of the "mixed
blessings" of echolocation; see Fenton, this volume) that reduce
reliance on echolocation without, however, loss of the potential
for making such fine discriminations as have been demonstrated
under experimental conditions.

The odontocete skull is unique among mammals, including
mysticetes, in its asymmetry, the result of enlargement of dorsal
elements on the right side. No reversal of this asymmetry has
been reported. The degree of asymmetry varies, reaching its most
extreme form in the sperm whale, *Physeter*, and pygmy sperm whales,
Kogia (Raven and Gregory, 1933).

Some years ago it was suggested that the puzzling asymmetry
of the odontocete skull was an adaptation related to the development of a sound-producing organ and represented a modification of
the skull having to do with echolocation (Wood, 1964, pp. 395-396).
The idea stemmed, in part, from dissection of a *Kogia* head, reveal-
ing in the right nasal passage a pair of tightly opposed horny lips
which appeared to constitute an admirable structure for pneumatic
sound production. This *museau de singe* (monkey's mouth) is presumed to be homologous with the *museau de singe* of the sperm whale,
a similar structure in the right nasal passage but located forward
of the spermaceti organ, not behind it as in *Kogia*. The right
nasal system of *Physeter* is also more specialized in other respects
(Raven and Gregory, 1933).

Schenkkan and Purves (1973) identify the *museau* as the
homologue of the external lips of the right naris in other
odontocetes, but on the basis of its "generally low level of
muscular control" and their presumption that as a sound generator
it would be very expensive in its use of air, they reject an
acoustic function for the *museau*, assigning instead a valving role
and attributing sound production to the larynx, not only in *museau*-
bearing forms but all odontocetes. We will return to this point
further on.

From detailed examination of the right nasal complex of the
sperm whale, Norris and Harvey (1972) argued that the *museau* is
indeed a sound generator and that the spermaceti organ acts as a
reverberation chamber in the production of the burst-pulse signals
of *Physeter*. Mackay (this volume) agrees that the spermaceti organ
of *Physeter* is involved in sound emissions but postulates a different mode of response to sounds produced by the *museau*. The work
of Karol et al. (1978) on a pygmy sperm whale also supports the

view that both the *museau* and spermaceti organ are components of
the sound production system in *Kogia*, although, considering the
topological relationship of the two structures, they propose a
wave guide function for the spermaceti organ.

It has been suggested (Norris and Harvey, 1972) that the
deeply basined forehead of *Physeter* which is covered over most of
its area by the posterior wall of the frontal sac (a diverticulum
of the right nasal passage), is indicative of a sound reflective
structure. This might also apply to the enlarged right side of
the bipartite basining of the *Kogia* skull in which the ethmoid
septum is deflected to the left (see Raven and Gregory, 1933).

It may be noted that the left nasal passage in physeterids is
considerably larger and follows a more direct course to the left
boney naris, which is correspondingly larger than the right. There
seems no reason to doubt that the left passage, as suggested by
Raven and Gregory, is for fast inhalation and exhalation, while
(contrary to Raven and Gregory) the right passage and associated
structures have a phonatory function.

The skulls of bottlenose whales, Ziphiidae, also exhibit
marked asymmetry, and the left boney naris is conspicuously larger
than the right, as shown in True (1910). In addition, a *museau* and
what is apparently a spermaceti organ have been reported in the
ziphiids *Ziphius cavirostris* (Norris and Harvey, 1972), *Mesoplodon
bidens* (Schenkkan, 1973), and *Hyperoodon ampullatus* (*museau*:
Schenkkan, 1973; spermaceti organ: Tomilin, 1957). The ziphiids
are also characterized by a maxillary crest which in all but one
.species forms a basin vertical to the long axis of the rostrum
(Moore, 1968). According to Fleischer (1978), both physeterids
and those ziphiids that have been examined lack a tympanic membrane,
having instead a thin, ceramic-like boney plate.

Along with the above likenesses it may be noted that the sperm
whale and some, at least, of the ziphiids are apparently the deepest and longest divers of all cetaceans (Caldwell et al., 1966;
Rice, 1978) and that the predominant food of both families is
cephalopods (Tomilin, 1957). We think it likely that the similar
acoustic modifications of the skull and ecological characteristics
of the two families are more than coincidental, especially since
they apparently are attributable to parallel evolution, not close
phylogenetic relationship (Miller, 1923).

Except for sounds made by the sperm whale (see Watkins, this
volume) almost nothing is known about the phonations of other
members of these families. Faint click trains have been recorded
from a *Kogia* by using a contact microphone in air (Caldwell et al.,
1966) and pulsed sounds have been recorded from a *Mesoplodon* at sea

(D. K. Ljungblad, pers. comm). Although Watkins presents evidence that *Physeter* click emissions he has recorded generally had more the characteristics of communication signals than echolocation pulses, it seems inherently unlikely, considering the complexity of the right narial passage and associated structures, along with the mode of feeding at great depths, that sperm whales lack a well developed sonar system. Further observations may clarify what appears to be an improbable limitation on the function of acoustic emissions by these animals. As Watkins suggests, we may be dealing with a different type of system.

The earliest known fossil delphinids, like the physeterids and ziphiids, have come from Lower Miocene strata. Although asymmetry of the skull and narial structures is present in dolphins, it is less extreme than in the aforementioned families. Sound production is still incompletely understood. The majority of workers have placed sound generation in the complex nasal system, although one early hypothesis was that clicks are produced in the nasal system and whistles by the larynx. The literature is briefly reviewed by Ridgway et al. (this volume) who then present compelling evidence that (in *Tursiops*) clicks and whistles originate in the nasal system and not, as maintained by Schenkkan and Purves (1973) in the larynx. The work of Ridgway et al. also clarifies the ability of *Tursiops* to simultaneously emit whistles and clicks, as previously reported by Lilly and Miller (1961) and Evans and Prescott (1962).

Schenkkan and Purves state that the *museau* of physeterids and ziphiids is homologous with the external lips of the right naris in other odontocetes. Norris (1969) has reported that in the common dolphin, *Delphinus delphis*, the lips of the nasal plugs (presumably what Schenkkan and Purves referred to as the external lips) are asymmetrical, with the right hand lip approximately twice the size of the other lip. This form of asymmetry has also been seen in a variety of other delphinids examined by one of us (W.E.E.).

Anatomically, the highly asymmetrical skull and other head structures of the physeterids especially, but also the ziphiids, appear to represent a more extreme departure from a primitive and more generalized condition than does the delphinid head. If the *museau* is a sound generator, as seems probable, it indicates specialization of one sound system. On the other hand, *Tursiops* has a bilateral, though still asymmetrical, system which is highly versatile in sound production and, in conjunction with neural processing, known to be capable of permitting very fine discriminations. A basically bilateral but asymmetrical arrangement of narial structures is present in other delphinids (see Schenkkan, 1973).

We suggest that the narial system and associated structures of the sperm and bottlenose whales are specialized for sonic

emissions adaptive for the ecology of these two families. The
Tursiops acoustic emission system may be more of a general purpose
nature, although variations among the delphinids (and other
odontocetes) may reflect differences in their ways of life.

THE ODONTOCETE BRAIN AND AUDITION

Basic to the echolocation capability of odontocetes is their
neuroprocessing of such sounds--the integration, analysis, and
interpretation of acoustic information contained in the echoes.
Neural processing requires neural tissue. It is well known that
all odontocetes, and some in particular, have very impressive
brains (see, for example, Morgane and Jacobs, 1972). Although
there is much variation in the size of odontocete brains, both
absolutely and by whatever index of cephalization one chooses to
use, in general the brains of these animals are characterized, in
gross morphology, by large size, with greater width and height than
length; great development of the cerebral hemispheres; and intricate convolutions of the cortex. Morgane and Jacobs (1972)
identify three major arcuate lobar formations in each cerebral
hemisphere, the limbic, paralimbic, and supralimbic lobes, in their
terminology. The paralimbic lobe, bounded principally by the
limbic cleft (sulcus splenialis) and paralimbic cleft (sulcus
entolateralis) is, according to Morgane and Jacobs, unique to
odontocetes and its homology unknown. The supralimbic lobe,
bounded mediad by the paralimbic cleft and laterad by the ectosylvan fissure, is the largest of the three divisions and is
subdivided into three arcuate lobules. Morgane and Jacobs state
that these lobules and defining sulci exhibit great variation in
position and configuration, not only in different species but
within the same species and even between right and left hemispheres
of the same individual, a condition characteristic of the so-called
"association" areas of the cerebrum. These are most extensive in
man but also relatively large in other higher primates and in
elephants (Kruger, 1966). Association areas are generally thought
to be concerned with the more complex mental processes.

More recently, Ladygina and Supin (1977) studied the location
of sensory projection zones of the cerebral cortex by electrophysiological mapping. According to them, the areas termed by Morgane
and Jacobs the paralimbic lobe and the superior (i.e., mediad)
lobule of the supralimbic lobe constitute the visual zone, while
most of the large intermediate lobule of the supralimbic lobe is
auditory area. Ladygina and Supin note that the location of projection areas in the bottlenose dolphin and harbor porpoise--and
probably, therefore, in other odontocetes--differs appreciably
from their location in terrestrial mammals studied. The visual
area is shifted rostrally, and the auditory area, which is the

largest, occupies most of what is usually designated as the supra-sylvian gyrus and not the ectosylvan gyrus, or temporal lobe, as in terrestrial mammals. In short, the greater part of what Morgane and Jacobs consider to be probable "associational" areas are, according to the work of Ladygina and Supin, visual and auditory projection zones. Considering both the anatomical and electro-physiological evidence, it is perhaps permissable to wonder if the importance of audition to odontocetes, especially in echolocation, may have brought about a cortical organization analagous to that represented by the association areas of other large brained but less acoustically oriented mammals.

If we accept Jerison's (1973) "principal of proper mass," which states that "the mass of neural tissue controlling a partic-ular function is appropriate to the amount of information proc-essing involved in performing the function," we might hypothesize that the high degree of encephalization of odontocetes reflects, to some considerable extent at least, their known and presumed acoustic attributes and capabilities. The only other hypothesis is that these animals--at least those with the biggest brains--possess an intelligence comparable to, if not exceeding, that of man (Lilly, 1961, 1963, 1967). But even Lilly has acknowledged that "Convincing scientific evidence of cetacean intelligence remains to be established, and arguments comparing levels of human and dolphin are philosophical not scientific" (Lilly, 1978).

In the remainder of this dissertation we will examine the evidence and arguments dealing with large odontocete brains and with odontocete bioacoustics and other factors that may be associated with the high degree of encephalization of these animals. This would seem to have a bearing on the ecological adaptiveness of echolocation in this particular group.

Paleontological and Anatomical Evidence

As pointed out by Jerison (1973), the history of the brain in whales is a history of early enlargement; "dramatically notable by the Miocene in the ancestors of living porpoises...The evolution of the whale was associated with the evolution of sensorimotor adaptations for maneuvering in water and, even more dramatically, the evolution of the auditory and sound producing systems for echo-location and for the perception of objects by their echoes" (pp. 346 and 351). Jerison notes, in part from paleontological evidence, the enlargement of the acoustic components of the eighth nerve, the inferior coliculi of the midbrain, the entire cerebral cortex, and, in association with the cortex, the cerebellum, all as parts of these adaptations.

In Oligocene odontocetes there was no sign of the asymmetry that now characterizes the skulls of toothed whales, or of the basining of the facial region that in some odontocetes is associated with the presence of a spermaceti organ that may play an important part in echolocation. Both of these features are apparent "in muted form" in early Miocene fossils (Whitmore and Sanders, 1976). Fleischer (1976) has presented evidence that the adaptation of the cochlea for high-frequency sound reception, and therefore echolocation, occurred in the Squalodontoidea during the late Oligocene. At the same time, these animals developed the multilayer acoustic shield at the forehead, in a process that is part of the telescoping of the odontocete skull described by Miller (1923). Fleischer considers this shield as particularly focussing the ultrasonic portion of the outgoing sound and insulating the hearing organs from the emitted pulses. In the Miocene, according to Fleischer, the hearing organ of the odontocetes was fully adapted for a sonar system.

There appears, then, to be at least a rough chronological correlation between the development of echolocation and the development of a large brain.

The Mysticeti, it may be noted, have symmetrical skulls and lack a spermaceti organ or melon (Harrison and King, 1965), as well as, apparently, anything like the narial asymmetries and specializations in odontocetes. Fleischer (1976), from a study of fossil material, concluded that "There is good evidence that neither the Archeoceti nor the Mysticeti ever possessed a sonar system. In contrast to the Odontoceti, the Mysticeti adapted their hearing organ to the perception of sound of very low frequencies." Mysticete sounds are typically very low frequency moans (Schevill, 1964); the complex "song" of *Megaptera* (Payne and McVay, 1961) appears to be a special case.

The ears of terrestrial mammals are inefficient sound receivers under water, and the earliest adaptations in the evolution of odontocete acoustic mechanisms must have involved alterations of the ancestral ears, designed for airborne sounds, to more efficient receivers of waterborne sounds. An aqueous medium, however, provides the potential for much better reception of sound, since the density differences between water and the bone and soft tissue of a submerged animal are a very small fraction of the differences between air and the bone and soft tissue of a terrestrial animal. Norris (1968) considers this the most important factor in the evolution of odontocete acoustic mechanisms, and cites Hickling (1962) who pointed out that we would expect environmental sounds to induce little vibration in terrestrial animals except at very localized points, while much vibration would be induced in aquatic organisms by environmental sounds, and at a

large variety of loci. Norris documents the extensive modification of the odontocete skull and head structures that he considers
most probably attributable ultimately to the much reduced density
difference encountered when the ancestral forms of the odontocetes
entered the water.

We will speculate that increase in brain size was a development that accompanied these modifications. An increment in hearing
sensitivity through an incremental adaption of the receiving system
for waterborne sounds may have resulted in an adaptive increase in
neural tissue for the integration, analysis, and interpretation of
the acoustic information now available. This, in turn, might permit the utilization of further adaptive refinements of the aural
structures, with, then, a further development of neural tissue for
processing the increase in available acoustic information, and so
on. At some point, perhaps quite early, a rudimentary echolocation
system appeared, and because of its adaptive value could have
accelerated the process, with the introduction now of neural components for a more and more refined control of sound emissions, along
with an increased capacity for processing the echo information as
well as sounds received from the environment. The development of
communication signals may have contributed to neural control and
processing capabilities, but the meager evidence for vocalizations
having apparent communication significance in delphinids (for
example, the repetitive whistle of a separated calf (McBride,
1940); the "mating call" of a bull *Tursiops* (Tavolga and Essapian,
1957); the "distress call" (Lilly, 1963; Wood 1973), and the
"signature whistle" concept (Caldwell and Caldwell, 1965, 1971,
1972)) suggests that they are of an "intuitive" nature, rather than
having the characteristics of formal language (the difference
between "Ouch" and "Fire is hot"). But this issue must remain open.

The Brains of Odontocetes (and Bats)

"Brains are metabolically expensive and don't get bigger
(phylogenetically) unless in some fashion they are more than paying for their upkeep" (Hockett, 1978). There is evidence that
odontocete brains as large as those in some modern delphinids
(taking into consideration body size) had evolved by early Miocene
times (Jerison, 1973). This suggests that the major period of
brain development occurred long ago and that any later modifications have been of a less radical nature.

Hockett (1973, p. 416) speaking of man, states that after
steady growth of the hominid brain for two million years, its size
seems to have leveled off about 50,000 years ago. Why did it stop?
Hockett suggests that mutations produced some crucial reorganiza-
tion of the workings of the brain whereby greater efficiency was

effected without continued metabolically expensive increase in size. Today, all our brains are fully human, and though differing in size the differences are irrelevant. Why, then, do sizes differ? Because, Hockett holds, they differed 50,000 years ago when size was important, and because since then there has been no selective reason anywhere for the size to change. The range in size of odontocetes, both as to body and brain, exceeds that of any other suborder of mammals, and the application of Hockett's statements to cetaceans is tenuous, but what he says may provide a valid insight.

It is a common tendency to think of large brains as indicating "intelligence," although there is no universally agreed upon definition of the term. Jerison (1973) avoids the usual difficulties by suggesting that "Biological intelligence may be nothing more (or less) than the capacity to construct a perceptual world." For us, this is the "real" world of which we are conscious. Other animals may have "real" worlds, too, but these worlds will differ according to how the animal's brain does the work of integrating sensory and motor events.

Jerison sees two important dimensions of biological intelligence. One is the extent to which sensory capabilities have been integrated with one another, and the second is the extent to which behavior in response to sensory information is flexible and adjustable to inconsistencies in that information.

It is easier to appreciate now, Jerison continues, why gross measures of the brain should be most closely related to biological intelligence. "The number of neurons and the complexity of their interconnections should reflect the degree to which sensory systems have been elaborated and interconnected; it might make little difference from the point of view of intelligence which systems have actually been emphasized in particular species."

After discussing other indices of encephalization and some of the difficulties encountered with them, Jerison describes a measure he developed that he calls the "encephalization quotient" or EQ. The EQ is related to "average" brain: body relations of the mammals as a whole. It is the ratio of actual brain size to expected brain size and is defined by the equation $EQ = \frac{E}{.12P^{2/3}}$, in which E is brain weight in grams, and P is body weight in grams. (For a detailed explanation of how this equation was developed the reader is referred to Jerison, 1973).

By this equation, the EQ of an "average" living mammal is unity. A higher EQ denotes a brain size larger than that expected for the size of the animal. The value of EQ for *Homo sapiens*, if E=1350 g and P=70,000 g, is 6.6.

It is interesting to apply Jerison's formula to odontocetes, more for the purpose of intercomparisons than for ascertaining the degree of departure from "expected" brain size. Table 1 gives brain and body weights and corresponding EQs for animals of a number of different species. We present these figures with a few qualifications. In some instances we have brain-body weights for only a single individual. Data for clearly immature animals could not be used since young animals have relatively larger brains for their body size than adults. In this regard *Inia* data are probably borderline; however, since the animals were females, and females apparently do not grow as large as males (see Layne, 1958), the EQ figures may not be far off.

Table 1. Brain weight (E), body weight (P) and encephalization quotient (EQ) of certain odontocetes. All weights in grams.

Species	E	P	EQ($\bar{x}$)
Tursiops truncatus (N=3)[1]	1511	99,000	5.9
Lagenorhynchus obliquidens (N=2)[1]	1211	105,500	4.5
Grampus griseus (N=1)[2]	2551	400,000	4.0
Delphinus delphis (N=9)[2]	802	69,000	4.0
Globicephala macrorhynchus (N=1)[2]	2798	522,000	3.5
Inia geoffrensis (N=2)[3]	618	62,400	3.0
Phocoenoides dalli (N=1)[1]	793	110,000	2.9
Orcinus orca (N=1)[2]	6138	2,409,000	2.9
Platanista gangetica (N=4)[4]	295	59,750	1.6

[1]Ridgway et al., 1966; [2]Ridgway and Brownson, 1979; [3]Gruenberger, 1970; [4]Pilleri and Gihr, 1970

If a correction for blubber weight (not an "average" mammal constituent) were incorporated, the EQs would be higher. Slijper (1962) states that blubber constitutes 30–35 percent of body weight in dolphins. On the other hand, cetacean data constituted only a small fraction of the mammalian brain-body weights used by Jerison in formulating his equation; it is likely that this in some degree counteracts the blubber weight factor and reduces any deviation from more accurately computed EQs. A different kind of error can result when dealing with very large animals; there is considered to be an upper limit on the allometric relationship between brain and body weight beyond which brain size does not keep pace with body size. Jerison thinks it probable that when selection pressures toward enlargement of the body result in body size above about 1000 kg (in land mammals, at least) they can act more or less independently of those toward enlargement of the brain. The body weight of the killer whale in Table 1 is such that the computed EQ is undoubtedly too low; we consider this in our discussion of the table.

These general comments aside, what are we to make of the range of EQ values shown? First, it is apparent that, with the exception of *Platanista*, all are considerably higher than Jerison's "average" mammal, with *Tursiops* having an EQ in the range of that of *Homo sapiens*. A second observation is that there is not a clear correlation with what is known about the echolocation capabilities of the different species. However, *Tursiops*, which has an excellent echolocation capability utilizing very broadband pulses, has a very broad geographic range, is found in varied types of aquatic habitats, and subsists on a wide variety of food organisms. It also adapts readily to captivity. For *Lagenorhynchus*, considering all of these factors, an EQ of around 4.5 appears appropriate. The other species, although demonstrating excellent echolocation capabilities, are more specialized with respect to their echolocation pulses, geographic range, habitats, food items, or a combination of these. Within the freshwater family Platanistidae, the higher EQ of *Inia*, even allowing for the possibly too high values in Table 1, may reflect the fact that it has functional eyesight, while *Platanista* does not.

The EQ of *Orcinus* is almost certainly too low on the basis of allometric considerations. We have noted above factors other than echolocation that seem to relate to EQ values. The killer whale has probably the greatest geographic range of any odontocete (from the Arctic to Antarctic), is found from shallow, restricted waters to the open sea, has a widely varied diet, and adapts very well to conditions of captivity. On the other hand, *Orcinus* preys on large fishes and also on other marine mammals, including whales, which are active noise sources. Locating food would appear to be less of a problem for killer whales than for most other odontocetes. Also, the computational capacity necessary to handle information in the bandwidth from 20 Hz to 30 kHz is certainly less than that needed for a 100 kHz bandwidth. However, we suspect that if an accurate EQ could be computed for *Orcinus* it would be well up the scale, possibly in the range of *Tursiops*. (See Appendix for relevant information on *Orcinus* and other species alluded to above.)

Eisenberg and Wilson (1978), using regression analysis, plotted cranial volume:body weight ratios for 225 species of bats of 14 families. Their data indicated that the aerial insectivores that rely almost completely on echolocation have the lowest brain

size to body weight ratios, while the frugivores of the suborder
Megachiroptera and those members of the microchiropteran family
Phyllostomatidae which have secondarily become fruit eaters, have
the largest. Insectivorous bats that discretely sample microhabitats or utilize a complex foraging strategy involving vision
and sonar to exploit micro-niches have intermediate ratios.
Eisenberg and Wilson modified Jerison's EQ formula by incorporating their extensive bat data (Jerison's sample included few
bats). The computed EQs correlated extremely well with trophic
roles, i.e., nectarivore and frugivore, carnivore, piscivore, and
so on in descending order of mean EQ values down to the aerial
insectivores. It was even possible to suggest that a species
suspected of being an aerial insectivore is instead, on the basis
of its EQ, a foliage gleaner.

With regard to the largest brains, Eisenberg and Wilson submit that "a foraging strategy based on locating relatively large
pockets of energy-rich food that are unpredictable in temporal and
spacial distribution necessitates the use of a complicated information storage and retrieval system involving input from several
sense organs."

The situation with respect to toothed whales is by no means
as clear cut as it seems to be with bats. We have presented
evidence for at least a rough chronological correlation between
the initial development of an echolocation system (quite different
from that of insectivorous bats, as discussed further on) and the
acquisition of a large brain. We suggest that other factors have
contributed to the variation in relative brain size seen in recent
toothed whales. In terms of Jerison's encephalization quotient,
higher values appear to reflect greater adaptability, as seen in
broad geographic range, penetration into a variety of ecologic
niches, a wide spectrum of food items, and adaptability to conditions of captivity. All of this, along with a very broadband
sonar, is descriptive of *Tursiops* which, so far as known, has the
highest odontocete EQ.

Integration of sensory inputs also represents a measure of
adaptability. We infer sensory integration in an animal that has,
for example, both good eyesight and a well developed sonar system;
the simultaneous use of vision and hearing seemed indicated by the
observations described early in this paper of the dolphins that
peered at, but also echolocated on, objects hung in an oceanarium
tank, and both looked at and echolocated on the fish being handed
out. In the discussion of Table 1, we suggested that the EQ of
Platanista, lower than that of the confamilial genus *Inia*, is at
least in part attributable to its lack of functional vision.

All of these factors, ecological and sensory, appear to bear
some relation to brain size in odontocetes. They indicate the
extent of variability of situations or circumstances the animal

may experience. Put another way, they determine the potential
scope of the animal's (in the species sense) perceptual world —
Jerison's concept of "biological intelligence."

While echolocation would seem to have a high adaptive value
for an aquatic vertebrate, there is hardly any evidence that it
has evolved in animals other than toothed whales. The sea catfish,
Arius felis, has been shown by Tavolga (1977) to have a rudimentary
acoustic orienting capability. Echolocation in pinnipeds and
penguins has been suggested by Poulter (1963, 1967), but attempts
to experimentally demonstrate such an ability in pinnipeds have
been unsuccessful (Evans and Haugen, 1963; Schusterman, 1966;
Scronce and Ridgway, this volume), and penguins have not been
studied in this respect. It remains possible that an echolocation
capability will be found in some of the Pinnipedia, such as those
species that are ice dwelling in the Arctic and Antarctic and have
a need to feed in total darkness and find breathing holes. At
present, however, there is no indication of the existence of a
complex aquatic biosonar system other than in odontocetes. Other
aquatic vertebrates seem to have taken the path of other sensory
systems or combinations of systems.

Even those dolphin species that have demonstrated a highly
developed sonar capability may use it less than we tend to suppose,
as pointed out with respect to the live fish experiment. Every
year a large number of dolphins and porpoises, including species
known to have an excellent sonar system, become entrapped in gill
nets, purse seines, and nets set off South African beaches to
protect bathers from sharks (Working Party on Marine Mammals, 1977,
pp. 100-107). The target strength of some, at least, of these nets
should make them readily detectable. While such entrapments may
result from inattention or concentrating on the pursuit of prey,
the fact remains that we have no idea how much of the remarkable
detection and discrimination capability of these animals is used
in the real world and how much of it is an untapped capacity
brought out by an artificial selection process called "training."

We cannot avoid further consideration of bats, which with
brains weighing only a fraction of a gram are skilled echolocators.
We have argued that, other factors notwithstanding, the large size
of the odontocete brain probably reflects in some considerable
degree the hypertrophy of neural tissue for processing acoustic
information, especially that contained in echoes, and for very
precise control of transient broadband sound emissions.* The idea

---

*The weakly electrogenic Mormyridae, freshwater fishes of Africa,
may, in the enormous development of their cerebellum, represent an
analogous example. Very brief polyphasic electric pulses, emitted
at a variable rate from a few to 130 per second (unless the fish
is "silent") provide an electrical field, any interruption of

has been expressed by others. Tomilin (1968), for example,
hazarded the guess that "The use of echolocation and the echolocation apparatus may have played as important a role in the formation
of the dolphin brain as did the hands, work, and articulate speech
in the development of the human brain." The minute brain of
insect-catching bats does not negate our hypothesis, but we need
to consider the bat brain and some of the differences between the
microchiropteran and odontocete echolocation systems.

One significant difference lies in the nature of the sound
emissions, the bats using FM sweeps (ignoring here the finer
distinctions among different species) and the odontocetes using
transient broadband clicks. We would expect that processing of
echoes from extremely brief broadband clicks emitted at rates up
to 600 per second or more would require a much greater neural
processing capability. Also, the sensorimotor control of such
pulses, including the ability to shift peak energy to different
parts of their frequency spectrum, would seem to be a much more
complex task (Au et al., 1974).

Although sound attenuation is greater in air than in water,
the air-target density match is such that echolocation and signal
processing should be much simpler than in an aquatic environment
where objects are generally more closely density-matched to water.

The bat system is highly specialized for a specific task (and
apparently has been for 50 or 60 million years); the odontocete
system appears to be more versatile and flexible. Also, the
interpretation of acoustic information from the environment may
have contributed to the neural processing capacity of cetaceans.

It is interesting that of the terrestrial/aerial vertebrates
known or suspected to use echolocation only the insectivorous bats
use time-frequency calls. All of the others, including the frugiv-
orous bats of the genus *Rousettus*, produce broadband pulses
(Fenton, this volume). In none of these is there any indication
of a highly developed sonar system; the broadband pulses, so far
as known, serve only for detection of obstacles in the path of the

which can be detected. By their electrosensory system mormyrids
can avoid obstacles and detect prey at a distance. The pulses
also appear to have communication significance (Moller and Baur,
1973). In contrast, the weakly electric (except for the electric
eel, which also has a powerful discharge) Gymnotidae produce mono-
phasic electric pulses at, in most species, a constant frequency.
Although able to detect obstacles and prey by their electric field,
gymnotids lack a large cerebellum. The "gigantocerebellum" of
mormyrids, which produce complex pulses under greater control, is
generally considered to be predominantly an association center for
the electrosensory system (Lissmann, 1958; Bennett, 1971).

echolocator. Fenton suggests that the use of broadband clicks for echolocation (in terrestrial settings) does not confer a strong adaptive advantage. He notes, however, that the facultative use of a broadband pulse form of echolocation by a wide variety of animals indicates that it does provide some advantages.

Phyletically, it would seem that echolocation by broadband clicks represents the "conventional" mode of biosonar in animals. The bats with their FM sweeps constitute a unique exception (unless the frequency-modulated chirps of some odontocetes have a significant sonar function; see Diercks' review, this volume).

There has evidently been little or no selective pressure to develop further the low-order broadband system that some terrestrial/aerial vertebrates possess. This is obviously not true of the odontocetes. The adaptive value of a sonar system for swift-swimming, deep-diving, fish or squid eaters seems apparent, especially given the sound-conducting properties of water. Considering the prevalence of a broadband form of echolocation, however rudimentary, in terrestrial/aerial vertebrates, it is not surprising that odontocete sonar is of the same type. Broadband clicks provide, in fact, distinct advantages in an aquatic environment because they are better for coping with reverberation — the reflection of sound off surface, bottom, and particulates in the water column.

We can assume that at some point in the history of odontocetes an incipient echolocation capability involving broadband pulse emissions appeared. There followed, then, evolutionary selection for sound generation, reception, and processing systems that would make the most of broadband emissions. Given the potential to develop a broadband system, including very fine sensorimotor control of their sound emissions and a highly developed neuroprocessing component, the odontocetes followed the evolutionary course that was most adaptive for them.

ACKNOWLEDGEMENTS

K. J. Diercks and R. T. Trochta kindly provided information and sketches of the live fish experiment. We are obliged to V. S. Gurevich for making us aware of and providing translations of most of the Soviet literature cited. The figure showing pulse wave forms and the drawings and layouts of the Appendix material were the work of illustrator Ann Dawes. Finally, we must express our appreciation to Jan Letourneau, Karie Wright, and Yvonne Moreno for their skill and patience in typing early drafts, and to Jan, especially, for the final, much revised, copy.

APPENDIX

Natural History and Echolocation Data
on Six Species of Odontocetes

ATLANTIC BOTTLENOSE DOLPHIN *(Tursiops truncatus)*

Natural History Data

Size: to 3.5 meters

Distribution: Found world wide in temperate and tropical waters. Most common in bays, rivers, estuaries, and mud flats. Also seen far at sea.

Food preference: Catholic feeders, predominantly fish but squid, octopus and a variety of large crustaceans also taken.

Swimming and diving: Will reach burst speeds of 18-20 knots for short periods, can sustain 4-5 knots for extended periods. Can travel 100 nautical miles/day. Dives to depths of over 350 meters have been recorded.

General information: Best studied of all marine delphinids. Adapts well to captivity. Has been bred and raised in captivity and some individuals have survived and thrived for over 20 years. Most of our available data on sound production in general and echolocation specifically is based on observations of this species. Good data are available on behavior physiology, nutritional requirements, also only species for which good data on visual acuity are available.

Echolocation Data

Signal type: pulse   Duration: 10-200 μsec

Peak frequency (kHz): 15-130

Measured source levels: Peak levels 155-204 dB re: 1 μPa (Ref. Evans, 1973; Babkin and Dubrovsky, 1971; Alexeyeva et al., 1971; Golubkov, 1972).

Detection: ability to detect spheres made of various materials, diameters from 3.3 cm to 7.6 cm at ranges of 6 meters to 73 meters. (Ref. Titov, 1972; Au et al., 1974).

Discrimination: Capability to discriminate objects of same size, different materials excellent, also shape discrimination. Can discriminate difference in steel spheres which differ in diameter by 4 mm (Ref. Evans, 1973; Dubrovsky et al., 1970; Norris et al., 1967; Fadeeva, 1973; Golubkov, Ivanenko, 1970; Dziedzic, Nachtigale, Murchison this vol; Evans and Powell, 1967).

Hearing: Range 75 Hz to 150 kHz, Maximum sensitivity 20-80 kHz (Ref. Johnson, 1966; Morozov et al., 1972).

COMMON DOLPHIN *(Delphinus delphis)*

Natural History Data

Size: to 2.6 meters

Distribution: World wide in temperate and tropical seas (water temperature range 10° to 28°C). Usually found in the deep open ocean outside depths of 200 meters, except frequently found in vicinity of banks and sea mounts in deep water. Probably the most numerous of all the cetaceans.

Food preference: Small schooling fishes and squid. Known to feed on anchovy, herring, a variety of mid-water fishes including myctophids and deep-sea smelt. In Eastern Pacific primarily a nocturnal feeder, taking advantage of photophobic organisms that migrate to the surface at night.

Swimming and diving: Very fast burst speeds to at least 24 knots. Can sustain speeds of 6 knots for long periods. Radio tagged individuals moved more than 110 nautical miles in 24 hours diving at night to depths in excess of 220 meters.

General information: This is one of the three species most heavily impacted by mortality associated with the tropical Eastern Pacific Yellowfin Tuna purse seine fishery. Has been maintained successfully in captivity for over 8 years and used extensively as an experimental animal, especially in USSR (Black Sea).

Echolocation Data

Signal type: pulse

Duration: 35-350 μsec

Peak frequency (kHz): 20-100 (Ref. Evans, 1973; Titov, 1972; Bel'kovich and Reznikov, 1971)

Measured source level (at 1 meter): peak level 140 dB re 1 μPa

Echolocation performance
Detection: 13 cm sculpin at 2.6 meters (Titov, 1972)

Discrimination: Several experimental demonstrations of ability to detect differences in shape, size, and material, e.g. spheres, cylinders, pyramids, 13 cm Herring from 13 cm metal "fish". (Ref. Titov, 1972; Bel'kovich *et al.*, 1969; Bel'kovich and Borisov, 1971; Demitrieva *et al.*, 1972; Bel'kovich, 1970; Rezvov *et al.*, 1973; Gurevich, 1968).

Hearing: Range measured 100 Hz to 280 kHz, maximum sensitivity at 60 to 100 kHz (Ref. Bel'kovich and Solntseva, 1970).

KILLER WHALE *(Orcinus orca)*

Natural History Data

Size: to 9 meters

Distribution: World wide from northern to southern polar waters. Throughout its range seems to prefer coastal areas and often enters bays, estuaries and river mouths in search of food. Sightings far at sea are not rare.

Food preferences: Known to feed on squid, large fish (tuna, salmon), sea turtles, seabirds, marine mammals, including large baleen whales.

Swimming and diving: Very fast swimmer, may be fastest of all the cetaceans. Sprint speeds in excess of 25 knots have been estimated. Can cruise for extended periods at speeds of 6-8 knots. Dives to depths of 250 meters have been observed under controlled conditions.

General information: Killer whales have now been successfully maintained under controlled conditions for several years. They are possibly most adaptable cetacean in captivity. They are very tractable and have been trained to participate in a variety of psycho-physical studies.

Echolocation Data

Signal type: pulse

Duration: 0.5-1.5 msec

Peak frequency (kHz): 14, range energy 100 Hz to 30 kHz
Measured source level (at 1 meter): 178 dB re 1 μPa (Diercks *et al.*, 1971)

Echolocation performance
Detection: Can detect a 5 cm diameter air-filled plastic cylinder at 6 meters.

Discrimination: At present there are no data available on the echolocation discrimination capability of this species (Evans, 1973).

Hearing: Range of hearing measured 500 Hz to 31 kHz, maximum sensitivity at 15 kHz (Ref. Hall and Johnson, 1972).

HARBOR PORPOISE *(Phocoena phocoena)*

Natural History Data

Size: to 1.8 meters

Distribution: Common in the North Atlantic with a northern limit of Iceland, White Sea, and the Davis Straights. Southern limit is Cape of Good Hope. Southern form may be different sub-species. Also in Eastern North Pacific near the United States and Japan. Inhabits the Black Sea, Sea of Azov and the Mediterranean Sea. Rarely in tropical waters. Usually in shallow water bays, rivers and estuaries.

Food preference: Main food items herring, small cod, sole; they also eat squid and crustaceans.

Swimming and diving: Does not stay submerged for more than three to four minutes. Seldom found in water deeper than 150 meters, so probably not a deep diver. Fast burst speeds when feeding to 15 knots or more. Cruising speed much slower, 2-4 knots.

General information: Has been successfully maintained at aquaria in United States and Europe (USSR). Majority of research on this species done in USSR and Europe (Denmark).

Echolocation Data

Signal type: pulse

Duration: 40-200 μsec

Peak frequency (kHz): 20-150 (Ref. Dubrovsky *et al.*, 1970; Møhl and Andersen, 1975)

Measured source level (at 1 meter): peak level 112 dB re 1 μPa

Echolocation performance

Detection: 0.2 diameter steel wire and nylon thread diameters of 0.9 mm. Also detection of 75 mm diameter cylinders of various materials at 8-11 meters (Ref. Busnel *et al.*, 1967; Zaslavsky *et al.*, 1969).

Discrimination: Can detect difference of 1.75 cm in diameter of steel cylinders at 2 meters (Ref. Titov, 1972).

Hearing: Range measured, 1 kHz-150 kHz, maximum sensitivity at 8, 32, 64 kHz (Ref. Andersen, 1970; Sukhoruchenko, 1973).

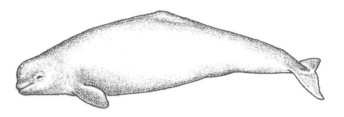

WHITE WHALE, BELUGA *(Delphinapterus leucas)*

Natural History Data

Size: to 4.3 meters

Distribution: Arctic waters of Atlantic, Pacific, and White seas. Seen as far south as 50°N, but this is rare. Frequents shallow bays and river mouths. Occasionally ascends rivers several hundred miles to feed and bear young. Also found offshore in southern part of range.

Food preference: Feeds on a variety of small fishes, including caplin salmon smolt, and small cod; also squid and a variety of benthic crustaceans.

Swimming and diving: Burst speeds to 15 knots probably faster. Slow cruising, 2-4 knots. In bays and rivers spends much time lolling at surface. Diving ability not known. Can easily breath-hold for more than 8 minutes. Feeds on organisms that are benthic at more than 300 meters.

General information: A very vocal cetacean and one of the first recorded. Possibly first species successfully kept in captivity. Has reproduced in captivity but has not successfully raised young. Several specimens have and are being maintained in United States, Canada and Europe. Echolocation ability and hearing only recently studied.

Echolocation Data

Signal type: pulse

Duration: 20-100 μsec

Peak frequency (kHz): 40, 80, 120
Measured source level (at 1 meter): Peak levels of 160-180 dB re 1μPa possibly capable of higher.

Echolocation performance
Detection: Detection of 10 cm^2 plastic square plate at 8 meters. Also can detect 5 mm ceramic cylinder at 5 meters.

Discrimination: Can discriminate flat square, triangle from three dimensional target (pyramid). Can discriminate a 3 cm x 3 cm square from 2 cm x 2 cm square. (Ref. Gurevich and Evans, 1976).

Hearing: Range measured 1 kHz to 123 kHz. Maximum sensitivity 60-65 kHz (Ref. White *et al*., 1978).

BOUTU, AMAZON RIVER DOLPHIN *(Inia geoffrensis)*

Natural History Data

Size: to 2.5 meters

Distribution: Found in the Amazon and Orinoco rivers and their tributaries. Will occasionally venture in the brackish bays at the mouth of the rivers.

Food preference: Eats mainly bottom dwelling fish, especially a species or armoured catfish. Also will eat freshwater crustacea.

Swimming and diving: In its native habitat speeds of 15 knots have been observed. Usually animal cruises at 2-3 knots. Experimental tests of diving ability have not been conducted, however can easily hold breath for 4 minutes.

General information: Has been successfully maintained for periods in excess of 10 years at several aquaria in the U.S., Europe, and Japan. Trains well, and is an excellent experimental animal.

Echolocation Data

Signal type: pulse

Duration: 15-100 μsec (Ref. Diercks *et al.,* 1971; Evans, 1973)

Peak frequency (kHz): 60-65
Measured source level (at 1 meter): peak level 146 dB re: 1 μPa (Ref. Evans, 1973)

Echolocation performance
Detection: 1.4 mm diameter wire at approximately 1 meter (Penner and Murchison, 1970).

Discrimination: Detection of 2.4 mm difference in diameter of cork/neoprene cylinders of same length (Ref. Diercks *et al.,* 1971; Evans, 1973).

Hearing: Range measured 1 kHz-105 kHz, Maximum sensitivity 30-50 kHz (Jacobs and Hall, 1972).

REFERENCES

Abramov, A. P., E. Sh. Ayrapet'yantz, V. I. Burdin, A. G. Golubkov, I. V. Ershova, A. R. Zhezherin, V. I. Korolev, Yu. A. Malyshev, G. K. Ul'yanov, and V. B. Fradkin 1971 Investigations of the dolphin's capabilities to differentiate volumetric targets according to their linear dimensions and material. Proceedings of the 8th All-Union Acoustical Conference, p. 3. (In Russian)

Alexeyeva, T. V., A. G. Golubkov, and I. V. Ershova 1971 On the problem of the active width of the spectrum of dolphins' echolocation signals. Trudy Akusticheskogo Instituta 17:99-103. (In Russian)

Andersen, S. 1970 Auditory-sensitivity of the harbour porpoise, *Phocoena phocoena*. In: Investigations of Cetacea, G. Pileri, Ed. Vol. 2, pp. 255-258. University of Berne, Switzerland

Au, W. W. L., R. W. Floyd, R. H. Penner, and A. E. Murchison 1974 Measurement of echolocation signals in the Atlantic bottlenose dolphin, *Tursiops truncatus* Montagu, in open waters. Jour. Acoust. Soc. Am. 56(4):1280-1290

Au, W. W. L., R. W. Floyd, and J. E. Haun 1978 Propagation of Atlantic bottlenose dolphin echolocation signals. Jour. Acoust. Soc. Am. 64:411-422

Ayrapet'yantz, E. Sh., A. G. Golubkov, I. V. Ershova, A. R. Zhezherin, V. N. Zvorykin, and V. I. Korolev 1969 Echolocation differentiation and properties of the emitted sounds of dolphins. Doklady Academii Nauka 188(5):1197-1199. (In Russian)

Babkin, V. P., and N. A. Dubrovsky 1971 Range and noise resistance of the bottlenosed dolphin echolocator during detection of various targets. Trudy Akusticheskogo Instituta 17:29-42. (In Russian)

Babkin, V. P., N. A. Dubrovsky, P. S. Krasnov, and A. A. Titov 1971 Material discrimination in spherical targets by the bottlenosed dolphin. Proceedings of the 8th All-Union Acoustical Conference, p. 5. (In Russian)

Bastian, J. R. 1967 The transmission of arbitrary environmental information between bottlenose dolphins. In: Animal Sonar Systems: Biology and Bionics, R.-G. Busnel, Ed. Vol. II, pp. 803-873. Laboratoire de Physiologie Acoustique, Jouy-en-Josas, France.

Bel'kovich, V. M.   1970   Mechanism of sound generation in Cetacea.
    In: Ekholocatsiya v Prirode, Nauka, Leningrad, pp. 297-310.
    (In Russian)

Bel'kòvich, V. M., V. I. Borisov, V. S. Gurevich, and N. L.
    Krushinskaya 1969 The ability of echolocation in *Delphinus
    delphis*. Zoologicheskii Zhurnal 48(6):876-884. JPRS 48780.

Bel'kovich, V. M., V. I. Borisov, V. S. Gurevich, and N. L.
    Krushinskaya 1969 Echolocational capabilities of the common
    dolphin (*D. delphis*). Zoologicheskii Zhurnal 48(6):876-883.
    (In Russian)

Bel'kovich, V. M., and V. I. Borisov 1971 Echolocational dis-
    crimination of the complex target by dolphins. Trudy
    Akusticheskogo Instituta 17:19-23. (In Russian)

Bel'kovich, V. M., V. I. Borisov, V. S. Gurevich, N. L.
    Krushinskaya, and I. L. Dmitrieva 1972 Echolocation dis-
    crimination of distance between the targets (angular resolu-
    tion) by dolphins. Proceedings of the Abstracts of the 5th
    All-Union Conference on Marine Mammals, Makhachkala, Part
    2:30-33. (In Russian)

Bel'kovich, V. M. and A. E. Resnikov 1971 New in dolphin echo-
    location. Priroda 7:71-75. (In Russian)

Bennett, M. V. O. 1971 Sensory systems and electric organs. In:
    Fish Physiology, Vol. 5, W. S. Hoar and D. J. Randall, Eds.
    Academic Press.

Berzin, A. A. 1971 The Sperm Whale. Pacific Scientific Research
    Institute of Fisheries and Oceanography (USSR). Trans. by
    Israel Program for Scientific Translations, publ. by U. S.
    National Technical Information Service, Springfield, Va.
    393 pp.

Busnel, R.-G., A. Dziedzic, and S. Andersen 1965 Seuils de
    perception du systeme sonar du Marsouin, *Phocoena phocoena*
    en fonction du diametre d'un obstacle filiforme. Comptes
    Rendu des Seances de l'Academie des Sciences, Paris
    260(1):295-297.

Busnel, R.-G., Ed. 1967 Animal Sonar Systems: Biology and
    Bionics. Two volumes. Laboratoire de Physiologie Acoustique,
    Jouy-en-Josas-78, France.

Busnel, R.-G. and A. Dziedzic 1967 Resultats metrologiques experimentaux de l'echolocation chez le *Phocoena phocoena*, et leur comparaison avec ceux de certaines chauves-souris. In: Animal Sonar Systems: Biology and Bionics, R.-G. Busnel, Ed. Vol. I, pp. 307-335. Laboratoire de Physiologie Acoustique, Jouy-en-Josas-78, France.

Caldwell, D. K. and M. C. Caldwell 1965 Individualized whistle contours in bottlenose dolphins. Nature 207:434-435.

Caldwell, D. K., M. C. Caldwell, and D. W. Rice 1966 Behavior of the sperm whale, *Physeter catodon* L. In: Whales, Dolphins, and Porpoises, K. N. Norris, Ed. Univ. of California Press

Caldwell, D. K., J. H. Prescott, and M. C. Caldwell 1966 Production of pulsed sounds by the pygmy sperm whale, *Kogia breviceps*. Bull. So. Calif. Acad. Sci. 65(4):246-248.

Caldwell, M. C., N. R. Hall, and D. K. Caldwell 1971 Ability of an Atlantic bottlenose dolphin to discriminate between, and potentially identify to individual, the whistles of another species, the spotted dolphin. Cetology No. 6, 6 pp.

Caldwell, D. K. and M. C. Caldwell 1972 Senses and communication. In: Mammals of the Sea: Biology and Medicine, S. H. Ridgway, Ed. pp. 466-502. Chas. C. Thomas, Springfield, Illinois.

Diercks, K. J., R. T. Trochta, C. F. Greenlaw, and W. E. Evans 1971 Recording and analysis of dolphin echolocation signals. Jour. Acoust. Soc. Am. 49(6, pt. 1):1729-1732.

Diercks, K. J., R. T. Trochta, W. E. Evans 1973 Delphinid sonar; measurement and analysis. Jour. Acoust. Soc. Am. 54:200-204.

Dmitrieva, I. L., N. L. Krushinskaya, V. M. Bel'kovich, and A. V. Shurkhal 1972 Echolocational discrimination of the shape of different subjects by dolphins. Proceedings of the Abstracts of the 5th All-Union Conference on Marine Mammals, Makhachkala, Part 2:76-80. (In Russian)

Dubrovsky, N. A., P. S. Krasnov, and A. A. Titov 1970 On the problem of emitting ultrasonic echolocational signals by harbor porpoise (*Ph. phocoena*). Akusticheskii Zhurnal 4:521-525. (In Russian)

Dubrovsky, N. A. and P. S. Krasnov 1971 Discrimination of the solid spheres according to their materials and dimensions by the bottlenosed dolphin (*T. truncatus*). Trudy Akusticheskogo Instituta 17:9-18. (In Russian)

Dubrovsky, N. A., P. S. Krasnov, A. A. Titov 1971 On the
    emission of ultrasound echolocation signals by the Azov Sea
    harbor porpoise. Akist. Zh. 16:521-525; Sov. Phys.-Acoust.
    16:444-447.

Dubrovsky, N. A. and G. L. Zaslavsky 1973 Temporal structure and
    directivity of sound emission by the bottlenosed dolphin.
    Proceedings of the 8th All-Union Acoustical Conference pp.
    56-59. (In Russian)

Eisenberg, J. F. and D. E. Wilson 1978 Relative brain size and
    feeding strategies in the Chiroptera. Evolution 32:740-751.

Evans, W. E. 1973 Echolocation by marine delphinids and one
    species of fresh-water dolphin. Jour. Acoust. Soc. Am.
    54(1):191-199.

Evans, W. E. and J. H. Prescott 1962 Observations of the sound
    production capabilities of the bottlenose porpoise: A study
    of whistles and clicks. Zoologica 47:121-128.

Evans, W. E. and Ruth Haugen 1963 An experimental study of the
    echolocation ability of the California sea lion *Zalophus
    californianus* (Lesson). Bull. So. Calif. Acad. Sci.
    62:165-175.

Evans, W. E. and B. A. Powell 1967 Discrimination of different
    metallic plates by an echolocating delphinid. In: Animal
    Sonar Systems: Biology and Bionics, R.-G. Busnel, Ed. Vol I,
    pp. 363-383. Laboratoire de Physiologie Acoustique,
    Jouy-en-Josas-78, France.

Evans, W. E. and J. Bastian 1969 Marine mammal communication:
    social and ecological factors. In: The Biology of Marine
    Mammals, H. T. Andersen, Ed. pp. 425-475.

Fadeeva, L. M. 1973 Discrimination of spheres with various
    echosignal structure by dolphins. Proceedings of the 8th
    All-Union Conference on Marine Mammals Part 1, p. 134. (In
    Russian)

Fleischer, Gerald 1976 Hearing in extinct cetaceans as determined
    by cochlear structure. Jour. Paleon. 50(1):133-152.

Fleischer, Gerald 1978 Evolutionary principles of the mammalian
    middle ear. Advances in Anatomy, Embryology, and Cell
    Biology, Vol. 55, Fasc. 5, 70 pp.

Ford, J. K. B. and H. D. Fisher 1978 Underwater acoustic signals of the narwhal (*Monodon monoceras*). Canadian Jour. Zool. 56:552-560.

Golubkov, A. G. and Yu. V. Ivanenko 1970 Study of the range resolution ability of the dolphin's echolocator. Proceedings of the 23rd scientific-technical conference of LIAP p. 65. (In Russian)

Golubkov, A. G., I. V. Ershova, I. V. Korolev, and Yu. A. Malyshev 1972 On energetical parameters of the Black Sea bottlenosed dolphin echolocational apparatus. Trudy LIAP 76:9-12. (In Russian)

Gruenberger, H. B. 1970 On the cerebral anatomy of the Amazon dolphin, *Inia geoffrensis*. In: Investigations on Cetacea V.II G. Pilleri, Ed. University of Berne, Switzerland.

Gurevich, V. S. 1969 On ability of the common dolphin (*D. delphinus*) to discriminate geometrical targets by means of echolocation. Vestnik Moskovskogo Universita 3:109-112. (In Russian)

Gurevich, V. S. and W. E. Evans 1976 Echolocation discrimination of complex planar targets by the beluga whale (*Delphinapterus leucas*). Jour. Acoust. Soc. Am. 60(Suppl. 1)S5-S6.

Hall, J. D. and C. S. Johnson 1971 Auditory thresholds of a killer whale *Orcinus orca* Linnaeus. Jour. Acoust. Soc. Am. 51(2):515-517.

Harrison, R. J. and J. E. King 1965 Marine Mammals. Hutchinson University Library, London.

Herald, E. S., R. L. Brownell, Jr., F. L. Frye, E. J. Morris, W. E. Evans and A. B. Scott 1969 Blind river dolphin: First side-swimming cetacean. Science 166:1408-1410.

Hickling, R. 1962 Analysis of echoes from a solid elastic sphere in water. Jour. Acoust. Soc. Am. 34(10):1582-1592.

Hockett, C. F. 1973 Man's Place in Nature. McGraw-Hill.

Hockett, C. F. 1978 In search of Jove's brow. American Speech 53(4):243-313.

Irvine, L. 1939 Respiration in diving mammals. Physiol. Rev. 19:112-134.

Jerison, H. J. 1973 Evolution of the Brain and Intelligence. Academic Press.

Karol, R., C. Litchfield, D. K. Caldwell, and M. C. Caldwell 1978 Compositional topography of the melon and spermaceti organ lipids in the pygmy sperm whale *Kogia breviceps*: Implications for echolocation. Mar. Biol. 47:115-123.

Kellogg, Remington 1928 The history of whales. Quart. Rev. Biol. 3:29-76, 174-208.

Kellogg, W. N. 1953 Ultrasonic hearing in the porpoise, *Tursiops truncatus*. Jour. Comp. and Physiol. Psych. 46(6):446-450.

Kellogg, W. N. 1958 Echo ranging in the porpoise. Science 128:982-988.

Kellogg, W. N. 1961 Porpoises and Sonar. Univ. Chicago Press.

Kellogg, W. N. and R. Kohler 1952 Reactions of the porpoise to ultrasonic frequencies. Science 116(3018):250-252.

Kellogg, W. N., R. Kohler, and H. N. Morris 1953 Porpoise sounds as sonar signals. Science 117:239-243.

Korolev, L. D., N. V. Lipatov, R. N. Resvov, M. A. Savel'ev, and A. B. Flenov 1973 Investigation of the spatial directivity diagram of the dolphin sound emitter. Proceedings of the 8th All-Union Acoustical Conference 1:33. (In Russian)

Kruger, Lawrence 1966 Specialized features of the cetacean brain. In: Whales, Dolphins, and Porpoises, K. S. Norris, Ed., pp. 232-252. Univ. of California Press.

Ladygina, T. F. and A. Ya. Supin 1977 Localization of sensory projection zones in the cerebral cortex of the bottlenosed dolphin, *Tursiops truncatus*. Zurnal Evolyutsionnoy Biokhimii i Fiziologii No. 6, pp. 712-718. (In Russian)

Layne, J. N. 1958 Observations on freshwater dolphins in the Upper Amazon. Jour. Mamm. 39(1):1-22.

Lilly, J. C. 1961 Man and Dolphin. Doubleday and Co.

Lilly, J. C. 1963 Critical brain size and language. Perspectives in Biology and Medicine 6:246-255.

Lilly, J. C. 1967 The Mind of the Dolphin. Doubleday and Co.

Lilly, J. C. 1978 Dolphin and the law. Ltr. to the Honolulu (Hawaii) Advertiser, 31 January

Lilly, J. C. and A. M. Miller 1961a Vocal exchanges between dolphins. Science 134(3493):1873-1876.

Lilly, J. C. and A. M. Miller 1961b Sounds emitted by the bottlenose dolphin. Science 133(3465):1689-1693.

Lissmann, H. W. 1958 On the function and evolution of electric organs in fish. Jour. Exp. Biol. 35:156-191.

McBride, A. F. 1956 Evidence for echolocation in cetaceans. Deep Sea Research 3:153-154.

McBride, A. F. and H. Kritzler 1951 Observations on pregnancy, parturition, and postnatal behavior in the bottlenose dolphin. Jour. Mamm. 32:251-266.

Miller, G. S., Jr. 1923 The telescoping of the cetacean skull. Smithsonian Misc. Collections 76(5):1-71.

Møhl, B. and S. Andersen 1973 Echolocation: high frequency component in the click of the harbour porpoise. Jour. Acoust. Soc. Am. 54(5):1368-1372.

Moller, P. and R. Bauer 1973 "Communication" in weakly electric fish, *Gnathonemus petersii* (Mormyridae). Interaction of electric organ discharge activities of two fish. Animal Behavior 21(3):501-512.

Moore, J. C. 1968 Relationships among the living genera of beaked whales. Fieldiana:Zoology 53(4):209-298.

Morgane, P. J. and N. S. Jacobs 1972 Comparative anatomy of the cetacean nervous system. In: Functional Anatomy of Marine Mammals, R. J. Harrison, Ed., Vol. 1, pp. 117-244. Academic Press.

Norris, K. S. 1968 The evolution of acoustic mechanisms in odontocete cetaceans. In: Evolution and Environment, E. T. Drake, Ed., pp. 297-324.

Norris, K. S. 1969 The echolocation of marine mammals. In: The Biology of Marine Mammals, S. Andersen, Ed., pp. 391-423. Academic Press.

Norris, K. S., J. H. Prescott, P. V. Asa-Dorian 1961 An experimental demonstration of echolocation behavior in the porpoise, *Tursiops truncatus* (Montagu). Biol. Bull. 20(2):163-176.

Norris, K. S. and G. W. Harvey 1972 A theory for the function of
     the spermaceti organ of the sperm whale. In: Animal Orienta-
     tion and Navigation, S. R. Galler et al., Eds., National
     Aeronautics and Space Administration, Washington, D.C.

Norris, K. S. and W. E. Evans 1967 Directionality of echoloca-
     tion clicks in the rough-tooth porpoise, *Steno bredanensis*
     (Lesson). In: Marine Bioacoustics, Vol. 2, W. N. Tavolga,
     Ed., pp. 305-314. Pergamon Press.

Norris, K. S., W. E. Evans, and R. N. Turner 1967 Echolocation
     in an Atlantic bottlenose porpoise during discrimination.
     In: Animal Sonar Systems: Biology and Bionics, R.-G. Busnel,
     Ed., Vol. I, pp. 409-437. Laboratoire de Physiologie
     Acoustique, Jouy-en-Josas-78, France.

Payne, R. S. and S. McVay 1971 Songs of humpback whales.
     Science 173(3997):585-597.

Penner, R. H. and A. E. Murchison 1970 Experimentally demon-
     strated echolocation in the Amazon River porpoise, *Inia
     geoffrensis* (Blainville). Naval Undersea Center (San Diego,
     Calif.) Technical Publication 187, Rev. 1, 26 pp.

Pilleri, G. and M. Gihr 1970 Brain-body weight ratio of
     Platanista gangetica. In: Investigations of Cetacea, Vol.
     II, G. Pilleri, Ed., University of Berne, Switzerland.

Poulter, T. C. 1963 Sonar signals of the sea lion. Science
     139:753-755.

Poulter, T. C. 1967 Systems of echolocation. In: Animal Sonar
     Systems: Biology and Bionics, R.-G. Busnel, Ed., Vol. I,
     pp. 157-183. Laboratoire de Physiologie Acoustique,
     Jouy-en-Josas-78, France.

Raven, H. C. and W. K. Gregory 1933 The spermaceti organ and
     nasal passages of the sperm whale (*Physeter catodon*) and
     other odontocetes. Am. Mus. Novitates No. 677, 18 pp.

Resvov, R. N., Ya. A. Savel'ev, and A. B. Flenov 1973 The study
     of dolphin's echolocator ability in detection of different
     targets, difference in range, and angle on vertical and
     horizontal planes. Proceedings of the 8th All-Union Acousti-
     cal Conference Part 1:134-135. (In Russian)

Rice, D. W. 1978 Beaked whales. In: Marine Mammals, Delphine
     Haley, Ed., pp. 89-95. Pacific Search Press.

Ridgway, S. H., N. J. Flanagan, and J. G. McCormick 1976
    Brain-spinal cord ratios in porpoises: possible correlation
    with intelligence and ecology. Psychon. Sci. 6:491-492.

Ridgway, S. H. and R. W. Brownson 1979 Relative brain sizes and
    cortical surface areas in odontocetes. In preparation.

Schenkkan, E. J. 1973 On the comparative anatomy and function of
    the nasal tract in odontocetes (Mammalia, Cetacea).
    Bijdragen Tot de Dierkunde 43(3):127-159.

Schenkkan, E. J. and P. E. Purves 1973 The comparative anatomy
    of the nasal tract and the function of the spermaceti organ
    in the Physeteridae (Mammalia, Odontoceti). Bijdragen Tot de
    Dierkunde 43(1):93-112.

Schevill, W. E. 1964 Underwater sounds of cetaceans. In:
    Marine Bio-Acoustics, W. N. Tavolga, Ed., pp. 307-316.

Schevill, W. E. and B. Lawrence 1953 Auditory responses of a
    bottlenose porpoise *Tursiops truncatus* to frequencies above
    100 kHz. Jour. Exp. Zool. 124(1):147-165.

Schevill, W. E., W. A. Watkins, and C. Raye 1963 Underwater
    sounds of pinnipeds. Science 141:50-53.

Schevill, W. E. and B. Lawrence 1965 Food-finding by a captive
    porpoise, *Tursiops truncatus*, Breviora 53:1-15.

Schevill, W. E., W. A. Watkins, and C. Ray 1969 Click structure
    in the porpoise, *Phocoena phocoena*. Jour. Mamm. 50(4):721-
    728.

Schusterman, R. J. 1966 Underwater click vocalization by a
    California sea lion: Effects of visibility. Psych. Rec.
    16:129-136.

Sergeant, D. E. 1978 Ecological isolation in some cetacea. In:
    Recent Advances in the Study of Whales and Seals, A. N.
    Severtsov, Ed., Nauka, Moskow.

Slijper, E. J. 1962 Whales. Basic Books, Inc., New York.

Tavolga, W. N. 1977 Mechanisms for directional hearing in the
    sea catfish (*Arius felis*). Jour. Exp. Biol. 67:97-115.

Tavolga, M. C. and F. S. Essapian 1957 The behavior of the
    bottlenosed dolphin (*Tursiops truncatus*): Mating, pregnancy,
    parturition, and mother-infant behavior. Zoologica 42, Part
    I, pp. 11-34.

Titov, A. A. 1972 Investigation of the sound activities and peculiarities of the Black Sea dolphin's echolocator. Ph.D. Thesis, Karadag. (In Russian)

Tomilin, A. G. 1957 Mammals of the U.S.S.R. and Adjacent Countries. Vol. IX Cetacea. Izdatel'sto Akademi Nauk, Moscow, U.S.S.R.

Tomilin, A. G. 1968 Factors promoting powerful development of the brain in Odontoceti: Trudy Vsesoyuznogo Sel'skokhozyaystvennogo Instituta Zaochnogo Obrasovaniya, No. 31, pp. 191-200. In Russian. JPRS No. 49777.

True, F. W. 1910 An account of the beaked whales of the family Ziphiidae in the collection of the United States National Museum, with remarks on some specimens in other American museums. U. S. National Museum Bulletin 73, 89 pp., 42 plates.

Watkins, W. A. and W. E. Schevill 1971 Underwater sounds of *Monodon* (Narwhal). Jour. Acoust. Soc. Am. 49(2), Pt. 2: 595-599.

Webster, F. A. 1963 Active energy radiating systems: The bat and ultrasonic principles II; acoustical control of airborne interceptions by bats. In: Proceedings of the International Congress on Technology and Blindness, Vol. 1, pp. 49-135.

Whitmore, F. C., Jr., and A. E. Sanders 1976 Review of the Oligocene Cetacea. Syst. Zool. 25(4):304-320.

Wood, F. G., Jr. 1952 Porpoise sounds. A phonograph record of underwater sounds made by *Tursiops truncatus* and *Stenella plagiodon*. Publ. by Marineland Research Laboratory, Marineland, Florida.

Wood, F. G., Jr. 1953 Underwater sound production and concurrent behavior of captive porpoises, *Tursiops truncatus* and *Stenella plagiodon*. Bull. Mar. Sci. Gulf and Caribbean 3:120-133.

Wood, F. G. 1964 General Discussion. In: Marine Bio-Acoustics, W. N. Tavolga, Ed., pp. 395-396. Pergamon Press.

Working Party on Marine Mammals 1977 In: Mammals in the Seas, Vol. 1, Report of the FAO Advisory Committee on Marine Resources Research. Food and Agriculture Organization of the United Nations, Rome.

Yablokov, A. V.   1962   The key to a biological riddle - the sperm whale at a depth of 2,000 meters.   Priroda 4:95-98.   (In Russian)

Yablokov, A. V., V. M. Bel'kovich, and V. I. Borisov   1972   Whales and Dolphins, Parts I and II.   Kity i Del'fini, Izd-vo Nauka, Moscow.   Translation JPRS 62150-1.   528 pp.

Zaslavsky, G. L.   1971   On directivity of sound emission in the Black Sea bottlenosed dolphin.   Trudy Akusticheskogo Instituta 17:60-70.   (In Russian)

Zaslavsky, G. L.   1972   Study of echolocational Signals of dolphins by means of two-channel registration system.   Biofisika 4:717-720.   (In Russian)

Zaslavsky, G. L.   1974   Experimental investigations of the spatial-temporal structure of the echolocational signals of dolphins.   Ph.D. Thesis.   (In Russian)

Zaslavsky, G. L., A. A. Titov, and V. M. Lekomtsev   1969   Investigations of the echolocational abilities of the harbor porpoise (*Ph. phocoena*).   Trudy Akusticheskogo Instituta 8:134-138.   (In Russian)

ADAPTIVENESS AND ECOLOGY OF ECHOLOCATION IN TERRESTRIAL (AERIAL) SYSTEMS

M. Brock Fenton

Department of Biology, Carleton University, Ottawa,

Canada, K1S 5B6

There are two clearly defined approaches to echolocation by terrestrial (aerial) animals based on the types of orientation sounds they use, namely broadband clicks or structured (time - frequency) calls that may be either broadband or narrowband. The broadband clicks are usually produced in pairs (Buchler and Mitz, this volume) and are associated with the location of obstacles in the paths of the echolocators, while the structured calls are used to identify specific features about the target and its position. To date, the only terrestrial animals that are known to use structured calls are bats in the suborder Microchiroptera; all other terrestrial echolocators, including the Megachiroptera, use broadband clicks (Table 1).

Echolocation is one method of obtaining information about one's surroundings under conditions of poor lighting and it shares many features with electrolocation, a system of orientation also effective in darkness (Hopkins 1976; 1977). Unfortunately, although it is clear that echolocation is polyphyletic in the animal kingdom, we have no clear idea of how it arose. Observations of bears, aardvarks and ratels using a loud snort and its associated echo to identify underground cavities (Kingdon 1977), suggests that the use of echoes to obtain information about the environment is not restricted to the 'conventional echolocators' (Table 1). Ewer (1968) proposed that echolocation calls are derived from communication signals, but there is no proof for this hypothesis. Data from the ontogeny of bat vocalizations (e.g. Gould 1975; this volume; Brown 1976; Brown and Grinnell, this volume) show no conclusive evidence of a transition from communication to echolocation calls in all of the species examined to date, although some workers have suggested otherwise (Gould 1971; Konstantinov 1973). Echolocating animals produce an

Table 1. Terrestrial (aerial) animals known or suspected to use echolocation (from a variety of sources including: Griffin 1958; Airapet'yants and Konstantinov 1973; Sales and Pye 1974; Buchler 1976; Novick 1977, and specific papers cited below).

animals	broadband calls	structured calls
Amphibia, Apoda, Caeciliidae		
Dermophis cf. septentrionalis[1]	X	
Aves, Caprimulgiformes, Steatornithidae		
Steatornis caripensis	X	
Apodiformes, Apodidae		
Collocalia vanikorensis[2]	X	
Collocalia hirundinacea[3]	X	
Collocalia brevirostris	X	
Collocalia maxima	X	
Collocalia fuciphaga	X	
Mammalia, Insectivora, Soricidae		
Sorex vagrans	X	
Sorex cinereus	X	
Sorex palustris	X	
Blarina brevicauda	X	
Cryptotis parva	X	
Tenrecidae		
Hemicentetes semispinosus	X	
Echinops telfairi	X	
Microgale dobsoni	X	
Chiroptera, Pteropodidae		
Rousettus spp.	X	
Microchiroptera, all families		
all species		X
Rodentia, Gliridae		
Glis glis	X	
Muridae		
Clethrionomys gapperi	X	
Mesocricetus auratus	X	
Rattus norvegicus[4]	X	
Primates, Hominidae		
Homo sapiens	X	

[1]Thurow and Gould 1977; [2]Griffin and Suthers 1970; n.b. the classification of swiftlets is a topic of some debate, and there are workers who classify these species of swiftlets in the genus Aerodramus, a name associated with a subgenus by other workers - see Pye, this volume; [3]Fenton 1975; [4]Chase, this volume.

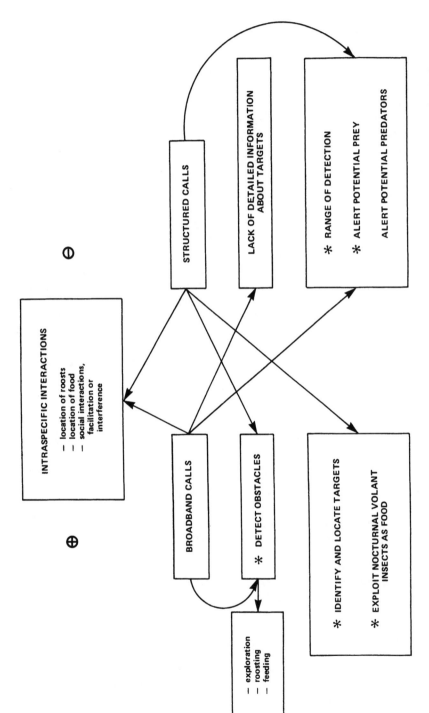

Figure 1. An outline of the advantages, disadvantages and mixed blessings associated with echolocation by terrestrial (aerial) animals. The asterisks indicate advantages or disadvantages documented by experimental evidence.

array of vocalizations which serve a variety of functions from mother-young interactions (e.g. Brown 1976; Barclay 1978; Gould, this volume) to mating (Barclay and Thomas 1979).

In short, we have no working hypothesis about how echolocation evolved. One way of clarifying our knowledge about this subject is to consider the advantages and disadvantages associated with echolocation. What follows here is a review of the sparse evidence on the adaptiveness of echolocation in terrestrial (aerial) animals, considering the advantages, disadvantages and mixed blessings associated with it (Fig. 1).

ADVANTAGES OF ECHOLOCATION

Detection of Obstacles

Broadband clicks or highly structured calls can be used by animals to detect the presence of obstacles in their paths. The size of the obstacles detected is partly a function of the wavelengths of the sounds produced (in broadband clicks, presumably the portion of the call containing the most energy) and the size and proximity of the target. Detection of obstacles independent of light conditions entails several important advantages to the animal. These benefits are clearer for flying echolocators such as bats and birds, than they are for terrestrial or fossorial species, partly because there are many more data for the fliers, especially the insectivorous bats.

Perhaps the principal advantage to all echolocators using broadband clicks, and a side benefit, albeit an important one for species using structured calls, is to make activity patterns independent of light conditions, allowing extensive activity at night or in poorly illuminated habitats during the day. For shrews echolocation allows poor resolution of habitat details but facilitates exploration activity (Buchler 1976). Buchler (1976) showed that Sorex vagrans tended to echolocate more when exploring unfamiliar territory than when using familiar areas. A direct benefit to shrews familiar with a large array of microhabitats could be increased access to food, an important point for an animal on a tight energy budget (Randolf 1973). Although I found no comparable data for other small terrestrial echolocators, similar constraints and advantages probably exist for their echolocation behaviour.

Activity independent of light conditions has made caves available to flying echolocators, providing benefits associated with feeding, breeding and predators. Most birds do not use fruit as food for nestlings because it does not provide adequate protein for rapid growth, so that nestlings reared on fruit grow more slowly and are in the nest longer, increasing the time they are exposed to predators. By nesting inside caves and reducing the threat of predation on nestlings, oilbirds (Steatornis caripensis) can exploit fruit, an abundant, low protein food, for their nestlings and themselves (Morton 1973).

Caves provide nesting or roosting sites that are relatively protected against many predators, although some arthropods prey on the eggs and young of some swiftlets (Collocalia spp.; Medway 1962). However, large populations of bats and/or swiftlets using the entrances to caves are an excellent food source for many predators, ranging from snakes to small Carnivora and a variety of birds of prey (Gillette and Kimbrough 1970). Thus exploitation of caves by large numbers of echolocators has inherent disadvantages typical of refuging populations (Hamilton and Watt 1970).

The large surface areas of the interiors of caves present an abundance of nest and roost sites, particularly for species such as swiftlets that can build their nests on relatively smooth walls, or animals such as Rousettus spp. that do not build nests. Since populations of some species of swifts appear to be limited by the availability of nest spaces (Lack 1956), the ability to echolocate and nest in caves may have fostered an increase in populations of echolocating swiftlets (Fenton 1975). Echolocation allows birds to come and go from their roosts at night, permitting Collocalia spp. access to the night skies where they could use vision to track insect prey (Fenton 1975).

Detection of obstacles by echolocation has permitted animals to sever the close ties of their activity patterns with ambient light levels, allowing nocturnal activity, increased opportunity for exploration and exploitation of caves as nesting and roosting sites. Caves may protect some animals, particularly young, from predators, allowing exploitation of abundant low protein food (Steatornis) and may have contributed to an expansion of populations by increasing the number of reproducing individuals in the population (Collocalia).

Locate and Identify Targets

Orientation sounds with specific time-frequency structures allow species in the Microchiroptera to gather specific information about targets in their paths. Specifically these calls permit location and tracking of flying insects (Griffin, Webster and Michael 1960; Simmons, Fenton and O'Farrell 1979) and determination of the general identity of the potential target (Goldman and Henson 1977; Schnitzler 1978). Using structured calls some bats can detect fish or other prey on or near the surface of water with enough precision to permit capture (Suthers 1965). Other species use echolocation to locate insects on vegetation or other surfaces, permitting capture by gleaning (Fenton and Bell 1979). Although the location and specific identification of targets is closely related to the ability to exploit populations of nocturnal insects as food (see below), it also allows accurate evaluation of details of roosts and other environmental factors.

The versatility of roosting and social behaviour known from the Microchiroptera, in addition to their diverse feeding habits, probably reflects the advantage conferred by an echolocation system that allows an animal to obtain specific, detailed information about targets, regardless of the light conditions.

Access to Volant Nocturnal Insects as Food

Insectivorous bats use structured calls to exploit nocturnal insects, particularly smaller ones, as food. These are a resource largely unavailable to animals that orient by vision, except under conditions where a predator can track prey against a constant, light-coloured background. Birds, including many species of owls and most Caprimulgiformes, regularly feed on nocturnal volant insects which they locate and track visually using the night sky as a background. Echolocating insectivorous bats and a variety of spiders exploit nocturnal flying insects as food, but their prey-catching prowess is not constrained by light.

Small nocturnal insects are an important food source, access to which provides predators with a rich, if not superabundant, food supply. A number of studies has shown that temperate insectivorous bats feed heavily on insects in proportion to their availability (Black 1974; Belwood and Fenton 1976; Bradbury and Vehrencamp 1976; Anthony and Kunz 1977) and none has provided evidence to support the idea that food is limiting for populations of insectivorous bats. In areas where 10 or 12, or even 25 species of insectivorous bats feed in the same place at the same time, there is no evidence of competition between species or individuals for food (Bell 1979 Fenton and Thomas 1979). The population levels of insectivorous bats are very high in some areas, suggesting that nocturnal insects are not in short supply in spite of astounding levels of predation associated with insectivorous bats (Davis, Herreid and Short 1962). These data all clearly demonstrate that nocturnal insects are an abundant food supply.

A reflection of the effect of this food resource is the fact that the majority of insectivorous bats are small (less than 40 grams as adults; Fig. 2), while all of the nocturnal insectivorous birds are much larger (over 40 grams as adults; Fenton 1974; Fenton and Fleming 1976). The preponderance of small insectivorous bats suggests that bats are very effective as small, nocturnal insectivores and much less effective as larger, nocturnal insectivores.

One of the principal advantages of echolocation is access to the food resource represented by small nocturnal insects and in

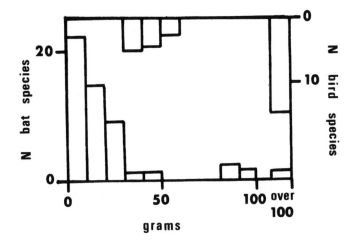

Figure 2. Size distribution (by weight) of the species of animalivorous and insectivorous bats and nocturnal birds which occur in Guyana (adapted from Fenton and Fleming 1976).

this context, I consider echolocation and flight to be the keys to the success of the Microchiroptera.

DISADVANTAGES OF ECHOLOCATION
Lack of Detailed Information about Targets

This disadvantage applies only to species using broadband clicks and is directly comparable to the lack of the ability for flapping flight in flying squirrels. The relative lack of resolution associated with broadband clicks is probably more a function of information processing by the animal than a limitation inherent in the signals themselves.

Range

One of the principal disadvantages of echolocation is its limited effective range in air, a situation potentially more hazardous for a flying than for a walking or running animal. Griffin (1971) discussed the role of attenuation in echolocation by bats and his data emphasized that bats using higher frequency calls (which provide better resolution of target information) suffered more from interference by attenuation. Pye (1971) demonstrated how rain and fog could hamper echolocation by increasing attenuation and providing additional clutter and phantom targets. It is clear that atmospheric conditions can strongly influence the effectiveness, especially the range of echolocation.

It is difficult to determine the effective range of
echolocation. Buchler (1976) suggested that for shrews distance
perception using echolocation was less than 1 m and Novick (1977),
in his summary of data on acoustic orientation by bats, provided
figures of 2 or 3 m for the effective range of echolocation of
insects and large obstacles by many Microchiroptera. Data on the
range of echolocation are usually associated with laboratory
studies of bats orienting towards preferred roosts (Novick 1977),
although Griffin, Webster and Michael (1960) provided data for
hunting Myotis lucifugus in the lab that showed that these bats
fixed on their insect prey at a range of about 0.8 m. Unfortunately,
the distance over which a bat or other echolocator fixes on its
target is the minimum effective range, not the maximum effective
range.

Vaughan's description of feeding behaviour of Hipposideros
commersoni (1977), the largest insectivorous bat, suggests an
effective range of less than 10 m in the field. Field observations
of Myotis volans feeding in Arizona suggest an effective range of
5 - 10 m, based on the distances at which individual bats fixed
on their insect prey (Fenton and Bell 1979). In the field, feeding
Myotis lucifugus and M. californicus appeared to locate prey at
much shorter distances (less than 1 m) and were able to make
repeated attempts to capture insects in a short distance, a
behaviour not observed in M. volans (Fenton and Bell 1979).

It is clear that different species of bats, sometimes in the
same genus, show different depths of perception through echolocation.
Brosset (1966) had suggested this when he proposed two basic
feeding strategies for insectivorous bats, long range and short
range. His observations on long range species were based mainly
on field observations of Taphozous peli, which appeared to locate
insect targets at 15 m, while his suggestions about short range
operators were based largely on the report of Griffin, Webster and
Michael (1960).

Intensity of echolocation calls probably influences the range
at which an animal can detect and identify a target. Griffin
(1971) estimated that the maximum effective range for Hipposideros
galeritus was no more than 2 m. This species produced a relatively
low intensity call, stronger than those typical of Nycteris spp.,
but much weaker than the high intensity calls associated with
Myotis spp., Eptesicus spp., or Rhinolophus spp.

When the effective range of echolocation is considered in
terms of body lengths of the animal involved, a problem of scaling
is obvious, further influenced by the speed at which the echolocator
is travelling. Larger, faster species will be adversely affected
by a short effective range (relative to their size) simply because
they may have reached a target before they have had time to

process information in the returning echoes. Slower, smaller
species will be less drastically affected in this context.
Considerations of food availability notwithstanding, the relative
dearth of larger insectivorous bats (Fig. 2) may be a reflection
of the limited range effectiveness of echolocation.

Larger insectivorous bats appear to have at least two strategies
for reducing the constraints associated with short range operation.
Hipposideros commersoni often hunts from a perch, flying out only
to capture vulnerable insects (Vaughan 1977), alleviating in part
the restrictions of range, while saving on energy spent in flight.
Production of loud orientation calls with fundamental frequencies
in lower ranges (15 - 20 kHz; Fig. 3) could also reduce the effects
of atmospheric attenuation, increase the effective range and
provide adequate details about larger insect targets appropriate
for larger bats.

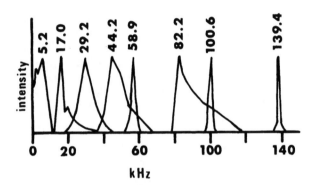

Figure 3. Fast Fourier Transform analysis of calls used for
echolocation by Collocalia hirundinacea (5.2 kHz), an unidentified
molossid in Cote d'Ivoire (17.0 kHz), Eptesicus fuscus (29.2 kHz),
Myotis lucifugus (44.2 kHz), Hipposideros cyclops (58.9 kHz),
Pipistrellus nanus (82.6 kHz), Rhinolophus landeri (100.6 kHz) and
Hipposideros ruber (139.4 kHz). In each case, at least 30 calls
were averaged by a Princeton Applied Research Model 4513 FFT to
produce these spectra.

Alerting Prey

The presence of ears that are sensitive to the echolocating
cries of bats in many moths (Roeder 1975) and some lacewings
(Miller 1975) demonstrates an important disadvantage for animals
which use echolocation for hunting, specifically, alerting potential

to the presence and approach of a predator. Roeder (1967) and Miller
(pers. comm.) have determined that moths and lacewings that can hear
bats have a 40% less chance of being captured by echolocating bats
than species lacking ears(see also Miller, this volume).

In the Lepidoptera ears are polyphyletic, appearing in different
places in different families (Table 2). The ears of moths are, like
those of other insects, tuned by their physical characteristics
including the diameter of, and the tension on, the tympanum
(Michelson 1971; Adams 1972). It is possible for an echolocating
bat to adjust its orientation calls to make them less conspicuous
to a moth, and thus to reduce the advantage to the moth (Fig. 4; see
also Fullard and Fenton, this volume). Crucial adjustments for a
bat in the context of foiling a moth include flight speed and the
intensity and frequency of orientation calls, modifications to
reduce the warning time available to the insect (Fenton and Fullard
1979).

Table 2. The locations of tympana in different families of moths
known to include tympanate species. Data from Roeder (1974; 1975),
Sales and Pye (1974), and James H. Fullard (pers. comm.)

| family | locations of ears | | |
	thorax	first or second abdominal segment	maxillary palp
Thaumetopoeidae	X		
Agaristidae	X		
Lymantriidae	X		
Noctuidae	X		
Pyralidae		X	
Arctiidae	X		
Notodontidae	X		
Geometridae		X	
Drepanidae		X	
Thyatiridae		X	
Thyretidae	X		
Sphingidae[1]			X

[1] only in the Choerocampinae

Some details of moth defense and possible bat counters have
been reviewed in the light of tuning characteristics of moths ears
(Fullard and Fenton, this volume). These data suggest that the
hearing ability of insects could exert an important selection
pressure on the echolocation behaviour of bats. It is important to

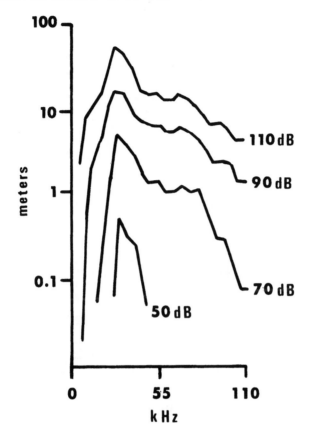

Figure 4. A representation of the maximum distances at which
a tympanate moth (Tmetopteryx bisecta ; Notodontidae) can detect
the echolocation cries of bats based on auditory threshold values.
These figures are derived by transforming audiogram values for the
moth with the data of Griffin (1971) on attenuation, and they
illustrate the effects of initial intensities of bat echolocation
calls (dB re 2 x 10^{-5} N/m^2; 10 cm), at 20% relative humidity.

note, however, that at least 65% of the bat species in any area,
and 100% in some areas, use FM echolocation calls that sweep through
the range of maximum sensitivity of tympanate moths (Fig. 5). Bats
which have countered moth defenses by adjusting their echolocation
calls appear to be consistently in a minority - one reason that
their counters are effective. The ultrasonic sounds produced by
some moths, usually arctiids, are an extension of insect hearing
defense against bats. These sounds cause some bats to reject
otherwise palatable food (Dunning and Roeder 1965), and are produced
just as the bat closes with the moth. The acoustic characteristics
of the clicks (power spectra and frequency-time structure) closely

match those of FM orientation calls produced by many bats as they
close with their insect prey. The clicks of the moths may be
initially processed as echoes by the bats' auditory systems,
resulting in disruption of information processing and producing a
Protean effect (Fullard, Fenton and Simmons 1979). A bat-
deterrent effect has also been suggested for the clicks of Peacock
Butterflies (Møhl and Miller 1976).

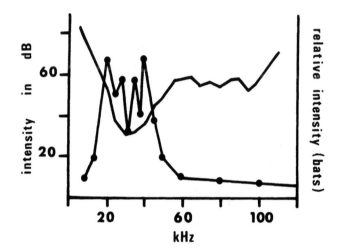

Figure 5. The audiogram of the African geometrid <u>Gorua apicata</u>
superimposed on an FFT analysis of 400 bat echolocation calls
recorded in the field in Cote d'Ivoire (dots). The moth is most
sensitive in the frequency range with the greatest energy of
echolocation calls.

In summary, echolocation calls may adversely affect the
hunting effectiveness of insectivorous bats by alerting potential
prey and providing them with time for escape maneouvers. The tuning
characteristics of moth ears allow bats to reduce the advantage
gained by the moth by adjusting the frequency and intensity of their
calls. Countering insect defensive strategies may represent an
important selective force on bat echolocation.

Alerting Predators

Trains of pulses emitted by an animal as it moves through an
area not only produce information about targets, but also provide
potential predators with specific details about the location and

course of the echolocator. Thus a shrew or bat producing
orientation calls could reveal its presence and course to a
predator and significantly reduce its chances of survival (assuming
that the predator can hear the calls).

There are relatively few predators which specialize on bats or
other echolocators, and the inability of birds to hear sounds above
20 kHz (Schwartzkopff 1973) precludes their exploitation of
ultrasonic echolocation cries to locate prey. Hawks and owls do
not appear to specialize on shrews as food, although these small
mammals comprise a consistently small portion of the diets of many
raptors and owls (e.g. Craighead and Craighead 1956), and they are
also taken by many mammalian predators. The Bat Hawk (Macheiramphus
alcinus) feeds heavily on bats (Fenton, Cumming and Oxley 1977)
which it locates using vision. Bat Hawks feed in groups of bats or
swiftlets that are leaving or returning to caves or other
roosts.

Carnivorous bats may be the predators best suited to exploit
the information available in the orientation calls of bats of
other animals. However, in spite of the suggestions that species
such as Vampyrum spectrum feed heavily on bats (e.g. Howell and
Burch 1974), one recent study of the diet of V. spectrum in Costa
Rica showed that bats did not constitute a large portion of its
diet (Vehrencamp, Stiles and Bradbury 1977).

There appear to be no data supporting the possibility that
the echolocation calls of other animals are used by predators as
aids to locating food. However, it may be significant that some
broadband echolocators cease production of their clicks when not
orienting in darkness (e.g. Collocalia hirundinacea; Fenton 1975),
while others continue to produce their cries, even when flying in
bright moonlight (e.g. Rousettus aegyptiacus; Thomas and Fenton 1978).

MIXED BLESSINGS

Information about the presence and course of an echolocator
provided by its calls is available to conspecifics. For example,
bats that increase their pulse repetition rates as they close with
insect targets reveal to eavesdroppers the presence of food. In
some settings listening for feeding buzzes could permit members of
the same colony to converge upon and exploit a rich patch of food,
but in other instances (e.g. territorial species), it may facilitate
interference with feeding behaviour of the individual that located
the food source.

In Arizona Fenton and Morris (1976) observed that some bats
occasionally responded to the playback of feeding buzzes by
increased activity around the speaker, suggesting that some

insectivorous bats exploit feeding calls of others as a means of
locating food. Some colleagues and I have conducted a number of
similar experiments involving the broadcast of feeding buzzes
produced by Myotis lucifugus in areas where conspecifics were flying.
The results mirror those of the earlier study (Fenton and Morris 1976)
since in most instances the playbacks elicited no response, but
on several occasions bat activity in the vicinity of the speaker
increased drammatically when the feeding buzzes were presented. In
southeastern Ontario M. lucifugus commonly forages in groups,
feeding for the most part on hatches of aquatic insects (Belwood
and Fenton 1976), and in this species in this setting it is
adaptive for individuals to be able to locate and exploit swarms of
insects.

The patchy distribution of many populations of nocturnal
insects may be best exploited by species that effectively communicate
information about the location of rich patches of food. In this
context the inherent communication properties of echolocation calls
may maximize use of nocturnal insects as food. However, territorial
insectivorous bats (e.g. Wallin 1961; Tuttle, Stevenson and
Rabinowitz 1978) can monitor intrusions by other bats into their
territories and distinguish between feeding and cruising trespassers.

In many instances insectivorous bats quickly locate and exploit
new roost opportunities (Brosset 1966; Fenton 1970), and roost
availability may limit populations in, and community structures of,
some bat faunas (Humphrey 1975). The echolocation calls of
individuals near or moving towards a roost may be used by others to
locate groups of conspecifics or roosts. Playback experiments we
have conducted in the vicinity of hibernation sites during swarming
behaviour have clearly demonstrated that M. lucifugus respond to the
calls of conspecifics and form large groups of flying bats in the
vicinity of the source of the calls. In this case we played recorded
echolocation calls of M. lucfugus about 50 m from the entrance to
an abandoned mine which is used each winter as a hibernation site
by about 5,000 bats. The echolocation calls (typical FM cries)
were presented to the M. lucifugus as they passed through the area
and within several minutes we consistently accumulated a column of
20 to 50 bats flying around in the vicinity of the speaker. This
response was not elicited by 40 kHz pure tones or by tape noise.

In this situation, bats were obviously using the echolocation
calls of others to locate groups of conspecifics. Vaughan and
O'Shea (1977) described the rallying behaviour typical of Antrozous
pallidus returning to a day roost in the morning. In that case
the bats produced distinctive 'directive' calls, and it is not clear
if the bat echolocation calls were also important in the responses
of individuals to the aggregation of conspecifics. There is no
evidence of a 'directive' call in the repertoire of M. lucifugus
(Barclay 1978).

Exploitation of orientation calls of conspecifics may be important in a number of bat social interactions. Again the calls may facilitate some interactions and increase the chances of interference in others. In some cases modifications to echolocation calls may enhance their effectiveness in communication. For example, the honking of one Noctilio leporinus at another was described by Suthers (1965) and we have observed this behavior in M. lucifugus, M. volans, and Eptesicus fuscus. In other cases calls produced by bats during social interactions may be drastically different from echolocation calls, probably to enhance their effectiveness (e.g. the copulation call of M. lucifugus; Barclay and Thomas 1979).

Mixed blessings in terms of the use of echolocation calls by conspecifics probably extend to other social interactions among insectivorous bats. Territorial behaviour has been reported from several species (e.g. Wallin 1961; Bradbury and Emmons 1974; Tuttle, Stevenson and Rabinowitz 1978) and individuals may be able to recognize their neighbours by their echolocation calls.

While working at the Rockefeller University Field Station near Millbrook, New York in July 1976, I occasionally observed chases involving pairs of Eptesicus fuscus. The chases sometimes involved physical contact, usually when the chasing bat struck the leader, and often vocalizations which were audible to me. Over a grassy clearing surrounded on three sides by deciduous woods a number of bats fed each evening, including M. lucifugus as they passed through en route to other areas, and up to 5 E. fuscus that remained in the vicinity of the clearing for at least 90 minutes after dark. One of the E. fuscus set up a patrol flight pattern in the clearing by flying a large four-leafed-clover pattern about 100 m across, over the other bats feeding in the area. Occasionally this bat left its flight path and pursued another E. fuscus, chasing it out of the immediate area before returning to its patrol path. The patrolling bat made no effort to chase away all of the other bats (M. lucifugus or E. fuscus) feeding below it. On the evening of 19 July 1976 I played the echolocation calls of an E. fuscus recorded 5 km away to the patrolling bat, causing it to leave its flight patrol and to chase the nearest E. fuscus. In 30 minutes I was able to elicit 7 such altercations, where the average number of altercations on 10 previous and 5 subsequent nights was 2 (range 0 to 3). Until the chases were initiated, all of the bats involved in the area appeared to have been producing 'typical' FM echolocation calls.

Since I cannot show that the patrolling bat was the same individual from night to night, and because I am not certain if the conspecifics feeding in the area were the same from night to night, it is not possible to provide a detailed interpretation of these observations. However, for the purpose of this discussion, the results strongly suggest that the patrolling bat responded to the echolocation calls of the intruder which it could distinguish from the familiar calls of other conspecifics feeding below it. Vision seems unlikely as an important factor in this situation as the patrolling bat was above the other bats and swooped down to chase intruders.

As we gather more information about the behaviour of echolocating animals in the field, it should become increasingly apparent that the orientation calls are used in social contexts and may be pirated. Interactions with conspecifics should demonstrate that echolocation offers a mixed blessing, an asset in some settings and a liability in others.

ADAPTIVENESS OF ECHOLOCATION

The relative lack of echolocation behaviour in terrestrial settings outside the Microchiroptera suggests that the use of broadband clicks for orientation does not confer a strong adaptive advantage to animals that practise it. On the other hand, the facultative use of echolocation by a variety of animals (e.g. man, and perhaps bears, aardvarks and ratels) demonstrates some advantages when lighting is poor or non-existant. Echolocation involving the use of structured calls, whether broadband or narrowband, represents a special adaptation (sensu Van Valen 1971) that has resulted in the occupation of a new adaptive zone, principally the exploitation of nocturnal insects as food.

STILL TO DO

There are a number of important questions to be addressed in the context of the adaptiveness of echolocation. How frequently do insectivorous bats turn off their echolocation and rely on other sensory input to locate food (e.g. Fiedler, Habersetzer and Vogler, this volume)? What is the importance of the sounds and smells of potential prey as cues for insectivorous bats? Do bats capable of producing structured calls occasionally use broadband clicks? How do individuals adjust the intensities of their calls in specific situations where a quiet or a loud call might be most effective?

Fortunately the number of researchers going into the field to observe the natural behaviour of bats and other echolocators is increasing along with the number of workers studying different aspects of echolocation in the lab. Furthermore, the availability of portable and relatively inexpensive apparatus for studying and monitoring the calls of echolocators has increased (Simmons, Fenton, Ferguson, Jutting and Palin 1979) allowing field workers to put themselves into the acoustic world of the animals they study.

Researchers wishing to study the behaviour of bats in the field can often exploit their feeding habits. Specifically, species in many families will feed at concentrations of insects around lights, while other species predictably visit feeding sites. In either situation, biologists can anticipate the response of the animal and study its feeding behaviour. However, it is important to remember that observing an animal producing a pattern of calls which we associate with echolocation does not prove that the animal was echolocating and not using other sensory cues to find its prey.

ACKNOWLEDGEMENTS

I am grateful to Robert Barclay, Gary Bell, Patricia Brown, Edward Buchler, James Fullard, Alan Grinnell and James Simmons for reading this manuscript and making helpful suggestions about it. L. Van Every kindly prepared Figure 1. My research on bats has been supported by National Research Council of Canada Operating and Equipment Grants.

REFERENCES

Adams, W.B. 1972. Mechanical tuning of the acoustic receptor of Prodenia eridania (Cramer) (Noctuidae). J. exp. Biol. 57: 297-304.

Airapet'yants, E. Sh., and A.I. Konstantinov. 1973. Echolocation in animals (translated from Russian). Israel Prog. Sci. Transl. Jerusalem: Keter Press.

Anthony, E.L.P. and T.H. Kunz. 1977. Feeding strategies of the little brown bat, Myotis lucifugus, in New Hampshire. Ecology 58: 775-786.

Barclay, R.M.R. 1978. Vocal communication and social behaviour of the little brown bat, Myotis lucifugus, (Chiroptera : Vespertilionidae). M.Sc. Thesis, Department of Biology, Carleton University, Ottawa, Canada.

Barclay, R.M.R. and D.W. Thomas. 1979. Copulation call of the bat, Myotis lucifugus; a discrete, situation specific communication signal. J. Mamm. 60: in press.

Belwood, J.J. and M.B. Fenton. 1976. Variation in the diet of Myotis lucifugus (Chiroptera : Vesperitilionidae). Can. J. Zool. 54: 1674-1678.

Bell, G.P. 1979. Summer habitat use and response to food patches by insectivorous bats in a desert community. M.Sc. Thesis, Department of Biology, Carleton University, Ottawa, Canada.

Black, H.L. 1974. A north temperate bat community: structure and prey populations. J. Mamm. 55: 138-157.

Bradbury, J.W. and L.H. Emmons. 1974. Social organization of some Trinidad bats. I Emballonuridae. Z. Tierpsychol. 36: 137-183.

Bradbury, J.W. and S.L. Vehrencamp. 1976. Social organization and foraging in emballonurid bats. I Field studies. Behav. Ecol. and Sociobiol. 1: 337-381.

Brosset, A. 1966. La biologie des chiroptères. Masson et Cie, Paris.

Brown, P.E. 1976. Vocal communication in the pallid bat, Antrozous pallidus. Z. Tierpsychol. 41: 34-54.

Buchler, E.R. 1976. The use of echolocation by the wandering shrew (Sorex vagrans). Anim. Behav. 24: 858-873.

Craighead, J.J. and F.C. Craighead. 1956. Hawks, owls and wildlife. Wildlife Management Institute.

Davis, R.B., C.F. Herreid II and H.L. Short. 1962. Mexican free-tailed bats in Texas. Ecol. Monogr. 32: 311-346.

Dunning, D.C. and K.D. Roeder. 1965. Moth sounds and the insect-catching behavior of bats. Science 147: 173-174.

Ewer, R.F. 1968. Ethology of mammals. Paul Elek Scientific Books, London.

Fenton, M.B. 1970. Population studies of Myotis lucifugus (Chiroptera : Vespertilionidae) in Ontario. Life Sci. Contr., R. Ont. Mus. no. 77: 1-34.

–   1974. The role of echolocation in the evolution of bats. Am. Nat. 108: 386-388.

–   1975. Acuity of echolocation in Collocalia hirundinacea (Aves : Apodidae), with comments on the distributions of echolocating swiftlets and molossid bats. Biotropica 7: 1-7.

Fenton, M.B. and G.P. Bell. 1979. Echolocation and feeding behaviour in four species of Myotis (Chiroptera). Can. J. Zool. 57: in press.

Fenton, M.B., D.H.M. Cumming and D.J. Oxley. 1977. Prey of Bat Hawks and availability of bats. Condor 79: 495-497.

Fenton, M.B. and T.H. Fleming. 1976. Ecological interactions between bats and nocturnal birds. Biotropica 8: 104-110.

Fenton, M.B. and J.H. Fullard. 1979. The influence of moth hearing on bat echolocation strategies. J. comp. Physiol. in press.

Fenton, M.B. and G.K. Morris. 1976. Opportunistic feeding by desert bats (Myotis spp.). Can. J. Zool. 54: 526-530.

Fenton, M.B. and D.W. Thomas. 1979. Dry season overlap in activity patterns, habitat use and prey selection by sympatric African insectivorous bats. Biotropica in press.

Fullard, J.H., M.B. Fenton and J.A. Simmons. 1979. Jamming bat echolocation: the clicks of archtiid moths. Can. J. Zool. 57: 647-649.

Gillette, D.D. and J.D. Kimbrough. 1970. Chiropteran mortality. IN About Bats, B.H. Slaughter and D.W. Walton (eds), Southern Methodist Univ. Press, Dallas, pp. 262-283.

Goldman, L.J. and O.W. Henson, Jr. 1977. Prey recognition and selection by the contrast frequency bat, Pteronotus parnellii parnellii. Behav. Ecol. Sociobiol. 2: 411-420.

Gould, E. 1971. Studies of maternal-infant communication and
     development of vocalizations in the bats Myotis and Eptesicus.
     Comm. Behav. Biol. Part A 5(5): 263-313.
     -          1975. Neonatal vocalizations in bats of eight genera.
     J. Mamm. 56: 15-29.
Griffin, D.R. 1958. Listening in the dark. Yale University Press.
     New Haven.
     -          1971. The importance of atmospheric attenuation for the
     echolocation of bats (Chiroptera). Anim. Behav. 19: 55-61.
Griffin, D.R. and R.D. Suthers. 1970. Sensitivity of echolocation
     in cave swiftlets. Biol. Bull. 139: 495-501.
Griffin, D.R., F.A. Webster and C.R. Michael. 1960. The echolocation
     of flying insects by bats. Anim. Behav. 8: 141-154.
Hamilton, W.J. III and K.E.F. Watt. 1970. Refuging. Ann. Rev. Ecol.
     Syst. 1: 263-286.
Hopkins, C.D. 1976. Stimulus filtering and electroreception:
     tuberous electroreceptors in three species of gymnotid fish.
     J. comp. Physiol. 111: 171-207.
     -          1977. Electric communication. IN How animals
     communicate. T. Sebeok (ed.), Indiana University Press,
     Bloomington, Indiana.
Howell, D.J. and D. Burch. 1974. Food habits of some Costa Rican
     bats. Revta. Biol. trop. 21: 281-294.
Humphrey, S.R. 1975. Nursery roosts and community diversity of
     Nearctic bats. J. Mamm. 56: 321-346.
Kingdon, J. 1977. East African mammals, an atlas of evolution in
     Africa, Volume IIIA, Academic Press, London.
Konstantinov, A.I. 1973. Development of echolocation in bats in
     postnatal ontogenesis. Period. Biol. 75: 13-19.
Lack, D. 1956. Swifts in a tower. Methuen, London.
Medway, Lord. 1962. The swiftlets (Collocalia) of Niah Cave,
     Sarawak. Ibis 104: 45-66.
Michelson, A. 1971. The physiology of the locust ear. II Frequency
     discrimination based upon resonances in the tympanum. Z. vergl.
     Physiol. 71: 63-101.
Miller, L.A. 1975. The behaviour of flying green lacewings, Crysopa
     carnea in the presence of ultrasound. J. Insect Physiol. 21:
     205-219.
Møhl, B. and L.A. Miller. 1976. Ultrasonic clicks produced by the
     peacock butterfly: a possible bat-repellent mechanism. J. exp.
     Biol. 64: 639-644.
Morton, E.S. 1973. On the evolutionary advantages and disadvantages
     of fruit-eating in tropical birds. Am. Nat. 107: 8-22.
Novick, A. 1977. Acoustic orientation. IN Biology of Bats, Volume
     III, W.A. Wimsatt (ed.), Academic Press, New York, pp. 74-287.
Pye, J.D. 1971. Bats and fog. Nature 229: 572-574.
Randolph, J.C. 1973. Ecological energetics of a homeothermic
     predator, the short-tailed shrew. Ecology 54: 1166-1187.
Roeder, K.D. 1967. Nerve cells and insect behavior, revised edition.
     Harvard Univ. Press, Cambridge.

–   1974. Acoustic sensory responses and possible bat
     evasion tactics of certain moths. Can. Soc. Zool. Ann. Meeting
     Proc., Univ. New Brunswick, Fredericton, pp. 71-78.

–   1975. Neural factors and evitability of insect
     behaviour. J. exp. Zool. 194: 75-88.

Sales, G. and J.D. Pye. 1974. Ultrasonic communication by animals.
     Chapman and Hill, London.

Schwartzkopff, J. 1973. Mechanoreception. IN Avian Biology, D.S.
     Farner and J.R. King (eds.), Volume 3, Academic Press,
     New York.

Schnitzler, H.-U. 1978. Die Detektion von Bewegungen durch
     Echoortung bei Fledermäusen. Verh. Dtsch. Zool. Ges. 1978:
     16-33.

Simmons, J.A., M.B. Fenton and M.J. O'Farrell. 1979. Echolocation
     and pursuit of prey by bats. Science 203: 16-21.

Simmons, J.A., M.B. Fenton, W.R. Ferguson, M. Jutting and J. Palin.
     1979. Apparatus for research on animal ultrasonic signals.
     Life Sci. Misc. Pub., R. Ont. Mus.

Suthers, R.A. 1965. Acoustic orientation by fish-catching bats.
     J. Zool. (London) 158: 319-348.

Thomas, D.W. and M.B. Fenton. 1978. Notes on the dry season
     roosting and foraging behaviour of Epomophorus gambianus and
     Rousettus aegyptiacus (Chiroptera : Pteropodidae). J. Zool.
     London 186: 403-406.

Thurrow, G.R. and H.J. Gould. 1977. Sound production in a caecilian.
     Herpetologica 33: 234-237.

Tuttle, M.D., D.E. Stevenson and A.R. Rabinowitz. 1978. Foraging
     site selection and territoriality of Myotis grisescens. Paper
     presented at the Fifth International Symposium on Bat Research,
     Albuquerque, New Mexico.

Van Valen, L. 1971. Adaptive zones and the orders of mammals.
     Evolution 25: 420-428.

Vaughan, T.A. 1977. Foraging behaviour of the giant leaf-nosed bat
     (Hipposideros commersoni). E. Afr. Wildl. J. 15: 237-249.

Vaughan, T.A. and T.J. O'Shea. 1976. Roosting ecology of the pallid
     bat, Antrozous pallidus. J. Mamm. 57: 19-42.

Vehrencamp, S.L., F.G. Stiles and J.W. Bradbury. 1977. Observations
     on the foraging behavior and avian prey of the Neotropical
     carnivorous bat, Vampyrum spectrum. J. Mamm. 58: 469-478.

Wallin, L. 1961. Territorialism on the hunting ground of Myotis
     daubentoni. Säugetierk. Mitt. 9: 156-159.

Chapter IV
Auditory Processing of Echoes

Chairman : L. Kay

co-chairmen

C. Scott Johnson and G. Neuweiler

- Underwater

Sound reception in the porpoise as it relates to echo-location.
   J.G. McCormick, E.G. Wever, S.H. Ridgway and
   J. Palin.

Behavioral measures of odontocete hearing.
   A.N. Popper

Electrophysiological experiments on hearing in odontocetes.
   S.H. Ridgway

Peripheral sound processing in odontocetes.
   K.S. Norris

Cetacean brain research : Need for new directions.
   T.H. Bullock

Important areas for future cetacean auditory study.
   C. Scott Johnson

- Airborne

Auditory processing of echoes : peripheral processing.
   G. Neuweiler

Organizational and encoding features of single neurons in the inferior colliculus of bats.
   G.D. Pollack

Auditory processing of echoes : representation of acoustic information from the environment in the bat cerebral cortex.
   N. Suga and W.E. O'Neill

SOUND RECEPTION IN THE PORPOISE AS IT RELATES TO ECHOLOCATION

James G. McCormick
Department of Otolaryngology
Bowman Gray School of Medicine
Wake Forest University
Winston-Salem, North Carolina 27103

E. G. Wever
Auditory Research Laboratories
Princeton University
Princeton, New Jersey 08540

S. H. Ridgway
Naval Ocean Systems Center
San Diego, California 92152

J. Palin
Auditory Research Laboratories
Princeton University
Princeton, New Jersey 08540

INTRODUCTION

Previous to the work of our group which was published in 1970 (McCormick, et al., 1970), all theories of hearing in Cetacea were based on dissections and experiments with dead specimens. Such experiments continue to be published in the literature to this day, and just as the earlier studies of dead material did, they only serve to confound the many students and investigators of Cetacean hearing, especially those who have little formal training in the science of physiological acoustics.

Our study was conducted with careful procedures and controls for artifacts, utilizing the electrophysiological method best suited for the study of middle and external ear mechanics of the

live animal--the A.C. cochlear potentials, which were discovered
and elucidated by our colleague Professor E. G. Wever.

As is true with most scientific works, the details of our
original paper of 1970 were read by few people, and retained by
even fewer people. Word of mouth--investigator to investigator--
has changed the context of our work so that many times we cannot
recognize our own work when we hear the "word of mouth" report
of our study at meetings. And so, many of our colleagues and
close friends have lost the essence of our publication.

On the other hand, we ourselves are all too late with the
new unpublished material which we present in this publication.
Furthermore, we have done our friends and colleagues an injustice
by not defining more specifically what we mean when we say the
porpoise ear receives sound by bone conduction. In a summation
statement, Dr. C. Scott Johnson, one of the outstanding pioneers
of the study of porpoise hearing, aptly pointed out the fact that
there are many theories and mechanisms of bone conduction in the
literature, thus presenting a semantic barrier to the understanding
of our work. In part, our new material presented here is directed
at rectifying this problem.

A second important point to be made here is that our work
supports and is in agreement with the heroic electrophysiological
research on the auditory system of the porpoise carried out by
Professors Theodore Bullock, Allen Grinnell, N. Suga, and their
colleagues in Japan (1968). Their study, although primarily
concerned with the central nervous system processing of auditory
signals by the porpoise, came before our work, and it was an
inspiration to some of our studies. We shared our new techniques
for anesthesia and major surgery on the porpoise (Ridgway and
McCormick, 1967) with Professor Bullock's group, and they in turn
shared their first electrophysiological experiences on the auditory
system of the porpoise with us.

Thirdly, this work is directed at clarifying the fact that we
find our experimental results in agreement with Professor Kenneth
Norris's theory of porpoise echolocation (1964). We feel that our
efforts, along with the work of Bullock and colleagues, support
the echo reception half of Norris's concept; and the classic work
of Bill Evans and his colleagues on the projection areas for sonar
in the porpoise supports the other half of Norris's hypothesis
(c.f. Diercks, et al., 1971).

Finally, recent work by Sam Ridgway and his colleagues has
added new credence to our speculation that the maintenance of air
over the round window membrane in the porpoise is important for
reception of sound. Ridgway's laboratory has produced evidence

for a mechanism in the porpoise which could easily carry out such
a requirement--the air pulse pressure system in the nares area of
the porpoise is directly connected to the porpoise middle ear
cavity via the eustacean tube (see Ridgway, Carder, Green, Gaunt,
Gaunt, and Evans this volume).

BONE CONDUCTION HEARING IN THE PORPOISE

Although it is generally accepted that the porpoise evolved
from a land mammal with an "air" ear (Romer, 1966), the results
reported by McCormick, et al. (1970) support the idea that the
meatus and ear drum of the porpoise no longer function in the
hearing process. Since the cochlear potential resulting from
point source vibration of the lateral meatus was not altered by
dissection and separation of the medial meatus pathway, one must
conclude that the meatus of the porpoise is no more important for
sound conduction than the surrounding tissue of the head. The
fact that relatively high pressure applications at the ear drum
failed to damp the cochlear potential argues strongly for the dis-
involvement of the ear drum with the hearing process. This disin-
volvement is also confirmed by the fine anatomical study of the
inner ear which revealed that the ear drum is not directly
connected to the malleus.

Since McCormick, et al., found that the Cetacean ear has no
direct functional or anatomical coupling between its degenerated
aerial system and the middle ear ossicular chain, they proposed
that it seemed appropriate to analyze this system in terms of
direct vibration modes of stimulation such as bone conduction and
tissue conduction. We believe that the evolution of the porpoise
ear has taken advantage of such a direct vibrational system, and
we should like to lead into a development of this thesis with a
presentation of bone stimulation theories. Wever and Lawrence
(1954) have formulated an excellent critique of bone stimulation
theories. As a springboard into an understanding of the operation
of the Cetacean ear, we take advantage of their discussion and
figures.

Two questions strike us as primary to our evolutionary theory
of the porpoise ear. Since we are proposing that the ear of the
land mammal lost its aerial function with adaptation to an aquatic
environment, how are the sensory hair cells of the cochlea stimulated in an aerial sound system, and is the stimulation phenomenon
the same for bone conduction?

The state of affairs for cochlear hair cell stimulation with
aerial sound hearing is well explained in the following quote from
Lawrence (1950, p. 1020):

"In order for the hair cells to function properly
they must be immersed in fluid. If these highly
specialized cells were like those of other tissues
in the body they could get their energy supply
from capillaries, but the very nature of these
cells activity precludes the presence of any
structure that might dilate, contract, or pulse.
The function of the hair cell is not only to be
stimulated mechanically but to be stimulated by
forces related only to the form of external
physical energy called sound. So of necessity
the hair cell is immersed in fluid from which
it can absorb nourishment without any extraneous,
disturbing influences to make it give off false
potentials. In this way the hair cell is very
happily situated; it can conveniently carry out
its metabolic processes without disturbance, and
it sits between two structures, the basilar
membrane and the tectorial membrane, so that if
the former can be made to move or vibrate a
squeezing of the hair cell will result."

Good experimental evidence has been marshalled in support
of the idea that the mode of the excitation of the cochlear endorgan in bone conduction is the same as that for aerial. The
critical studies are outlined by Wever and Lawrence (1954) and by
Hood (1962). Bekesy (1932) was the first to conclude that cochlear
excitation is the same for bone and aerial stimulation. He found
that the perception of a bone conducted tone of 400 cycles per
second could be cancelled out by aerial sound of the same frequency
but 180° out of phase. Lowy (1942) applied Bekesy's technique of
phase cancellation while measuring the cochlear potential of cats
and guinea pigs with stimulation in the frequency range of 250 to
3000 cycles per second. Cancellation obliterated the cochlear
potential, and moving the recording electrode to different positions along the cochlear boundary did not alter the cancellation
phenomenon. In a more comprehensive study, Lowy's findings were
extended and confirmed by Wever and Lawrence (1954). They found
total vectorial cancelling of air- and bone-conducted stimuli in
the frequency range of 100 to 15,000 cps.--providing that all the
stimuli were below the level of overloading. The development of
our theory leads us to the consideration of two different theories
of bone conduction--one theory for the low frequencies and another
theory for the high frequencies. Therefore it is significant to
note here that Wever and Lawrence (1954) find the same mode of
excitation of the organ of Corti for low frequency bone stimulation
as for high frequency bone stimulation.

The foregoing arguments make it clear that an evolutionary

change of the porpoise ear to a direct vibrational system does not
in itself call for a modification of the nature of hair cell stimulation found in the land mammal ear. However, I believe the case
was not so simple for evolution with regard to the mode of bone
stimulation to be utilized by the porpoise ear. There are two
different theories of bone conduction for the air mammal ear--
translational and compressional. There are pros and cons with
these two different systems which have definite significance for
the development of the function of the porpoise ear as we know it
today. To convey our point we must first explain the two theories.

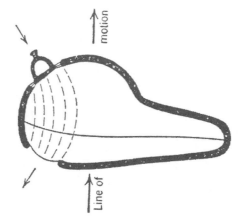

Figure 1. The translatory mode of bone conduction. When the
petrous bone is moved along the line indicated by the heavy arrows
the cochlear fluid is displaced toward the round window and the
basilar membrane is depressed. From Wever and Lawrence (1954).

The principles of translational bone conduction are depicted
in Figure 1. This mode calls for a relative movement between the
footplate of the stapes and the otic capsule. Vibration of the
skull is believed to set the otic capsule in motion relative to
the footplate of the stapes; thus the cochlear fluid is displaced
toward the round window and the basilar membrane is depressed.
This type of bone conduction exploits the same stapes-oval window
"piston-cylinder" system which functions in aerial hearing.

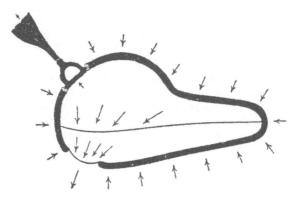

Figure 2. The compressional mode of bone conduction, as conceived
by Herzog. When the cochlear capsule is compressed from all sides
its fluid contents are pressed out, mainly taking the path of
least resistance through the round window. From Wever and Lawrence
(1954).

     Compressional bone stimulation, proposed by Herzog (1930),
does not call for a functioning of the stapes "piston-cylinder"
system. Instead it is postulated that the skull when stimulated
with a bone conduction receiver sets up compressional waves which
tend to force the fluid in the cochlear capsule out the round
window (see Figure 2). Compression of the cochlear capsule is
thought to result in fluid being forced out the round window for
two reasons. First, the mass of the stapes in the oval window on
the vestibular side of the basilar membrane functions as a load
which favors the release of the compressional pressure at the
round window where there is no loading mass. Secondly, the 5/3
vestibular to tympanic fluid ratio is believed to favor a depression of the basilar membrane towards the round window during otic
capsule compression.

     What empirical evidence supports the translational and com-
pressional theories of bone conduction? The first contribution
to an understanding of bone conduction theory was made by Bekesy
in 1932--later expanded in 1948. As illustrated in Figure 3
Bekesy discovered that when the head is driven with a bone stimulator at low frequencies it vibrates as a unit mass in a translational manner. However, as the frequency of the bone stimulation
is raised through 1000 cycles per second the skull begins to
vibrate in a compressional mode--with forehead stimulation, the
forehead and occiput began to get out of phase with one another.
At 1500 cycles per second the front and back regions of the head
were 180° out of phase.

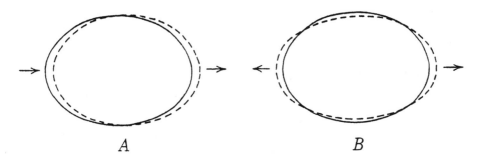

Figure 3. From Wever and Lawrence (1954): The form of motion of
the head in bone conduction according to Bekesy. For low tones
the motion is as represented at A, with the head moving as a
whole, and for high tones it is as represented at B, with opposite
surfaces moving in contrary directions.

Bekesy's experimental results call for the operation of a
translational system up to 1000 cycles per second, a combination
translational-compressional system to 1500 cycles per second, and
compressional mode alone for the high frequencies above 1500
cycles per second. In accordance with translational and compressional theories these findings predict an operational stapes
"piston-cylinder" system for low frequency bone stimulation, and
inoperation of the stapes for high frequency bone stimulation. A
few ingenious experiments demonstrate this point while further
elucidating the nature of bone stimulation:

Stacy Guild (1936), in concert with a low frequency translational theory, found 512 cycles per second bone stimulation most
effective when applied to an osseous pathway in the same direction
as the stapes oval window system. Guild determined this through
thousands of post-mortem histological examinations of temporal
bones. The temporal bones were obtained from patients in the
Johns Hopkins Hospital that had had audiometric bone and aerial
hearing tests before death.

In a consideration of the "nerve deafness" bone stimulator
tests in which a bone stimulator is applied to the side of the
head, Guild found that when all osseous pathways to the otic capsule were intact except for the sub-aditus trabeculae there was a
marked degree of impairment of hearing by bone conduction. The
sub-aditus trabeculae run between the medial part of the posterior
wall of the external auditory canal and the inferolateral aspect
of the prominence of the horizontal semicircular canal. Patients
with unmolested trabeculae did not show a deficit in bone stimulation hearing.

Most remarkable of all, Guild found a patient with chronic otitis media who heard better by bone conduction than by air conduction--this patient's sub-aditus trabeculae were not fractured.

In direct support of the operation of the stapes in low frequency bone stimulation Guild (1936, p. 747) notes the following: "The combination of anatomic arrangements is such that sound waves transmitted to the otic capsule via the subaditus trabeculae approach the organ of Corti from a direction and in a manner more nearly like sound waves through the air conduction mechanism than do those that come to the otic capsule via the other osseous pathways."

Using 435 cycles per second bone stimulation, Barany (1938) corroborated Guild's hypothesis with an entirely different experimental approach. As portrayed in Figure 4, Barany masked one ear of human subjects with white noise while he applied a point source bone stimulator to different locations of the head. By varying the phase and loudness of an aerial sound to the unmasked ear until cancellation of the bone stimulation perception occurred, Barany was able to determine the effective loudness of bone stimulation to different loci on the head. As Guild's work predicts, the most effective bone stimulation was obtained with

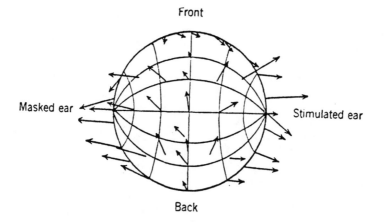

Figure 4. Effect of the point of application of a bone-conduction stimulus to the surface of the head. The head is seen from above, with the forehead at the top of the drawing. The left ear was masked and the effective loudness of the sound was measured in the right ear. The origins of the arrows represent the points of stimulation, their lengths represent the relative loudness, and their directions represent the phase relations. From Barany's work referenced in Wever and Lawrence (1954).

applications in the same direction as that in which the stapes operates.

Happily for both theories of bone stimulation, Smith (1943) has demonstrated the degree of involvement of the stapes in bone conduction from 100 cycles per second to 10,000 cycles per second. He recorded the cochlear potentials of cats, first with the stapes in its natural condition, and secondly with the stapes immobilized by a string around it connected to a pin in the bulla. The string could be released to return the stapes to its normal condition. "Translational" low frequency reduction of the cochlear potential occurred with stapes immobilization, and high frequency "compressional" cochlear potential reduction did not occur during stapes immobilization.

Smith has included in Figure 5 the changes of cochlear responses for air conduction as well as bone conduction. Since air conduction utilizes the same stapes piston system as translational bone conduction it is also reduced--adding to the evidence for stapes involvement in translational bone conduction.

Returning to the work of McCormick, et al. (1970), what now do these theories of bone conduction say for the operation of the porpoise ear? Although, as we have seen, the introduction of a land ear into an aquatic environment does not call for a new mode of stimulation of the organ of Corti, it does restrict the auditory endorgan to what we believe is a relatively inefficient method of stimulation for the high frequencies, i.e., a compressional system.

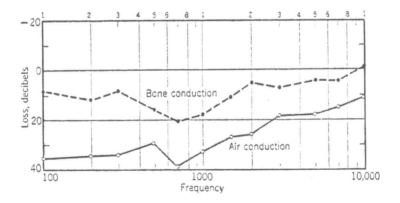

Figure 5. Changes of cochlear responses for air and bone conduction caused by a mechanical fixation of the stapes. From observations in the cat by Smith (1943).

The most effective machinery developed by the land ear for
transmission of sound energy to the inner ear is the stapes-piston
cylinder system. Even at the level of the fish it becomes
apparent that the development in evolution of a "piston-cylinder
system" is a distinct advantage for the stimulation of an auditory
endorgan. Vallancien (1963, p. 540) notes that: "Fish possessing
a Weberian apparatus respond to higher frequencies and lower
intensities than fish not so equipped."

The porpoise makes considerable use of high frequency hearing.
In order to perform this and also to exploit the piston-cylinder
system of the land mammal ear, we propose that the evolutionary
development of the porpoise ear extended the operation of translational system over the entire frequency sensitivity range of the
animal.

With regard to a compressional mode of bone stimulation,
Wever and Lawrence (1954, p. 226) have noted that "this mode of
vibration can be realized in a solid only if it is contained in a
denser solid whose flexions compress it equally from all sides."
Although these requirements are partially met in the human ear and
other land mammal ears, it is apparent to us from a consideration
of the gross anatomy of the porpoise ear that evolution has
built a system which will not respond to compressional movements
of the skull. The porpoise has a bulla which is completely
separated from the skull. The bulla is connected to the skull by
cartilage, connective tissue, and fat--thus anatomically isolating
it from the skull proper.

Less obvious, but very important for a translational system,
the bulla of the porpoise ear is much denser than the skull, and
the walls of the cochlear capsule are quite thick. Boenninghaus
(1904) is the first we know of to note that the Odontocete bulla
is much more dense than that of the land mammal.

The principles set down by Wever and Lawrence (1954, p. 266)
and the anatomical relationships just described for the porpoise
suggest acoustic isolation of the ear from the skull. The body
mapping study confirmed the acoustic isolation of the ear from
the skull: Cochlear potential readings taken with bone stimulation of the upper jaw and points over the dorsal portions of the
skull were consistently, from animal to animal, much lower than
signals elicited from sensitive areas on the lower jaw. The
vector stimulation of the teeth of the upper jaw as opposed to
that on the lower jaw also corroborated the idea of acoustic
isolation of the bulla.

In our 1970 publication we only showed a summary figure of
our auditory body mapping studies (see Figure 6). Some of the

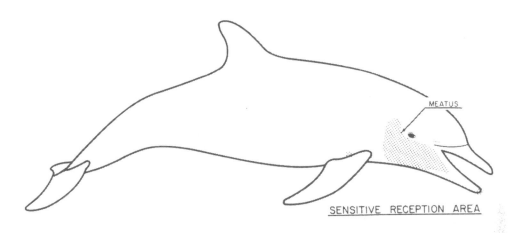

MEATUS

SENSITIVE RECEPTION AREA

Figure 6. An outline of the body of a dolphin (<u>Tursiops truncatus</u>) showing by the three dotted areas the regions of greatest sensitivity to sounds. (After McCormick, et al., 1970)

detailed data of our experiment are presented for the first time in Figures 7A, 7B, 8A, 8B, and 9.

The fact that the axis of the stapes of the porpoise was noted to project on an imaginary line through the aspect of the lower jaw which was found to be a sensitive reception area in the body mapping study suggests the operation of Barany's phenomena for the porpoise. As shown in Figures 10A and 10B the bone stimulation of the lower jaw of the porpoise was most effective when directed back toward the bulla.

Such a high frequency reproduction of Barany's results (the body mapping studies were all done with stimulation above 20kHz) would not be expected in the human since a high frequency "compressional system" would not be selective for areas of the body linked to the functioning of the stapes piston-cylinder system. The porpoise CNS auditory "body mapping" at 30 kHz of Professor Bullock's group (1968) agrees very well with our cochlear potential mapping of the porpose at 20K Hz (McCormick, et al., 1970). See Figures 11A, 11B, and Table One. We have no higher frequency data to compare with the Bullock study's 65 kHz data.

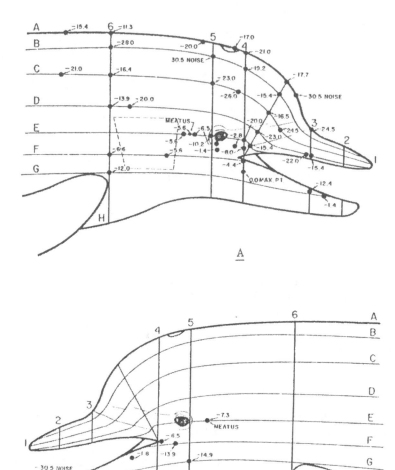

Figure 7. Cochlear potentials for 23,277 Hz point source bone stimulation plotted in db relative to the highest reading for porpoise number one (Tursiops truncatus). (A) is ipsilateral and (B) is contralateral to the round window recording electrode.

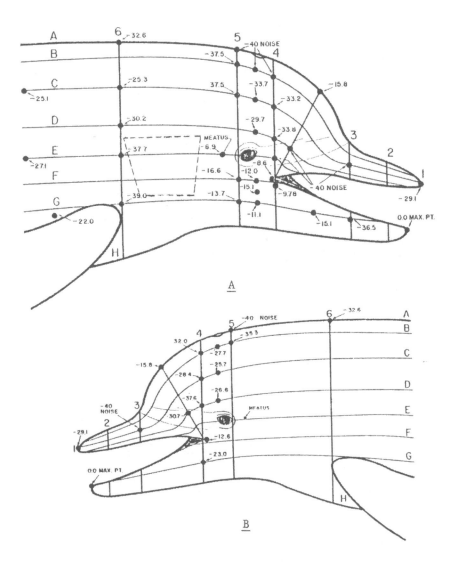

Figure 8. Cochlear potentials for 23,420 Hz point source bone stimulation plotted in db relative to the highest reading for porpoise number two (<u>Tursiops truncatus</u>). (A) is ipsilateral and (B) is contralateral to the round window recording electrode.

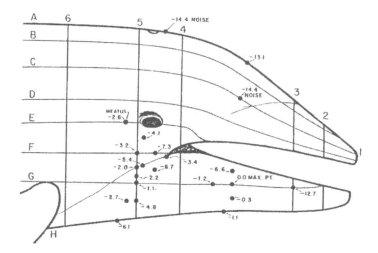

Figure 9.   Cochlear potentials for 19,762 Hz point source bone
stimulation plotted in db relative to the highest reading for
porpoise number three (Lagenorhynchus obliquidens).

Table 1.   Averages of lowest collicular evoked potential thresholds
found in Stenella to sounds presented via hydrophone pressed
against different parts of the head.   Thresholds are shown in dB
relative to the most sensitive spot on the side of the contralateral
lower jaw

	High frequencies (60 kHz/s up)	Low frequencies (17-30 kHz/s)
Under tongue	+ 9 dB	+ 21 dB
Ventral surface under lower jaws	+ 9	+ 5
Contralateral lower jaw	0	0
Ipsilateral lower jaw	− 5	+ 1
Contralateral melon	−16	− 19
Ipsilateral melon	−11	− 17
Midline of melon	−14	− 19
Contralateral external meatus	−23	− 8
Ipsilateral external meatus	−22	− 14
Blowhole	−22	− 16
Rostrum	−42	− 27
		(no sensitivity)

After Bullock, et al. (1968).

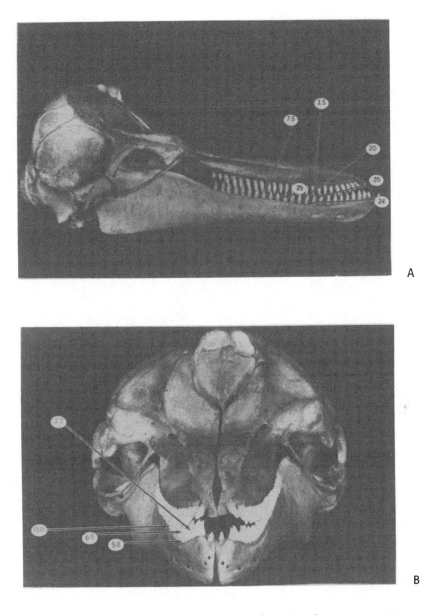

Figures 10A and 10B. Two photographs of a <u>Tursiops truncatus</u>
skull are used to illustrate the cochlear potentials resulting
from stimulating the teeth of the lower jaw of a live <u>Tursiops</u>
<u>truncatus</u>. The arrows represent the angle at which a bone stim-
ulator was applied to the teeth. The numbers in the white ovals
are the cochlear potentials (in microvolts) obtained during the
particular stimulation trial. The stimulation frequency was
20,305 Hz.

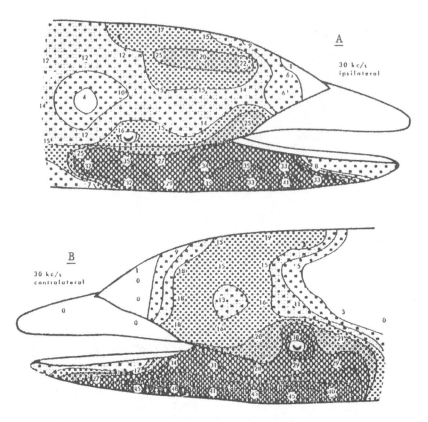

Figure 11A and 11B. Pattern of sensitivity of an individual
Stenella to sound produced by a hydrophone pressed against the
head surface at the points shown. The numerical values represent
attenuation at threshold; therefore, the largest numbers represent
greatest sensitivity. Contour lines are drawn at intervals of 5
db in sensitivity. Recording was from the inferior colliculus.
Note that sensitivity was greatest along the side of the contra-0
lateral mandible (except for under the tongue), and on the ipsilateral melon. After Bullock, et al., (1968).

    The 18 dB drop in the cochlear potential observed with damp-
ing of the ossicular chain during high frequency stimulation
certainly argues further for an operational ossicular chain at
high frequency, and thus a high frequency translational system in
the porpoise. It will be recalled that in Smith's experiment a
restriction of the ossicular chain of the cat had no effect on the
high frequency cochlear potential—as would be predicted for a
high frequency "compressional" land mammal ear.

The drops of the cochlear potential observed with damping of the bulla suggest that movement of the bulla in response to sound energy is critical for the hearing process. Since the malleus is fused to the bulla, the ossicular chain must function as a rigid spring mounted inside the tympanic cavity. When sound energy passing through the body tissues sets the bulla in motion, the inertia and stiffness of the ossicular chain probably cause the stapes to move relative to the oval window of the petrous bone-- the petrous bone being part of the moving bulla. Disconnecting the malleus of the porpoise from the bulla as we did in 1970 should have little effect on the sound reception of the porpoise since the middle ear chain would still function as an inertial mass lagging the motion of the bulla and petrous bone.

REFERENCES

Barany, E. A., 1938, Contribution to the physiology of bone con-
    duction, Acta Oto-laryngology, Supp. 26:223.
von Békésy, G., 1932, Zur theorie des hörens bei der schallaufnahme
    durch knochenleitung, Ann. Physik., 13:111.
von Békésy, G., 1948, Vibration of the head in a sound field and
    its role in hearing by bone conduction, J. Acoust. Soc. Amer.,
    20:749.
von Békésy, G., 1960, "Experiments in Hearing", McGraw-Hill, New
    York.
Bullock, T. H., Grinnell, A. D., Ikezono, E., Kameda, K., Katsuki,
    Y., Nomoto, M., Sato, O., Suga, N., and Yanagisawa, K., 1968,
    Electrophysiological studies of central auditory mechanisms
    in Cetaceans, Zeitschrift für vergleichende Physiologie, 59:
    117.
Boenninghaus, G., 1904, Das ohr des zahnwales, zugleich ein beitrag
    zur theorie der schalleitung, Zoologische Jahrbucher Abthei-
    lung für Anatomie und Ontogene der Tiere, 19:189. (English
    translation by E. G., Wever).
Diercks, K. J., Trochta, R. T., Greenlaw, C. F., and Evans, W. E.,
    1971, Recording and analysis of dolphin echolocation signals,
    J. Acoust. Soc. Amer., 49:1729.
Guild, S. R., 1936, Hearing by bone conduction: the pathways of
    transmission of sound, Annals of Otology, Rhinology, and
    Laryngology, 17:207.
Herzog, H., 1926, Das knochenleitungsproblem; theoretische erwäg-
    ungen, Zeits. f. Hals-Nasen-Ohrenheilk, 15:300.
Herzog, H., 1930, Die mechanik der knochenleitung im modellversuch,
    Zeits, f. Hals-Nasen-Ohrenheilk, 27:402.
Hood, J. D., 1962, Bone conduction: a review of the present posi-
    tion with a special reference to the contributions of Dr.
    George von Békésy, J. Acoust. Soc. Amer., 34:1325.

Lawrence, M., 1950, Recent investigations of sound conduction, part
    I. The normal ear, Annals of Otology, Rhinology, and Lar-
    yngology, 59:1020.
Lowy, K., 1942, Cancellation of the electrical cochlear response
    with air and bone conducted sound, J. Acoust. Soc. Amer.,
    14:156.
McCormick, J. G., Wever, E. G., Palin, J., and Ridgway, S. H., 1970,
    Sound conduction in the dolphin ear, J. Acoust. Soc. Amer.,
    48:1418.
Norris, K. S. 1964, Some problems of echolocation in Cetaceans,
    in: "Marine Bio-acoustics, Vol. 1", W. N. Tavolga, ed.,
    Pergamon Press, New York.
Norris, K. S., 1969, The echolocation of marine mammals, in: "The
    Biology of Marine Mammals", H. T. Andersen, Academic Press,
    New York.
Ridgway, S. H., 1972, Homeostasis in the aquatic environment, in:
    "Mammals of the Sea: Biology and Medicine", S. H. Ridgway,
    ed., Charles C. Thomas, Illinois.
Ridgway, S. H., and McCormick, J. G., 1967, Anesthetization of
    porpoises for major surgery, Science, 158:510.
Ridgway, S. H., McCormick, J. G., and Wever, E. G., 1974, Surgical
    approach to the dolphin's ear, J. Exp. Zool., 188, N° 3:265.
Ridgway, S. H., Carder, D. A., Green, R. F., Gaunt, A. S., Gaunt,
    S. L. L., and Evans, W. E., 1979, Electropyographic and
    pressure events in the nasolaryngeal system of dolphins
    during sound production, this volume.
Romer, A. S., 1966, "Vertebrate Paleontology",3rd Edition, The
    University of Chicago Press, Chicago.
Smith, K. R., 1943, Bone conduction during experimental fixation of
    the stapes, J. Exp. Psychol., 33:96.
Tavologa, W. N., 1965, Review of marine bio-acoustics, state of the
    art: 1964, Technical report: NAVTRADEVCEN, U. S. Naval Train-
    ing Device Center, Port Washington, N.Y.
Vallancien, B., 1963, Comparative anatomy and physiology of the
    auditory organ in vertebrates, in: "Acoustic Behavior of
    Animals", R. G. Busnel, ed., Elsevier Press, New York.
Wever, E. G., 1959, The cochlear potentials and their relation to
    hearing, Annals of Otology, Rhinology and Laryngology, 68:975.
Wever, E. G., and Lawrence, M., 1954, "Physiological Acoustics",
    Princeton University Press, New Jersey.
Wever, E. G., McCormick, J. G., Palin, J., and Ridgway, S. H., 1971,
    The cochlea of the dolphin, Tursiops truncatus, Proc. National
    Acad. Sciences, 68:2381.
Wever, E. G., McCormick, J. G., Palin, J., and Ridgway, S. H., 1971,
    The cochlea of the dolphin, Tursiops truncatus. I. General
    morphology, Proc. National Acad. Sciences, 68:2381.
Wever, E. G., McCormick, J. G., Palin, J., and Ridgway, S. H., 1971,
    The cochlea of the dolphin, Tursiops truncatus. II. The bas-
    ilar membrane, Proc. National Acad. Sciences, 68:2708.

Wever, E. G., McCormick, J. G., Palin, J., and Ridgway, S. H., 1971,
    The cochlea of the dolphin, Tursiops truncatus. III. Hair
    cells and ganglion cells, Proc. National Acad. Sciences,
    68:2908.
Wever, E. G., McCormick, J. G., Palin, J., and Ridgway, S. H., 1972,
    Cochlear structure in the dolphin, Lagenorhychus obliquidens,
    Proc. National Acad. Sciences, 69:657.

BEHAVIORAL MEASURES OF ODONTOCETE HEARING

Arthur N. Popper

Department of Anatomy, Georgetown University

Washington, D.C. 20007

Much of what is now known about hearing mechanisms in vertebrates has come from studies using behavioral and psychophysical
techniques to ask questions regarding auditory detection and processing. These studies have included determination of detection
capabilities for a variety of signals presented in a quiet environment and in the presence of other signals. Such studies are
valuable in determining the resolving and analysis capabilities
of the auditory system. Other investigations have been directed
at determining an animal's ability to discriminate between signals
that differ in a variety of parameters, including intensity, frequency, spectral components, position in space, or temporal relationships. Such studies provide insight into detection capabilities
of an animal, as well as provide information on mechanisms of
acoustic processing at different levels of the auditory system.

While psychophysical studies have proven particularly fruitful
in enhancing our understanding of the auditory system in terrestrial
mammals, it has not been until the past 10 to 15 years that carefully controlled investigations have been performed on odontocetes.
Certainly the technical difficulties associated with any behavioral
study of a large mammal, especially involving sound in water, have
hampered such investigations. However, information about the
auditory system of these animals is of considerable interest since
it has become clear that many odontocete species are exceptionally
well adapted for analysis of their environment using acoustic signals. The studies discussed below, in spite of very limited data
on intra- and inter-specific differences, clearly demonstrate that
the auditory system in representatives of several odontocete families is extraordinarily well adapted for the use of sound in intra-
and inter-animal communications.

AUDITORY SENSITIVITY

Measures of pure tone auditory sensitivity provide insight into the sound detection capabilities of a species, most significantly in terms of the range of frequencies that can be heard and the minimum sound level detectable (or threshold) at each of the frequencies.

Behavioral and physiological studies of auditory sensitivity have been done in a number of odontocete species, starting with investigations by Kellogg (1953) and Schevill and Lawrence (1953a, 1953b) who first demonstrated that the bottle-nosed dolphin, Tursiops truncatus, could detect sounds about 100 kHz. Subsequent to these studies, and particularly since Johnson's study in 1967, there have been a number of investigations using careful behavioral, psychophysical and acoustical techniques to measure hearing sensitivity in five odontocete species (Fig. 1). Additional studies, using electrophysiological techniques to record evoked potentials from the inferior colliculus (Bullock et al., 1968), have provided data on range of sensitivity in several additional species (broken lines in Fig. 1). Care must be taken, however, in comparing the behaviorally and physiologically determined thresholds since data determined physiologically are relative to an arbitrary threshold rather than 1 microbar (dyne/cm^2). Consequently, while it is appropriate to compare data determined with the two approaches in terms of the frequency range over which sounds can be detected, comparisons of absolute sensitivity cannot be made.

The species for which behaviorally determined data are available (Fig. 1) include the bottle-nose dolphin, Tursiops truncatus (Johnson, 1966, 1967), the Amazon River dolphin, Inia geoffrensis (Jacobs & Hall, 1972), Orcinus orca (Hall & Johnson, 1972), the common dolphin, Phocoena phocoena (Andersen, 1971), and the beluga whale, Delphinapterus leucas (White et al., 1978). Auditory sensitivity for each of these species except Phocoena were measured using similar behavioral techniques. In each case, thresholds were determined using an operant conditioning paradigm where the animal had to make a discrete and quantifiable response when it detected a sound. In each study, attempts were made to assess thresholds in a relatively quiet environment. However, it must be pointed out that Johnson (personal communication) now suggests that thresholds determined behaviorally at frequencies above 50 kHz may actually not be absolute values, but instead, may have been limited by masking thermal noise.

Although there are several trends that can be discussed regarding the data in Figure 1, it is apparent that there is substantial agreement among all species regarding the shape of the sensitivity curves. The one exception to these data is reported in a study by Supin and Sukhoruchenko (cited in Ayrapet'yants and Konstantinov,

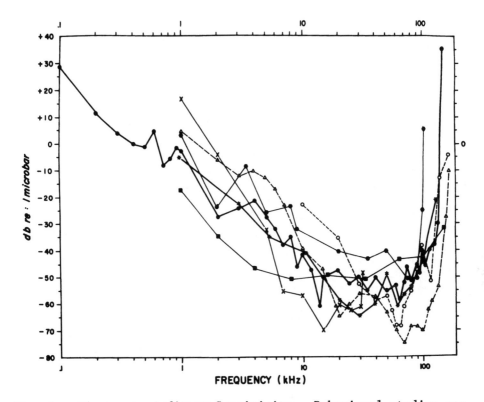

Fig. 1. Odontocete Auditory Sensitivity. Behavioral studies are indicated by a solid line and electrophysiological studies by a broken line. Symbols: * -Delphinapterus leucas (composite of two animals, White et al., 1978); ✳ Inia geoffrensi (Jacobs & Hall, 1972); × -Orcinus orca (Hall & Johnson, 1972);▇ -Phocoena phocoena (Andersen, 1971); o -Stenella caeruleoalba (Bullock, et al., 1968); Δ-Tursiops gilli (Bullock et al., 1968); ●-Tursiops truncatus (Johnson, 1967).

1974). These workers, using a "galvanic skin response" to measure sensitivity to tones, reported that Phocoena was able to detect sounds from 4 to 180 kHz with a maximum sensitivity of about -52dB (re: 1 μbar) at 128 Hz. This is in contrast to Andersen's findings (Fig. 1) of best sensitivity between 8 and 30 kHz for Phocoena, and a sensitivity of -35 dB at 128 kHz. The significant differences in the data obtained by Andersen and the Russian workers may be methodological, but this cannot be said for certain since little information is available regarding the procedures used by Supin and Sukhorunchenko.

Several points may be made regarding data on pure tone sensitivity for odontocetes. The range of frequencies detectable by all the species studied (including species investigated electrophysiologically) with the exception of Orcinus, is wider than that of non-echolocating terrestrial mammals. There are considerable inter-specific differences in auditory sensitivity among the odontocetes studied behaviorally, with Tursiops, Delphinapterus, Stenella and Phocoena having a wider range of sensitivity than Inia or Orcinus. Best sensitivity, however, is seen in Orcinus, which was able to detect signals as low as −72 dB (re: 1 μbar) at 16 kHz (Hall & Johnson, 1972). Since all studies, with the exception of one on Delphinapterus by White et al. (1978), were with a single specimen, we do not know for certain whether the inter-specific variation seen in the behavioral data reflects true species differences in sensitivity, or whether the data would be less variable were studies done with more than one specimen. Sensitivity for two specimens of Delphinapterus, one male and one female, did show some substantial differences, particularly at higher frequencies where the male could detect lower intensity sounds than the female. However in the region of best sensitivity for both animals (30–90 kHz), the female was about 4 dB more sensitive than the male (White et al., 1978). It is significant that the variability in hearing between the two specimens of Delphinapterus is substantially less than inter-specific variability between the five odontocetes shown in Figure 1. If we can extrapolate to the other species, we might tentatively suggest that intra-specific variability among odontocetes will be substantially less than the inter-specific variability seen in Fig. 1. Thus, the variation most likely represents real species differences in the ability to detect pure tones.

AUDITORY DISCRIMINATION

Behavioral studies of pure tone sensitivity provide information about the general range of sounds detectable by an animal. However, such data tell little about the actual functioning of the auditory system or about the capabilities of an animal for the detection of biological signals used in communications. Communicative signals in many vertebrates, as in dolphins, are complex and may include considerable variation in frequency, intensity, pulse patterns and duration. Only with studies of an animal's ability to handle these functions is it possible to accurately assess its ability to detect and analyze signals that hold relevance to inter- or intra-specific communications.

Frequency Discrimination

Odontocete sounds cover a broad frequency spectrum (e.g. Evans, 1973). The energy in the echolocation clicks may cover a 20–50 kHz band, with the specific spectrum varying with the species and the echolocating situation (Au et al., 1973; Evans, 1973). Communicative signals generally contain energy below 20 kHz and consist of pure tone whistles which are often frequency modulated. It has been suggested that analysis of target information during echolocation involves spectral analysis, and the differences in frequency spectra between emitted and returned signals would provide the animal with information about the nature of the target (e.g. Norris, 1969). Consequently, the ability to analyze frequency would hold considerable importance to a dolphin in echolocation, as it would in social communication where whistles with different frequency components may be of communicative significance.

Data from several experiments have demonstrated that Tursiops can discriminate small changes in frequency. These data are also supported by physiological and morphological studies showing that the peripheral and central regions of the auditory system are well adapted for frequency analysis. Studies by Bullock et al. (1968) have demonstrated that the inferior colliculus in anesthetized animals was very responsible to small changes in frequency. Investigations of the morphology and innervation of the cochlea in several different delphinid species by Wever et al. (1971, 1972) show that the cochlea is anatomically suited for good frequency analysis. Wever and his colleagues suggest that the very large number of ganglion cells innervating the cochlea in several delphinid species, particularly as compared with man (although there are about the same number of sensory hair cells in both species), is possibly indicative of a good frequency analysis system.

Direct studies of discrimination capabilities in Tursiops have been performed using behavioral measures of the discrimination limens (DL's), or the minimum difference between two frequencies that an animal can discriminate successfully at a particular frequency. The first study (Jacobs, 1972) found that the relative DL's (DL/F where F is frequency in Hertz) ranged from 0.28% to 1.4% for frequencies from 1 to 90 kHz. The best relative DL's occurred between 5 and 20 kHz where the animal could discriminate signals that were 0.3% to 0.4% apart. This work was replicated and extended by Herman and Arbeit (1972) and by Thompson and Herman (1975),who found that the best discrimination for Tursiops was for frequencies from 2 to 53 kHz where the relative discrimination limens were from 0.2% to 0.4%. Frequency discrimination was consistently better than 0.89% at all other frequencies studied except at 1 and 140 kHz where the relative DL's rose to 1.4%. Significantly, the data for the studies by Jacobs and by Herman and his colleagues on two different animals are similar to one another, with the only major difference being the range of best discrimination capabilities. This difference may possibly be explained, in part, by the poor health of Jacob's animals, which died before the completion of his study.

While not directly comparable to the data for Tursiops due to methodological differences, data on frequency discrimination have also been reported by Sukhoruchenko (1973) for Phocoena phocoena.

Sukhoruchenko reported relative discrimination limens of 0.1% to
0.29% from 3 to 225 kHz. It is striking that while these data are
similar to the data for Tursiops below about 53 kHz, the DL's for
Phocoena are essentially flat over the whole range of hearing.
This finding is contradictory to all other studies of mammalian
frequency discrimination, including those for Tursiops, where the
relative DL's function rises steeply towards the end of the animal's
range of auditory sensitivity (Fay, 1974).

Intensity Discrimination

The ability to discriminate sound intensity may be of considerable significance to odontocetes if amplitude variation of spectral components of echolocation clicks is important in target detection and discrimination. Few data are available regarding the
intensity discrimination capabilities of odontocetes, although some
data lead to the suggestion that several species may be able to
discriminate signals that differ by as little as 1 dB (Bullock
et al, 1968; Johnson, 1967), an ability which is on the same order
as that in humans (Reisz, 1928). The only study of intensity discrimination was done by Burdin et al. (1973), who determined, in
behavioral experiments, that Tursiops could discriminate between
two broad-band (50-100 kHz) white noise signals that differed in
amplitude by 4% to 7%. Experiments only indirectly related to intensity discrimination in odontocetes indicate a discrimination
ability of about 1 dB (Evans, 1973), which is comparable to data
from preliminary behavioral studies (Johnson, 1967) and physiological recordings from the inferior colliculus (Bullock et al., 1968).

Temporal Analysis

Several investigations have been performed on the ability of
Tursiops to compare sound quality and to determine the temporal
resolving ability for sounds of different durations. Johnson
(1968) measured auditory sensitivity for pure tones of different
durations and found that sensitivity in Tursiops, as in man, becomes
poorer as the duration of the pulsed pure tones is decreased below
100 to 200 msec. The data for Tursiops closely resembles that for
humans. In both species, resolution of very short signals improves
substantially when a series of short-duration signals come in rapid
succession, permitting temporal summation of the signal by the
auditory system.

Discrimination of signals of different durations has been investigated by Yunker and Herman (1974), who found that Tursiops
could discriminate between two tones that differed in duration by
as little as 8% (e.g., between 300 and 324 msec signals). While
Yunker and Herman obtained similar results for a variety of different tonal signals, they did not use any signals shorter than 300
msec in their study. Thus, it remains to be seen whether temporal

resolution would stay the same or change for signals more closely approximating the duration of echolocation-like pulses found in delphinids.

Critical Bands and Masking

Most "normal" behavioral situations involve detecting and discriminating signals in noisy environments rather than detecting signals in relatively quiet environments such as a laboratory tank (which may also contain a good deal of thermal noise about 50 kHz, resulting in masking at these frequencies) (Johnson, personal communication). Consequently, it is of considerable importance that any animal be able to detect signals in the presence of noise. In the case of odontocetes, it is also necessary that the animals be able to discriminate their own signals from those of nearby sound producing conspecifics.

The mechanism involved in discrimination of signals in noise, and for spectral analysis of signals in mammals, is generally considered to center around the concept of the critical band. This concept suggests that the mammalian auditory system is functionally analogous to a series of narrow band-pass filters (Green and Swets, 1966), each of which passes signals only within its own frequency range. This effectively eliminates noise that is not within a particular band of signals, thereby enhancing the signal-to-noise ratio around the signal. Studies of the critical band attempt to determine how finely tuned the filters in the ear are through measurement of the ability of an animal to detect the presence of a signal that is imbedded in noise. Investigations with humans have demonstrated the presence of 24 non-overlapping critical bands within the approximate frequency range from 50 kHz to 16 kHz (Scharf, 1970). These critical bands may represent distinct loci on the basilar membrane and correspond to constant distances along the membrane (Greenwood, 1961).

Johnson (1968) measured critical ratios (a function related to the critical band) (Reisz, 1928) for Tursiops by determining differences in threshold levels for selected pure tone frequencies as a function of spectrum level of the masking noise at the same frequency. He found evidence for up to 40 critical bands in Tursiops. The critical ratios for Tursiops are similar to those for the ringed seal (Pusa hispida) (Terhune and Ronald, 1975), humans (Hawkins and Stevens, 1959) and the domestic cat (Watson, 1963). However, Tursiops appears to have narrower critical ratios than these other species at frequencies where corresponding data are available.

SOUND LOCALIZATION

A fundamental auditory task is the determination of the position in space of a sound source (Erulkar, 1972; Mills, 1972) and,
in fact, it has been suggested that adaptation for sound localization has had broad significance in the evolution of the mammalian
auditory system (Masterton et al., 1969). The process of sound
localization is of particular importance for odontocetes since it
would be used in the determination of the position of other animals
or, in part, in the location of a target during echolocation.
While investigations of odontocete sound localization have been
limited, recent experiments have demonstrated an excellent ability
to do localization (McDonald-Renaud, 1974; Renaud and Popper, 1975).

Renaud and Popper (1975) studied localization in an open body
of water with an animal that was trained not to move its head during
the behavioral localization task (head movement enhances detection
abilities). Renaud and Popper measured the ability of <u>Tursiops</u> to
discriminate between targets that were coming from the animal's midline, as opposed to targets located off of the midline. This determination of minimum audible angle was done with 3 msec duration
pure tone pulses for frequencies from 6 to 100 kHz. The minimum
audible angle from 20 to 90 kHz was between 2° and 3°, indicating
that an animal could successfully discriminate between sounds that
were on its midline from sounds coming from 2-3° off of the midline.
At 6, 10, and 100 kHz, the minimum audible angle was between 2.8°
and 4.0°.

Since 3-msec duration pure tone pulses are not very similar
to echolocation sounds, Renaud and Popper (1975) measured minimum
audible angles using trains of echolocation-like clicks, each of
which was a 35 μsec transient with energy centered at 64.35 kHz.
The minimum audible angle for this signal was 0.9°, leading to the
suggestion that the broad energy spectrum and/or rapid onsets in
this signal enhanced the localization ability of the animal.

Sounds are often not located along an animal's midline, but
instead come from anywhere around the body. Consequently, Renaud
and Popper (1975) determined the minimum audible angle using sounds
centered at different points on the animal's horizontal plane.
When the center point for the signals were changed to 15° and 30°
to the side of the animal the minimum audible angle for the pure tone
pulses was 1.3 to 1.5° for the 15° azimuth, while the MAA dropped
to 5° when the azimuth was 30°, suggesting a slight improvement in
localization ability when sounds are centered away from the animal's
horizontal midline as compared to sounds on this midline.

The ability to localize sounds above and below an animal is
of particular importance for aquatic and flying animals where
behaviorally significant sounds may come from anywhere in three

dimensions. Consequently, Renaud and Popper (1975) studied the
vertical localization ability of Tursiops and found data indicating
an ability that is comparable to discrimination thresholds on the
horizontal plane. However, Renaud and Popper (1975) pointed out
that the mechanism used for localization on the vertical plane
may be different from that used for horizontal localization by
Tursiops.

It is difficult to compare the localization for Tursiops determined by Renaud and Popper (1975) with other odontocetes due to
significant methodological differences between experiments. Di-
yachenko and colleagues (cited in Ayrapet'yants and Konstantinov,
1974) reported that Tursiops could localize correctly in 90% of
trials when sound came from speakers 1.5° apart. While these were
long duration sounds, and experimental details were lacking, the
data compare reasonably well with those of Renaud and Popper.
Dudok Van Heel (1959, 1962) studied Phocoena phocoena and found
discrimination ability of 11° for a 3.5 kHz pure tone and 8° for
a 6 kHz pure tone. Andersen (1971), also using Phocoena, found a
discrimination ability of 3° for a 2 kHz pure tone. However, both
experiments with Phocoena used unrestrained animals, and stimulus
control was weak. Further, the frequencies used were far below the
range of best auditory sensitivity of Phocoena. Thus, it was
possible for these animals to "scan" the environment with their
heads during the position discrimination task, thus improving
directional discrimination.

DISCUSSIONS AND CONCLUSIONS

It is apparent, in spite of limited ata, that several species
of odontocete have exceptionally good hearing capabilities that is,
as far as is now known, rivaled by few other mammals. Since less
data are available for bats, it is still not possible to state
whether the auditory capabilities in bats and odontocetes run parallel courses. However, we can compare behavioral auditory data
between odontocetes and another group of marine mammals, the pinnipeds. Data for a number of pinniped species demonstrates excellent hearing capabilities, although no way on a par with that
of most odontocetes. For example, in measures of pure tone sensitivity, the California sea lion (Zalophus californianus) was found
to only be able to detect sounds to about 50 kHz (Schusterman et
al., 1972), and sensitivity at most frequencies was substantially
poorer than for Tursiops. While the harbor seal (Phoca vitulina)
is capable of detecting signals over almost as wide a range as
Tursiops (Møhl, 1968), sensitivity for this species, and that for
several other pinnipeds with similar hearing ranges, is substantially poorer than for Tursiops above 30 kHz. Frequency discrimination capabilities compared between Tursiops and other animals
in-water show that this species has as good, if not better, dis-

crimination ability than man below 2000 Hz (Thompson & Herman,
1975) and exceeds that of Phoca vitulina (Møhl, 1967), and Zalophus
(Moore & Schusterman, 1978) by about 2½ times. Similarly, the
ability to discriminate intensity appears to be somewhat better in
Tursiops than in Zalophus (Moore & Schusterman, 1976). While data
are scanty for the pinnipeds, it appears that the critical ratio
for Tursiops is slightly narrower than for Pusa hispida (Terhune
& Ronald, 1975). However, the differences in critical ratios between Tursiops, Pusa, and several other mammalian species may not
be significant at frequencies for which there are overlapping data
(Gurevich, 1970).

Finally, the ability to discriminate the position of a sound
source in space appears to be considerably better in Tursiops
than in other aquatic mammals including Zalophus californianus
(Moore & Au, 1975; Moore, 1975; Gentry, 1967) and Phoca vitulina
(Møhl, 1964; Terhune, 1974).

While it is not now reasonable to make any sort of conclusive
statement, it does appear that the excellent auditory capabilities
seen in many odontocetes may be related to the use of high frequency
sounds for echolocation and determination of information about the
nature of a target using acoustic reflection. It is, however, im-
perative that additional studies be undertaken with odontocetes,
making use of behavioral techniques, to extend our knowledge of
hearing abilities of these species. While replicated data on
frequency discrimination in Tursiops and on hearing sensitivity in
Delphinapterus leucas indicate that the general trend in sensitivity and discrimination shown by the data for all species is real,
it will still be of considerable value to have additional replicated
data on hearing capabilities of a single species, such as Tursiops,
as well as an extension of the data to additional species. It would
be of particular interest to obtain psychophysical data for a
species of delphinid, such as Orcinus, in which hearing and/or the
range of the frequency components of the echolocation click, are
different from that in Tursiops.

Clearly, the excellent ability to detect, and presumably
analyze sounds, as demonstrated by the behavioral data currently
available, correlates well with the sophisticated echolocation
ability of these species. Without the refined ability to detect
and analyze sounds it is highly likely that such an echolocation
ability would not be possible in these species.

ACKNOWLEDGMENTS

Portions of the work presented here were supported by the
Office of Naval Research, Contract N00123-74-C-0660. Preparation
of the manuscript was supported in part by Research Career Develop-

ment Award NS-00312 from the National Institute of Neurological and Communicative Disorders and Stroke.

REFERENCES

Andersen, S. Auditory sensitivity of the harbour porpoise Phocoena phocoena. In: Investigations on Cetacea. (Ed. by G. Pilleri) Benteli Ag. Vol. 3. pp. 255-259 (1971a).

Andersen, S. Directional hearing in the harbour porpoise Phocoena phocoena. In: Investigations on Cetacea. (Ed. by G. Pilleri) Benteli Ag. Vol. 3. pp. 260-264 (1971b).

Au, W.W.L., Floyd, R.W., Penner, R.H. & Murchison, A.E. Measurement of echolocation signals of the Atlantic bottlenose dolphin, Tursiops truncatus Montagu, in open water. J. Acoust. Soc. Amer., 56: 1280-1290 (1974).

Ayrapet'yants, E.S. & Konstantinov, A.I. Echolocation in Nature. Joint Publication Research Service. #JPS 63326-1-2 (1974).

Bullock, T.H., Grinnel, A.D., Ikezono, E., Kameda, K., Katsuki, Y., Nomoto, M., Sato, O., Suga, N. & Yanagisawa, K. Electrophysiological studies of the central auditory mechanisms in cetaceans. Z. vergl. Physiol., 59: 117-156 (1968).

Burdin, V.I., Markov, V.I., Reznik, A.M., Skoriyakov, V.M. & Chupakov, A.G. Determination of the differential intensity threshold for white noise in the bottlenose dolphin (Tursiops truncatus Barabasch). In: Morphology and Ecology of Marine Mammals (Ed. by K.K. Chapskii & V.E. Sokolov) Wiley, NY p. 112 (1973).

Dudok van Heel, W.H. Audio-direction findings in the porpoise Phocoena phocoena. Nature 183: 1063 (1959).

Dudok van Heel, W.H. Sound and cetacea. Netherlands J. Sea Res. 1, 407-507 (1962).

Erulkar, S.D. Comparative aspects of spatial localization of sound. Physiol.Rev., 52: 237-260 (1972).

Evans, W.E. Ecolocation by marine delphinids and one species of freshwater dolphin. J. Acoust. Soc. Amer., 54: 191-199 (1973).

Fay, R.R. Auditory frequency discrimination in vertebrates. J. Acoust. Soc. Amer., 56: 206-209 (1974).

Gentry, R.L. Underwater auditory localization in the California sea lion (Zalophus californianus). J. Aud. Res., 7: 187-193 (1967).

Gourevitch, G. Detectability of tones in quiet and in noise by rats and monkeys. In: Animal Psychophysics (Ed. by W.C. Stebbins) Appleton-Century-Crofts, NY pp. 67-98 (1970).

Green, D.M. & Swets, J.A. Signal Detection Theory and Psychophysics. Wiley, NY (1966).

Greenwood, D.N. Critical bandwidths and the frequency coordinates of the basilar membrane. J. Acoust. Soc. Amer., 33: 1344-1356 (1961).

Hall, J.D. & Johnson, C.S. Auditory thresholds of a killer
   whale Orcinus orca Linnaeus. J. Acoust. Soc. Amer., 51:
   515-517 (1972).

Hawkins, J.E. & Steven, S.S. The masking of pure tones by white
   J. Acoust. Soc. Amer., 22: 6-13 (1950).

Herman, L.M. & Arbeit, W.R. Frequency discrimination limens in
   the bottle-nose dolphin: 1070 KC/S. J. Aud. Res., 12: 109-120
   (1972).

Jacobs, D.W. Auditory frequency discrimination in the Atlantic
   bottlenose dolphin, Tursiops truncatus Montagu: A preliminary
   report. J. Acoust. Soc. Amer., 52: 696-698 (1972).

Jacobs, D.W. & Hall, J.D. Auditory thresholds of a freshwater
   dolphin, Inia geoffrensis Blainville. J. Acoust. Soc. Amer.,
   51: 530-533 (1972).

Johnson, C.S. Auditory thresholds of the bottlenose dolphin
   (Tursiops truncatus Montagu). U.S. Naval Ordinance Test
   Station NOTSTP 4178, 25 pp. (1966).

Johnson, C.S. Sound detection thresholds in marine mammals.
   In: Marine Bio-Acoustics II. (Ed. by W.N. Tavolga) Pergamon
   Press, NY pp. 247-260 (1967).

Johnson, C.S. Masked tonal thresholds in the bottlenosed porpoise.
   J. Acoust. Soc. Amer., 44: 965-967 (1968).

Kellogg, W.N. Ultrasonic hearing in the porpoise, Tursiops
   truncatus (Mont.). Aquatic mammals, 4: 1-9 (1953).

Masterton, B., Heffner, H., & Ravizza, R. The evolution of
   human hearing. J. Acoust. Soc. Amer. 45: 966-985 (1969).

McDonald-Renaud, D.L. Sound localization in the bottlenose
   porpoise, Tursiops truncatus (Montagu). Ph.D. Thesis, University
   of Hawaii (1974).

Mills, A.W. Auditory localization. In: Foundations of Modern
   Auditory Theory (Ed. by J.V. Tobias) Academic Press, NY Vol.
   II. pp. 303-348 (1972).

Möhl, B. Preliminary studies on hearing in seals. Vidensk.
   Medd. fra Dansk Naturh. Foren., 127: 283-294 (1964).

Möhl, B. Frequency discrimination in the common seal and a
   discussion of the concept of upper hearing limit. In: Underwater Acoustics. (Ed. by V.M. Alberts) Plenum, NY Vol. 2.
   pp. 43-54 (1967).

Möhl, B. Auditory sensitivity of the common seal in air and
   water. J. Aud. Res., 8: 27-35 (1968).

Moore, P. Underwater localization of click and pulsed pure tone
   signals by the California sea lion (Zalophus californianus).
   J. Acoust. Soc. Amer., 57: 406-410 (1975).

Moore, P. & Au, W. Underwater localization of pulsed pure tones
   by the California sea lion (Zalophus californianus). J. Acoust.
   Soc. Amer., 58: 721-727 (1975).

Moore, P. & Schusterman, R.J. Discrimination of pure tone intensities by the California sea lion. J. Acoust. Soc. Amer.,
   60: 1405-1407 (1976).

Reisz, R.R. Differential sensitivity of the ear for pure tones.
Phys. Rev., 31: 867-875 (1928).

Renaud, D.L. & Popper, A.N. Sound localization by the bottlenose
Porpoise, Tursiops truncatus. J. Exp. Biol., 63: 569-585 (1975).

Scharf, B. Critical bands. In: Foundations of Modern Auditory
Theory (ed. by J.V. Tobias) Academic Press, NY Vol. I. pp.
157-202 (1970).

Schevill, W.E. & Lawrence, B. Auditory response of a bottlenosed
porpoise, Tursiops truncatus, to frequencies above 100 KC.
J. Exp. Zool., 124: 147-165 (1953a).

Schevill, W.E. & Lawrence, B. High-frequency auditory responses
of a bottlenosed porpoise, Tursiops truncatus (Montagu).
J. Acoust. Soc. Amer., 25: 1016-1017 (1953b).

Schusterman, R.J., Balliet, R.F. & Nixon, J. Underwater audiogram
of the California sea lion by the conditional vocalization
technique. J. Exp. Anal. Behav., 17: 339-350 (1972).

Schusterman, R. J. & Moore, P. A. The upper limit of underwater
auditory frequency discrimination in the California sea lion.
J. Acoust. Soc. Amer., 63: 1591-1595 (1978).

Sukhoruchenko, M.N. Frequency discrimination in dolphins (Phocoena
phocoena). Fiziol. AH. SSR. IM. M. Sechenova, 59: 1205-1210
(1973).

Terhune, J.M. Directional hearing of a harbor seal in air and
water. J. Acoust. Soc. Amer., 56: 1862-1865 (1974).

Terhune, J.M. & Ronald, K. Masked hearing thresholds of ringed
seals. JH. Acoust. Soc. Amer., 58: 515-516 (1975).

Thompson, R.K.R. & Herman, L.M. Underwater frequency discrimination in the bottlenosed dolphin (1-140 kHz) and human (1-
8 kHz). J. Acoust. Soc. Amer., 57: 943-948 (1975).

Watson, C. Masking of tones by noise for the cat. J. Acoust.
Soc. Amer., 35: 167-172 (1963).

Wever, E.G., McCormick, J.G., Palin, J. & Ridgway, S.H. The
cochlea of the dolphin Tursiops truncatus: General morphology.
Proc. National Acad. Sci., 68: 2381-2385 (1971).

Wever, E.G., McCormick, J.G., Palin, J. & Ridgway, S.H. Cochlear
structure in the dolphin, Lagenorhynchus obliquidens. Proc.
National Acad. Sci., 69: 657-661 (1972).

Yunker, M.P. & Herman, L.M. Discrimination of auditory temporal
differences in the bottlenose dolphin and by the human.
J. Acoust. Soc. Amer., 56: 1870-1875 (1974).

ELECTROPHYSIOLOGICAL EXPERIMENTS ON

HEARING IN ODONTOCETES

Sam H. Ridgway

Naval Ocean Systems Center

San Diego, Ca.   92152   U.S.A.

INTRODUCTION

Although electrophysiological techniques have not been extensively employed in the study of cetacean hearing, there are more
than a dozen significant studies that will be briefly reviewed in
this paper. These techniques have been useful for measuring hearing thresholds by evoked responses and classical conditioning
(several animals each of Phocoena and Tursiops). The auditory
cortex has been located at the dorsum of the cerebral hemisphere
only 1.5 to 3.0 cm lateral to the midline and adjacent to visual
and somatosensory cortex. The midbrain auditory structures of
dolphins are specialized for ultrabrief, ultrasonic, fast-rising,
closely spaced sounds like echolocation clicks, and temporal resolution is rapid. Frequency-modulated tones produce maximum responses
both at midbrain locations and at cortical recording sites, suggesting the importance of such signals to the animals. Methods and
equipment are now available for broader application of electrophysiological techniques without expenditure of valuable animals.
These techniques can contribute further insight concerning how
odontocetes process sound and what characteristics of the acoustic
information package are most important to the animal.

Previous Studies

Among cetaceans only the small toothed whales, commonly called
dolphins or porpoises, have been studied by electrophysiological
techniques. The advent of safe and effective anesthesia techniques
(Ridgway and McCormick, 1967) promised to open the way for a
thorough physiological assessment of the unusually large brains of

these aquatic mammals. However, because of several factors, including protective legislation and economics, many types of investigations, such as single-unit recording and mapping of the cerebral
cortex, have not been possible except in the Soviet Union.

Since much of the Soviet literature is not generally available
in the West, Bullock and Gurevich (1979) have recently completed a
review. Ladygina and Supin (1970) published the first Soviet investigations in this field. They studied the auditory evoked potential
(AEP) with electrodes placed in small holes drilled in the skull of
unanesthetized dolphins. Since 1970, the Soviet investigators have
produced an impressive list of reports, primarily from the laboratory
of A. Ya. Supin. Much of this is apparently reviewed and updated in
a recent book by Supin, Mukhametov, and Ladygina (1978) that is
not yet available in English translation. This paper will mention
a few of the Soviet studies that appear to be of major significance
and will also discuss some Western studies that predate most of the
Soviet work.

LITERATURE REVIEW AND COMMENTS

Bullock et al (1968) were the first to employ electrophysiological techniques in the study of cetacean hearing. They studied the
AEP from midbrain auditory structures, and among other things they
provided the first physiological evidence for the lower jaw sound
path postulated by Norris (1964). This is a landmark paper that
deserves careful reading. Conclusions of major importance are as
follows: 1) Temporal resolution of successive sounds was found to
be extremely rapid. 2) There was no evidence of facilitation of
response to the second of two identical stimuli as had been observed
in echolocating bats. 3) Dramatic changes in AEP waveform, but not
amplitude, were produced by distortion of the sound field by holding a sheet of paper between the sound source and the lower jaw.
4) Gentle stroking of the lower jaw or splashing water on the jaw
masked response to sounds, but electrical stimulation of the skin
of the jaw did not produce responses or mask AEPs. 5) Small changes
in stimulus frequency resulted in changes of AEP amplitude and waveform. 6) Frequency-modulated (FM) tones were more likely to produce
a large AEP than were constant frequency tones. In one case, a 100-
kHz tone produced a weak AEP, while a 135-kHz tone produced none.
An FM pulse starting at 100 kHz and sweeping upward to 135 kHz
caused a large AEP. The effect was also present for tones sweeping
down in frequency. 7) This investigation showed that the AEP audiogram was similar to the behavioral curve reported by Johnson (1966)
for Tursiops, but with a slightly lower upper limit of about 135
kHz. 8) An AEP to a given tone pip could be masked only by a relatively narrow band of frequencies surrounding that stimulus (Bullock
et al, 1968).

Bullock and Ridgway (1971 and 1972) recorded and telemetered
electrical activity from the inferior colliculus or nucleus of the
lateral lemniscus and from a few locations in the cerebral cortex.
The recording electrodes were placed stereotaxically with the
animal under anesthesia. The experimental animals were trained
to emit on command a series of clicks at a rate of 15 to 300 per
second. (These clicks are similar to those used in echolocation.)
The voluntary clicks resulted in AEPs. However, the clicks of
highest intensity often evoked quite modest AEPs while clicks of
much weaker intensity gave maximal AEPs, suggesting important
differences in click composition. Artificial echoes showed recovery
to be more rapid (0.5 ms including the duration of the click)
following the animal's own click than following an artificial conditioning tone burst of equal evoking power.

The typical midbrain AEP was found to be specialized for ultrasonic, ultrabrief, fast-rising, closely spaced sounds like the echolocation clicks. No potential was evoked if the frequency was below
5 kHz or if the rise time was above 5 ms. At several cerebral locations (mainly in the posterior lateral temporal cortex), however,
long latency, long duration, slowly recovering potentials were
evoked by frequencies below 5 kHz, whether fast or slowly rising
acoustic envelopes were used (Bullock and Ridgway, 1972).

Another electrophysiological technique, cochlear potential
recording, was employed by McCormick et al (1970) in studies of
sound conduction in the dolphin ear. Since a study employing cochlear potentials will be discussed by McCormick in this volume, the
subject will not be further mentioned.

Lende and Akdikmen (1968) and Lende and Welker (1972) did some
limited investigations on cortical mapping in Tursiops. A small
somatic motor area was found in the frontal pole (Lende and Akdikmen, 1968). These investigators, however, found it necessary to
remove most of the melon and much of the blowhole and nasal sac
system for access to frontal cortex. Since these ablated structures
probably require fine motor control to produce the varied complex
sounds of odontocetes (see Ridgway et al this volume), the nasal
system will probably occupy a large area of motor cortex. Thus,
its removal precludes studies of its representation.

Ladygina and Supin (1970 and 1977) used AEPs to locate sensory
projection zones in the cerebral cortex of Phocaena and Tursiops.
They reported that visual and acoustic regions in both species are
shifted to the dorsal surface of the hemisphere and that all projection zones are adjacent to each other (cortical projection areas
forming a single complex are also found in Monotremata and Edentata).
The auditory cortex was found approximately 1.5 to 3.0 cm lateral
to the sagittal suture of the skull. Thus, as compared to other
mammals, there is an apparent shifting of the auditory area from

the temporal to the parietal lobe. Zanin et al (1978) duplicated
some of these findings. It should be mentioned, however, that a
tonotopic map of the cochlea, based on sound frequency areas on the
cortex, has not been done. Additional areas of auditory sensitivity
may be found in temporal cortex which is less accessible in cetaceans (Cf. Bullock and Ridgway, 1972).

Popov and Supin (1976 and 1978) in studies of Phocaena and
Tursiops concentrated on changes in AEPs from the auditory cortex
produced by varying the characteristics of sound stimulation. They
found distinct changes in the cortical AEPs in response to switching
the sound on and off compared to changes in the duration of the
signal and to changes in the frequency of the signal. The threshold
for AEPs in the auditory cortex of Tursiops was found to be close
to the behavioral hearing threshold.

Voronov and Stosman (1977) studied the frequency-threshold
characteristics of subcortical elements of the central auditory
system of Phocaena; they used 200-μs clicks and 5-ms tones. Both
"on" and "off" responses were measured for the tone stimuli. In
all structures examined, maximum sensitivity was found in the 100-
to 120-kHz range (with preception extending to 200 kHz), the same
as that found by Sukhoruchenko (1971) using classical conditioning
of the galvanic skin response (GSR). The frequency bandwidth at a
level 10 dB over the lowest thresholds was 40 to 50 kHz in the
cochlear nuclei, 70 to 90 kHz in trapezoid body, 70 to 80 kHz in
Superior olive, 45 to 55 kHz in lateral lemniscus, and 35 to 45 kHz
in inferior colliculus and medial geniculate. There was an additional maximum of sensitivity from 20 to 80 kHz in the lateral
lemniscus (bandwidth 30 to 40 kHz), inferior colliculus (bandwidth
40 to 50 kHz), and medial geniculate (bandwidth 40 to 50 kHz).
"Off" responses had a lower threshold up to about 150 kHz. From
150 to 200 kHz both "on" and "off" responses had the same threshold,
which showed an elevation of 10 to 20 dB per 10 kHz in frequency
increase.

Supin and Sukhoruchenko (1970 and 1974) and Sukhoruchenko
(1971 and 1973) used a classical conditioning method to measure
hearing thresholds. After restraining the experimental animal with
straps in a small rubber-lined tank, the investigators recorded
respiration, heart action, and the GSR. They paired a tone (the
conditioned stimulus) with an electric shock (the unconditioned
stimulus). Ten to fifteen pairings of the tone and shock were
usually sufficient to develop a conditioned reflex (respiratory
pause or GSR). Every tenth to fifteenth tone was reinforced, bringing about a stable response that did not extinguish. By attenuating
the tone and noting the disappearance of the response, a complete
audiogram was done in a single experiment or within two or three
experiments. This made it possible to test a reasonable number of
experimental animals.

The technique was extended (Supin and Sukhoruchenko, 1974) to measure differential frequency thresholds. An FM tone at an intensity approximately 40 dB higher than the threshold of audition in the frequency band under test was used as the conditioned stimulus. After a conditioned reflex was developed to FM tones as distinct from unmodulated tones, the range of the frequency modulation was lowered until the reflex disappeared (threshold). In Phocaena, the differential frequency threshold was found to be 10 Hz at 3 kHz increasing to 300 Hz at 190 kHz. Above 22 kHz, the differential thresholds were approximately constant at 0.1 to 0.2% at any given frequency. Thompson and Herman (1975) using behavioral techniques found values as low as 0.2 to 0.3% between 6 and 20 kHz, but at 140 kHz their value (for Tursiops) was 1.4% (seven times coarser than the Soviet findings) and no responses were obtained above 150 kHz. These findings are especially noteworthy since they provide strong evidence for frequency analysis over the entire range of hearing in Phocaena and Tursiops.

Recent discoveries have shown that electroencephalographic (EEG) recordings made in the far field (i.e., scalp in humans, vertex of the skull, mastoid process) can be used to obtain AEPs that reflect activity even at the brain stem level. Thus, a great deal of information about different levels of the auditory system can be collected without placing electrodes in the brain (Galambos and Hecox, 1977).

Seeley et al (1976) used such recordings in a technique for rapidly assessing the hearing of Tursiops. A single monopolar electrode was placed into the vertex of the skull 5.8 cm posterior

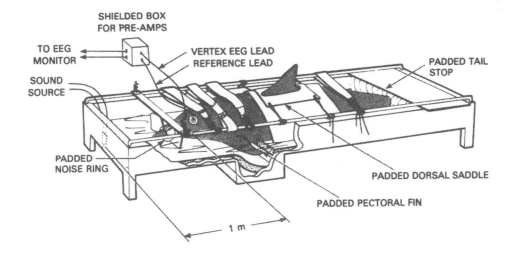

Figure 1. Tursiops positioned in restraint apparatus to assess hearing using the AEP (Seeley et al, 1976).

to the caudal margin of the blowhole. A reference electrode was
built into a suction cup (7.5 cm in diameter) and attached with
active vacuum to the skin surface of the head rostral to the first
electrode. The experimental animal was restrained with straps in
a foam, plastic-padded box. The box was partially filled with
water that just covered the dorsal surface of the snout (figure 1).
The animal was kept wet by frequent misting from a garden sprayer
during recording sessions of 4 to 7 hr.

Because of the proximity of the massive nasal muscles to the
brain (see Green et al this volume) and to the recording electrodes,
each breath at average intervals of approximately 20 s resulted in
artifact several times larger than the EEG. In addition, other
gross body movements produced large artifact. Seeley et al (1976)
electronically removed this movement artifact by using an artifact-
inhibiting system (figure 2). The system provided a 13.5-s delay
between data recording and computer analysis of average transients.
In this time period the stimulus epochs containing artifact were
eliminated, making it possible to obtain an AEP with as few as 16
stimuli; however, 48 stimuli were more commonly used (figure 3).

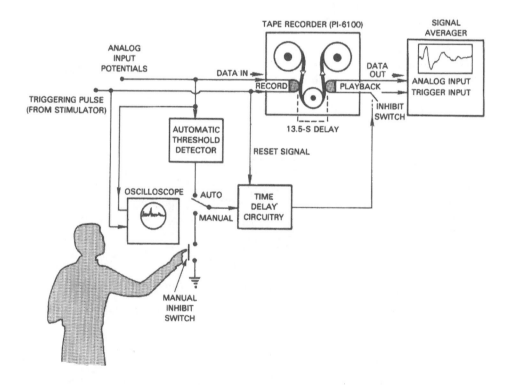

Figure 2.   Artifact-inhibiting system used by Seeley et al (1976).

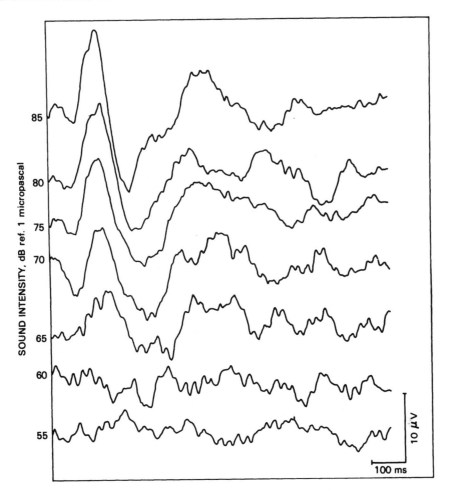

Figure 3. Effect of change in stimulus intensity on auditory
evoked potentials in an unanesthetized Tursiops. Each potential
represents an average of 48 sound stimuli at 50 kHz. Upward
deflection indicates negativity of the vertex relative to the
reference electrode (Seeley et al., 1976).

   This technique was used to obtain audiograms from seven
Tursiops of both sexes and various ages. Most of these audiograms
were similar to the behavioral curve of Johnson (1966), but one
31-year-old female was virtually deaf, showing a 50 to 70 dB higher
threshold over most of the frequency range.

   Ridgway and Seeley (1979) used implanted telemetry with electrodes positioned over the dura at the vertex and 9.0 cm lateral
over temporal cortex to record an EEG continuously from an unrestrained dolphin. Four different sound stimuli were used for AEP
studies: 1) a 500 ms tone with slow (10 ms) rise and fall,

2) a pulse train with the same envelope as the tone, i.e., 10-ms rise and fall, containing 100 20-μs pulses 5 ms apart, 3) a single 20-μs pulse, and 4) an FM tone sweeping up in frequency from 20 to 130 kHz (slow rise and fall). Each stimulus was presented at intervals of 5 s, and 64 responses were averaged for each plot.

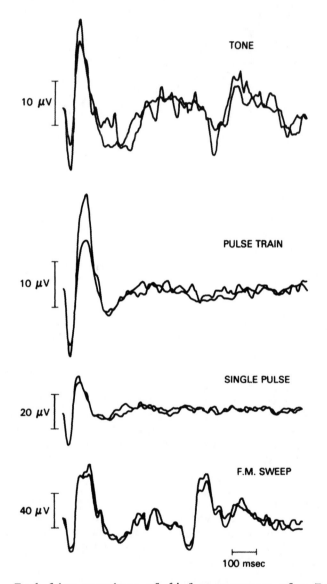

Figure 4.   Each line consists of 64 1-s averages of a _Tursiops_ EEG in response to a different sound stimulus. Upward deflection indicates negativity of the vertex relative to an electrode over the right temporal cortex (Ridgway and Seeley, 1979).

All of the responses began with a positive component peaking at about 30 ms followed by a negative component at about 100 ms. The tone burst and FM stimuli showed a strong "off" response after the stimuli at 500 to 600 ms. It is interesting to note that the pulse train of the same duration did not show this response. The AEPs produced by FM sweeps were much larger than with the other stimuli employed. These recordings from the cortex are consistent with the earlier findings of Bullock et al. (1968) at the subcortical levels. Many dolphin whistles are long FM sweeps, and the consistent findings that this type of signal produces large responses indicates that the FM signals are probably very important to the animals. Thus, they perhaps deserve intensive studies equivalent to the attention given the brief clicks known to be important in echolocation.

FUTURE STUDIES

Future studies at NOSC will employ improvements in the techniques reported by Seeley et al. (1976) for recording AEPs at various levels of the central auditory system of odontocetes. Auditory processing at the brain stem reflected by the averaged brainstem response, sensory processing at the thalamus and cortex reflected by longer latency AEPs, and cognitive processing that may be reflected in endogenous potentials such as the P300 or "Ah Ha" response are three major areas of study that could yield considerable insight concerning how the odontocetes process sound and what characteristics of the acoustic information package are important to the animal.

REFERENCES

Bullock, T. H., Grinnell, A. D., Ikezono, E., Kamedo, K., Katsuki, Y., Nomoto, M., Sato, O., Suga, N., and Yanagisawa, K., 1968, Electrophysiological studies of central auditory mechanism in cetaceans, Zeitschrift für vergleichende Phys. iologie, 59:117.

Bullock, T. H., and Gurevich, V. S., 1979, Soviet literature on the nervous system and psychobiology of cetaceans, International Review of Neurobiology, (in press).

Bullock, T. H., and Ridgway, S. H., 1971, Evoked potentials in the auditory system of alert porpoises (Cetacea) and sea lions (Pinnipedia) to their own and to artificial sounds, Proc. Int. Un. Physiol., IX:89.

Bullock, T. H., and Ridgway, S. H., 1972a, Neurophysiological findings relevant to echolocation in marine mammals, in: "Animal Orientation and Navigation", S. R. Galler, K. Schmidt-Koenig, G. J. Jacobs, R. E. Belleville, eds., NASA, U.S. Govn't Printing Off., Washington, D.C.

Bullock, T. H. and Ridgway, S. H., 1972b, Evoked potentials in the central auditory system of alert porpoises to their own and artificial sounds, J. Neurobiol., 3:79.

Galambos, R. and Hecox, K, 1977, Clinical applications of the brain stem auditory evoked potentials in man,"Psychopharmacology Correlates of Evoked Potentials", Vol. 2, J. E. Desmedt, ed., Krager, Basel.

Johnson, C. S., 1966, Auditory thresholds of the bottlenosed porpoise (Tursiops truncatus Montagu), N.O.T.S. TP 4178.

Ladygina, T. F., and Supin, A. Y., 1970, The acoustic projection in the dolphin cerebral cortex, Fiziol. Zh. SSSR im. I. M. Sechenova, 56:1554.

Ladygina, T. F., and Supin, A. Y., 1977, Localization of sensory projection zones in the cerebral cortex of the bottlenosed dolphin, Tursiops truncatus, Zh. Evolyutsionnoy Biokhimii I Fiziol.

Lende, R. A., and Akdikmen, S., 1968, Motor field in the cerebral cortex of the bottlenosed dolphin, J. Neurosurgery, 29:495.

Lende, R. A., and W. I. Welker, 1972, An unusual sensory area in the cerebral neocortex of the bottlenose dolphin, Tursiops truncatus, Brain Research, 45:555.

McCormick, J. G., Wever, E. G., Palin, J., and Ridgway, S. H., 1970, Sound conduction in the dolphin ear, J. Acoust. Soc. Amer., 48:1418.

Norris, K. S., 1964, Some problems of echolocation in cetaceans, in: "Marine Bio-Acoustics", W. N. Tavolga, ed., Pergamon Press, New York.

Popov, V. V., and Supin, A. Y., 1976, Determination of the hearing characteristics of dolphins by measuring induced potentials, Fiziol. Zh. SSSR im. I. M. Sechenova, 62:550.

Popov, V. V., and Supin, A. Y., 1976, Responses of the dolphin auditory cortex to complex acoustic stimuli, Fiziol. Zh. SSSR im. I. M. Sechenova, 62:1780.

Popov, V. V., and Supin, A. Y,, 1978, Electrophysiological studies of the auditory system of the Tursiops truncatus, in: "Morskiye Mlekopitayushchiye. Resul'taty i Metody Issledovan- iya",Izdatel'stvo Nauka, Moscow.

Ridgway, S. H., Carder, D. A., Green, R. F., Gaunt, A. S., Gaunt, S. L. L., and Evans, W. E., 1979, Electromyographic and pressure events in the nasolaryngeal system of dolphins during sound production, this volume.

Ridgway, S. H., and McCormick, J. G., 1967, Anesthetization of porpoises for major surgery, Science, 158:510.

Ridgway, S. H., and Seeley, R. L., 1979, Auditory evoked responses of the dolphin cortex to four different sound stimuli, in preparation.

Seeley, R. L., Flanigan, W. F., and Ridgway, S. H., 1976, A tech- nique for rapidly assessing the hearing of the bottlenosed porpoise Tursiops truncatus, Naval Undersea Center TP 552, San Diego, California.

Sukhoruchenko, M. N., 1971, Upper limit of hearing of dolphins
        with reference to frequency, Tr. Akust. Inst. Moscow, 17:54.
Sukhoruchenko, M. N. 1973, Frequency discrimination of dolphin
        (Phocoena phocoena), Fiziol. Zh. SSSR im. I. M. Sechenova,
        59:1205.
Supin, A. Y. and Sukhoruchenko, M. N., 1970, The determination of
        auditory thresholds in Phocoena phocoena by the method of
        skin galvanic reaction, Tr. Akust. Inst. Moscow, 12:194.
Supin, A. Y., and Sukhoruchenko, M. N., 1974, Characteristics of
        acoustic analyzer of the harbor porpoise Phocoena phocoena,
        in: "Morfologiya, Fiziologiya i Akustika Morshkikh Mlekopita-
        yushchikh", V. Y. Sokolov, ed., Izdatel'stvo Nauka, Moscow.
Supin, A. Y., Mukhametov, L. M., Ladygina, R. F., Popov, V. V.,
        Mass, A. M., and Polyakova, I. G., 1978,"Electrophysiological
        Studies of the Dolphin's Brain", V. E., Sokolov, ed.,
        Izdatel'stvo Nauka, Moscow.
Thompson, R. K. R., and Herman, L. M., 1975, Underwater frequency
        discrimination in the bottlenose dolphin (1-140 kHz) and in
        human (1-8 kHz), J. Acoust. Soc. Amer., 57:943.
Voronov, V. A., and Stosman, I. M., 1977, Frequency-threshold
        characteristics of subcortical elements of the auditory
        analyzer of the Phocoena phocoena porpoise, Zh. Evolyutsionnoy
        Biokhimii I Fiziol., 6:719.
Zanin, A. V., Bibikov, N. G., Vodyanaya, E. G., 1978, Evoked pot-
        entials of the Tursiops under the stimulation with short
        sound signals, in: "VII-aya Vsesoyuznayz Konferentsiyz po
        Morskim Mlekopitayushchim", Simpheropol'.

PERIPHERAL SOUND PROCESSING IN ODONTOCETES

Kenneth S. Norris

Coastal Marine Laboratory
University of California - Environmental Studies
Santa Cruz, California 95064, U.S.A.

A decade ago two divergent theories describing peripheral sound processing in odontocetes had been formulated. Their differences were unresolved. One view held that sounds were generated in the larynx, radiated through the soft tissue of the throat or transmitted up through the skull and rostrum. Reception was thought restricted to the region of the external auditory meatus, a pinhole in most odontocetes. Sounds gathered at this region were thought to be transmitted to the middle ear via the narrow and sometimes occluded external auditory canal and the tympanic ligament (Fraser and Purves, 1959 ; Purves, 1966). The second view suggested quite radical modification of both sound reception and transmission paths. Sounds were envisioned as being produced by extra-laryngeal structures adjacent to the nasal passages in the forehead, and transmitted and transduced into the water via the fatty melon of the odontocete forehead, or through the mesorostral canal of the odontocete snout. The external ear canal was viewed as non-functional, at least for high frequency sounds, and instead sound was postulated to enter the lower jaw through a thinned posterior portion of the mandible (the acoustic window, or pan bone) and to be transmitted via the fat filled mandibular canal to the tympan-operiotic complex of the middle ear. (Evans and Prescott, 1962 ; Norris, 1964 ; Norris, 1968). These modifications were presumed to have answered the multiple problems attendant upon the re-entry of a terrestrial mammal into the aquatic environment. Major selective pressures involved in such change resulted from partial loss of sensitivity, partial loss of binaurally based directionality, and partial loss of discrimination. These problems sprang from impedance missmatches between the terrestrial ear and water, complexity of the aquatic acoustic environment, and velocity related problems (sound travels approximately 5 times as fast in water as air). Also involved was the new found need to recycle air for phonation in an aquatic animal living much of its life below the surface, in deep water.

Much new evidence now suggests that the second view is generally correct. Major evolutionary changes of the sorts envisioned seem to have occurred. Only the broad outlines are known, however, and many simple functions are poorly understood or remain to be investigated.

THE MELON

The porpoise forehead, anterior to the blowhole, is, in most odontocete species, formed of a convex fatty tissue mass lying on the dorsal surface of the bony rostrum (Fig. 1). Smooth fibrous blubber covers its exterior surface and grades inward to an oil-

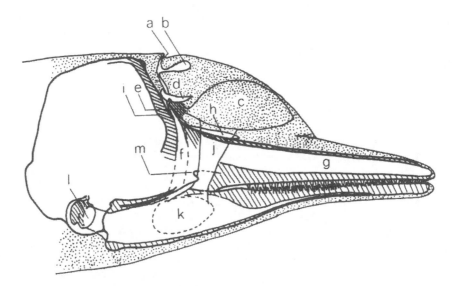

Fig. 1.   Structures in the delphinid odontocete head : a, blowhole ;
          b, vestibular sac ; c, melon ; d, tubular sac ; e, nasal
          valve with lip entering tubular sac ; f, internal nares ;
          g, mesorostral (or ethmoid) cartilages ; h, premaxillary
          nasal sac ; i, cribriform plate ; j, proposed wave guide
          from melon to outer tissue of mandible, passing anterior to
          antorbital notches ; k, acoustic window in mandible, and
          the site of overlying fatty window ; l, attachment point of
          mandibular wave guide onto bulla ; m, antorbital notch.

rich fatty core that lies over the center of the rostrum and the dorsal groove of the mesorostral canal. Embracing the melon is the rostral muscle that inserts on the rostrum and terminates in the fibrous tissue on either side of the melon. In the mid-sagittal plane the melon is an oblate spheroidal mass lying on the rostrum posterior to the beak, and in Delphinus delphis ends in the tissue of the right nasal plug lip. Behind it lies the complexity of the sacs of the dorsal narial canals, and the blowhole. To some extent the vestibular and pre-maxillary sacs embrace the posterior melon, and are thought to contribute to the shaping of emitted sound fields, and to aid in isolation of the middle ears from emitted sounds (Norris, 1968 ; Giro and Dubrovsky, 1974).

The melon has been proposed as a sound focussing lens that may serve to shape emitted sounds prior to their entry into water. Reflection from air sacs and bone, both impedance missmatched to the melon tissue, has been proposed as a focussing means (Norris, 1964), as has a differential transmission velocity topography within the melon (Norris and Harvey, 1974). The melon is probably deformable through the action of the embracing muscles and this is clearly the case in the Beluga Delphinapterus leucas, whose melon seems to be deformed foreward and thinned when the animal ensonifies objects of interest with click trains (G. Carlton Ray, personal communication).

CHEMICAL COMPOSITION OF MELON AND JAW FAT

A transverse section of the melon reveals regular and obvious topographic variations in the fatty tissue from the surface to the core. The surface blubber is very tough, with fat droplets tightly held in a connective tissue matrix in most odontocete species. Medially this changes until at the core the translucent fat is so oil-rich and lightly held that slight pressure will express abundant clear oil from the tissue. Through my sponsor at the time, the Office of Naval Research, I was able to interest Dr. Donald Malins, a lipid biochemist, in the problem. His work, and that of his colleague Usha Varanasi quickly revealed unique biochemical topography in the odontocete melon and jaw lipids. Interest among biochemists in this unique system spread and several laboratories in North America and Europe have contributed studies. This work has linked biochemical structure, molecular branching and chain length as determinants of acoustic transmission velocity. Lipids in "acoustic tissues" are different from those in surrounding tissues. They contain triacylglycerols and wax esters rich in short and medium chain (C_5-C_{12}) acids, while blubber is composed primarily of long chain triacylglycerols (C_{14}-C_{22}). In delphinids and monodontids, isovaleric and longer iso acids are present in large amounts in areas implicated in acoustic transmission (Varanasi and Malins, 1971 ; Wedmed et al., 1973 ; Litchfield and Greenberg, 1974 ;

Varanasi, Feldman and Malins, 1975 ; Malins and Varanasi, 1977 ;
Blomberg, 1978). Certain areas, such as the acoustic window and
inner melon, contain almost exclusively 1, 3 diisovaleroyl-2-
acylglycerols and isovaleroyl wax esters. This topography has been
associated with a sound transmission velocity topography in the
melon ; the central melon being of lower transmission velocity than
the outer areas and encasing blubber (Norris and Harvey, 1974).

A sharp biochemical difference has been found between the
acoustic fatty tissues of delphinids and the same tissues in
physeterids, ziphiids and platanistids (Litchfield, Greenberg and
Mead, 1976). This difference accounts for the different odors
emanating from fatty tissues of the two assemblages, and seems to
me to be phylogenetically related. Ziphiids and physeterids are
typified by waxy fats of little odor, while delphinid fats are
typically pungent with the clinging odor of isovalerate derivatives.

Interestingly, the unique fat composition and distribution
variations mentioned above closely define tissues that have been
implicated in acoustic processing in both the lower jaw and melon,
and are sharply different from body fat (Litchfield, Greenberg and
Mead, 1976). Thus it seems valid to speak of "acoustic tissues"
as I and others have done.

EXPERIMENTAL STUDIES

Contact hydrophone studies, in which small receiving hydrophones can be pressed or attached by suction to various loci on the
porpoise head have shown that impulsive or click-type sounds are
emitted in structured fashion from the anterior melon and the tip
of the upper jaw (Diercks et al., 1971 ; Evans, 1973 ; Norris and
Harvey, 1974). High frequencies and intensities are focussed foreward so that for Tursiops truncatus, all sound above 100 kHz is
typically restricted in a beam to a few degrees of arc, centering
directly ahead of the rostrum, and upward (Evans, 1973). A similar
field has been described for the roughtooth porpoise, Steno bredanensis (Norris and Evans, 1967), and a field emphasizing lower
frequencies with a small high frequency component has been outlined
for the killer whale, Orcinus orca (Schevill and Watkins, 1966).
Sound fields emanating from the side of the delphinid head are
restricted to lower frequencies, and the entire emitted beam of
all of these species drops both in frequency and intensity nearly
to zero at 90° to the longitudinal exis. This corresponds closely
to the common observation of workers recording in the field, that
when delphinids pass a recording station, significant sound levels
are only recorded when animals face or are passing the hydrophone.

These fields result from sources that have been localized
behind the melon (see Ridgway et al., this volume). On anatomical

grounds, Evans and Prescott (1962), Norris (1966) and Evans and
Maderson (1973) postulated a nasal plug source as the most likely
locus. In 1968 I suggested that from the stand-point of impedance
matching the most likely mechanism was a lip vibrating in an air
column, with the generated sound propagating inside the lip itself
passing into the melon and hence to the water, thus avoiding the
major expected energy loss if vibration of the air column itself
was the major source. Evans (1973) has proposed a friction model
in which a lip rubs over another surface, releasing energy as the
rubbed tissue folds and restitutes elastically.

Evans and Prescott (1962) were able to produce click trains
in an excised porpoise head by forcing air through the bony nares
from the basicranial space. The sequence of events in chirp production in Stenella longirostris was visualized by Norris, Dormer,
Pegg and Liese (1971) using x-ray cinematography, using a living
animal trained to phonate while in air. Phonation was precisely
correlated with the passage of small blebs of air upward past the
nasal plugs. Such air passage inflated the vestibular sac. Periodic
emptying of this sac with the return of air into the basicranial
space without opening of the blowhole was postulated as the air
cycling system for porpoises phonating below the surface. Air sac
inflation was noted to provide adequate sound mirrors for sound
projection. This work has recently been expanded and extended by
Dormer (1974). The elegant electromyographic work of Ridgway and
his colleagues in this volume is consonant with these x-ray studies
and extends them with precise timing of the muscular events in
phonation, and clearly shows that in those studies at least, no
sounds were being produced by the larynx since its muscles were
inactive during phonation. Even though Purves (1966) was able to
produce sound from the porpoise larynx by insertion of a Galton
whistle into the tip of the laryngeal spout and application of air
under pressure, I do not feel there is any evidence that clearly
implicates the larynx in porpoise sound production, though future
studies may so implicate it. Thus far, then, electromyographic,
x-ray, and sound field measurements all implicate a source at the
level of the nasal plugs. Of course, some sounds are produced by
odontocetes at the blowhole valve by expressing air past it into
the water, and other sounds are made by jaw clapping and aerial
behavior, but these are not of concern to us here.

Only one direct measurement, taken for other purposes, shows
sound focussing in the melon. Norris and Harvey (1974) attempting
to find a pathway by which received sound might enter the melon and
reach the cochlea, ensonified an excised bottlenose porpoise
(Tursiops truncatus) head with 25 sec. clicks from a click generator
at various angular positions around the preparation in the horizontal plane. With a small receiving hydrophone implanted approximately
in the center of the melon sound, levels above calibration

level for an all water path were produced when the source was 20°
to either side of the midline and horizontal to the axis of the
melon. When the sound source was rotated ahead of the preparation,
sending sound directly into the apex of the melon, a reduction in
received sound was obtained.

That the melon shows sound transmission velocity topography
sufficient to account for beam production was shown by Norris and
Harvey (1974) who measured velocity at many points on tissue slices
comprising in composite, the entire melon. A low velocity core
along the longitudinal axis of the melon was found to grade into
a higher velocity tissue closer to the melon surface.

THE SPERMACETI ORGAN

The spermaceti organ is found in varying form in the sperm
whale Physeter catodon, the pygmy sperm whales, Kogia sp, and in
the various Ziphiid whales. Its function has long been a subject of
speculation. Suggestions have ranged from a "force pump" (Raven and
Gregory, 1933), a hydrostatic organ (Clarke, 1978), to a sound
generation system (Norris and Harvey, 1972 ; Mackay, this volume).

In the sperm whale it is an enormous elongate spermaceti oil-
filled structures that may comprise 1/4 the length of a large adult
whale. Bounded on both ends by air sacs that are diverticula of the
right nasal passage, and underlain by a channellike passage of the
same nasal passage, it may overhang the lower jaw by as much as
10 % of an adult animal's length. Anteriorly, the organ grades into
the tissue of the upper lip of a broad horizontal pair of lips that
block the narial passage, the museau du singe, or Monkey's Muzzle.
This strange structure is thought by some to be a sound generator
(Norris and Harvey, 1972 ; Mackay, this volume).

Recently four of the various views about its function have
been published, and their differences remain unresolved. Norris
and Harvey (1972) noting the peculiar reverberent pulses often
emitted by the whale, in which an initial strong pulse decays
through a series of 7 or more pulses in a diminuendo through about
25 m sec., postulated that the organ is a reverberation chamber in
which an initial sound is produced at the museau producing a reverberent signal between the two end air sacs which act as sound
mirrors. A model using design criteria determined from dissections,
produced signals closely similar to those of the whale. They were
able to use the structure of clicks to check the model, using a
click sequence of a small whale that surfaced alongside a vessel of
known length. Clicks were used to predict whale length and checked
against the actuality of the sighted animal. The correspondence was
close. Nonetheless, as Mackay indicated (this volume) compensating
errors seem to have been involved in this calculation. The velocity

of spermaceti used in the Norris and Harvey model postulated a
closed organ pipe model for reverberation. The spermaceti organ
is more probably likened to an open pipe because its rear wall is
bounded by the compliant proximal sac. A halving of wavelength
approximately compensates for the velocity error. Mackay suggests
that air bubbles in the spermaceti sample may have given the peculiar velocity measure, which was made by Dr. Diercks of the University of Texas. This author suggests that this may be correct and
that the source of such bubbles may have been the harpoon explosion
that killed the animal. Mackay (this volume) proposed a bugle model,
noting that sperm whale sounds are not notably directional as
compared to those of other odontocetes. It seems to me that lack
of strong directionality is to be expected of a signal that peaks
between 6-12 kHz, and that, like the killer whale Orcinus orca,
(Schevill and Watkins, 1966) there may be high frequency components
to the sound that have yet to be recorded, and that may be closely
focussed foreward, and may be of much lower intensity than the
remainder of the signal. Mackay reports that when a fathometer was
playing into the spermaceti organ of his specimen a reverberent
pulse resulted. Only further studies using living animals seem likely
to resolve these different views. In my view, other recent suggestions seem unconvincing for various reasons. The overriding reason
for skepticism is that in each case the model put foreward has not
been directed by any observed behavior by the animal itself that
would suggest these adaptive systems. For example, it seems gratui-
tous to me to construct a model for the spermaceti organ as a
hydrostatic structure where there is no evidence that sperm whales
habitually hover in nature, as Clarke (1979) suggests they may do.

Kozak (1974) postulates a "video acoustic system" in which
the spermaceti organ is a sound receptor, using the serious fluid-
filled blisters on the posterior wall of the proximal sac as a
receptor field. The idea is that clicks are generated by the museau
and return, as echos to a receptor field on the sac wall, to provide
an acoustic image of objects in the environment. Nervous innervation,
ending as free nerve endings, as shown by staining the tissues of
the blister-covered wall is presented as supportive evidence. No
transduction system is demonstrated that would convert received
sound into nerve impulses and sensation and the innervation seems
scattered and slight. The geometry of the spermaceti organ seems
better suited to sending than receiving in that it narrows anteriorly
and enters the upper lip of the museau, and is surrounded laterally
and above by a thick meshwork of ligamentous cables. Nonetheless,
Bullock et al. (1968) demonstrated high sound sensitivity for the
porpoise melon, and no transduction system seems to exist there,
in a region primarily innervated by the trigeminal nerve, which has
not previously been involved in sound processing. The sperm whale,
like other odontocetes shows a good separation of emitting and
receiving structures, isolation being achieved primarily by air sac

barriers, and the sperm whale shows a well developed lower jaw reception system that seems clear evidence of reception at that locus. Nonetheless, I am not yet ready to dismiss Kozak's ideas entirely, but await further data. Clarke (1979) postulates that the sperm whale uses the spermaceti organ to establish neutral buoyancy during dives, by changing the temperature of the contained spermateci which concurrently undergoes slight density changes. My basic doubts about the idea are several. First, the spermaceti organ seems ill designed to be a heat exchange organ. Blood flow within the spermaceti is very modest indeed, the spermaceti case being so clear of all obstructions that whalers could bail its contained spermaceti oil out with buckets. As Schenkkan and Purves (1973) point out, vascular heat exchange requires cooling of blood and hence either cooling of the entire animal, or a heat loss mechanism that has yet to be demonstrated.

To bolster the argument for the organ as a heat exchange structure, Clarke postulates that the whale draws water into its right nostril, cooling the spermaceti from below. Once again, this seems a very inefficient design for heat exchange, in which the approximately 300 liters of waxy oil (Schenkkan and Purves, 1973) lie on top this flattened air passage, in the nearly cylindrical organ whose crossectional diameter is a meter or more. Such inhalation of water, if it can be done past the museau du singe, which seems to be a valve whose closure is directed against such entry, seems a cumbersome and peculiar adaptation for an animal that may phonate with the same structure.

The time course of density change under even the most extreme rates of heat exchange is slow enough that significant portions of many dives would find density change in process when thermoclines are reached. The occurence, or even existence of such thermoclines, is variable in the world oceans, and finally, there is no evidence for whales hovering, but instead for a fanning out of schools underwater, and continued foreward movement (Watkins, 1977). Watkins (this volume) found that not all animals phonate but some may be silent for fairly long periods during dives, but that some phonation in sperm whale schools is the norm. If the silent animals are hydrostatically equilibrating, what are the phonating animals in the same school doing ?

Another view is contained in an anatomical work by Schenkkan and Purves (1973). Their studies lead them to propose that the spermaceti organ assists in evacuation of the lungs prior to a deep dive and that it serves as an air reservoir for phonation at extreme depths. Perhaps because their preparation for the sperm whale was a fetal animal they were unable to conceive that the right nasal passage of the animal can express air into the left passage through a connecting vestibule, and thus allow air recycling at depth, and also phonation during dives that on occasion last more than an hour. They postulate phonation in the sperm whale larynx, a view as we note from the work of Ridgway (this volume) not substantiated for other odontocetes. I should note however, that the sperm whale is very different anatomically from the delphinids used by Ridgway. Schenkkan and Purves' contention that the spermaceti organ assists in lung evacuation seems without support. First, there is an enormous developmental investment by the animal in the spermaceti organ, which extends a quarter the length of the animal, and which involves many very intricate structures. It is hard for me to beleive that it was developed as a means for lung evacuation when other diving whales do not have it, and evacuate their lungs into the trachea and air sacs of the head by simple provision of muscular means of lung compression and of structures that provide varying degrees of resistance to pressure. During diving the collapsible rib cages of odontocetes direct air into the somewhat more pressure resistent trachea, and the air sacs of the skull. This process of evacuation probably involves muscular activity at alveolar sphincters, intercostal muscles and elsewhere. But it seems unreasonable to invoke the modest buoyancy of the spermaceti organ as a major feature of such evacuation. Changes in animal orientation underwater would reduce or reverse such an affect in any case.

Suffice it to say, there exists no definitive evidence of how sperm whales use their spermaceti organs. I feel direct experiments will be needed involving living animals, for us to learn its function. To me, not surprisingly, postulates for a phonation and sound processing function seem the most likely, and the best supported by knowledge of the animal in nature.

THE ODONTOCETE SKULL

Various attempts have been made to correlate skull contours with emitted sound fields (Evans, Sutherland and Beil, 1964 ; Dubrovsky and Zaslavski, 1970). Most of these studies have sprung from observation of the marked asymmetry found in most Delphinid, Physeterid and Ziphiid skulls, and assume that the bone-tissue interface serves as a sound reflector. This view may go back to my own early work (Norris, 1964) which pictured such a relationship. I now feel that the effects are mostly secondary ones, with the bony tissues supporting soft tissue structures that are directly involved in sound processing. In this view the bones give shape and rigidity to soft structure processing systems, involving muscles, fatty sound processing conduits, air film reflectors, and soft tissue sound generation structures. The mesorostral canal and the mandible have sound processing functions. The canal seems to provide a channel for sound conducted in its contained cartilage and may be involved in shaping the content and form of emitted signals from the rostral tip.

At its medial end the mesorostral cartilage enters the isthmus between the two nasal plugs, near the site of sound production in Delphinids, and at its distal end, its cartilage actually extends beyond the ivory-like encasing premaxillary bone to end in a rounded tip just beneath the skin. As noted earlier, this structure does carry high frequency sounds (Evans, 1973 ; Purves, 1966) as determined by sound recordings made near or on it.

THE MANDIBLE

It seems clear now that environmental sounds enter the odontocete mandible by passing through overlying blubber through what I have termed the "acoustic window", penetrating the very thin "pan bone" of the rear mandible, enter the mandibular fat body and are transmitted to the middle ear. Considerable controversy exists about the route by which such sounds enter the cochlea (see McCormick et al., this volume ; Fleischer, this volume).

The odontocete lower jaw and its adjacent structures have been greatly modified for this sound transmission function. First, the zygomatic arch which overlies the posterior portion of the mandible has been reduced from a very stout bony strut in many terrestrial mammals, to a thin flexible bony strand in most odonto-cetes (not platanistids). It is presumed that such reduction inter-ference by the bone in received sound fields (Norris, 1968). The oval fatty tissue of the acoustic window is either completely free of muscle strands, or nearly so, and it directly overlies the thinnest part of the mandible, or pan bone (Fig. 2). Its tissue has been found to be composed of "acoustic fat" like that of the melon and mandibular canal, and not found elsewhere in the odontocete body (Varanasi and Malins, 1971). Sound apparently may also enter the mouth of odontocetes, presumably during the last stages of feeding or biting, since high sensitivity to sound has been found for the base of the porpoise tongue (Bullock et al., 1968). Here there is no bony barrier since the mandible is excavated and the mandibular fat body directly underlies the tissue at the inner angle of the jaw and mouth cavity. Sounds, hitting the animal outside the mouth however, must traverse the pan bone to reach the mandibular fat body, and this bone, which is very thin (to 0.2 mm) has been found to vary widely between species in both thickness and thickness topography. A sound processing function has been postulated for this topography and the physical relations that allow partial sound ana-lysis by movements of this bone across complex sound fields have been presented (Norris, 1968). In effect, the pan bone of most odontocetes flares at a small angle to the mid-sagittal axis of the animal, and sounds returning from directly foreward hit the bone in many odontocetes at an angle that can be expected to attenuate them somewhat. Ten degrees or more to either side, transmission into the fat body, which will vary according to wavelength, can be expected

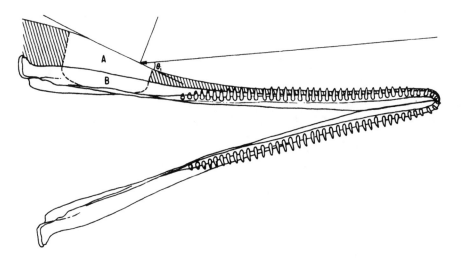

Fig. 2.   Delphinid lower jaw, showing the tissue overlying the jaw
          with its fatty "acoustic window", and the angle of
          incidence (θ) of sound coming from directly ahead of the
          animal.

to increase as normality is approached. An optimum angle, less
than normal, is expected because sounds hitting the jaw at right
angles can be expected to transmit poorly down the jaw because
reflections would be involved. Optimum transmission between 15-30°
of incidence is expected because the path to the bulla is direct.

As the animal moves its head during scanning (Kellogg, 1960 ;
Norris et al., 1961) the animal will scan across sound fields in
which both transmitted intensity and frequency composition change
can be expected. Once inside the mandibular fat body, which is a
well defined subcylindrical mass of fat flaring posteriorly, sounds
may be transmitted directly to the tympanic bulla, to which the fat
body attaches. The transmission route is direct and relatively
short in most odontocetes, though some, such as the Platanistids,
show complexity that is not understood. The fat body, it has been
postulated (Norris and Harvey, 1974), may function as a tuned
receiver, related to frequencies of most interest, as does the
human external ear canal.

Some experimental evidence exists that supports these conten-
tions. Studies of cochlear microphonics or evoked midbrain potentials in relation to projected sound have been performed (McCormick
et al., 1970 ; Bullock et al., 1972). Both studies show sensitivity to brief clicks or pips of sound to be highest over the lower
jaw, and as much as 6 times as high there as for sound projected at

the external auditory meatus. Bullock et al. (1968) found further
that the tongue was highly sensitive, and anomalously, that the
porpoise melon was 5.5 times as sensitive as the external auditory
canal to clicks. They also found that jaw sensitivity to these
sounds was highest in the contralateral colliculus, indicating a
normal path through the cochlea, with a crossover before reaching
the collicular level. On the other hand, reception in the melon
proved best in the ipsilateral colliculus, suggesting either a
crossover prior to reaching the cochlea, as for example if sound
passed through the melon and down the subcutaneous fat over the
antorbital notches to the opposite mandibular fat body, or deeper
in the acoustic system of the brain. Such superficial passage over
the antorbital notches was tested by Norris and Harvey (1974) by
implantation of small hydrophones in the proposed sound path and
ensonification of the opposite side the melon. Such transmission was
not found to occur. Also possible is some form of reception in the
melon itself, perhaps mediated by the trigeminal nerve. The
observation remains a mystery.

Angular changes in received intensity have been demonstrated
by hydrophone implantation inside the mandibular fat body and subsequent ensonification, using freshly dead porpoises. A partial
null for sounds projected at 0 directly ahead of the animal, was
found using hydrophone implants in the mandibular fat body, as was
predicted by structure. Highest received levels were found between
20-25° to the side of the animal's longitudinal axis, also as
predicted. Two fold increases in intensity were found for implants
in the fat body just anterior to its insertion on the bulla. This
indicates a structural intensification of received sound, like
that in a hearing trumpet, or the external ear canal of man (Norris
and Harvey, 1974).

Finally, hydrophone implants inside the peribullary air sinus
showed no reception at all in an ensonified animal, indicating that
this diverticulum of the eustacian canal indeed isolates the middle
ear eomplex to sounds, except those reaching the ear by tissue
pathways (Norris and Harvey, 1974).

Taken in its entirety, the odontote head seems adapted to
sound emission from the forehead and upper jaw, with the melon and
air sacs providing a means for internal sound processing (Giro
and Dubrovsky, 1974) and impedance matching prior to emission into
the water. Air sacs especially, seem positioned to isolate emission
sources from reception locations. Reception for clicks at least,
seems isolated in the lower jaw and tongue (except for the anomalous
observation of such reception in the melon, by Bullock et al.,
1968). In both emission and reception, sounds are processed by
structures composed of special acoustic fats not found elsewhere.
They seem to function by differential transmission velocity topography, and by means of tissue interfaces of various sorts. The entire

system seems to represent an adaptive response to the problems
faced by a terrestrial animal entering water, including extreme
impedance missmatching, and radically different environmental transmission velocities, which all but nullify normal aerial directional
hearing.

REFERENCES

Blomberg, J., 1978, Functional aspects of odontocete head oil lipids
    with special reference to pilot whale head oil, Prog. Chem.
    Fats and other Lipids, 16:257.
Bullock, T.H., Grinnell, A. D., Ikezono, E., Katsuki, Y., Nomoto, M.,
    Sato, O., Suga, N., and Yanigasawa, K., 1968, Electrophysio-
    logical studies of the central auditory mechanisms in cet-
    aceans, Z. vergl. Physiol., 59:117.
Bullock, T. H., and Ridgway, S. H., 1972, Evoked potentials in the
    central auditory system of alert porpoises to their own and
    artificial sounds, J. Neurobiol., 3:79.
Clarke, M. R., 1978, Buoyancy control as a function of the sperma-
    ceti organ in the sperm whale, J. Mar. Biol. Assoc., United
    Kingdom, 58:27.
Clarke, M. R., 1979, The head of the sperm whale, Scientific Amer.,
    1:128.
Diercks, K. J., Trochta, R. T., Greenlaw, R. L., and Evans, W. E.,
    1971, Recordings and analysis of dolphin echolocation signals,
    J. Acoust. Soc. Amer., 49:1729.
Dormer, K. J., 1974, The mechanism of sound production and measure-
    ment of sound processing in delphinid cetaceans, Ph. D. Diss.,
    University of California, Los Angeles.
Dubrovsky, N. A., and Zaslavski, G. L., 1975, Role of the skull bones
    in the space-time development of the dolphin echolocation sig-
    nal, Sov. Phys. Acoust., 21 (Trans. Amer. Inst. Physics.).
Evans, W. E., 1973, Echolocation by marine delphinids and one species
    of fresh water dolphin, J. Acoust. Soc. Amer., 54 (1):191.
Evans, W. E., and Prescott, J. H., 1962, Observations of the sound
    production capabilities of the bottlenose porpoise: a study
    of whistles and clicks, Zoologica, 47:121.
Evans, W. E., Sutherland, W. W., and Beil, R. G., 1964, The direction-
    al characteristics of delphinid sounds, in: "Marine Bio-
    Acoustics", W. N. Tavolga, ed.
Evans, W. E., and Maderson, P. F. A., 1973, Mechanisms of sound pro-
    duction in delphinid cetaceans: a review and some anatomical
    considerations, Amer. Zool., 13:1205.
Fraser, F. C., and Purves, P. E., 1959, Hearing in whales, Endeavour,
    18:93.
Giro, L. R., and Dubrovsky, N. A., 1974, The possible role of supra-
    cranial air sacs in the formation of echo ranging signals,
    A. Kurstichevkiy Zhurnal, 20:706.

Kellogg, W. N., 1960, Auditory scanning in the dolphin, Psychol. Rec., 10:25.

Kozak, V. A., 1974, The "video acoustical system" of the sperm whale, Trans.: Zhur. Evolyutsionnoi Biokhimii i Fiziologii, 10(3): 276.

Litchfield, C., and Greenberg, A. J., 1974, Comparative lipid patterns in the melon fats of dolphins, porpoises, and toothed whales, Comp. Biochem. Physiol., 478:401.

Malins, D. C., and Varanasi, U., 1977, Acoustic pathways in the cetacean head: assessment of sound properties through the use of a new microtechnique, (abst.), Proc. 2nd Conf. on Biol. of Marine Mammals, 36.

McCormick, J. G., Wever, E. G., Palin, J., Ridgway, S. H., 1970, Sound conduction in the dolphin ear, J. Acoust. Soc. Amer., 48:1418.

Norris, K. S., 1964, Some problems of echolocation in cetaceans, in: "Marine Bioacoustics", Pergamon Press, New York.

Norris, K. S., 1968, The evolution of acoustic mechanisms in odontocete cetaceans, in: "Evolution and Environment", E. T. Drake, ed., Yale University Press, New Haven.

Norris, K. S., Presscott, J. H., Asa-dorian, P. V. and Perkins, P., 1961, An experimental demonstration of echolocation behavior in the porpoise, Tursiops truncatus (Montagu), Biol. Bull., 120:163.

Norris, K. S., Evans, W. E., and Turner, R. N., 1966, Echolocation in an Atlantic bottlenose porpoise during discrimination, in: "Les Systèmes Sonars Animaux, Biologie et Bionique, Tome I.", R. G., Busnel, ed., Laboratoire de Physiologie Acoustique, INRA-CNRZ, Jouy-en-Josas.

Norris, K. S., and Evans, W. E., 1967, Directionality of echolocation clicks in the rough-tooth porpoise, Steno bredanensis (Lesson), in: "Marine Bio-Acoustics, Vol. 2", W. N. Tavolga, ed., Pergamon Press, New York.

Norris, K. S., Dormer, K. J., Pegg, J., and Liese, G. J., 1971, The mechanisms of sound production and air recycling in porpoises: a preliminary report, in: "Proceedings of the 8th Annual Conference Biol. Sonar and Diving Mammals".

Norris, K. S., and Harvey, G. W., 1972, A theory for the function of the spermaceti organ of the sperm whale (Physeter catodon L.), National Atmos. and Space Admin., Special Publication, 262:397.

Norris, K. S., and Harvey, G. W., 1974, Sound transmission in the porpoise head, J. Acoust. Soc. Amer., 56:659.

Purves, P. E., 1966, Anatomical and experimental observations on the cetacean sonar system, in: "Les Systèmes Sonars Animaux, Biologie et Bionique, Tome I", R. G. Busnel, ed., Laboratoire de Physiologie Acoustique, INRA-CNRZ, Jouy-en-Josas.

Raven, H. S., and Gregory, W. K., 1933, The spermaceti organ and nasal passages of the sperm whale (Physeter catodon) and other odontocetes, Amer. Mus. Novitates, 677:1.

Schenkkan, E. J. and Purves, P. E., 1973, The comparative anatomy of the nasal tract and the function of the spermaceti organ in the Physeteridae (Mammalia, Odontoceti), Bijdragen tot de Dierkunde, 43:93.

Schevill, W. E., and Watkins, W. A., 1966, Sound structure and directionality in Orcinus (Killer Whale), Zoologica, 51:71.

Varanasi, U., and Malins, D. G., 1971, Unique lipids of the porpoise (Tursiops gilli): Differences in triacyl glycerols and wax-esters of acoustic (mandibular canal and melon) and blubber tissues, Biochem. and Biophys. Acta, 231:415.

Varanasi, U., Feldman, H. R., and Malins, D. C., 1975, Molecular basis for formation of lipid sound lens in echolocating cetaceans, Nature, 255:340.

Watkins, W. A., 1977, Acoustic behavior of sperm whales, Oceanus, 50-58.

Wedmid, Y., Litchfield, C.,Ackman, R. G., Sipos, J. C., Eaton, C. A., and Mitchell, E., 1973, Heterogeneity of lipid composition within the cephalic melon tissue of the pilot whale (Globicephala melaena), Biochem. and Biophys. Acta, 326:439.

CETACEAN BRAIN RESEARCH: NEED FOR NEW DIRECTIONS

Theodore Holmes Bullock

Neurobiology Unit, Scripps Institution of Oceanography, and Department of Neurosciences, School of Medicine, University of California, San Diego, La Jolla, CA 92093

Whereas neuroscience is a rapidly moving field, characterized by the application of many new techniques, progress in understanding the greatest brains has slowed down in the west and is only forging ahead in a few Soviet laboratories in particular directions. The virtual absence of work on the cetacean brain in the U.S.A. is attributable to a combination of cost, the Marine Mammal Protection Act, and public sentiment against exploiting these animals. Two things need to be said here, in this historic international forum. One is that I support the Marine Mammal Protection Act and believe most of my colleagues do; I am working to extend similar protection to primates. The other is that new research is needed and justified especially on the brain and behavior, albeit with a high threshold of justification, that is, for questions of particularly high significance and to which answers are likely.

Doomed animals such as those that beach themselves to die of natural causes, those that are harvested in Japan and other countries for food and those that are accidentally trapped in commercial fishermen's nets, should not be wasted when humane studies could contribute to our meager knowledge of the unique brain. Hopefully, there will be far fewer such animals within a short time, as commercial fishing is stopped. In the meantime, a few of the hundreds still landed would permit the first application of modern methods of brain anatomy and some important new physiology and pharmacology.

The Soviet work referred to is not well known elsewhere. Actually several hundred active scientists in the USSR are publishing currently on the cetacean brain and behavior. Bullock and Gurevich (1979) have reviewed the literature since 1960; they conclude that there is a need for Western workers to study selected items in

Russian and for Soviet authors to publish full papers, with methods and data, in the international journals more often than they do. Otherwise the animals they have used are not able to contribute as much as they could to the accumulation of knowledge about them. Bullock and Gurevich also point out the concentration of Soviet attention to certain scientific areas, leaving other problems relatively undeveloped. Even in the areas of maximum attention some of the work bears confirmation or re-examination with newer methods.

In the remaining paragraphs, I will touch upon a few of the areas where new work is especially needed.

ANATOMY OF THE BRAIN

Comparative neurology is now in a renaissance with major new findings and reinterpretations, due largely to a battery of recently discovered techniques that permit tracing connections between distant structures, thus establishing their real place in the system, and allowing comparisons on a more rational, informed basis.

Almost no modern methods have been applied to the study of cetacean brain anatomy, especially the powerful axon transport techniques. Probably very few workers have even had well fixed, i.e. perfused brains for classical methods. A special feature of the new axon transport techniques for tracing pathways is that hours following injection of the innocuous marker substance (e.g. an amino acid or a peroxidase) suffice for its distribution along axons, so that the methods are useable on doomed animals that will survive only for a day or so. Injection under anesthesia would employ routine technique used in human stereotaxic surgery to place a long needle at a chosen site in the brain.

The reason that major new insights into the evolution of the mammalian brain are to be expected is that cetaceans have evolved separately so far that structures such as the olives, the cerebellum, and the cerebralneocortex are not now satisfactorily compared to the structures of the same names in other mammalian orders. Some structures are unaccountably small such as the auditory dorsal cochlear nucleus and major parts of the limbic system, including the hippocampus. Others are of prodigious size and very likely have hitherto unrevealed specializations such as the inferior colliculus which is 12 times as big, in a dolphin brain of 780 gms., as the homologous structure in the human.

PHYSIOLOGY OF THE BRAIN

I will mention only one technique still hardly used on cetaceans, although Seeley, Flanigan and Ridgway (1976) demonstrated its feasibility. This is the averaged evoked potential recorded through the skull, without entering the brain. An enormous literature shows its usefulness in studying the human brain. It can be a quick and objective way of plotting response against frequency of sound for example, or the effect of interval in pairs of clicks or the best parameters, rate and span of FM. In recent experiments with Ridgway and colleagues we have found that, as in the human, responses of several brain stem centers early in the auditory pathway can be studied, as well as later responses of higher centers. Evoked potentials might be able to show that two stimuli are discriminable by the brain, or that a continuum of slightly different stimuli are categorized into two distinct classes, like phonemes, or that one hemisphere does this more than the other. There are special waves associated in the human with recognition and cognitive processes; these have not yet been identified in any non-human species. Clearly this powerful tool should be used to assess many aspects.

By extending the technique to intracranial electrodes, such as are in standard use in human stereotaxic recording, a large additional domain of insights based on more localized and sensitive recording, can be obtained, as demonstrated first by Bullock and Ridgway (1971, 1972a,b) and further by a Soviet group (reviewed in Supin et al., 1978), and on localized electrical stimulation as demonstrated by Lilly (1958, and see Lilly and Miller, 1962). In the absence of mortality data from the Soviet experience the method must be considered to have appreciably more risk in dolphins than in humans although this may in fact be much lower now than some years ago, especially if the electrodes are only left in place for some days. It is well suited to the class of animals mentioned in Section II which are expected to survive only a day or so. A long list of deep as well as localized surface structures become accessible and many of their functional relations can be revealed by electrical recording. Given the special interest and value of understanding the functioning cetacean brain it will probably be worthwhile to employ multichannel recording, current source density analysis, focal and diffuse electric stimulation, cryoprobes and controlled focal release of drugs to bring out the comparisons and contrasts with brains in other mammals.

This is not the place to give background and protocol for specific experiments. The point of this note is to say that, with all due caution, and appropriate due process, modern techniques for brain research should be applied to cetaceans because the appreciation and understanding of their unique status in respect to nervous system achievement amply justify the humane imposition of low risk procedures.

REFERENCES

Bullock, T. H., and Gurevich, V. S., 1979, Soviet literature on the
    nervous system and psychobiology of cetaceans, International
    Review of Neurobiology, (in press).
Bullock, T. H., and Ridgway, S. H., 1971, Evoked potentials in the
    auditory system of alert porpoises (Cetacea) and sea lions
    (Pinnipedia) to their own and to artificial sounds, Proc.
    Int. Un. Physiol. Sci., IX:89.
Bullock, T. H., and Ridgway, S. H., 1972a, Neurophysiological find-
    ings relevant to echolocation in marine mammals, in: "Animal
    Orientation and Navigation", S. R. Galler, K. Schmidt-Koenig,
    G. J. Jacobs, R. E. Belleville, eds., NASA, U.S. Govn't
    Printing Off., Washington, D.C.
Bullock, T. H., and Ridgway, S. H., 1972b, Evoked potentials in the
    central auditory system of alert porpoises to their own and
    artificial sounds, J. Neurobiol., 3:79.
Lilly, J. C., 1958, Electrode and cannulae implantation in the
    brain by a simple percutaneous method, Science, 127:1181.
Lilly, J. C., and Miller, A. M., 1962, Operant conditioning of the
    bottlenose dolphin with electrical stimulation of the brain,
    J. Comp. Physiol. Psychol., 55:73.
Seeley, R. L., Flanigan, W. F., and Ridgway, S. H., 1976, A tech-
    nique for rapidly assessing the hearing of the bottlenosed
    porpoise Tursiops truncatus, Naval Undersea Center TP 552,
    San Diego, California.
Supin, A. Y., Mukhametov, L. M., Ladygina, T. F., Popov, V. V.,
    Mass, A. M., and Polyakova, I. G., 1978, "Electrophysiologi-
    cal Studies of the Dolphin's Brain", V. E. Sokolov, ed.,
    Izdatel'stvo Nauka, Moscow.

IMPORTANT AREAS FOR FUTURE CETACEAN AUDITORY STUDY

C. Scott Johnson

Naval Ocean Systems Center

San Diego, California 92152

INTRODUCTION

We have now accumulated enough information on cetacean hearing
to discover some anomalies, that is, experimental results that we
do not understand as well as we would like to. Three of these
anomalies will be discussed. In each case more data is necessary
in order to clarify the question. Until this is accomplished,
speculation as to the basis of the problems is of limited value.

AUDITORY THRESHOLDS AND CRITICAL RATIOS

In the frequency region above about 50 kHz, cetacean hearing
appears to be limited by thermal noise (C. S. Johnson, 1979). This
is an important result because it means that at least for the frequencies above 50 kHz, and quite possibly for those below as well,
we need not perform the laborious duty of determining absolute
thresholds. Since the thresholds are always thermal noise limited
in healthy animals, the only relevant numbers are the steepness of
the cut-off at the upper limit of hearing and the critical ratios.

The critical ratios referred to here are as defined by Fletcher
(1940) and are estimated by measuring auditory thresholds of pure
tones superimposed on a background of continuous broad-band noise.
Unfortunately, critical ratios have only been determined for one,
T. truncatus (C. S. Johnson, 1968), and those values are in some
question (c. s. Johnson, 1979). The reason for this is that the
threshold values predicted using the thermal noise spectrum and the
measured critical ratios differ from those measured on several
cetacean species by about an order of magnitude, i.e., 10 dB. This
means that either the critical ratio values are too large by a

factor of ten, which is quite a large discrepancy, or these animals can hear 10 dB below the noise in a critical band. This is obviously an anomalous situation that can only be resolved by doing more masked threshold experiments.

FREQUENCY MODULATED SIGNALS

The following account relates one instance of unexplained behavior related to the use of FM signals.

This author worked closely with Jacobs (1972) on the reported frequency discrimination experiment. The original procedure was to train the animal to respond by pushing on one lever when it heard an intermittently pulsed tone of constant frequency, and to a second lever when tones of two different frequencies were pulsed alternately. The duration of the individual pulses was a few tenths of a second and the series of pulses was played for a few seconds. In the one case the animal was presented with a da da da da etc. signal and the other a da dee da dee etc. signal. Once this task was learned the separation in frequency between the alternating tones could be reduced until the discrimination failed and the just noticeable frequency determined.

This procedure was tried for several weeks using two different animals and no matter how different we made the frequencies of the two tones, neither animal performed above the chance level in making the discrimination. Many different pulse lengths and overall stimulus durations were tried to no avail. Finally, in desperation, the signals were changed from pulsed tones of equal or different frequencies to a constant frequency tone versus a sine wave modulated FM signal. By the end of the first day following the change in the signals, the animal was performing the discrimination with near perfection! Unfortunately at the time it was not possible to explore this problem further and it has not been investigated since.

It is obvious from this experience that the type of modulation, in this case pulsed versus sine wave, is of great importance to the animal and emphasizes the importance of investigating the significance of LPM and other types of modulation.

CRITICAL TIME INTERVAL

Soviet scientists, Belkovich and Dubrovsky (1976), p. 91, and Velmin and Dubrovsky (1976), have reported some very interesting results indicating that something anomalous may be happening when pulse echo separations are in the range of 200 to 300 μsec.

The first reference above describes an experiment in which two T. truncatus were trained to discriminate two click trains; one of single pulses and one of pairs of pulses. The second pulse of the

pair train was varied in both time separation from the first pulse
and in amplitude relative to the first pulse. Their results show
that the animals' performance was good in making the discrimination
as the separation of the paired pulses decreased from 2,000 µsec
until a separation of about 500 µsec. The scores of both animals
decreased rapidly with decreasing pulse separation to about 90 µsec.
At this point the score of animal number 1 began improving rapidly
as pulse separation decreased while that of animal number 2 continued to decline. The data imply that as the pulse separation
decreased the acoustic stimuli perceived by the animals changed,
e.g., from pulse pairs to a time separation pitch signal or other.
Animal number 1 caught on to the change while animal number 2 did
not. Similar behavioral effects have been reported before (C. S.
Johnson, 1971).

    In the experiment reported by Velmin and Dubrovsky (1976), a
single T. truncatus was required to discriminate between two trains
of paired pulses. One of the pulse trains always had a pulse
interval of 100 µsec while the interval in the second train was
varied from 150 µsec to 2,000 µsec. The results show the animal's
performance scores at 100 percent until the pulse separation reached
200 µsec. Performance dropped with increasing separation above to
200 µsec to 0 percent at about 320 µsec and all larger separations.
As in the first experiment it appears that the stimulus changed
between 200 and 300 µsec and the animal didn't catch on to the
change.

    What is happening here? The distance from the blow hole of a
T. truncatus, approximate location of the sound source, to the tip
of the nose and back to the ear bones is about 75 cm. This is
presumably the minimum round trip distance encountered by the animal
in echolocating. A 75 cm round trip distance gives a pulse-echo
time interval of 500 µsec as the minimum pulse-echo interval of
value to the animal while echolocating. If what is happening is a
transition from perception of individual echo pulses to the time
separation pitch for range determination since such short (300 µsec)
pulse-echo intervals are never encountered while echolocating. It
would be very useful to both verify these results and to extend
them to intervals of greater than 2,000 µsec (a corresponding
target range of 140 cm).

    Another important question is: under what conditions can we
compare data from passive listening multiple pulse experiments with
that obtained from active echolocation experiments?

REFERENCES

Bel'kovich, V. M., and Dubrovskiy, N. A., 1976, "Sensory Bases of
    Cetacean Orientation",Nauka, Leningrad (English translation
    JPRS L/7157).
Fletcher, H., 1940, Auditory patterns, Rev. Mod. Phys., 12:47.
Jacobs, D., 1972, Auditory frequency discrimination in the Atlantic
    bottlenosed dolphin, Tursiops truncatus, Montagu: a prelim-
    inary report, J. Acoust. Soc. Amer., 52:696.
Johnson, C. S., 1971, Auditory masking of one pure tone by another
    in the bottlenosed porpoise, J. Acoust. Soc. Amer., 49:1317.
Johnson, C. S., 1968, Masked pure tone thresholds in the bottle-
    nosed porpoise, J. Acoust. Soc. Amer., 44:965.
Johnson, C. S., 1979, Thermal noise limit in cetacean hearing,
    submitted for publication in J. Acoust. Soc. Amer.
Velmin, V. A., and Dubrovskiy, N. A., 1976, On the auditory analysis
    of pulsed sounds by dolphins, Doklady Akademii Nauka SSSR,
    225:470.

AUDITORY PROCESSING OF ECHOES: PERIPHERAL PROCESSING

Gerhard Neuweiler

J.W.Goethe-Universität, Zoologisches Institut

Frankfurt, Germany

COCHLEA - THE FREQUENCY ANALYZER OF THE AUDITORY SYSTEM

In mammals sounds hitting the ear release a travelling wave on the Basilar Membrane (BM). The BM acts as a bank of mechanical low pass filters where high sound frequencies produce a maximal wave amplitude in the basal part of the BM and low ones in the apical region. The place of maximal vibration is considered the site of the BM where that frequency is represented. From base to apex the width of the BM gradually increases and the thickness decreases resulting in a continuous decline of the stiffness of the BM which correlates to the frequency representation along the course of the BM. From base to apex gradually lower frequencies are represented, each octave covering an equal distance on the BM.

This uniform pattern of logarithmic frequency representation in the mammalian cochlea has been recently challenged by physiological and anatomical studies in bats. Already in 1964, 1966 and 1967 A.Pye found that in some bat species, notably Rhinolophus, thickness and width of BM are regionally differentiated and depart from a smooth gradual change. In search of the 83 kHz filter in horseshoe bats Bruns (1976) described several structural specializations of the basal and outer (pectinata) part of BM (e.g. thickenings of the BM and secondary spiral lamina, minimal width of the BM) which abruptly disappear at the 4.3 mm site (from base) of the BM. By frequency mapping he showed that the filter frequency of about 83 kHz is represented exactly at that site where the sudden discontinuity of the stiffness gradient occurs (Fig. 1).

Furthermore these studies (Bruns 1976a, b) disclosed that the narrow frequency band from 82 - 86 kHz containing most relevant

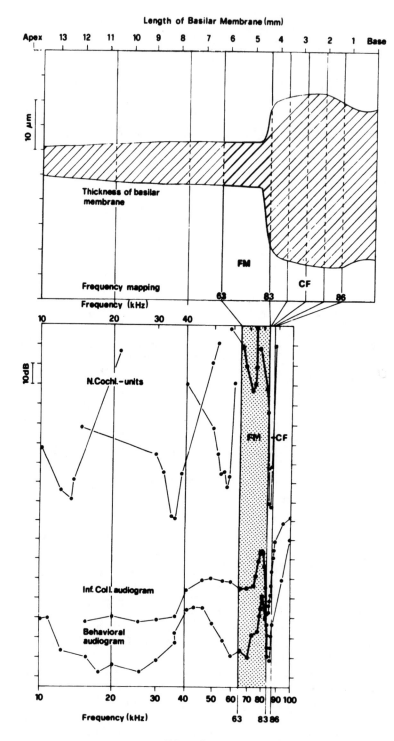

Fig. 1.

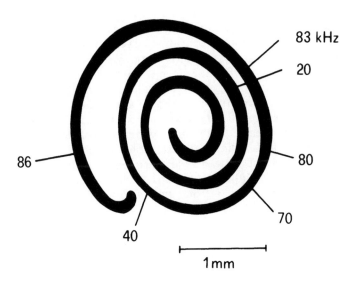

Fig. 2. Disproportional representation of frequencies on the basilar membrane in the greater horseshoe bat (<u>Rhinolophus ferrum-</u>
<u>equinum</u>). Basilar membrane is projected into a transmodiolar plane.
The frequency range of the constant frequency part of echoes from
82 to 86 kHz is vastly expanded. (From Bruns and Schmieszek, in
press).

Fig. 1. (Opposite page) The filter tuned to the constant frequency part of the echoes in the cochlea of the greater horseshoe bat.
Upper graph: Thickness of basilar membrane from base to apex (upper abscissa). Lower abscissa: Frequency representation on the basilar membrane as determined by frequency mapping. CF: frequency
range of the constant frequency part of the echoes. FM and shaded
area: frequency range of the final frequency modulated part of the
echoes. Lower graph: Tuning curves of cochlear nucleus units, inferior collicular and behavioral audiogram in horseshoe bats. Ordinate: Thresholds in relative dB-units. Note: Precise correlation
of sharp tuning in the 83-86 kHz range with its representation on
the thickened part of basilar membrane. (Bruns and Schmieszek,i.p.).

echo informations is represented on the BM in a vastly expanded fashion covering a length of the BM otherwise used for a complete octave, e.g. 80 - 40 kHz. This basal region of the BM in horseshoe bats may be called a frequency magnifier or an acoustic fovea (see Fig. 2, Schuller and Pollak, in press).

An analysis of the basilar membrane fine structure (Bruns, in prep.) disclosed that within the basal, specialized "foveal" region the inner and the outer part of the BM form rigid plates only weakly coupled mechanically by a thin sulcus. Capacity of independent vibration of the outer parter is further facilitated by a loose connection of the outer BM to the cochlear bony wall. As a result for stimulus frequencies above 80 kHz (and not for lower frequencies!) the inner and outer part of the BM vibrate out of phase (Wilson 1977, Wilson and Bruns, in prep.). These antiphasic movements may create radial fluid motions in the organ of Corti partially cancelling out longitudinal travelling wave vibrations. By this interaction of longitudinal and transverse motions deflection peaks of the BM caused by a travelling wave may become narrowly localized. Thus antiphasic movements of the two uncoupled BM-parts may assist in a precise frequency analysis (second mechanical filter, guard-ring principle of Wilson 1977).

All these structural specializations of the BM and its anchoring system together with additional specializations of receptor arrangements and spacings (Bruns, in press) along with the antiphasic vibration mode for the filter frequency clearly demonstrate that the horseshoe bat has a mechanically specialized part of the BM resulting in a narrow band pass filter matched to the constant frequency component of the echoes and also possesses a frequency magnifier or acoustic fovea built into the cochlea for the filter frequencies from 82 to 86 kHz. Bruns (in prep.) incorporated all these data into a model of cochlear function based on Steele's vibration model for the mammalian cochlea. This specialized and sophisticated case of mechanical adaptation to hearing in a narrow frequency band and acute frequency discrimination strongly supports the notion that in the mammalian auditory system the frequency spectrum is mechanically analyzed by the cochlea without additional neuronal filtering. The frequency resolution of an auditory system is that of its basilar membrane.

These structural data fairly well corroberate cochlear microphonic (CM) recordings. In all bat species commonly investigated, the CM-audiograms match the frequency bands of the specific echolocation sounds. In "long CF/FM" - bats the CM-audiogram includes a narrow filter for the CF - frequency (Pollak et al., 1972). Suga et al. (1975) elaborated on the CM of Pteronotus parnellii, a "long CF/FM" - bat. As demonstrated by slow decay times of cochlear microphonics they described ringing of a mechanical resonator tuned to 61.1 kHz, i.e. 0.5 to 1.0 kHz below the best frequency of summated auditory nerve responses. A similar resonator has been described in horseshoe bats (Schnitzler et al., 1976). However, in these bats only a small dip of no more than 14 dB was recorded in the cochlear microphonic audiogram 0.5 to 1.0 kHz below the frequency of the constant frequency component in the heard echoes. Since part of the experiments were performed in anaesthetized bats and since the site of the recording electrode greatly influences the results, cochlear microphonic recordings have to be interpreted cautiously.

All microphonic studies in Pteronotus including detuning experiments of the 62 kHz - filter by overstimulation (Pollak et al., in press) are in favor of a resonance system whereas the vibration measurements of the basilar membrane in Rhinolophus indicate other possible mechanisms which, however, also may show ringing qualities.

The horseshoe bat's basilar membrane is certainly a specialized but not an exceptional case. Morphological differentiations of the basilar membrane are recorded for other bat species and other mammals from different parts of the BM and a detailed inspection with precise frequency mapping methods may disclose specialized regions and differentiated frequency representation in frequency ranges specifically relevant for a species' behavioral repertoire. We should depart from the notion that in all mammalian basilar membranes the frequency range heard by a species is represented in a uniform logarithmic array.

An interesting comparative case is Pteronotus parnellii, a "long CF/FM" - bat with a CF-frequency of 62 kHz and a correspondingly tuned narrow filter in its cochlear microphonic audiogram (Pollak et al., 1972). However, no morphological specializations focusing on a certain site of the basilar membrane have been recorded so far (M. Henson 1978). Since no frequency mapping is available for the cochlea of Pteronotus a detailed comparison with that of Rhinolophus is not yet possible, but it is conceivable that these two unrelated bat species achieved the same filtering precision by different cochlear mechanical adaptations.

NEUROANATOMY
A) INNERVATION DENSITIES OF THE COCHLEA AND TONOTOPY

Apart from dolphins, bats have the largest number of spiral ganglion cells. The organ of Corti is highly innervated; in Myotis lucifugus the ratio of ganglion to hair cells is 55300/3500 (700 inner hair cells and 2800 outer hair cells) or 15:1 (Ramprashad et al., 1978). The density distribution along the basilar membrane is not uniform and reaches a maximum of 22:1 at the end of the lower basal turn. In the "long CF/FM" - bat Pternotus parnellii the in-

nervation appears particularly dense at the beginning and at the
end of the basal turn with sparse fibers inbetween (M.Henson 1978)

In horseshoe bats the total number of spiral ganglion cells
only amounts to 16000 as in the case of Plecotus austriacus (Firbas 1972). The innervation density in the basal region is low
(500 fibers/mm length of BM) and suddenly increases to a maximum
of 1750 fibers/mm at the transition from the first to the second
half turn where the frequencies of the final FM-sweep of the echolocation sounds are represented (Bruns and Schmieszek, in press).
80 to 90 % of all afferent fibers run to the inner hair cells.
About 8 to 16 fibers converge onto one inner hair cell in the
"acoustic fovea" - region, 24 fibers in the region where the frequencies of the FM-sweep are represented and each inner hair cell
in the region representing lower frequencies is innervated by
about 15 fibers. Although the innervation of the "acoustic fovea"-
region seems to be low it is actually high related to the frequency

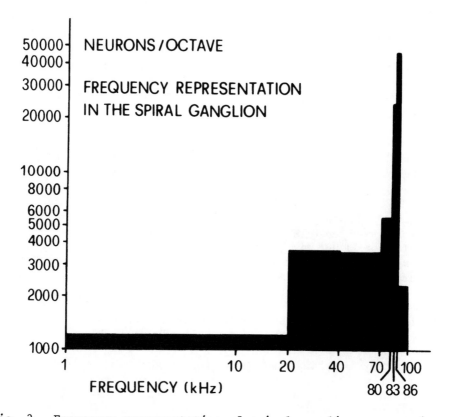

Fig. 3. Frequency representation of spiral ganglion neurons in
greater horseshoe bats. Ordinate: number of neurons/octave; abscissa: best frequency of neuron as determined by its afferent origin
on the basilar membrane. (From Bruns and Schmieszek, in press).

range covered. Since the frequency range from 82 to 86 kHz spreads over most of the lower basal turn 3360 spiral ganglion cells or 21% of the total cell number innervate the basilar membrane representing this narrow frequency range (Fig. 3.). This example illustrates that functional conclusions based on cochlear hair cell and innervation densities can only be made when the frequency representation on the basilar membrane has been mapped.

In all auditory nuclei of bats and other mammals so far studied neurophysiologically and/or neuroanatomically a tonotopic arrangement of the neurons' best frequencies has been reported. In our lab for instance we have anatomical and/or physiological data in the horseshoe bat's spiral ganglion, cochlear nuclei, superior olivary complex, inferior colliculus, medial geniculate body and the auditory cortex (also see Ostwald 1978) unequivocally demonstrating tonotopic organization. Within this tonotopic order the frequency band of 82 to 86 kHz always is overrepresented. As already mentioned one fifth of the afferent fibers come from that region of the basilar membrane where 82 to 86 kHz are represented. Thus the peripheral "acoustic fovea" is faithfully reflected in the afferent fiber population (Fig. 3.), wired through and quantitatively maintained throughout all nuclei of the central auditory system up to the auditory cortex. This is a striking analogy to the foveal representation in the visual central nervous system (Schuller and Pollak, in press). Thus in the horseshoe bat, central nervous overrepresentation of a narrow frequency band always is the result of a corresponding blow-up of that frequency range on the basilar membrane.

It would be most interesting to investigate whether the central overrepresentation of behaviorally relevant frequency bands and its tracing back to a corresponding overrepresentation and ordering in the peripheral auditory system could be applied not only to horseshoe bats but to other bat species and perhaps to all mammals also. The first crucial test for this hypothesis would be the frequency map of the basilar membrane and a quantitative count of innervation densities in Pteronotus parnellii.

If the hypothesis of the peripheral origin of frequency overrepresentation could be generalized, overrepresentation would become a mere special case of the more general fact that all auditory nuclei are organized in a tonotopic fashion, always originating in in the corresponding orderly representation of the frequencies on the basilar membrane.

B) ASCENDING AUDITORY PATHWAY

Neuroanatomical studies of the Chiropteran auditory system have long been neglected. Schweizer (1978) has studied the cyto-

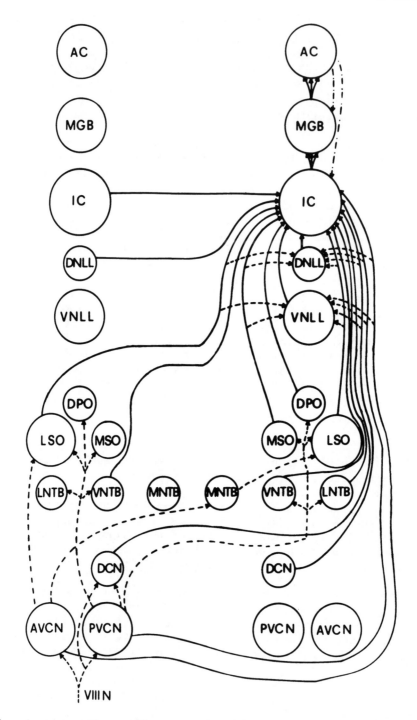

Fig. 4. Ascending auditory pathway in bats (Legend opposite page).

architecture and projections to and from the inferior colliculus
(IC) in horseshoe bats, Rhinolophus ferrumequinum, and Zook et al.
(in press) have made a similar study in Pteronotus parnellii and
Artibeus jamaicensis. Both authors used horseradishperoxidase (HRP)-
tracing methods.

In all three bat species the auditory pathways do not diverge
from the principle mammalian scheme. Except for the medial nucleus
of the trapezoid body (MNTB, a relay nucleus for the projection
from the contralateral anteroventral cochlear nucleus AVCN to the
lateral superior olivary nucleus LSO) all peripheral nuclei project
to the inferior colliculus resulting in 15 different inputs from
the auditory periphery to the inferior colliculus. These inputs come
from: contralateral anteroventral, posteroventral and contra- and
ipsilateral dorsal cochlear nuclei, from contra- and ipsilateral
ventral and ipsilateral lateral nuclei of the trapezoid body, from
ipsi- and contralateral lateral and ipsilateral medial superior
olivary nucleus, from ipsilateral dorsal periolivary nucleus, from
contra- and ipsilateral dorsal and ipsilateral ventral nuclei of
the lateral lemniscus and finally from contralateral inferior colliculus (Fig. 4.).

In contrast the higher auditory centers receive only input
from one auditory nucleus each, the medial geniculate body (MGB)
receives ascending input from the inferior colliculus and the auditory cortex (AC) from the medial geniculate body. The auditory cortex projects back into the medial geniculate body and into the inferior colliculus but not further peripherally.

In all bats the auditory nuclei are largely hyperthrophied
including auditory cortex and medial geniculate body, contrary to
earlier findings (Poljak 1926). The most hyperthrophied nuclei in
echolocating bats are the anteroventral cochlear nucleus and the
medial nucleus of the trapezoid body pointing to the functional importance of the lateral superior olivary nucleus in bats.

Fig. 4. Scheme of the ascending auditory pathway in Rhinolophus
ferrumequinum. Left ipsi- and right column contralateral nuclei.
Below IC not all interconnections between the nuclei are shown.
Dashed lines: pathways not directly projecting to IC. Broken lines:
Descending projections from auditory cortex. AC auditory cortex;
AVCN anteroventral cochlear nucleus; DCN dorsal cochlear nucleus;
DNLL dorsal nucleus of lateral lemniscus; DPO dorsal periolivary
nucleus; IC inferior colliculus; LNTB lateral nucleus of trapezoid
body; LSO lateral superior olivary nucleus; MGB medial geniculate
body; MNTB medial nucleus of trapezoid body; MSO medial superior
olivary nucleus; PVCN posteroventral cochlear nucleus; VNLL ventral
nucleus of lateral lemniscus; VNTB ventral nucleus of trapezoid body;
VIII N auditory nerve. (Adapted from Schweizer 1978).

The wiring diagram of the three species studied reveals a remarkable divergence of the afferent auditory information into 15
peripheral nuclei and a subsequent convergence of all peripheral
outputs upon one single nervous center, the inferior colliculus
(Fig. 4.). Auditory information is already frequency analyzed in
the cochlea and the frequency information is allotted to specific
auditory nerve fibers. The complete set of frequency analyzed information is projected upon both ventral cochlear nuclei and upon
the dorsal cochlear nucleus directly or via the posteroventral
cochlear nucleus. As will be shown in the chapter on specialized
peripheral neurons complex neuronal processing starts already in
the cochlear nuclei and continues on the next stages, the trapezoid
body , superior olivary complex and the lemniscal nuclei. About the
specific nature of processing performed in the different nuclei
little is currently known.

A good example of our scant knowledge can be illustrated by
considering the superior olivary complex, the most peripheral nuclei where binaural interaction may occur. According to a theory of
Irving and Harrison (1967) the presence of a large lateral superior
olivary nucleus and the absence of the medial one is related to
small head size and the use of binaural intensity differences for
high frequency sound localization. On the other hand a large medial superior olivary nucleus is related to a large head size and
the use of binaural time differences for sound localization and
subsequent tracking of the sound direction by eye movements. Accordingly they described echolocating bats as not possessing a medial
superior olivary nucleus. But Rhinolophus ferrumequinum, Molossus
ater and Molossus molossus as well as Pteronotus parnellii (Schweizer 1978, Zook et al., in press) do have a well developed medial
superior olivary nucleus. From his HRD-labeling data Schweizer
(1978) concludes that the prominence of either the medial or lateral nucleus is related to the frequency range heard by a mammal
since in the lateral nucleus high frequencies are preferentially
represented whereas the medial superior olivary nucleus shows ret-
rograte labeled cells from inferior collicular regions where low
frequencies are represented.

Both Zook et al. and Schweizer obtained remarkably similar results with some minor exceptions: Zook et al. describe an intermediate lemniscal nucleus in Pteronotus which may be homologous to
the dorsal part of the ventral nucleus of lateral lemniscus described in Rhinolophus. In Pteronotus the anteroventral cochlear
nucleus is reported to project bilaterally to the next nuclei
whereas in Rhinolophus it seems to project only contralaterally.
Since in Artibeus labeled cells were found bilaterally throughout
the olivary complex Zook et al. concluded that this bat does not
have a medial superior olivary nucleus that is considered to project only to the ipsilateral inferior colliculus. This issue requires more detailed comparative studies.

Both studies contain rich informations on the cytoarchitecture of the nuclei so that neurophysiologists now can aim at describing functionally the local subdivisions and morphologically identified neuron types.

The convergence of all peripheral outputs upon the inferior colliculus assigns a central rank in neuronal processing to this midbrain nucleus. The direct and the parallel multistage projections from the cochlear nuclei to the colliculus reminds one of Licklider's delay-line theory describing an analyzing mechanism in the time domain (Tobias, 1970).

In an echolocating bat, auditory information has to be transformed in appropriate motor commands governing the flight musculature. The linkage between auditory and motor system in horseshoe bats is represented by a large projection from the inferior colliculus and from the auditory cortex (Radtke, 1979) to the lateral pontine nuclei. The latter project to the enormously hyperthrophied paraflocculus and to the vermis of the cerebellum.

Collicular projections are also sent to the formatio reticularis and to the central grey matter where vocalization of echolocation sounds can be electrically elicited (Suga et al., 1973). The auditory cortex is also connected to the grey matter and to other nuclei related to vocalization, e.g. the area cinguli, the amygdala, nucleus caudatus and an area in the frontal cortex (Radtke, 1979). The interaction of auditory and vocalization system will be dealt with in the last chapter.

BINAURAL ANALYSIS OF ECHOES

It is widely believed that bats rely on binaural intensity differences for determining horizontal angular positions of a sound source. There exist many evoked potential and single unit studies from superior olivary complex to higher nuclei demonstrating that the discharge rate may encode azimuth differences of a loudspeaker position as small as 0.5 to 2.5° (Grinnell and Hagiwara, 1972, Jen and Suga, 1974; Shimozawa et al. 1974; Jen, 1978).

Schlegel (1977 and in prep.) experimentally tested the binaural interaction of stimulus intensities with the use of earphones in collicular, lemniscal and superior olivary neurons of horseshoe bats. In all neurons recorded, the tuning curves were the same for ipsi- and contralateral stimulation. 70% of all units recorded were excited by contralateral and inhibited by ipsilateral stimulation (E/I-types of neurons). This result confirms to the classical concept of binaural intensity difference coding. However the discharge of the E/I-neurons not only depends on the binaural intensity differences but also on the absolute intensity level rendering the discharge information ambiguous. This equivocality has to be eliminated

at a higher interacting level. Assuming that a 5% change of discharge rate is detected by the brain these neurons could encode
azimuth differences of about 1°.

By periodically amplitude modulating the pure tone intensities
diametrically in the ipsi- and contralateral earphone, Schlegel
(in prep.) simulated a sound source moving horizontally around the
head of the horseshoe bat at different speeds (4°/sec to 1000°/
sec). He tested the reaction of superior olivary, lemniscal and collicular neurons to movement of direction and to its speed. Most
neurons faithfully encode the azimuth angle of the sound source
irrespective of movement direction and speed (proportionalists in
the classification of Schlegel). Some neurons responded only to a
narrow angular sector with a marked "best angle". These neurons
also encode the speed of the sound source movement by increasing
discharge rates for faster movements ("speed-encoders"). They do
not discriminate the direction of the movement. However, there are
other speed encoders which are vigorously excited by an angular
movement from the contra- to the ipsilateral side and are immediately inhibited by movement in the opposite direction (movement direction encoder, see Fig. 5.).

It has to be emphasized that most of these speed and direction
sensitive neurons also encode the direction of a stationary sound
source, sometimes only in a limited horizontal angular range. Yet
there are neurons which do not respond to any stationary sound
source at all but only discharge when the sound source is moving
in the horizontal plane (movement detectors, Fig. 5.). This detailed study shows that directional information and directional movement of a sound source are intricately analyzed and processed at
peripheral levels of the auditory system.

Although most units respond to several parameters there exist
many specialists responding preferentially or even exclusively to
a single parameter such as movement or a certain direction of movement, demonstrating once more that specialized (feature) detectors do exist at peripheral levels.

Similar studies on binaural time differences are much in de-
mand. In this context the few data already available in our lab
suggest that the generally accepted statement of bats using only
binaural intensity differences for directionality encoding may have
been a hasty conclusion.

COMPARATIVE ASPECTS
A)   NEURAL PROCESSING IN FM - AND "LONG CF/FM"- BATS

The numerous bat species use a wide range of different echolocation
sounds. Although there exists a continuum of structural features

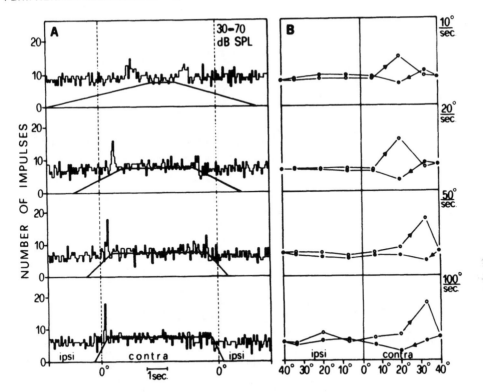

Fig. 5. Inferior collicular neuron selectively sensitive to direction of angular movement in Rhinolophus ferrumequinum. A) PST-histograms to tone pips at the unit's best frequency. Sound source is
apparently moved from 40°ipsi- to 40° contralateral and reverse by
diametrically changing the intensity from 30 to 70 dB SPL between
both earphones until complete reversal (70 to 30 dB) during a time
course as indicated by the slopes of the trapezoid. As steeper the
slopes as faster the directional movement (right hand ordinate).
Broken vertical line marks midline position (i.e. equal intensities
in both earphones) of sound source. B) Discharge rate of the unit
as in A but plotted against angular position in the horizontal plane. Open circles: movement from ipsi- to contralateral, black dots:
movement from contra- to ipsilateral side. Note: unit preferentially responds during movement and unit reacts by strong excitation to
sound source movement from ipsi- to contralateral and a shallow inhibition by movement in the reverse direction. (From Schlegel, in
prep.).

one may classify the bat species as FM - bats using broadband downward frequency modulated signals (often with harmonics), "short
CF/FM"-bats emitting an FM-sweep preceeded by a pure tone (CF) of
up to 8 msec duration, and "long CF/FM"-bats emitting a narrow FM-
sweep preceded by a pure tone of 10 to 100 msec duration (Grinnell,
1973).

Even at the level of the cochlear nucleus remarkable differences of response patterns to pure tones appear between FM- and "long
CF/FM"-bats (Fig. 6.). In horseshoe bats many cochlear neurons have
very complex response patterns displaying simultaneous excitatory
and inhibitory interactions of various durations and elicited by a
single pure tone stimulation. Frequently longlasting inhibition of
spontaneous activity is elicited by the onset of a pure tone. Thus
inhibition is a dominating part of neural activity in the cochlear
nucleus of horseshoe bats (Neuweiler and Vater, 1977) as in many
other mammals.Half of the neuron sample recorded were phasic on.

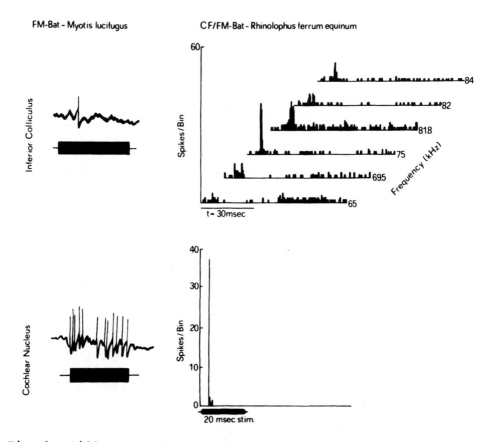

Fig. 6. Differences of neuronal response patterns to pure tones
between FM- and "long CF/FM"- bats. In FM-bats (left column) cochlear nucleus units mostly respond tonically whereas in "long CF/FM"-
bats (right column) they display a variety of response patterns, a
a large fraction responds phasic on. More than 90% of inferior collicular neurons in FM-bats respond phasic on whereas in "long CF/FM-
bats most of them respond in complex patterns dependent on stimulus
frequency. (Adapted from Suga, 1964 and Möller et al., 1978).

In contrast no such phasic on - neurons were found in the cochlear nucleus of the FM-bat Myotis lucifugus and the units recorded
uniformly responded by a tonic excitatory pattern enduring as long
as the stimulus lasted (Suga, 1964). Specialized neurons responding
only to certain stimulus parameters have not been described in the
cochlear nucleus of FM-bats whereas in horseshoe bats different specialized cochlear neurons have been recorder (Neuweiler and Vater,
1977).

In the inf.colliculus a similar striking difference of neural
processing between the two groups of bats reappears. In the horseshoe bats collicular neurons show response patterns similar to neurons in the cochlear nucleus with a trend to even greater complexity
and longlasting inhibitory effects, a feature of neuronal processing
seemingly neglected by auditory physiologists. The percentage of
phasic on - neurons drops to about 30 to 40% of some 600 neurons recorded in Rhinolophus ferrumequinum. However, in the FM - bats
Mollossus ater and Molossus molossus, 95% of 329 collicular neurons
recorded were not spontaneously active and responded only with phasic on to pure tones (Vater et al.,1979).

Again in complete contrast to the "long CF/FM" - bats in FM-
bats a large majority of collicular neurons so far recorded are phasic on - neurons with no spontaneous activity. One is impressed by
the uniformity of response patterns in FM - bats and the wide diversity of neuronal responses in "long CF/FM" - bats (Fig. 6.).

The same is true for inferior collicular responses in the FM-
bats Tadarida brasiliensis (78% phasic on-responders; Pollak et al.,
1978) and Myotis lucifugus (95% phasic on-responders; Suga, 1969).
Thus in neuronal peripheral processing a dichotomal trend appears
between FM - bats on the one hand and the "long CF/FM"- bats on the
other hand. Whereas the neuronal activities of cochlear nucleus and
inferior colliculus of the latter group agrees with that in other
mammals the neural processing in the FM - bat group is more uniform
and specialized. The prevalence of phasic on - neurons and the lack
of spontaneous activity above the cochlear nucleus level is the result of strong dominant inhibition and may be related to the possible timing capacities of such neurons (see also next review by
G.Pollak).

The benefit of emitting a short CF-component in echolocation
sounds is still a puzzling question. In the auditory nuclei of
"short CF/FM"-bats the frequencies of the CF - part are never overrepresented as they are so extravagantly in "long CF/FM"-bats. Grinnell and Hagiwara (1972) provide neurophysiological evidence that
the short CF - part facilitates the response to the terminal FM -
component of the echo. However, in some other bats sometimes also
emitting a short CF-part no such facilitation was observed in single unit recordings (Vater and Schlegel 1979).

B) OPTIMAL ADAPTATIONS

Since the introduction of optimal filter theory into experimental echolocation research by Simmons (1971) and Glaser (1974) the search of neuronal correlates of optimal adaptations to specific echolocation sounds became topical. As already mentioned, audiograms reach minimal thresholds in frequency ranges of the species' echolocation sounds. Even in closely related species, e.g. Molossus molossus and Molossus ater the minimal thresholds of audiograms are 5 kHz apart related to the main frequency range of the sounds emitted at 35-25 kHz in M. ater and 45-30 kHz in M. molossus (Vater et al., 1979).

In a detailed study in the above mentioned species Vater and Schlegel (1979) showed that most collicular neurons respond equally sensitive to CF- and FM-signals of the same duration. However, about one third of the units recorded responded with up to 6o dB lower thresholds to a signal mimicking the FM-component of the echolocation sound. FM-signals with longer sweeps than those emitted were less effective. These results are in agreement with those found in Tadarida brasiliensis by Pollak et al. (1978).

Thus there exists an adaptation, though suboptimal, of parts of the auditory system to the structure of the echolocation sound. Possible neuronal mechanisms for this adaptation to FM - echolocation sounds are discussed in Vater and Schlegel (1979).

Optimal receivers detect signals they are matched to much better in noise than other receiving systems. In fact this improvement in signal to noise - detectability is the major asset of a matched filter. The signal to noise ratio of its output should improve by a factor of 2 TB (T effective duration, B effective bandwidth of signal; Glaser, 1974). In our lab signal to noise ratios of threshold responses in collicular neurons were measured with pure tone signals in Rhinolophus ferrumequinum, a "long CF/FM"-bat, and with frequency downward modulated signals in Molossus ater and M. molossus, so called FM - bats. Both signals mimick the species-specific echolocation signals.

Threshold signal to noise ratios of these units range between +4o to - 20 and rarely reach values predicted by matched filter theory. However, in Molossid bats FM-sweeps are better detected in noise than pure tones by about 10 dB, thus indicating an adaptation of the receiver system to the echo signals, though again suboptimal (Fig. 7.).

Additionally in Molossus, collicular neurons detect a masked signal of long duration much easier than a short one which nicely relates to Simmons' finding that Tadarida, another FM - bat, lengthens its echolocation sounds in the presence of noise. The better detectabi-

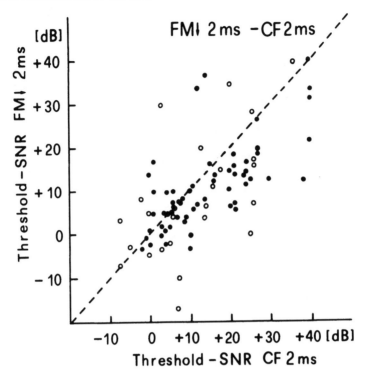

Fig. 7. Comparison of threshold - signal/noise - ratios (Threshold-SNR) to downward sweeping frequency modulated (FM) and constant frequency (CF) signals of equal duration (2 msec) for inferior collicular neurons in two FM-bats, Molossus molossus (open circles) and Molossus ater (black dots). Broken line represents equal values for both CF- and FM- signal. CF-signals were at best frequency, all FM-signals (modulation heights 40, 20 or 10 kHz) were centered at best frequency. Note: a majority of units has lower threshold - SNR for FM- than for CF-signals. (From Vater and Schlegel, 1979).

lity of long sweeps is readily explained by the higher energy content of the long sweep.

SPECIALIZED PERIPHERAL NEURONS

In auditory research the dispute on the concept of specialized (feature) detectors continues and has been recently revived by the spectacular findings in the auditory cortex of the mustache bat (Suga and his associates, see also review by Suga and O'Neill in this book). All detector concepts imply a hierarchical neuronal processing with unspecific neurons in the periphery and more and more selective neurons in the next stages converging onto the highly specific "detector" neuron in the cortex.

As a matter of fact there is ample evidence that specialized neurons do exist in peripheral nuclei and that their number probably increases at higher levels. There are specialized neurons in the superior olivary complex only responding when a stimulus is moving in the horizontal plane from ipsi- to contralateral side; it will not respond to a stationary or oppositely moving sound source (Schlegel, in prep.; see also Fig. 5.). There are neurons in the cochlear nucleus of horseshoe bats only responding to a pure tone component of the echolocation sound, they will not react to any noise band as narrow as 120 Hz (Neuweiler and Vater 1977). There are neurons in the inferior colliculus which do not respond to a pure tone but vigorously fire when the carrier frequency is modulated (Schuller, 1979). Apparently specialization (or selectivity or special rejection mode) is achieved parallelly within peripheral and central stages.

The auditory system deals with continua of bioacoustic dimensions such as frequencies, frequency modulations, narrow noise bands, amplitude modulations and so on (see also Scheich, 1977). Specialized neurons respond exclusively or preferentially to certain sectors or combinations of these continual dimensions and reject other signals.

The main issue is the problem if these specialized neurons are matched to behaviorally significant signals only or if specialized neurons exist along all bioacoustic dimensions irrespective of the known existance of correspondant signals in the acoustic world of a species. To be sure, there is ample evidence in mammals and in birds that specialized neurons matched to behaviorally relevant or species-specific signals do exist. However, since we hardly offer any other complex stimuli other than behaviorally relevant ones to an auditory system we can not expect to find specializations other than those that we asked for.

In FM- as well as in "long CF/FM"- bats, specialized neurons have been recorded not matching the structure of the specific echolocation sounds (Suga, 1973; Neuweiler and Vater, 1977; Vater et al., 1979). Recently it was shown that some bat species may vary the time structure of their echolocation sounds to a greater extent than breviously found and one might argue that these behaviorally unspecified specialized neurons may match to echolocation sounds or to communication calls hitherto undiscovered. In no case has the specifity of these neurons been tested with real signal structures emitted by the bat. Thus these neurons may be still considered as evidence of specialization in the auditory nervous system not implying immediate behavioral relevance. Apparently the brain has the capacity of discriminating complex stimuli in a much larger matrix surpassing the limited range of what we call behaviorally relevant.

Moreover recent research in awake squirrel monkey indicate
that detector neurons in the auditory cortex change their selective
(or rejective) qualities in time to an unexpected extent (Manley
et al., 1978). Apparently those detector neurons have much more dynamic properties than we like them to have for our theories. Unfortunately due to convenience of the currently adopted neurophysiological data collecting and processing methods longlasting response
variability of neurons is hardly recorded and induces neurophysiologists to look at neuronal properties statically. In this context
the few studies at hand already indicate that auditory neurons may
dramatically change their response characteristics related to stimulus history. We should design dynamic auditory detector hypothe-
sises by not only playing around in the threedimensional space of
a brain, i.e. in the frequency-domain, but by expanding our think-
ing into the time dimension. Any scheme of the auditory system or
a subsystem not integrating its functions and functional states in
time is unrealistic.

All data presented here indicate that specialized neurons
exist parallel at peripheral and central levels of the neural auditory system and along all bioacoustic dimensions. Accordingly we
expect a matrix of specialized neurons far beyond those behaviorally relevant. Within the large range of specialized neurons those
may become overrepresented by experience or by heritage which match
to frequently encountered auditory signals. The outlines presented
here are compatible with the concept of a "focal property selective
system" forwarded by Scheich (1977).

THE MOVING PREY DETECTING SYSTEM OF THE HORSESHOE BAT
- A CASE OF MULTIPLE AND CONVERGING SPECIALIZATIONS

The long constant frequency (CF) component of 10 to 100 msec
duration emitted by horseshoe bats is an excellent signal for detecting frequency (FM) and amplitude (AM) modulations caused by
sound reflections from moving wing beating prey. The emitted frequency of 81 to 85 kHz (interindividual variation) is simply a carrier frequency for those complex modulated echo informations.

The ear of the horseshoe bat has a built-in mechanical filter
narrowly tuned to the individual carrier frequency of the echoes
(Fig. 1 and 2, Neuweiler, 1970; Bruns 1976a, b; Schuller and Pollak, in press; see also chapter 1 of this review). The narrow frequency band of about 82 to 86 kHz containing the biologically relevant signals from moving prey is represented on the basilar membrane in a magnified, vastly expanded fashion as a kind of an acoustic fovea (Fig. 2., Bruns, 1976b; Schuller and Pollak, in press).

This foveal arrangement results in an extremely sharp tuning
with Q-values up to 500 in the afferent influx for the frequency

band of 82 to 86 kHz. This sharp tuning is preserved throughout the auditory system with Q_{10dB}-values of single units frequently as high as 400 to 500 (highest Q-values in other mammals and in FM-bats: 20 to 60). In horseshoe bats precise tuning has been found in the auditory nerve (Neuweiler and Vater, 1977), in the cochlear nucleus (Suga et al. 1976), in the inferior colliculus (Neuweiler, 1970; Möller et al. 1978) and in the auditory cortex (Ostwald, 1978). However, it has to be emphasized that the extremely sharp tuning in the filter band is only due to mechanical specializations of the horseshoe bat's cochlea. There occurs no additional neuronal tuning, as expressed by Q-values of the units recorded. As in the eye the fovea and its functional consequences are a result of peripheral receptor organization.

In an evoked potential study, Schuller (1972) discovered that the inferior colliculus responds to frequency modulation depths as small as 10 Hz as long as the carrier frequency lies between 82 and 86 kHz. Schuller postulated that the horseshoe bat takes advantage of the fine FM- and AM-sensitivity for detecting wing beating prey. In a recent behavioral study Goldmann and Henson (1977) proved that this hypothesis is correct in another "long CF/FM"-bat, Pteronotus parnellii.

Collicular single units faithfully encode frequency modulations of the carrier frequency with modulation rates of up to 500 Hz and modulation depths as low as 10 Hz. These neurons resolve frequency changes periodically repeated every 3 - 5 msec (Fig.12 in the next review by Pollak; Schuller, 1979). Schuller also described a great sensitivity to amplitude modulations down to modulation indeces of 3-5%. Whereas sensitivity to amplitude modulations is a general feature of collicular neurons, sensitivity to low frequency modulation depths is restricted to the neurons with best frequencies in the filter band possessing extremely narrow response areas. The FM-sensitivity is the result of this narrow tuning originating in the frequency expanded part of the cochlea, i.e. the acoustic fovea (Schuller, 1979).

Consequently high sensitivity to FM-signals in neurons with best frequencies between 82 and 86 kHz is already encountered in the cochlear nucleus (Vater, in prep.). Responsiveness to frequency modulation rates is even better in the anteroventral cochlear nucleus reaching a maximum of 900 Hz compared to maximal 500 Hz in the inferior colliculus. Interestingly FM-signals are differentially processed in the different subdivisions of the cochlear nucleus. Tonically (primary like) responding neurons, preferentially occurring in the anteroventral cochlear nucleus most faithfully encode timestructures of a sinusoidally frequency modulated signal, whereas phasically responding , built-up and complex neurons,mostly occurring in the dorsal cochlear nucleus, are less suitable or uncapable of following periodic modulations (Vater, in prep.).

Specialized neurons were also found in the inferior colliculus as well as in the cochlear nucleus. Many collicular neurons preferentially or exclusively encoded FM - signals only at low or medium sound intensity levels (10 to 60 dB SPL) and rejected FM - signals of higher intensities. This feature was not related to upper threshold tuning curves and may be interpreted as an adaptation to echo listening (Pollak and Schuller, in prep.). Other neurons in both cochlear and collicular nuclei only responded to FM-stimuli but failed to fire to the carrier frequency at any intensity or to any amplitude modulation.

These detailed recent studies amply prove that the capacity to respond to minute frequency and amplitude modulations again is due to peripheral adaptations and may be fully explained by the narrow tuning areas originating in the cochlea, at least as far as FM-encoding is concerned. Selective responsiveness to modulations of the carrier frequency caused by potential wing beating prey also starts at the level of the cochlear nucleus.

There are more general adaptations of the periphery facilitating hearing of the pure tone echo component in horseshoe bats. Under natural conditions the bat never hears a single echo but has to listen to at least two overlapping signals, the emitted sound and the echo (or a time smeared sequence of echoes).

In collicular neurons of horseshoe bats as of other mammals the response to a second signal is easily inhibited by a preceeding one. However there is an interesting exception to this rule: in neurons with best frequencies within the filter region (82-86 kHz) this inhibition of the response to the second tone (i.e. echo) is completely missing when the frequency of the preceeding signal (i. e. the emitted sound) lies in the frequency band of 78 to 82 kHz which actually corresponds to the frequencies emitted by a hunting horseshoe bat. Yet the frequencies of the emitted echolocation sounds not only prevent inhibition but in addition enhance the response to the subsequent echo by increasing firing rates and decreasing thresholds up to 20 dB (Fig. 8.; Möller, 1978). It has to be emphasized that lack of inhibition and facilitation is only elicited by the narrow frequency band of 78 to 82 kHz, actually emitted by a flying horseshoe bat.

We have always suspected that vocalization may neuronally condition auditory neurons to echo - listening. Recently Schuller (in press) has shown that, indeed, some collicular neurons only respond to a pure tone or only encode FM-signals (echoes from wing beating prey) when the horseshoe bat is vocalizing. Acoustic stimulation with the same stimuli but without vocalization does not elicit any response in these neurons (Fig. 10). This fascinating interaction of vocalization and auditory system shall be dealt with in the last chapter.

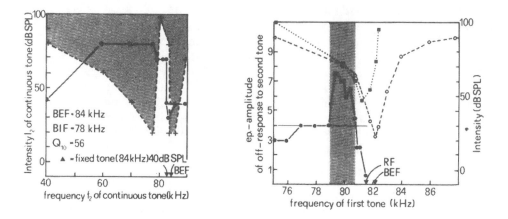

Fig. 8. Lack of inhibition and enhancement of response by two tone stimulation in inferior colliculus neurons of Rhinolophus ferrumequinum. Left figure: response to a fixed tone (echo) is inhibited by a continuous tone (shaded area) except to a narrow frequency range of 78-82 kHz, corresponding to the frequencies actually emitted by a flying horseshoebat. Solid line and black dots: tuning curve of the unit; BEF Best Excitatory Frequency; BIF Best Inhibitory Frequency. Right Figure: Frequency range of a first tone (hatched column) causing facilitation of response to a second fixed tone as shown by evoked potential amplitudes (solid line, black dots). Fixed tone : 60 dB SPL, frequency at Reference Frequency (RF), i.e. frequency of the echo, preceeded by a first tone of variable frequencies (abscissa) and 90 dB SPL. Delay time between both stimuli 10 msec, stimulus duration 30 msec each. Horizontal left bar marks e.p.amplitude for fixed tone alone. Right hand ordinate: thresholds of e.p. to single tone stimulation. Broken line and open circles: Threshold curve for on-responses. Dotted line and black squares: Threshold curve for off-responses. Note: Frequency range causing facilitation coincides with that causing lack of inhibition (left figure). (From Möller, 1978).

Finally the flying horseshoe bat has to eliminate its own travelling speed from the movement detecting system, since the own flight speed shifts the carrier frequency of the echoes to higher values and out of the center frequency of the cochlear filter. This uncoupling of the movement detector from the bat's own movements is achieved by Doppler effect compensation (Schnitzler, 1968). The bat, after listening to the echo frequency , decreases the frequency of the next emitted sound in such a way that the carrier frequency of the echoes is kept close to the center frequency of the cochlear filter at 83 kHz. The bat maintains the echo frequency with

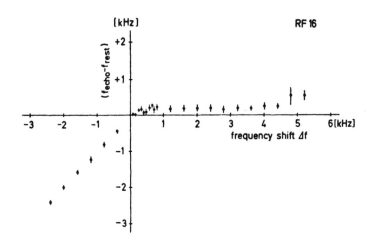

Fig. 9. Doppler-shift compensation in horseshoe bats (Rhinolophus ferrumequinum). Ordinate: Difference between heard echo frequency (f_{echo}) and frequency emitted (f_{rest}). Abscissa: experimentally induced frequency shifts of the echo (Δf). For frequency shift up to 4 kHz above f-rest the bat keeps the echofrequency close to f-rest by lowering the emitted frequency (horizontal sequence of dots in the upper right quadrant). Frequency shifts below f-rest are not compensated for (lower, left quadrant). (From Schuller et al.,1974).

an accuracy of ± 50 Hz (Schuller et al., 1974; Fig. 9.). This sophisticated feed-back system for locking the echo carrier frequency to the center frequency of the receiver (f_{rest} in Fig. 9.) only functions when emitted sound and returning echo overlap by at least 15 msec (Schuller, 1977). The capacities of the feed-back control system have been elegantly described by Schuller et al. (1975). Ingeniously the feed back system reacts so sluggishly that it only compensates frequency shifts of the complete echoes and never reacts to short shifts occurring within the pure tone part which actually are the signals the bat needs to hear.

All these many specializations of mechanical and neuronal nature enumerated in this chapter occur between cochlea and inferior colliculus and converge into a highly sophisticated echolocaction system specialized in detecting moving prey. The adaptations to moving prey detection so far described are: 1) sharply tuned mechanical filter in the cochlea matched to the 83 kHz carrier frequency of the echo; 2)an acoustical fovea in the cochlea matched to the filter frequency range and subsequent overrepresentation of the filterfrequencies throughout the auditory nervous system; 3) high sensitivity to minute frequency modulations of the echo carrier frequency; 4) lack of inhibition and facilitation of neuronal respon-

ses to echoes by frequencies of preceeding emitted sound (Fig. 8.);
5) conditioning of auditory neurons to FM-stimuli by vocalizing an
echolocation sound; 6) Dopplershift compensation (Fig. 9.).

A final question has to be answered. Many insectivorous bats
hunt insects on the wings. Why do only a few of them (e.g. mustache bat and horseshoe bat) emit long pure tones exploiting the encoding capacities of this signal for movements?

For a bat hunting in open air any echo received may indicate
a potential prey (or predator!) since obviously no nonflying object
will be on air. Thus for such a bat there is no need of differentiating living prey against stationary objects.

In contrast, a bat searching for flying insects close to foliage or any other dense background will receive a sequence of time
smeared echoes destroying the signal structure (e.g. FM-sweeps)
and deteriorating the detectability of the prey. However, time
smeared echoes of a pure tone signal retain the signal structure,
namely a pure tone, as long as the background is stationary whereas
any moving object in front of the dense target will "pop up" in the
pure tone as a complex small modulation. Even when the background
is randomly moving (e.g. foliage) wing beating prey may be easily
discriminated against the random modulations of the echo (caused by
background movements) by the periodicity of modulations coming from
beating wings. The latter argument may be also the reason why these
bats use very long pure tones in search flights.

Thus it appears that the use of a long pure tone component in echolocation is an adaptation for hunting flying insects close to dense
backgrounds (see also Glaser, 1974 and Brosset, 1966). Of course
this movement detecting system only works when the receiver is adapted to the pure tone as is the case in horseshoe bats and in mustache bats (Fig. 1.).

INTERACTION OF VOCALIZATION AND AUDITORY SYSTEM

Vocalization centers may control auditory input in two ways:
a) peripherally by eliciting contraction of middle ear muscles just
prior to or during vocalization or
b) directly by exerting some neuronal influence onto auditory neurons.

Both ways have been shown to exist: the peripheral one by Henson (1965) and Suga and Jen (1975) and the central one by Suga and
Schlegel (1972) and Suga and Shimozawa (1974). In Myotis grisescens
the response of lateral lemniscal neurons to vocalized echolocation
sounds was attenuated up to 25 dB compared to responses to an identical play-back. Since this attenuation was never observed in sum-
mating potential recordings of the auditory nerve (N_1) this neuronally conveyed damping influence elicited by vocalization commands

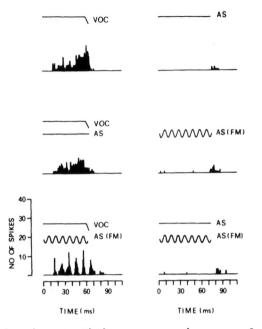

Fig. 10. Vocalization conditions responsiveness of collicular neurons in Rhinolophus ferrumequinum to frequency modulated signals.
PST-histograms to a vocalized echolocation sound alone (VOC); to
an artificial pure tone at the same frequency and intensity as the
vocalized pure tone part of the echolocation sound (AS); to a
simultaneous presentation of vocalization and the pure tone (VOC+
AS) ; to a sinusoidally frequency modulated tone with carrier frequency equal to that of the vocalized tone and a modulation depth
of + 500 Hz (AS(FM)); to a combination of vocalization and frequency modulated signal (echo) (VOC+AS(FM)); and to a combination of
a pure tone at the frequency of the vocalized pure tone part and
the frequency modulated signal (AS+AS(FM)). Note: Frequency modulation is strongly encoded during an ongoing vocalization. The
same neuron fails to respond at all when stimulated with the same
signals but without vocalization (graphs in the lowest row).
(From Schuller, in press.).

has to occur somewhere after the auditory nerve and before the inferior colliculus. This neural attenuation serves a general protection of the ear from selfstimulation by vocalization.

Recently Schuller (in press) reported fascinating specific and
facilitatory influences of vocalization commands on auditory neurons. In the inferior colliculus many units responded differently
to a vocalized and to an identical nonvocalized signal. This vocal
effect was not due to middle ear muscle activities. Among those neurons influenced by vocalization commands were some specialized to

respond to an FM-signal (mimicking an echo returning from a wing beating insect) only when the bat vocalized (Fig. 10.). If the same neuron was stimulated with the same stimulus set in the nonvocalizing bat it did not respond at all. In fact the vocalized signal could not be replaced by any other identical or slightly varied signal in order to evoke the FM-encoding capacity of the neuron. To avoid any misinterpretation it has to be emphasized that only few neurons showed this specific behavior.

Apparently these neurons are sensitized to FM-stimuli by activities of the vocalization system. This enhancing effect is lost when the FM-stimulus is presented 40 to 60 msec after onset of vocalization or after the emitted sound is terminated. Since the duration of the vocalized echolocation sound could not be manipulated it is not clear if the sensitizing effect is exerted as long as vocalization endures or if it operates during a fixed time window of 40 to 60 msec triggered by onset of vocalization.

These new results open up an entirely new aspect of auditory physiology. The interaction of vocalization and auditory system has to be studied neurophysiologically and neuroanatomically in great detail. These linking pathways are not only of interest for a Doppler-shift compensating or any other echolocating bat but also for each acoustically communicating mammal and especially for humans. The few results already available should us make careful in transforming responses of auditory neurons into behavioral relevance.

ACKNOWLEDGEMENTS

This research was supported by DFG Br 593/2, Ne 146/9 and A.v.Humboldt - Foundation.
I thank Rosa Veltlin and Beatus Habkern for valuable assistance in preparing this review.

REFERENCES

Brosset, A., 1966, La biologie des Chiropteres, Masson, Paris.
Bruns, V., 1976a, Peripheral auditory tuning for fine frequency analysis by the CF-FM bat, Rhinolophus ferrumequinum. I.Mechanical specializations of the cochlea. J.comp.Physiol.A 106, 77-86.
Bruns, V., 1976b, Peripheral auditory tuning for fine frequency analysis by the CF-FM bat, Rhinolophus ferrumequinum. II.Frequency mapping in the cochlea. J.comp.Physiol.A 106, 87-97.
Bruns, V., in prep., Adaptations for frequency sharpening in the cochlea of the Greater Horseshoe Bat. II. Functional morphology of the basilar membrane and its anchoring system.

Bruns, V., and Goldbach, M., in prep., Adaptations for frequency sharpening in the cochlea of the Greater Horseshoe Bat. I.Receptors and tectorial membrane.

Bruns, V., and Schmieszek, E., in prep., Regional variation in the innervation pattern in the cochlea of the Greater Horseshoe Bat.

Firbas, W., and Einzinger, H., 1972, Über das Ganglion spirale der Chiroptera. Z.Säugetierk. 37, 321-326.

Glaser, W., 1974, Zur Hypothese des Optimalempfangs bei der Fledermausortung. J.comp.Physiol. 94, 227-248.

Goldman, L.J., and Henson, O.W., 1977, Prey recognition and selection by the constant frequency bat, Pteronotus p. parnellii. Behav.Ecol.Sociobiol. 2, 411-420.

Grinnell, A.D., 1973, Neural processing mechanisms in echolocating bats, correlated with differences in emitted sounds. J.Acoust. Soc.Am., 54, 147-156.

Grinnell, A.D., and Hagiwara, S., 1972, Adaptation of the auditory system for echolocation: studies of New Guinea bats. Z.vergl. Physiol. 76, 41-81.

Henson, M.M., 1978, The basilar membrane of the bat, Pteronotus p. parnellii. Amer.J.Anat. 153, 143-158.

Henson, O.W., 1965, The activity and the function of the middle-ear muscles in echolocating bats. J.Physiol. 180, 871-887.

Irving, R., and Harrison, J.M., 1967, The superior olivary complex and audition: a comparative study. J.comp.Neurol., 130, 77-86.

Jen, P.H.-S., 1978, Electrophysiological properties of auditory neurons in the superior olivary complex of echolocating bats. J.comp.Physiol., 128, 47-56.

Jen, P.H.-S., and Suga, N., 1974, Coding of directional information by neurons in the superior olivary complex of bats. J.Acoust. Soc.Amer., 55, S52.

Manley, J.A., and Müller-Preuß, P., 1978, Response variability of auditory cortex cells in the squirrel monkey to constant acoustic stimuli. Exp.Brain Res., 32, 171-180.

Möller, J., 1978, Response characteristics of inferior colliculus neurons of the awake CF-FM bat, Rhinolophus ferrumequinum. II.Two-tone stimulation. J.comp.Physiol. A, 125, 227-236.

Möller, J., Neuweiler, G., and Zöller, H., 1978, Response characteristics of inferior colliculus neurons of the awake CF-FM bat, Rhinolophus ferrumequinum. I. Single-tone stimulation. J.comp. Physiol. A, 125, 217-225.

Neuweiler, G., 1970, Neurophysiologische Untersuchungen zum Echoortungssystem der Großen Hufeisennase Rhinolophus ferrum equinum. Z.vergl.Physiol., 67, 273-306.

Neuweiler, G., and Vater, M., 1977, Response patterns to pure tones of cochlear nucleus units in the CF-FM bat, Rhinolophus ferrumequinum. J.comp.Physiol. A., 115, 119-133.

Ostwald, J., 1978, Tonotope Organisation des Hörcortex der CF-FM-Fledermaus Rhinolophus ferrumequinum. Verh.Dtsch.Zool.Ges., 1978, p.198.

Poljak, S., 1926, Untersuchungen am Octavussystem der Säugetiere
    und an dem mit diesem koordinierten motorischen Apparaten des
    Hirnstammes. J.Psychol.Neurol. 32., 170-231.

Pollak, G., Henson, O.W., and Johnson, R., in press, Multiple spe-
    cializations in the peripheral auditory system of the CF-FM
    bat, Pteronotus parnellii. (J.comp.Physiol.).

Pollak, G., Henson, O.W., and Novick, A., 1972, Cochlear micropho-
    nic audiograms in the pure tone bat Chilonycteris parnellii
    parnellii. Science, 176, 66-68.

Pollak, G., Marsh, D., Bodenhamer, R., and Souther, A., 1978, A
    single unit analysis of inferior colliculus in unanesthetized
    bats: response patterns and spike count functions generated
    by CF and FM sounds. J.Neurophysiol., 41, 677-691.

Pollak, G., and Schuller, G., in prep., Encoding properties of col-
    licular neurons for sinusoidal frequency modulations at dif-
    ferent intensities in Rhinolophus ferrumequinum.

Pye, A., 1964, A comparative anatomical study of the auditory appa-
    ratus of selected members from the orders Chiroptera and
    Rodentia. Ph.D. Thesis, London Univ.

Pye, A., 1966, The structure of the cochlea in Chiroptera. I.Micro-
    chiroptera: Emballoruroidea and Rhinolophoidea. J.Morph. 118,
    495-510.

Pye, A., 1967, The structure of the cochlea in Chiroptera. III.Mic-
    rochiroptera: Phyllostomatidae. J.Morph. 121, 241-254.

Radtke, S., 1979, Struktur und Verschaltung des Hörcortex der Gros-
    sen Hufeisennase (Rhinolophus ferrumequinum). Staatsex.Thesis,
    Univ. Frankfurt.

Ramprashad, F., Money, K.E., Landolt, J.P., and Laufer, J., 1978,
    A neuro-anatomical study of the cochlea of the little brown
    bat, Myotis lucifugus. J.Comp.Neurol. 178, 347-364.

Scheich, H., 1977, Central processing of complex sounds and feature
    analysis, in "Recognition of complex acoustic signals" Th.H.
    Bullock ed., Dahlem Konferenzen, Berlin.

Schlegel, P., 1977, Directional coding by binaural brainstem units
    of the CF-FM bat, Rhinolophus ferrumequinum. J.comp.Physiol.
    118, 327-352.

Schlegel, P., in prep., Coding of moving sounds by binaural brain-
    stem units in the Greater Horseshoe Bat, Rhinolophus ferr.

Schnitzler, H.-U., 1968, Die Ultraschall-Ortungslaute der Hufeisen-
    Fledermäuse in verschiedenen Orientierungssituationen. Z.vergl.
    Physiol. 57, 376-408.

Schnitzler, H.-U., Suga, N., and Simmons, J.A., 1976, Peripheral
    auditory tuning for fine frequency analysis by the CF-FM bat,
    Rhinolophus ferrumequinum. III. Cochlear microphonics and
    auditory nerve responses. J.comp.Physiol. A, 106, 99-110.

Schuller, G., 1972, Echoortung bei Rhinolophus ferrumequinum mit
    frequenzmodulierten Lauten. Evoked potentials im Colliculus
    inferior. J.comp.Physiol. 77, 306-331.

Schuller, G., 1974, The role of overlap of echo with outgoing echo-
    location sound in the bat Rhinolophus ferrumequinum. Naturw.

61, 171-172.

Schuller, G., 1979, Coding of small sinusoidal frequency and amplitude modulations in the inferior colliculus of "CF-FM" bat, Rhinolophus ferrumequinum. Exp. Brain Res. 34, 117-132.

Schuller, G., in press, Vocalization influences auditory processing in collicular neurons of the CF-FM bat, Rhinolophus ferrumequinum. (J. Comp. Physiol.).

Schuller, G., Beuter, K., and Rübsamen, R., 1975, Dynamic properties of the compensation system of Doppler shifts in the bat Rhinolophus ferrumequinum. J. Comp. Physiol. 97, 113-125.

Schuller, G., Beuter, K., and Schnitzler, H.-U. 1974, Response to frequency shifted artificial echoes in the bat Rhinolophus ferrumequinum. J. Comp. Physiol. 89, 275-286.

Schuller, G., and Pollak, G., in press, Disproportionate frequency representation in the inferior colliculus of horseshoe bats: evidence for an acoustic fovea. (J. Comp. Physiol.)

Schweizer, H., 1978, Struktur und Verschaltung des Colliculus inferior der Großen Hufeisennase (Rhinolophus ferrumequinum). Diss.-Thesis, University of Frankfurt, Germany.

Shimozawa, T., Suga, N., Hendler, P., and Schuetze, S., 1974, Directional sensitivity of echolocation system in bats producing frequency modulated signals. J. Expl. Biol. 60, 53-69.

Simmons, J. A., 1971, Echolocation in bats: signal processing of echoes for target range. Science 171, 925-928.

Suga, N., 1964, Single unit activity in the cochlear nucleus and inferior colliculus of echolocating bats. J. Physiol. 172, 449-474.

Suga, N., 1969, Classification of inferior collicular neurons of bats in terms of responses to pure tones, FM sounds, and noise bursts. J. Physiol. 200, 555-574.

Suga, N., 1973, Feature extraction in the auditory system of bats, in: "Basic Mechanisms in Hearing", A. R. Møller, ed., pp. 675-744. Academic Press, New York and London.

Suga, N., and Jen, P.H.-S., 1976, Disproportionate tonotopic representation for processing CF-FM sonar signals in the mustache bat auditory cortex. Science, 194, 542-544.

Suga, N., and Jen, P.H.-S., 1975, Peripheral control of acoustic signals in the auditory system of echolocating bats. J. Expl. Biol. 62, 277-311.

Suga, N., Neuweiler, G., and Möller, J., 1976, Peripheral auditory tuning for fine frequency analysis by the CF-FM bat, Rhinolophus ferrumequinum. IV. Properties of peripheral auditory neurons. J. Comp. Physiol. A, 106, 111-125.

Suga, N., and Schlegel, P., 1972, Neural attenuation of responses to emitted sounds in echolocating bats. Science, 177, 82-84.

Suga, N., and Shimozawa, T., 1974, Site of neural attenuation of responses to self-vocalized sounds in echolocating bats. Science, 183, 1211-1213.

Tobias, J. V., Ed., 1970, Foundations of modern auditory theory. Academic Press, New York and London.

Vater, M., in prep., Responses of cochlear nucleus neurons to comp-
    lex stimuli in the CF-FM bat, Rhinolophus ferrumequinum.
Vater, M., and Schlegel, P., 1979, Comparative auditory neurophysio-
    logy of the inferior colliculus of two Molossid bats, Molossus
    ater and Molossus molossus. II. Single unit responses to fre-
    quency-modulated signals and signal and noise combinations.
    J. Comp. Physiol. A, 131, 147-160.
Vater, M., Schlegel, P., and Zöller, H., 1979, Comparative auditory
    neurophysiology of the inferior colliculus of two Molossid bats,
    Molossus ater and Molossus molossus. I. Gross evoked potentials
    and single unit responses to pure tones, J. Comp. Physiol. A,
    131, 137-146.
Wilson, J. P., 1977, Towards a model for cochlear frequency analysis,
    in: "Psychophysics and Physiology of Hearing", E. F. Evans and
    J. P. Wilson, Eds., Academic Press, New York and London.
Zook, J. M., and Casseday, J. H., in press, Identification of audi-
    tory centers in lower brain stem of two species of echolocating
    bats: evidence from injection of horseradish peroxidase into
    inferior colliculus.
Zook, J. M., Siegel, B. M., and Casseday, J. H., in prep., Projec-
    tions to inferior colliculus in the mustache bat, Pteronotus
    parnellii parnellii as seen by retrograde transport of horse-
    radish peroxidase.

ORGANIZATIONAL AND ENCODING FEATURES OF SINGLE NEURONS IN THE

INFERIOR COLLICULUS OF BATS

George D. Pollak

Department of Zoology, University of Texas

Austin, Texas 78712 U.S.A.

INTRODUCTION

The past five years have witnessed a remarkable increase in the number of studies concerned with the neural basis of echolocation. These studies have dealt with features of the peripheral auditory system (Bruns 1976a,b; Henson 1978; Pollak et al. 1979; Schnitzler et al. 1976; Suga et al. 1975), the lower central parts of the auditory pathway, such as the cochlear nucleus (Neuweiler and Vater 1977) and superior olivary complex (Jen 1978), as well as with higher regions such as the inferior colliculus (Möller et al. 1978a,b; Schuller, 1979a; Schuller and Pollak 1979) and cortex (reviewed by Suga). But beyond the sheer proliferation, the reports have provided much more sophisticated and detailed insights into the principles that underly both the structural and functional organization of the Chiropteran auditory system than were known before. In the previous section, Dr. Neuweiler reviewed the data from studies of the cochlea and lower brain stem centers of the horseshoe bat whereas in the following section Drs. O'Neill and Suga have summarized their findings from the cortex of the long CF/FM mustache bat. In this chapter, I shall attempt to bridge the intervening gap by reviewing the results obtained from experiments on the inferior colliculus.

Recent neurophysiological studies have focused on two types of echolocating bats. One type is characterized by bisonar cries which are composed primarily of loud but brief downward sweeping frequency modulated components. I shall refer to this group as the FM bats. Among these are bats of the genus _Myotis_ and _Tadarida_, the two genera which have been most intensively studied.

The second type are the long CF/FM bats so called because their echolocation calls are composed of a long constant frequency (CF) component followed by a brief downward sweeping FM portion. The best known of the long CF/FM bats are the greater horseshoe bat, Rhinolophus ferrumequinum, and the mustache bat, Pteronotus parnellii. Since the behavior as well as the neurophysiology differ in a number of significant ways between the FM as compared to the long CF/FM bats, I shall treat them in separate sections. In the section below I shall first review the results from neurophysiological studies conducted on FM bats and in the subsequent section results from recent studies concerned with the processing of the CF component in long CF/FM bats will be presented.

FM BATS
Tuning Curves

One of the fundamental characteristics of all auditory neurons is the tuning curve. The tuning curve defines the range of frequencies capable of evoking discharges in a given neuron and thus is a reflection of the range of supra-threshold excitatory synaptic inputs, originating in the cochlea and auditory nerve, that are mapped upon an individual neuron. The majority of tuning curves obtained from collicular neurons are more or less triangular in shape with the frequency at the apex of the triangle requiring the lowest intensity to evoke discharges. (Grinnell 1963a; Suga 1964b; Pollak et al. 1978). This frequency will hereafter be called the best frequency (BF) of the neuron.

The tuning curves of the population of collicular neurons differ with regard to their widths (Suga 1964b; Pollak et al. 1978). Some are quite wide and receive excitatory inputs from a large region of the cochlea while others are narrower reflecting a more restricted excitatory input from the cochlea. The sharpness of a tuning curve is typically expressed as a Q value, defined as the BF divided by the bandwidth of the tuning curve at a level 10 db above the BF. Hence, a high Q value means that the tuning curve is very sharp while a small Q value means that the tuning curve is broad.

In FM bats, such as Myotis and Tadarida, the majority of collicular neurons have relatively broad tuning curves (Suga 1964b; Pollak et al. 1978). The Q values typically range from 5 to 10 or 15 with only a very small number of neurons having Q values above 20, i.e., only a small segment of the population is sharply tuned.

Auditory neurophysiologists commonly interpret the sharpness of tuning as an indication of how well the system can discriminate small frequency differences. A neuron having a small

Q value is thought to be poorly suited for distinguishing or recognizing small frequency differences because the cell responds to a wide range of frequencies. Thus, an organism whose nervous system is composed only of broadly tuned neurons is considered to have a lesser ability to discriminate small frequency differences than is an animal having a large population of very sharply tuned neurons. Although this interpretation may be true for some neurons, I shall present evidence in a later section which demonstrates that in some neurons at least, the tuning curve has no predictive value for ascertaining a neuron's frequency selectivity.

Tonotopic Organization

Whereas the tuning curve is a measure of the extent to which the cochlea maps upon an individual neuron the manner in which BFs vary as a function of position within the IC provides an index of how the cochlea maps upon the entire expanse of this nuclear region. Studies of FM bats indicate that the IC is tonotopically organized where the mapping of frequency is expressed as a systematic shift in the BF. In both Myotis (Grinnell 1963a) and Tadarida (Pollak et al. 1978) low frequencies are represented in the dorsal portions of the IC while higher frequencies are represented in progressively more ventral parts of the colliculus. Frequency, then, is mapped upon the IC in a systematic fashion which, in principle, is typically mammalian and bears a striking resemblance to the tonotopic organization of the colliculus of other mammals that have been studied (Aitkin 1976).

Discharge Patterns Evoked by Tone Bursts

When tone bursts at the frequency of the neuron's BF are presented, a wide variety of discharge patterns can be evoked among the population of collicular neurons (Suga 1973; Pollak et al. 1978; Vater et al. 1979; Vater & Schlegel 1979). These patterns however fall into two general categories: 1) phasic patterns and 2) tonic or sustatined patterns. Within the category of phasic responders, three basic types are recognized (Pollak et al. 1978): phasic on neurons, phasic on-off neurons, and phasic burst neurons. Several varieties of sustained patterns can also be distinguished. The two basic types are found in roughly the following proportions: phasic 78%, sustained 22%.

I shall not discuss further the discharge features evoked by tone bursts. The reason is that very different results obtain when FM bursts, simulating the natural orientation cries, are presented (Pollak et al. 1977c, 1978). These differences are reflected in two general ways. First, there is a change in the proportion of units having phasic as compared to tonic discharge

patterns. Whereas about 20% of the neurons have tonic patterns
when driven with tone bursts, this pattern never occurs with FM
signals. Second, a unit having a phasic pattern to both FM and
tone bursts may exhibit radical differences in latency, synchrony
and firing rate changes with intensity depending upon which signal is presented. In short, the encoding features of a neuron
may be, and usually are, considerably different with FM signals
as compared to its discharge characteristics observed with tone
bursts. The important point is that since loud FM bats use, for
the most part, brief FM signals for echolocation the response
features evoked by tone bursts cannot be invoked in explanations
of how neurons encode information extracted from echoes.

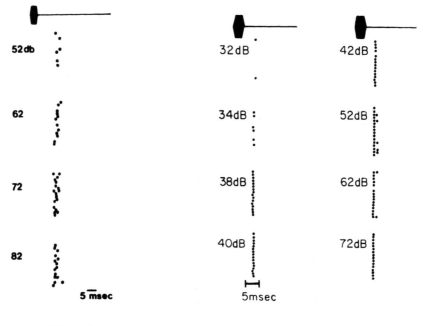

Fig. 1   Fig. 2

Fig. 1. pELR pattern evoked by 4.0 msec FM bursts that swept downward from 50-25 kHz. Threshold was about 47 dB. Latencies were
52 dB, 14.9 ± .98 msec; 62 dB, 13.87 ± 1.08 msec; 72 dB, 13.23 ±
.73 msec; 82 dB, 13.06 ± .65 msec.

Fig. 2. An example of a pCLR pattern. Signals were 4.0 msec FM
bursts that swept downward from 50-25kHz. Threshold was 32dB. Latencies were 34 dB, 7.83 msec ± 245 μsec; 38 dB, 7.20 msec ± 150
μsec; 72 dB, 6.90 msec ± 117 μsec. Note overall latency change
determined 2dB above threshold (34dB) and at 40 dB above threshold
(72dB) was only 0.93 msec. From Pollak et al. 1978.

Discharge Patterns Evoked by FM Signals

Brief downward sweeping FM signals evoke predominately phasic on discharges from collicular neurons in Mollosids (Pollak et al. 1978; Vater and Schlegel 1979). The phasic on patterns, however, are not homogeneous and can be divided into two subgroups. First, are neurons we have called phasic erratic latency responders (pELRs). The feature which characterizes these neurons is the large shifts in latency when an FM signal having a particular sweep, intensity and duration is repeatedly presented. (Fig. 1). The second type of phasic on pattern evoked by FM signals is quite dramatic and the significance of the response characteristics for echolocation is much better understood than for any other type of neuron. We call these phasic constant latency responders (pCLRs) (Suga 1970; Pollak et al. 1977c; Vater and Schlegel 1979) and they are distinguished by four features (Fig. 2): 1) a phasic on response pattern where the neuron fired at most two spikes to an FM burst; 2) a reliable firing to every or nearly every stimulus presentation for intensities 2-6 dB above threshold or greater; 3) a highly consistent latency of initial firing for a wide range of intensities; and 4) a relatively constant mean response latency where the average latency would change at most by 1.75 msec over a 40 dB intensity range.

It is of some interest to compare the response features of pCLRs when driven with tone bursts with the response features evoked by brief FM signals. Consider first that some pCLRs are unresponsive to any tonal stimulation while others will discharge to tone bursts (Pollak et al. 1978). Second, many, but not all, of the pCLRs that are responsive to tone bursts exhibit a pELR pattern to tone bursts but have a pCLR pattern to FM signals (Pollak et al. 1977c). In other words, in many units the discharges fall into tight registry with FM signals whereas when tone bursts are presented the firing latencies are erratic from stimulus to stimulus. Third, and most importantly, all pCLRs that can be driven with tonal stimulation have wide tuning curves (Fig. 3) where the Q values range from about 3-8 (Bodenhamer et al. 1979). The commonly accepted interpretation of such wide tuning is that these neurons are not frequency selective. However, and this point is critical, we can demonstrate that the tight registration of latencies is a direct consequence of an exceptionally high degree of frequency selectivity which only becomes apparent when the stimuli are FM bursts.

When FM signals are presented the pCLRs appear to respond to only one frequency or a very narrow band of frequencies in the FM signal. When such a brief FM burst is repeatedly presented, the neuron is fired by, and thus locks onto, the same frequency component or narrow frequency band each and every time and, as a consequence, the latencies fall into tight registration.

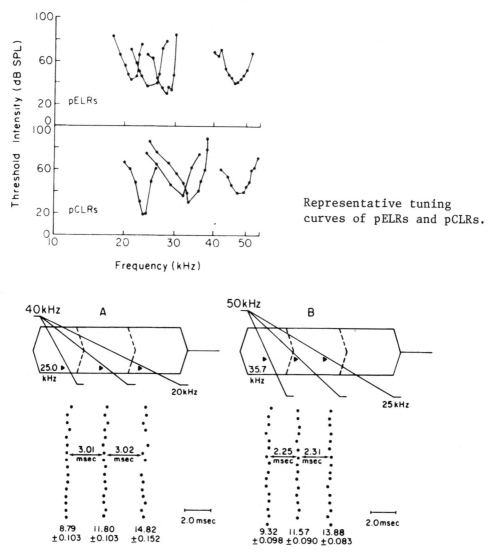

Representative tuning curves of pELRs and pCLRs.

Fig. 3. pCLR sensitivity to a particular EF. Each dot column represents spikes elicited by 16 presentations at a given duration. The dot columns have been aligned directly beneath the FM signals to show how they line up with the proposed EF in each sweep (arrow-heads). Each dot column's mean response latency and standard deviation is shown below that column (in msec). In A, the response latency increased by 3.01 msec. when signal duration increased from 4.0 to 8.0 msec and by 3.02 msec when the duration increased from 8.0 = 12.0 msec. The 25 kHz constant shifted by 3.01 msec for each change in duration & thus the neuron is "following" the 25 kHz component of the FM pulse. A similar "following" is shown for another pCLR in B but this unit was following the 35.7 kHz component. From Bodenhamer et al. 1979.

This feature is illustrated by two pCLRs in Fig. 3. The
rationale of this experiment is that if these neurons are synchronized to one frequency component in an FM burst, then the
excitatory frequency (EF) will shift in time as the duration of
the FM signal is increased and the discharge latency should
shift by an equivalent amount of time. That this is indeed the
case is demonstrated by the two units in Fig. 3.

The implications of these features for the processing of
echolocation signals are considerable. These units are highly
selective for both temporal and spectral attributes of the
biosonar signals and are likely candidates for encoding both
target distance or range as well as the spectral composition of
the echo which could be important for target identification.
Below, I shall discuss these features in greater detail.

The fact that pCLRs record the temporal event for a particular frequency in an FM burst means that these neurons can
serve as precise time markers. That is to say, such neurons
should encode the time of occurrence of a particular frequency
component in the emitted pulse and the time of occurrence of
the same frequency component in the returning echo. The period between the occurrences of a frequency in the pulse and echo is a
function of the pulse-echo delay, i.e., target distance. As shown
in Fig. 4, pCLRs can encode the temporal separation between two FM
signals with remarkable precision and, as a consequence, are almost
certainly conveying distance information to higher auditory centers.

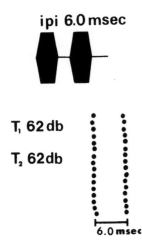

ipi 6.0 msec

FM 50-25 kHz
4 msec signals

T₁ 62 db

T₂ 62 db

Fig. 4. An example of the
precision with which pCLRs
can encode the temporal separation between two FM signals.

The timing precision illustrated by the pCLR in Fig. 4 is a
necessary but not a sufficient demonstration to warrant a conclusion that pCLRs are encoding range information. The two signals in Fig. 5 were both of equal intensity and were presented
at only one time interval. Since range must be evaluated at a
variety of distances and with a wide variety of echo intensities,
it is important to demonstrate that discharge characteristics,
similar to those shown in Fig. 4, can be elicited with signals
simulating the spectrum of pulse-echo combinations that would
occur during echolocation.

Two features of the pCLRs are important in this regard.
First, is that the EF observed for a pCLR is invariant with intensity (Bodenhamer et al. 1979). Thus, the same frequency in both
the loud pulse and fainter echo will cause the neuron to fire and
this feature will serve to preserve the timing accuracy. The
second feature concerns the fast recovery times of pCLRs where
the presentation of a loud initial signal has almost no influence upon the response to a much fainter second signal following shortly thereafter (Pollak et al. 1977a,b,c). These

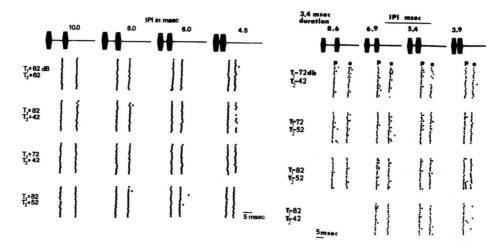

Fig. 5. Recovery in two collicular neurons. All signals were FM
bursts sweeping downward for 50-25 kHz. The intensity of the first
and second signals are shown as T_1 and T_2 respectively. Both neurons are phasic constant latency responders (pCLRs). Note the
fast recovery & the precision with which the response intervals
follow the inter-pulse-interval (IPI). Each pair of dot columns
was generated by 16 repetitions of a simulated pulse-echo-combina-
tion.

features, then, allow the pCLRs to encode the temporal period
between the pulse and echo under most of the conditions occuring during echolocation, as shown in Fig. 5.

The precision with which pCLRs encode the temporal interval
between two FM bursts, however, is influenced by the relative
intensities of the two signals. Louder signals evoke discharges
at shorter latencies than do fainter signals. It follows then
that a loud pulse followed shortly thereafter by a fainter echo
will, in any given pCLR, evoke two spikes but the discharge interval will be slightly longer than the temporal interval of the
signals. In this regard, it is most important to note that one
feature of the pCLRs is that their discharge latencies increase
on the average by only 0.75 msec over a 30 dB range of intensities
(Pollak et al. 1977c). In other words, if the bat emits a pulse
where the intensity reaching the cochlea is 82 db and receives
an echo 5 msec later having an intensity of 52 dB, the most
common pCLR will respond with discharges separated by 5.75 msec.
The unit, then, would encode a distance of 98 cm instead of the
true distance of 86 cm, an error of 12 cm. The error depends
only upon the pulse-echo intensity difference, the smaller the
difference the less the timing error. The important point is that
the possible range of errors is sharply limited because of the
small latency variation with intensity.

Such errors in an individual neuron are not difficult to re-
concile with the ranging precision required of bats. The argument is that the timing error depends only upon pulse-echo
intensity differences and since the bat encounters these errors
throughout life, it must somehow learn to make appropriate
corrections. In short, the bat could easily associate each pulse-
echo intensity difference with a given error and thereby correct
for it.

The spectrum of the biosonar signals could also be encoded by
the pCLRs. Consider that each pCLR discharges to only one frequency component in the FM sweep and that pCLRs have been found
with EFs from 24.5 to 45 kHz. Since the IC is tonotopically
organized, it follows that the pCLRs tuned to the various frequencies in the echolocation cries are topographically organized
with neurons discharging to the lower frequencies being situated
more dorsally while neurons discharging to the higher frequencies
in the biosonar signals being situated more ventrally.

The echoes bats receive from different targets vary in spectral
fine structure depending on the nature of the target. These echoes
may resemble filtered replicas of the emitted pulse with some targets acting primarily as high pass filters, others as low pass filters
and others as variable notch filters (Griffin, 1967). Such echo differences are important cues for target evaluation and since each pCLR is

triggered by only a small frequency band, a population of these cells, each tuned to a slightly different EF and arranged in a topographic manner within the IC, should be capable of encoding the fine structure of these echoes by a place code within the brain and hence be of value for encoding target features.

In summary, the pCLR's sensitivity to a particular EF is valuable to FM bats in at least two ways. First, it leads to a tight firing registration enabling the pCLRs to serve as pulse-echo time markers, and thereby encode target range information. Second, since each pCLR is fired by only a narrow frequency band of the FM signal, populations of pCLRs should be able to encode echo spectral characteristics and thus convey information about the physical nature of targets as well as target range.

Further Coding of Target Range

A very recent report by Feng et al. (1978) is highly suggestive of the way in which range information may be processed by higher centers. These workers recorded from individual neurons in the intercollicular nucleus, lying between the inferior and superior colliculi, of the loud FM bat, Eptesicus fuscus. They found neurons that are unresponsive to any tonal or FM stimuli but which would vigorously discharge when a pair of FM bursts were presented at a certain range of pulse-echo delays. The interpretation given to these results was that "target range may thus be encoded by a neural "place" mechanism incorporating different delay-tuned neurons in the central nervous system, presumably at levels above the inferior colliculus".

This is a most exciting hypothesis which, I believe, is likely to be the case. However, the preferred pulse-echo delay times exhibited by intercollicular neurons were quite broad. The width of the delay-tuning was typically 5-8 msec which would not satisfy the ranging requirements of echolocating bats.

Although I disagree with the authors' interpretation of their data, this should not distract from the importance of the finding. At the very least, neurons selective for a range of pulse-echo delays must be important for eliminating the clutter produced by echoes from nearer and more distant objects and may allow the bat to focus upon one, primary target of interest. Furthermore, the demonstration that a population of neurons discharge only to pairs of FM bursts having preferred temporal separations, opens the possibility that other neurons might have delay-tuning properties much sharper than the intercollicular neurons and these could be of value for the accurate encoding of target range, as Feng et al. originally proposed and Suga and O'Neill suggest in the following chapter.

LONG CF/FM BATS

Within the past several years considerable interest has been directed toward the long CF/FM bats. This interest has been generated by the demonstrations that these animals possess remarkable behavioral abilities as well as striking anatomical and physiological features which seem to correlate and provide a system designed to process the CF component of their biosonar cries.

The aspect of these bats that, I believe, has had a profound influence upon those who work with these animals is they emit a pure tone component (Novick 1963, 1971; Schnitzler 1968; Grinnell 1970; Grinnell & Hagiwara 1972). Following from this feature is the assumption that these bats are measuring the relative velocity between themselves and their targets by evaluating the amount of Doppler shift in the echo CF component of their echoes. When this is coupled with the fact that long CF/FM bats Doppler-compensate (Schnitzler 1968; Simmons 1973; Schuller et al. 1974), thereby holding echoes within a narrow band of frequencies only ± 50-100 Hz wide, one is led to conclude that they are interested in pure tones.

This is a most seductive idea. Tone bursts have traditionally been and continue to be the most widely used stimuli in studies of the auditory system and this is true of both vertebrates and invertebrates. Some of the historical reasons as well as the rationale for using tone bursts have been reviewed by Capranica (1972) and need not be repeated here. The point is that the long CF/FM bats emit pure tones and possess nervous systems which are thought by many investigators to be constructed for the evaluation of small frequency differences existing between two tonal signals. These bats, then, appear to be a mammalian subject for which tone bursts are the biologically relevant signals and hence are the stimuli of choice for neurophysiological investigations of the auditory system.

It seems to me, at least, that this is a major conceptual as well as a major strategic point, one which is of some interest to examine in greater detail. In many reports dealing with long CF/FM bats the argument is made that a CF signal is well suited for measuring Doppler shifts, i.e., relative velocity, due to the narrow bandwidth of the signal but, for the same reason, is poorly suited for target characterization, localization and target ranging (Simmons et al. 1975; Simmons 1979; Suga 1977, 1978). More recently, many workers have suggested that the filter system could also be important for prey recognition and identification by encoding the modulation patterns imposed on the echo CF component as it strikes and is reflected from the beating wings of a small insect (Schuller 1972, 1979a; Schuller & Pollak, 1979; Schnitzler 1970, 1978; Henson et al., 1974; Henson & Goldman 1976; Goldman & Henson 1977). Curiously, some investigators hold that the long

CF/FM bats are extracting both a precise estimate of relative velocity and modulation pattern information from the echo CF component
(Suga et al. 1975; Suga 1978). What makes this interpretation curious is that the two hypotheses are, in fact, incompatible. The
evaluation of relative velocity is based on the assumption that the
echo CF is a pure tone, and consequently of very narrow bandwidth,
whereas the prey recognition hypothesis is based on the _fact_ that
echoes reflected from a small insect have periodic amplitude and
frequency modulations (AM and FM) which increase the echo bandwidth
and introduce a significant ambiguity into any velocity measurement.

Goldman and Henson (1977) have presented evidence strongly
indicating that the long CF/FM mustache bat chooses which among
a population of insects it will pursue and capture on the basis
of wing beat patterns. Furthermore, recent neurophysiological
studies have shown that neurons in the cochlear nucleus (Suga et
al. 1975; Suga and Jen 1977; Vater personal communication), inferior colliculus (Schuller 1972; 1979a; Pollak & Schuller 1978), and
auditory cortex (Ostwald, 1979) of long CF/FM bats are exceptionally well suited for encoding both AM and FM modulation patterns.
Considerations, then, both of the type of echoes the bats receive
during echolocation as well as neurophysiological features of
auditory neurons seem to support the idea that the primary function
of the filter system is for target characterization rather than for
relative velocity estimates.

In the following sections I shall review the results obtained
from recent single unit studies of the horseshoe bats' inferior
colliculus conducted in collaboration with Gerd Schuller. It is my
intention to show the advantage that accrues to Doppler-shift compensation for target characterization in light of the organizational features of the IC together with the manner in which collicular
neurons encode FM and AM modulation patterns.

Tuning Curves

Two general types of tuning curves are obtained from collicular
neurons in horseshoe bats: those that are broadly tuned and those
that are sharply tuned (Möller et al. 1978). The broadly tuned
neurons typically have BFs from 9 kHz to about 75-77 kHz. These
neurons rarely have Q values greater than 20 and in this respect
are similar to the collicular neurons in FM bats. In marked contrast, neurons tuned to the filter frequencies, i.e., 78-86 kHz,
are usually very sharply tuned and have Q values between 50-200
with some as sharp as 400-500. The demarcation point is quite
striking, as shown in Fig. 6, and it is this feature of the neurons
tuned between about 77-88 kHz that justifies the term "filter
neurons".

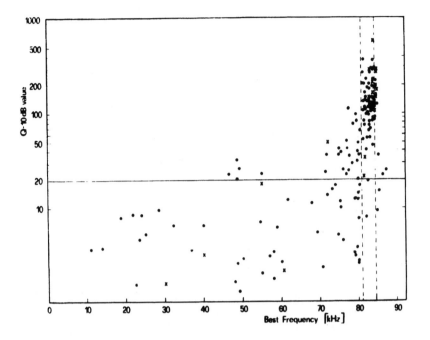

Fig. 6. Q values of collicular neurons recorded from horseshoe bats. From Möller et al. 1978.

Tonotopic Organization of the Horseshoe Bats' Inferior Colliculus

The finding that collicular neurons can be grouped into filter and non-filter neurons is also reflected in the tonotopic organization of the IC. The IC can be functionally divided into a dorsal region, where the BFs range from 9kHz to 75-76 kHz, and a ventral region where the filter frequencies are found (Fig. 7).

The tonotopic arrangement in the dorsal region is typically mammalian and is similar to that found in the IC of FM bats. Proceeding from the dorsal to the more ventral region, the neurons encountered have progressively higher BFs to a depth of approximately 800-1200 microns below the collicular surface. At that point there is an abrupt change in the topographic mapping of the cochlea upon the IC. For the final 1000-1200 microns of depth the tonotopy assumes a quasi-cortical appearance where almost all neurons along a particular vertical axis have the same or nearly the same BFs (Figs. 7 & 8).

Only filter frequencies are represented in the ventral, quasi-cortical region of the IC. For this reason, I shall hereafter refer to this region of the IC as the filter region. The tonotopic organization of the filter region is in the anterior-posterior axis

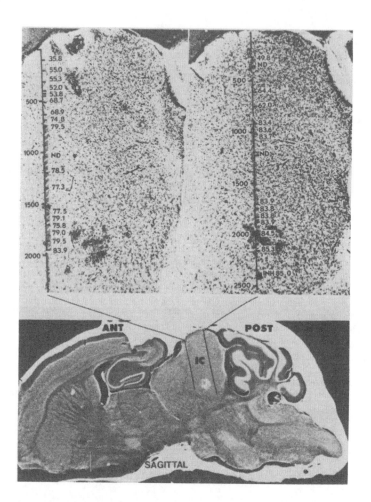

Fig. 7. Sagittal sections through the inferior colliculus of a horseshoe bat showing the BFs of the neurons encountered in two electrode penetrations. The regions, in the anterior-posterior axis, from which these sections were obtained can be seen from lower sagittal section of one entire brain. Note the progressive increase in BFs from dorsal surface to a depth of 700-900 microns and the relatively constant BFs for neurons at lower depths. Also note that the BFs in the filter region were all 77-79.5 in the anterior region whereas in the posterior region the BFs were 83.4-83.9. All electrode tracts were retracted from Alcian Blue Dye marks deposited from the electrode. Some neurons were not driveable (ND) with acoustic signals.

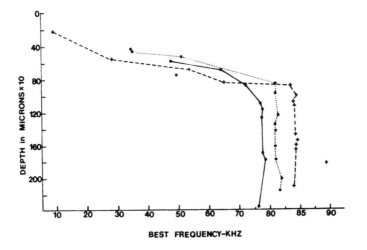

Fig. 8. Change in BFs with depth for three penetrations of the horseshoe bat's inferior colliculus. Note how similar the BFs in the filter region are within a given electrode penetration. These are different penetrations from those shown in Fig. 7.

where low frequencies (i.e., 77-80 kHz) are represented in the anterior portion while progressively higher frequencies (i.e., 81-86 or 87 kHz) are represented in the more posterior regions (Fig. 7).

It is important to point out that the arrangement described above holds true only for the central portion of the IC. A very different organization is found in both the medial and lateral margins of the IC. The tonotopy of the most medial edge of the IC is somewhat of a compromise between the dorsal and filter regions. The BFs in this region proceed from low to high with increasing depth but only the filter frequencies are represented. The BFs systematically increased with depth in this region but the range of BFs extends only from 78-89 kHz.

The most lateral region of the IC has yet a different organization. In this margin there is also a systematic progression of BFs with depth, from low to high, but the filter frequencies are only sparsely represented. This region, however, is truly striking but not for the tonotopy. Rather the neurons existing in vertical arrays within the lateral margin represent a homogeneous population in that they all have a high rate of spontaneous activity and, almost without exception, they respond to tone bursts with a pronounced inhibition of activity.

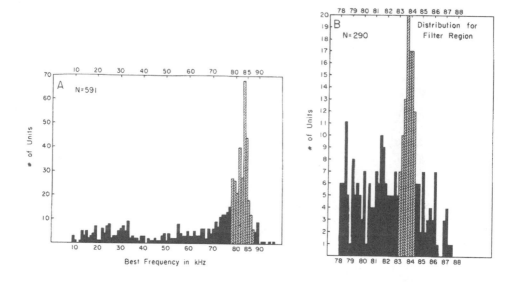

Fig. 9. (A) Histogram showing the number of units tuned to the frequencies comprising the audible range of horseshoe bats. The BFs are grouped and plotted in intervals of 1.0 kHz. (B) Histogram showing the distribution of units tuned to the filter frequencies where the BFs are grouped and plotted in intervals of 200 Hz. Note the precipitous increase in the number of neurons tuned between 83-84.5 kHz (cross-hatching). From Schuller and Pollak, 1979.

Proportional Representation of Audible Frequencies Within the IC

Previous studies have shown that approximately 50% of the neurons in any auditory center are tuned to the filter frequencies (Suga et al. 1976; Schlegel 1977; Möller et al. 1978). This result was verified in the IC by Schuller and Pollak (1978; 1979) (Fig. 9A). However, we had a systematic sample of the entire IC, having obtained the BFs of almost 600 neurons, and we therefore were able to plot the distribution of BFs for the filter frequencies in intervals of 200 Hz. The results, shown in Fig. 9B, are quite striking and we believe, are of fundamental importance. It is clear from this plot that the 83.0-84.5 kHz region has a much greater representation than do the other filter frequencies. This region of over-representation gains additional significance because the bats used in these studies had resting frequencies of 83.5-83.8 kHz. From studies of Doppler-shift compensation (Simmons 1973; Schuller et al. 1974) it is well established that horseshoe bats adjust their emitted CF components such that the echo frequencies are stabilized or held at the bat's reference frequency, a frequency 100-300 Hz above the emitted or resting frequency. Conse-

quently, when Doppler-compensating these bats hold their echo CF
component between 83.6-84.1 kHz. The advantage of DSC now becomes
apparent. By manipulating the frequencies of their emitted CF components, horseshoe bats confine echoes to a very narrow frequency
band and this behavior insures the bat that echoes will be processed
by an exceptionally large number of sharply tuned filter neurons.

Doppler shift compensation then, can be thought of as, a mechanism for "foveation". In the visual system, eye movements are used
to keep an image focused on the fovea. Horseshoe bats, on the
other hand, move their voices, i.e., adjust the frequency of the CF
component, in order to stabilize echoes on the "acoustic fovea". The
"acoustic fovea" is the greatly expanded and highly specialized region of the cochlear partition devoted to the filter frequencies
(Bruns 1976a,b) whereas the over-represented 83.0-84.5 kHz band is
the "acoustic foveal" region of the brain.

The purpose of foveation, in vision, is for visual acuity.
Similarly, it appears that DSC, or "acoustic foveation", is
utilized for acoustic acuity; to enhance the resolving power for
fine target features by devoting a disproportionately large
number of neurons to the processing of the echo CF component.

The target features that the bat is attempting to resolve,
most likely, are the frequency and amplitude modulation patterns
which are created as the CF component is reflected from the
beating wings of a small insect. These modulation patterns
could well serve as a source of information for distinguishing
flying insects from inanimate objects as well as for identifying
which species the echo is reflected from.

Discharge Patterns to Frequency Modulated Signals

A few reports of evoked potentials (Schuller 1972) and recodings
from cochlear nucleus (Suga et al. 1975; Suga and Jen 1977) neurons
suggest that the nervous system of long CF/FM bats is well suited
for encoding frequency modulation patterns. The ability to encode FM
patterns was firmly established in a recent study by Schuller (1979a).
He recorded the discharge patterns of neurons in the inferior colliculus of horseshoe bats when presented with sinusoidally frequency
modulated signals (SFM). The truly spectacular synchronization of
discharges to the modulation waveform, even when the depth of modulation is as small as $\pm$ 25 Hz, is illustrated in Fig. 14.

Neurons having firing patterns tightly locked to the modulating
waveform are commonly encountered in the colliculus. However, most
collicular neurons are quite selective with regard to some parameter
of the modulating waveform to which they will synchronously discharge.
This feature is perhaps best illustrated by considering how discharge
synchrony is influenced by signal intensity.

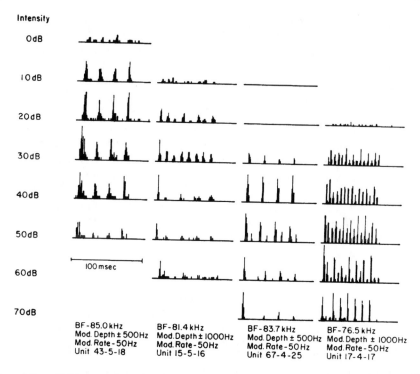

Fig. 10. Effect of stimulus intensity upon the locked discharges to SFM signals in four collicular neurons. The neuron on the right (unit 17-4-17) had tightly locked firings at all intensities above threshold while the three other units each locked best to only a small range of intensities.

Intensity influenced locking behavior in two broad ways. First are neurons in which the discharge registrations are about equally secure at all intensities above threshold (Fig. 10). The second type of neuron exhibited the sharpest locking and most vigorous responding only at low or moderately low intensities and had a significant decline or complete absence of locking at higher intensities (Fig. 10). Stated differently, these units discharged in registry with the modulating waveform only over a preferred range of intensities. But notice from Fig. 10, that the preferred intensity range differs slightly from neuron to neuron. The population, then, forms a continuous gradient where some units encode the modulation pattern equally well at all intensities above threshold whereas others are more selective and only encode the modulation waveform if the echo falls within a narrow intensity slot.

Before considering how collicular neurons encode different modulation rates and depths, I should like to digress for a moment in

order to make an important point: the encoding features for SFM signals cannot always be predicted from the way in which the same neuron responds to tone bursts. In Fig. 11, for example, are shown
the responses to SFM signals centered at 85 kHz and the responses
to pure tones at 85 kHz, the BF of this neuron. Notice first that
the threshold for the 85 kHz tone burst was between 20-30 dB SPL
and that the discharges increased monotonically with intensity.
Both of these features were considerably different when SFM signals
were presented. The threshold for SFM was somewhat lower than 0 dB
and the firing rate, as seen in the amplitude of the histogram peaks,
first reached a maximum at 20 dB and then decreased, becoming neglig-
ably small at 50 dB.

The second parameter of the modulation waveform we varied was
the modulation rate which would simulate the various wing beat frequencies of different insects. As was the case for intensity, neurons exhibited a continuous gradation in their abilities to encode
modulation rate. Some neurons could accurately encode modulation
rates from 50 Hz to 500 Hz whereas most other units showed a preference for a range of rates (Fig. 12). In general the preferences were
expressed in a continuous shift among units in their capacity to encode progressively higher modulation rates. There is also evidence
suggestive of a synchronization in the opposite direction. In Fig.

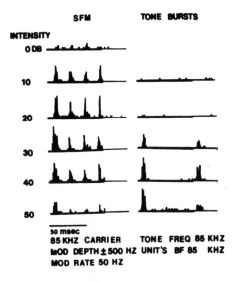

Fig. 11. Differential behavior to SFM and tone burst stimulation.
Notice that threshold for SFM was about 25 dB lower than to tone
bursts at the neuron's BF. The neuron also responded monotonically
with intensity to tone bursts but was clearly non-monotonic to SFM.

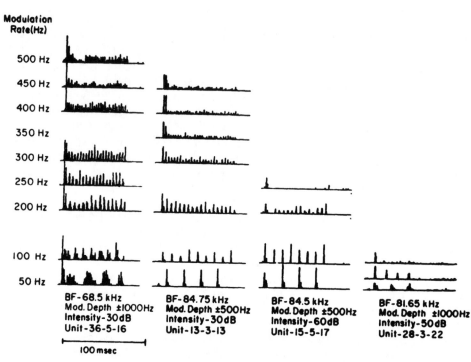

Fig. 12. Abilities of four collicular neurons to follow SFM signals having different modulation rates. Unit 36-5-16 (left) followed rates from 50-500 Hz with well synchronized firings while unit 28-3-22 followed only up to 75 Hz. Spike scale for unit 28-3-22 is compressed for clarity.

13 for example, is a neuron which locked poorly to modulation rates lower than 100 Hz but synchronized with reliable firings to modulation rates between 100 to about 250 Hz. Notice that the modulation depth was only ± 100 Hz.

Another perturbation imposed upon the echo CF is the depth of modulation. The depth can vary considerably and depends upon the size of the wings, wing beat frequency and the angle of the insect with respect to the bat. It is only fair to say that we have much less data concerning modulation depth than we have about intensity and rate. However, the data obtained suggests that, in this case also, neurons can be divided into two categories. One category discharges synchronously with the waveform at almost all depths while the second category are more selective for a range of depths. An example of the first type is shown in Fig. 14.

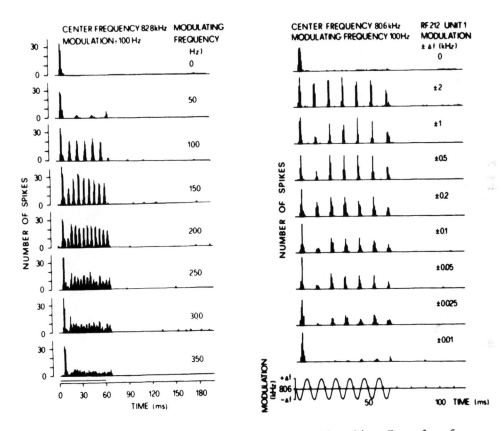

Fig. 13. Neuron exhibitory preference for higher modulation
rates. The neuron responded
poorly to 50 Hz rates but discharged with vigorous well synchronized firings at rates from
100-200 Hz and declined at higher
rates. The modulation depth was
± 100 Hz. From Schuller, 1979a.

Fig. 14. Example of a
neuron able to respond
with locked discharges
for modulation depths
ranging from ± 2 kHz to
± 25 Hz. From Schuller,
1979a.

The Effects due to Variations of the Carrier Frequency

All of the results presented above were obtained when the carrier frequency was set at the neuron's BF. However, the center frequency of the echo will not necessarily correspond to the BF of all neurons excited by the signal. Consider, for example, that upon Doppler compensating, a horseshoe bat holds the echo carrier around 84 kHz and that the FMs vary by +500 Hz around the carrier. Under these conditions, units having BFs of 84 kHz will be excited but so will other units having BFs both above and below 84 kHz. Consequently, it becomes of some interest to assess how filter neurons encode modulation patterns for a variety of different carrier frequencies. I shall refer to the locked responses across carrier frequency at a particular intensity as the SFM response area and the frequency range of tone bursts capable of evoking responses at a given intensity as the unit's response area (RA).

In Fig. 15 is shown how a typical neuron responds to SFMs having different carrier frequencies. The most striking feature of the SFM response areas of filter units is the very sharp boundaries on both the high and low frequency sides. This is especially evident in units which locked in a symmetrical manner around a carrier set at the unit's BF. In units having this type of symmetric SFM response area, the height of the locked discharge peaks correlated closely with the extent to which the SFM signals encroached upon the RA; the greater the encroachment the larger the response and if the signal failed to enter the RA, no discharges were evoked.

Other units have asymmetric SFM response areas where the locking, as judged by the amplitudes of the peaks in the histograms, was clearly superior on the low, or in other units, on the high frequency side of the BF. An asymmetric SFM response area where the neuron favored frequencies above the BF is shown in Fig. 16.

Response Patterns Evoked by Sinusoidally Amplitude Modulated (SAM) Signals

Neurons in the IC of horseshoe bats encode amplitude modulation patterns with discharges synchronized to the modulation waveform in a manner similar to the way in which they encode SFM patterns. An example of the degree of sensitivity for even very small modulation indices is shown in Fig. 17.

Intensity influenced the locking to SAM signals in a manner similar to that of SFM signals. Some units would lock to the SAM waveform over a wide range of intensities. However, most collicular units exhibited a preference for an intensity range

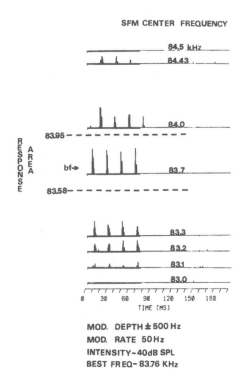

SFM CENTER FREQUENCY

MOD. DEPTH ± 500 Hz
MOD. RATE 50 Hz
INTENSITY - 40dB SPL
BEST FREQ - 83.76 KHz

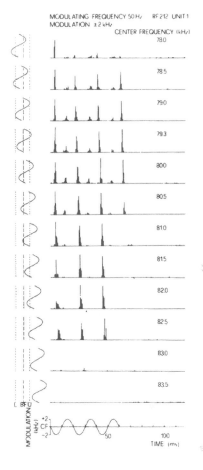

Fig. 15. Symmetric SFM response
area. The BF was 83.76 kHz
and the modulation depth was ±
500 Hz. When the carrier was
83.10 kHz, the SFM signal just
clipped the lower limit of the
RA (dashed lines) and elicited
some locked discharges. The locked firings became increasingly
vigorous as the carrier frequency progressively entered more of
the RA. As the carrier rose above the BF, the histogram peaks
decreased in amplitude and reached an upper limit for evoking responses when the carrier was 84.43 kHz. The unit failed to respond when the SFM carrier was 84.5 kHz, a signal that just failed
to enter any part of the RA.

Fig. 16.

Fig. 16. Asymmetric SFM response area. The bottom line describes
the time course of the modulation whereas the drawing to the left
of each histogram depicts the position of the modulation cycle
within the limits of the response area. Notice that when the
carrier was 78.0 kHz and crossed the neurons BF, hardly any firings
were evoked. L: lower limit of RA, U: upper limit of RA. From
Schuller, 1979a.

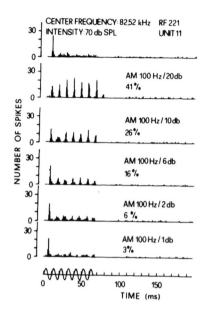

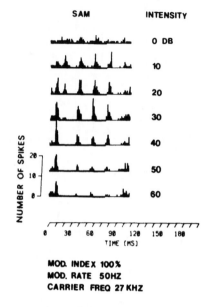

Fig. 17. Responses to sinusoidally amplitude modulated (SAM) signals. AM is indicated as the difference of the minimum and maximum intensities in decibels as modulation index in percent. Modulation
rate was 100Hz. From Schuller, 1979a.

Fig. 18. Neuron showing intensity preference for SAM stimulation.

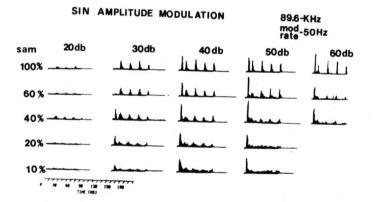

Fig. 19. Interactive effect of intensity and modulation index
(percent modulation) upon locked firings evoked by SAM stimuli.
See text for further explanation.

over which they would discharge in registry with the SAM waveform.
This intensity preference is illustrated by the neuron in Fig. 18.

In many units the ability to lock onto the modulation waveform at various depths was strongly influenced by the signal
intensity. The interaction of intensity and modulation index
is shown in Fig. 19. This unit responded with sharp, well
synchronized discharges at all intensities, but only when the
modulation index was 100%. At smaller modulation indices, of
60% and 40%, the locking was most secure and peaks of the
histograms were highest at intensities of 30-50 dB, and a pronounced decline in responding was observed at 60 dB. At a modulation index of 20%, locking was seen only at 30-40dB and disappeared when the signal was 50 dB.

Response Selectivity

The majority of collicular neurons responded to tone bursts
as well as SFM and SAM signals, in the case of the latter two
signals with synchronized discharges, but there is a sizeable
population that were more selective. The selectivity can be
grouped into three categories. First, were neurons that were
unresponsive to pure tones or SAM stimuli but responded to SFM
with firings in tight registration with the modulation pattern.
A few units were unresponsive to tone bursts and SFM stimuli but
locked to the envelope of SAM signals.

Second were neurons that responded to tone bursts but
were selective in the sense that they would respond with locked
firings either to SFM stimuli or to SAM stimuli but not to both.

In the third category are neurons whose response properties,
at least superficially, seem to be poorly suited for encoding
modulation patterns. Included here are neurons that would respond to tone bursts, as well as SAM and SFM stimuli, but in no
case were the discharges locked to the modulation waveform. Indeed, they responded to the SFM and SAM signals as they did to
tone bursts. The other type of neurons that can be placed in
the third category are those that could not be driven with any
type of acoustic stimulation, including tone bursts, SFM, SAM
nor by combinations of SFM-SAM, and were detected only because
of their spontaneous activity. It might seem puzzling that
neurons in the third category are considered as being highly
selective for some stimulus but there is now evidence to suggest
that some neurons of this type may be highly selective but the
selectivity is only expressed while the animal is actively engaged in echolocation. I shall consider the evidence for this
suggestion in the next section.

Single Unit Activity Recorded from Echolocating Bats

Recently, the activity of filter neurons has been monitored while horseshoe bats were emitting pulses and receiving electronically generated echoes played back from a loudspeaker (Schuller 1979b; Schuller & Pollak unpublished observations). The number of units recorded from behaving bats is quite limited and the results should be considered to be of a preliminary nature. The data, from at least some neurons, however, are of considerable interest and are instructive for guiding future studies. Below I shall first discuss the activity evoked by the emitted cries and correlate that activity with the discharge patterns evoked by electronically generated tone bursts. Subsequently, activity evoked by electronically generated echoes during periods of pulse echo overlap will be presented.

The emission of an orientation cry typically evokes discharges from filter neurons, or at least from neurons in the filter region of the IC. The discharge patterns can be tonic, phasic or inhibitory and will often, but not always, be similar to the pattern evoked by tone bursts at the BF or BIF. Of particular interest is that some filter neurons that would not discharge to tone bursts, SFM or SAM stimuli, fired bursts of spikes to each emitted orientation cry. Whether these discharges resulted from some combination of internally generated activity which primed or facilitated the responses to the emitted sounds or whether the discharges were evoked solely by internally generated activity is not known. On the other hand, a few neurons of a similar type, i.e., not driveable with acoustic signals, failed to respond when echolocation calls were emitted and present what seems at the moment to be a paradox.

Most neurons would discharge upon pulse emission but the events associated with pulse emission would also generate a potent, underlying inhibition in the majority of cells studied. The inhibition prevented the neuron from responding to echoes that returned during periods of overlap. As soon as the pulse had ended the normal response to the electronically generated echo would be evoked. In other words, the response to an echo which partially overlapped with the emitted cry would be evoked only during the non-overlapping portion of the echo.

The above data seemingly indicate that the majority of filter neurons do not respond to echoes during periods of pulse-echo overlap which, if true, would be startling. However, it should be stressed that the frequency relationship between the emitted CF components and the echoes was substantially different in experiments cited above from that occurring under most normal conditions. While hunting insects, horseshoe bats Doppler compensate and thereby listen to a cry having a frequency substantially lower than the frequency of the echo CF

portion. Consequently, the sounds heard in the neurophysiological experiments represent a different set of stimulus
conditions than would normally occur.

In this context, a recent study by Möller (1978), of the
lateral inhibitory regions of filter neurons, seems to be highly
significant. Möller found that filter neurons tuned around
the bat's reference frequency had inhibitory surrounds flanking
and overlapping with the excitatory tuning curve. The important
point in this study is that a gap in the flanking inhibitory
region exists just <u>below</u> the bat's reference frequency and extends to frequencies 500-4000 Hz below the neurons BF. From
these data it follows, then, that two signals, an emitted
pulse at the resting frequency and an echo having roughly the
same frequency, would create a set of conditions whereby the
emitted cry would inhibit the responses to the echo. But the
surround inhibitory pattern also suggests that during Doppler-
compensation, when the emitted CF is lowered in frequency and
falls within the flanking gap of the inhibitory surround region,
no inhibition would be present and neurons would be free to respond to the echoes, even during periods of pulse echo overlap.

The most intriguing and surely the most important result
from the experiments with echolocating bats comes from a study
by Schuller (1979b). He showed that some neurons which do not
respond to SFM stimuli or others which do respond with discharges
that are <u>not</u> locked to the modulation waveform become synchronized
to the modulated signal but only if the bat is actively emitting
echolocation cries and only during periods of pulse-echo overlap.
These rather remarkable features are illustrated by the two
neurons shown in Fig. 20.

The unit in Fig. 20A responded to emitted cries (vocalizations) with a tonic firing pattern but was almost unresponsive to
SFM signals (AS(FM)). When the bat emitted orientation sounds and
the same SFM signal was simultaneously presented, the unit fired
with well defined discharges synchronized to the modulation waveform (Voc +AS(FM)). That active pulse emission was required
to evoke the locked firings is demonstrated by the control condition where a tone burst having the same frequency as the emitted
cry was presented simultaneously with the SFM signal (AS + AS
(FM)). Notice that almost no discharges were evoked with a simulated orientation cry. Finally, presenting an 83 kHz tone burst
during pulse emission resulted in a tonic firing pattern but one in
which the discharges were not in synchrony with the modulating waveform (VOC + AS), showing that the SFM stimulus was indeed necessary
to evoke locked firings.

A similar phenomenon is shown for another unit in Fig. 20B.

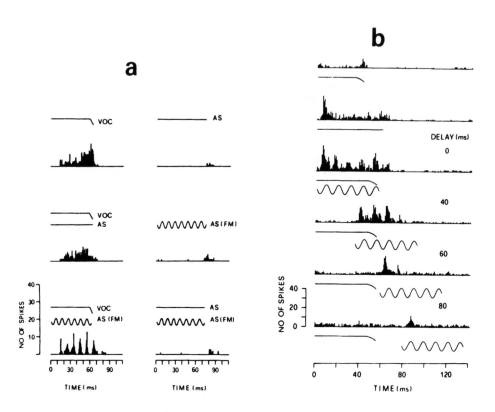

Fig. 20 Discharges of two single neurons evoked by emitted orientation cries (Voc) and electronically generated tone bursts (AS)
and frequency modulated signals (AS(FM)). (A): Neuron whose firings were locked to the modulation waveform only when orientation
cries were emitted. See text for further explanation. (B): Neuron illustrating the necessity of pulse-echo overlap for evoking
locked discharges. In top figure is the firing pattern to emitted
cries: The CF component was 83.3 kHz. The figure below shows response pattern to 84.3 kHz tone burst at 70 dB. In the subsequent
figure are shown the firing patterns evoked by vocalizations
coupled with SFM signals delayed by various times with respect to
the beginning of the emitted cry. Notice that locked discharges
were evoked only during the periods of pulse-echo overlap. From
Schuller 1979b.

This neuron also demonstrates that for locked firings to be evoked the echo must overlap with the emitted CF component.

What Makes Filter Neurons so Special?

The majority of reports concerned with the neurophysiology of long CF/FM bats have focused on those features associated with the processing of the CF component. The reasons, of course, derive from the highly dramatic nature of the behavior and the truly spectacular features of the nervous system that have evolved for processing this component of the biosonar signals. The descriptive term most often associated with the filter system is that it is specialized or even highly specialized. Implied in the usage of such a term is the idea that something unique derives from the specialization, that filter neurons have encoding properties so extraordinary as to set them apart from the neurons found in the nervous systems of other animals. To be sure, the system is "specialized," but I believe it worthwhile to define and to clarify a bit more carefully exactly what is meant by the term "specialized".

The three characteristics that have had an immediate and striking impact on all workers are the remarkably sharp tuning of the filter neurons, the great over-representation of the filter frequencies and the unusual tonotopic organization of the filter regions. What these features share in common is that they are basically structural in nature. The sharp tuning, for example, derives from the morphological features of that portion of the cochlea which responds to the filter frequencies (Bruns 1976a,b). But judging from a recent study by Bruns and Schmieszek (1979), there is little unique in the neural innervation of the cochlea. During evolution, apparently the most basal portion of the basilar membrane was, in essence, "stretched" out and fortified with a number of structural modifications. What this means is that a much longer strip of the basilar membrane is now devoted to each filter frequency than would have been the case if the membrane had not been "stretched". The mechanical properties of the "stretched" portion are sharpened by structural modifications. Within the elongated frequency strip, the density of the afferent neural innervation is reduced relative to that of the adjacent area but the decrease is more than compensated for by the increase in length devoted to each frequency. Thus, the absolute number of auditory nerve fibers devoted to each filter frequency is increased but the afferent neural innervation pattern is less dense than, but otherwise similar to that occurring in the nonfilter region. In short, it is the structural arrangement of the cochlear filter region that imparts the sharp tuning to a population of, otherwise, typically mammalian auditory neurons.

The results of recent experiments in the IC also lead me to conclude that the specializations are largely architectural in nature. The filter region is somewhat amplified within the central nervous system in that the proportion of neurons devoted to the filter frequencies is greater in the central nervous system than in the cochlea. Furthermore, the neurons deriving their synaptic inputs from the filter region of the cochlea assume their own domains and tonotopic organization within the acoustic centers of the brain. This arrangement serves to segregate the processing of the CF from the processing of the FM component of the biosonar signals, as originally postulated by Simmons (1973) and more recently also demonstrated in the auditory cortex of the mustache bat (Suga and Jen, 1977). In this sense the system is indeed unique and highly specialized but these features say little about the encoding abilities of individual neurons.

It has been suggested in a number of reports that the sharp tuning endows filter neurons with the ability to encode SFM modulation patterns (Suga et al. 1975; Suga & Jen 1977, Suga 1978). The rationale is that an FM signal sweeping in and out of a very narrow response area would cause the neuron to discharge only when the signal entered the RA, the result being discharges synchronized to the period of the waveform. In this view, the ability to synchronize to the modulating envelope is a direct consequence of the sharp tuning and thus is a specialized property of filter neurons.

This view, however, has not proven to be the case. Neurons tuned to most of the audible frequency region of horseshoe bats exhibit discharges to SFM signals that are as tightly locked to the modulation waveform as are those of the filter neurons. Furthermore, some sharply tuned filter neurons, as well as broadly tuned nonfilter neurons, fail to lock to SFM stimulation. The nonfilter neurons that do lock to the period of the SFM envelope have broad tuning curves but the locking is still evoked when the spectrum of the signals is confined entirely to, and sweeps back and forth within, the neuron's RA. This feature is also observed for filter neurons, as can be seen in Fig. 15[1]. The lock-

---

[1] The ability to lock to very small modulation depths, i.e., +10–40 Hz, is seen only in filter neurons and has never been observed in neurons having BFs below about 78 kHz (Schuller 1979a). A number of investigators consider the locking to very small modulation depths to be a specialized property of filter neurons and have suggested that locking to such low depths is associated with the sharp tuning of the filter neurons (Suga and Jen 1977; Schuller 1979a). However, it should be pointed out that there does not appear to be a correlation between the sharpness of tuning and the lowest modulation depth capable of evoking synchronized discharges. Apparently, neurons having

ing then is not so much a consequence of the sharp tuning but
rather seems much more to be due to the sequence and potency of
synaptic events evoked by the temporal and spectral features of
the signals. This, in principle, is the same argument advanced
for the pCLRs.

If, as I have argued, the sharp tuning does not confer extraordinary encoding abilities upon individual neurons, what advantage
accrues to having a population of sharply tuned elements? Perhaps
the functional significance of the sharp tuning is to confine and
spatially segregate the activity evoked by the emitted CF from
the activity evoked by the echo CF component (Suga et al. 1976;
Suga 1978). In conjunction with the pattern of the inhibitory
surround region observed for many filter neurons, which is a
specialized adaptation, the emitted CF would be prevented from
masking the responses to the echoes during periods of pulse-echo
overlap, a feature that is surely of primary importance to long
CF/FM bats.

The point I wish to make is simply this: there are a number of specialized structural and architectural features within
the auditory systems of long CF/FM bats which set these animals
apart from all other vertebrates. However, the response patterns
evoked by tone bursts are strictly mammalian (Möller et al., 1978
a; Pollak & Schuller, in prep.). Furthermore, except for the
locking to the very small modulation depths, the response features evoked by SFM and SAM signals are not fundamentally different in filter as compared to nonfilter neurons. This argument
is equally valid for the preferences of both filter and nonfilter neurons for the intensities, modulation depths and modulation rates to which each neuron will fire with locked discharges.

Although the encoding abilities of most individual filter
neurons appear not to be "specialized" to any extraordinary degree, the system with strong justification, can be said to be
functionally specialized. The functional specialization is imposed upon the system by the structural modifications in the
cochlea and by the architectural arrangement of elements within
the central nervous system. Each element receives input from a
sharp peripheral filter which severely restricts the extent of
the neuron's receptive field. Thus, a strip of basilar membrane
is projected upon a vertical array of collicular neurons with
each neuron in the array being selective for one or another signal

such encoding abilities must have a minimum Q value of about 40 (filter neurons encoding depths as low as $\pm$10 Hz can have Q values of 40
-45) but neurons having even sharper tuning curves, i.e., 100-300,
seem not to derive any additional advantage for encoding low modulation depths (Schuller 1979a; Pollak and Schuller, in prep).

parameter and with the extent of activity generated in each array
being sharply bounded in neural space. These features are then
expressed as a unique spatio-temporal pattern of activity which,
as a Gestalt, creates within the brain a neural image of a flying insect whose path happened to cross that of an echolocating
horseshoe bat.

Functional Organization of the Inferior Colliculus

There now exists sufficient information which can be fitted
together to provide a rough picture of how an acoustic image is
formed within the brain. Below I shall attempt to correlate the
known behavior modifications long CF/FM bats exhibit during echolocation with the architectural arrangement and the various encoding properties of collicular neurons and present what is likely
to be the form that the above mentioned spatiotemporal patterns of
activity might assume.

It seems reasonable to suppose that upon detecting a target,
the bat first needs to determine if the target is animate or is
inanimate, e.g., a falling leaf or background objects. A basis
for such a discrimination resides in the temporal patterns of the
responding elements; a flying insect can, presumably, be recognized
as such by the synchronized discharges locked to the modulation
pattern of the echo whereas the echo from an inanimate object will
be encoded in a manner similar to a tone burst.

After ascertaining the nature of the target, it may well
be that horseshoe bat seeks to determine whether or not the insect is one that is palatable and worthy of pursuit and capture.
In this context, the two manipulations the bat performs on the
emitted CF component during the goal directed portion of echolocation
seem to be important. The first type of manipulation is the
systematic lengthening of the emitted CF component that occurs
after target detection and continues for a considerable time
as the bat tracks its target (Novick & Vaisnys 1964; Novick
1971). One effect of the pulse lengthening is to increase the
number of modulation cycles present in each echo. The second
manipulation concerns Doppler-shift compensation where the bat
stabilizes the echoes within a narrow frequency band around the
reference frequency. The significance of this behavior is that
a disproportionately large number of filter neurons are tuned
to a 1.5 kHz band around the bat's reference frequency. It
appears, then, that horseshoe bats actively create a signal which
is designed to insure that an echo having a sufficient number of
modulation cycles will be received and processed by an exceptionally large number of sharply tuned neural elements, most of which
are well suited to encode some component of the modulation pattern.

Insights into the spatial extent of activity evoked by a
modulated echo can be obtained from considering the tonotopic organization together with the features of the SFM response areas.   If
one assumes that filter units respond to frequency modulations in
a natural echo as they do to SFM signals, then the characteristics
of SFM response areas suggest that a modulated echo should produce
more or less equally vigorous activity over a large expanse of
neural tissue. That is, the discharge rate and synchrony should
be maximal in units tuned to the echo center frequency and having
symmetric SFM response areas.   However, the activity should also be
maximal in other units having BFs above and yet in other units
having BFs below the center frequency of the echo because these
neurons have asymmetric SFM response areas.   Furthermore, the
tuning properties of the filter neurons indicate that the active
region will have sharply confined boundaries in the anteri.oposterior
axis of neural space (Fig. 21).

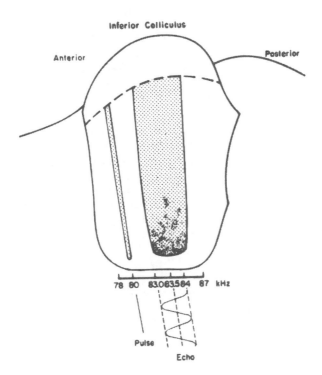

Fig. 21.   Schematic diagram to visualize the spatial extent of
activity evoked by the CF of an emitted pulse and the CF of an echo
having frequency modulation of $\pm$ 500 Hz after the bat has compensated
for a 3.5 kHz Doppler shift in the echo CF component.   The reference
frequency is assumed to be 83.5 kHz.

Within the active region, the echo will evoke discharges synchronized to the modulation waveform. Moreover, the frequency
changes occuring in the echo will, by virtue of their variations
with time, also cause a corresponding shifting, back and forth, of
the locus of activity as the spectrum enters and leaves the sharp
response areas of elements tuned to different frequencies. Thus,
the frequency modulation pattern is coded both by the temporal
sequence of discharges in a sharply bounded subpopulation of
filter neurons and in the spatial shifting of activity within
that subpopulation over time.

Due to changes in position, orientation and speed of either
the bat or its target, subsequent echoes will differ more or
less in carrier frequency, modulation pattern and intensity from
the previous echo(es). In principle, each echo will be encoded in
a similar manner. However, the preferences of many filter units
for selective ranges of intensity as well as modulation rate and
depth suggest that the changes in echo parameters will cause some
neural elements to drop out, new elements to be recruited while
others will simply change response vigor and/or firing registration to reflect the changes in echo characteristics. In short, the
properties of filter neurons endow the system with the ability to
encode the pertebations imposed on the echo CF component with the
sum total of spatio-temporal activity being a dynamic pattern
which differs from echo to echo.

Given the capacity to encode target features, the question
arises as to whether the information derived from one echo is
sufficient for target recognition or whether several echoes are
required. In this regard, it should be pointed out that there is
considerable variation of both wing length and wing beat frequency
among insect species. Both of these features have profound effects
on the frequency and amplitude modulations in the echo CF component and must result in substantial differences among the spatio-
temporal patterns of activity generated by echoes reflected from
different insects. Schuller (1979a) has discussed this issue in
considerable detail and from a number of considerations has concluded that the possibility for recognition from one echo, while
not impossible, is unlikely. On the other hand, the system encodes
modulation patterns with great precision and diversity. It, therefore, would be surprising if horseshoe bats could not associate
the acoustic image derived from the profile of activity integrated
over several echoes with a particular population of insect.

References

Aitkin, L.M. Tonotonic organization at higher levels of the auditory pathway. In: Internatural Review of Physiology: Neurophysiology II. R. Porter (Ed). University Park Press, London. Vol. 10, pp. 249-280 (1976).

Bruns, V. Peripheral auditory tuning for five frequency analysis by the CF-FM bat, Rhinolophus ferrumequinum. I. Mechanical specialization of the cochlea. J. Comp. Physiol. 106, 77-86 (1976a).

Bruns, V. Peripheral auditory tuning for fine frequency analysis by the CF-FM bat, Rhinolophus ferrumequinum: II Frequency mapping of the cochlea. J. Comp. Physiol. 106, 87-97 (1976b).

Bodenhamer, R., Pollak, G.D. and Marsh, D.S. Coding of fine frequency information by echoranging neurons in the inferior colliculus of the Mexican free-tailed bat. Brain Res. (in press).

Capranica, R. Why auditory neurophysiologists should be more interested in animal sound communication. Physiologist 15, 55-60 (1972).

Feng, A.S., Simmons, J.A., and Kick, S.A. Echodetection and target ranging neurons in the auditory system of the bat, Eptesicus fucus. Science 202, 645 (1978).

Friend, J.H., Suga, N. and Suthers, R.A. Neural responses in the inferior colliculus of echolocating bats to artificial orientation sounds and echoes. J. Cell. Physiol. 67, 319-332 (1966).

Goldman L.J. and Henson, O.W., Jr. Prey recognition and selection by the constant frequency bat, Pteronotus p. parnellii. Behav. Ecol. Sociobiol. 2, 411-419 (1977).

Griffin, D.R. Discriminative echolocation by bats. In Les Systems Sonars Animaux. R.G. Busnel (Ed.) INRA-CNRZ, Jouy-en-Josas, France, Vol. 1, pp. 273-300 (1967).

Grinnell, A.D. The neurophysiology of audition in bats; intensity and frequency parameters. J. Physiol., London 167, 38-66 (1963a).

Grinnell, A.D. The neurophysiology of audition in bats: temporal parameters. J. Physiol., London 167, 67-96 (1963b).

Grinnell, A.D. Comparative auditory neurophysiology of neotropical bats employing different echolocation sounds Z. Vergl. Physiol. <u>68</u>, 117-153 (1970).

Grinnell, A.D. and Hagiuvara, S. Adaptations of the auditory system for echolocation: studies of New Guinea bats. Z. Vergl. Physiol. <u>76</u>, 41-81 (1972).

Henson, M.M. The basilar membrane of the bat, <u>Pteronotus</u> <u>p</u>. <u>parnellii</u>. Am. J. Anat. <u>153</u>, 143-158 (1978).

Henson, O.W. Jr., Pollak, G.D., Johnson, R.A., and Goldman, L.J. Specialized properties of the auditory system in the bat <u>Pteronotus</u> <u>p</u>. <u>parnellii</u>. Anat. Rec. <u>107</u>, 373 (1974).

Henson, O.W., Jr. and Goldman, L.J. Prey detection and physiological aspects of the cochlea in the bat <u>Pteronotus</u> <u>p</u>. <u>parnellii</u>. Anat. Rec. <u>184</u>, 425 (1976).

Jen, P.H.-S. Electrophysiological properties of auditory neurons in the superior olivary complex of echolocating bats. J. Comp. Physiol. <u>128</u>, 47-56 (1978).

Möller, J., Neuweiler, G., and Zöller, H. Response characteristics of inferior colliculus neurons of the awake CF-FM bat, <u>Rhinolophus</u> <u>ferrumequinum</u>. I. Single tone stimulation. J. Comp. Physiol. 125, 217-225 (1978a).

Möller, J. Response characteristics of inferior colliculus neurons of the awake CF-FM bat, <u>Rhinolophus</u> <u>ferrumequinum</u>. II. Two tone stimulation. J. Comp. Physiol. <u>125</u>, 227-236 (1978b).

Neuweiler, G. and Vater, M. Response patterns to pure tones of cochlear nucleus units in the CF-FM bat, <u>Rhinolophus</u> <u>ferrumequinum</u>. J. Comp. Physiol. <u>115</u>, 119-133 (1977).

Novick, A. Orientation in neotropical bats. II. Phyllostomatidae and Desmodontidae. J. Mammal. <u>44</u>, 44-56 (1963).

Novick, A. Echolocation in bats: Some aspects of pulse design. Amer. Sci. <u>59</u>, 198-209 (1971).

Novick, A. and Vaisnys, R. Echolocation of flying insects by the bat, <u>Chilonycteris</u> <u>parnellii</u>. Biol. Bull. <u>128</u>, 297-314 (1964).

Pollak, G.D., Marsh, D., Bodenhamer, R., and Souther, A. Echo-detecting characteristics of neurons in the inferior colliculus of unanesthetized bats. Science <u>196</u>, 675-678 (1977a).

Pollak, G.D., Bodenhamer, R., Marsh, D.S., and Souther, A.

Recovery cycles of single neurons in the inferior colliculus of un-anesthetized bats obtained with frequency modulated and constant frequency sounds. J. Comp. Physiol. <u>120</u>, 215-250 (1977b).

Pollak, G.D., Marsh, D.S., Bodenhamer, R. and Souther, A. Characteristics of Phasic on neurons in the inferior colliculus of unanesthetized bats with observations relating to mechanisms for echo ranging. J. Neurophysiol. <u>40</u>, 926-942 (1977c).

Pollak, G.D., Marsh, D.S., Bodenhamer, R., and Souther A. A single unit analysis of inferior colliculus in unanesthetized bats: response patterns and spike-count functions generated by constant frequency and frequency modulated sounds. J. Neurophysiol. <u>41</u>, 677-691 (1978).

Pollak, G.D., and Schuller, G. Tonotopic organization and response patterns to frequency modulated signals in the inferior colliculus of Horseshoe bats. Soc. Neurosci. Abstr. <u>4</u>, 9 (1978).

Pollak, G.D., Henson, O.W. Jr. and Johnson, R.A. Multiple specializations in the peripheral auditory system of the CF-FM bat, <u>Pteronotus parnellii</u>. J. Comp. Physiol. (in press).

Schlegel, P. Directional coding of binaural brainstem units of the CF-FM bat, <u>Rhinolophus ferrumequinum</u>. J. Comp. Physiol. <u>118</u>, 327-352 (1977).

Schnitzler, H.-U. Die Ultraschall-Ortungslaute der Hufeisen-Fledermäuse (<u>Chiroptera</u>, <u>Rhinolophidae</u>) in verschiedenen. Orientier-ungesituationen. Z. Vergl. Physiol. <u>57</u>, 376-408 (1968).

Schnitzler, H.-U. Echoortung bei der Fledermäus <u>Chilonyteris rubiginosa</u>. Z. Vergl. Physiol. <u>68</u>, 25-38 (1970)

Schnitzler, H.-U. Die Detektion von Bewegungen durch Echootung bei Fledermäusen. Verh. Dtsche Zool. Gesc. 16-33 (1978).

Schnitzler, H.-U., Suga, N., and Simmons, J.A. Peripheral auditory tuning for fine frequency analysis by the CF-FM bat, <u>Rhinolophus ferrumequinum</u>. J. Comp. Physiol. <u>106</u>, 99-110 (1976).

Schuller, G. Echoortung bei <u>Rhinolophus</u> ferrumequinum mit frequenz-modulierten Lauten. Evoked potentials in colliculus in-ferior. J. Comp. Physiol. <u>77</u>, 306-331 (1972).

Schuller, G. Coding of small sinusoidal frequency and ampli-tude modulations in the inferior colliculus of the "CF-FM" bat, <u>Rhinolophus ferrumequinum</u>. Exp. Brain. Res. <u>34</u>, 117-132 (1979a).

Schuller, G. Vocalization alters responsivness of auditory neurons in CF-FM bat, Rhinolophus ferrumequinum. J. Comp. Physiol. (1979b, in press).

Schuller, G., Beuter, K., and Schnitzler, H.-U. Response to frequency shifted artificial echoes in the bat, Rhinolophus ferrumequinum. J. Comp. Physiol. 89, 275-286 (1974).

Schuller, G. and Pollak, G.D. Disproportionate frequency representation in the inferior colliculus of Horseshoe bats: Evidence for an "acoustic fovea". J. Comp. Physiol. (in press).

Simmons, J.A. The resolution of target range by echolocating bats. J. Acoust. Soc. Amer. 54, 157-173 (1973).

Simmons, J.A., Howell, D.J., and Suga, N. Information content of bat sonar echoes. Amer. Sci. 63, 204-215 (1975).

Simmons, J.A., Fenton, M.B., and O'Farrell, M.J. Echolocation and pursuit of prey by bats. Science 203, 16-21 (1979).

Suga, N. Single unit activity in the cochlear nucleus and inferior colliculus of echolocating bats. J. Physiol. (London). 172, 449-474 (1964a).

Suga, N. Recovery Cycles and responses to frequency modulated tone pulses in auditory neurons of echolocating bats. J. Physiol. (London) 175, 50-80 (1964b).

Suga, N. Echo-ranging neurons in the inferior colliculus of bats. Science 170, 449-452 (1970).

Suga, N. Feature extraction in the auditory system of bats. In: Basic Mechanisms in Hearing. (Ed. A.R. Möller) pp. 675-742, Academic Press, New York (1973).

Suga, N. Amplitude-spectrum representation in the Doppler-shifted CF processing area of the auditory cortex of the mustache bat. Science 196, 64-67 (1977).

Suga, N. Specialization of the auditory system for reception and processing of species-specific sounds. Fed. Proc. 37, 2342-2354 (1978).

Suga, N., Simmons, J.A., and Jen, P.H.-S. Peripheral control of acoustic signals in the auditory system of echolocating bats. J. Exp. Biol. 62, 277-311 (1975).

Suga, N. and Schlegel, P. Coding and processing in the nervous system of FM signal producing bats. J. Acoust. Soc. Amer. 84, 174-190 (1973).

Suga, N., Neuweiler, G., and Möller, J. Peripheral auditory tuning for fine frequency analysis by the CF-FM bat, Rhinolophus ferrumequinum. IV. Properties of peripheral auditory neurons. J. Comp. Physiol. 106, 111-125 (1976).

Suga, N. and Jen, P.H.-S. Disproportionate tonotopic representation for processing CF-FM sonar signals in the mustache bat's auditory cortex. Science 194, 542-544 (1976).

Vater, M., Schlegel, P. and Zöller, H. Comparative auditory physiology of the inferior colliculus of two mollosid bats, Molossus ater and Molossus molossus. I. Gross evoked potentials and single unit responses to pure tone stimulation. J. Comp. Physiol. (1979a, in press).

Vater, M. and Schlegel, P. Comparative auditory physiology of the inferior colliculus of two mollosid bats, Molossus ater and Molosus molosus. II Single unit responses to frequency modulated signals and signal/noise combinations. J. Comp. Physiol. (1979b, in press).

AUDITORY PROCESSING OF ECHOES: REPRESENTATION OF ACOUSTIC INFORMATION FROM THE ENVIRONMENT IN THE BAT CEREBRAL CORTEX

Nobuo Suga and William E. O'Neill

Department of Biology, Washington University

St. Louis, Mo., U.S.A.

Clues to understanding information processing by the auditory system are hidden in the differences in response properties of single neurons within and among individual auditory nuclei. In order to explore neural mechanisms for signal processing, we thus focus our research on differences in response properties among neurons as well as response properties per se. All the acoustic information of use to an animal is contained in the activity of its auditory nerve. In mammals, all nerve fibers with identical best frequencies show nearly the same response pattern and frequency-tuning curve. In bats, all these primary auditory neurons respond to both emitted orientation sounds (also called pulses) and echoes by discharging multiple action potentials to each. The role of the central auditory system is to process this activity and send appropriate information to regulate the motor system, if necessary.

If all neurons in a given central auditory nucleus responded to both emitted orientation sounds and echoes as do primary auditory neurons, they would be concerned with coding, but not encoding or decoding information. In the central auditory system, single neurons with identical best frequencies typically show wide variation in their response properties; this is true for characteristics such as (1) response pattern (e.g., Rose, Greenwood, Goldberg & Hind, 1963; Suga, 1971). (2) impulse-count function (e.g., Grinnell, 1963a; Suga, 1965a, 1977), (3) frequency-tuning curve (e.g., Grinnell, 1963a; Suga, 1965a, b, 1973), (4) response to FM sounds (e.g., Suga, 1965a, b, 1973), (5) recovery cycle (e.g., Grinnell, 1963b, 1970; Suga & Schlegel, 1973), and (6) directional sensitivity (e.g., Grinnell & Hagiwara, 1972; Knudsen & Konishi, 1978). Such wide variations for given response

589

properties are apparently related to encoding or decoding of
certain parameters of acoustic signals (for echolocating bats,
more specifically, echo patterns associated with certain physical
aspects of the environment).

The most recent research on the functional organization of
the auditory system clearly indicates that neurons with response
properties different from those at the periphery are the elements
which establish the various specialized and beautiful functional
organizations found at the higher levels of the auditory system:
that is, groups of neurons represent certain important aspects of
the acoustic environment in orderly patterns, usually along axes
or in coordinate systems. For instance, neurons with especially
narrow frequency-tuning curves contained within a population
showing a wide variation in the shape of the curves are associated
with the fine tonotopic representation found in the auditory
cortex of the mustached bat Pteronotus parnellii rubiginosus
(Suga & Jen, 1976; Suga, 1977). These same neurons show a wide
variation in impulse-count function and are arranged in an orderly
way to produce a systematic amplitopic representation in the same
area of the auditory cortex of this bat (Suga, 1977). Broad
variation in directional sensitivity curves is apparently essential
to form the neural map of auditory space discovered in the dorso-
lateral mesencephalic nucleus of the barn owl Tyto alba (Knudsen
& Konishi, 1978). Furthermore, neurons showing a broad range of
responses to FM sounds (in particular, FM-specialized neurons)
and a broad spectrum of recovery cycles are apparently responsible
for creating the various response properties of delay (or range)
sensitive neurons found in the mustached bat (Suga, O'Neill &
Manabe, 1978; O'Neill & Suga, 1979). Our most recent finding is
that these delay-sensitive neurons are systematically arranged
along a target-range axis in the cerebral cortical plane (Suga
& O'Neill, 1979). The two most important findings in acoustic
signal processing are that a broad distribution of response properties is created by excitatory and inhibitory neural interactions and that neurons with different response properties form
different aggregates and are arranged in specific ways in certain parts of the brain depending upon their function.

The following will summarize our recent findings on the
functional organization of the auditory cortex of the mustached
bat. We will minimize the detailed discussion of mustached bat
behavior because it will be covered in other parts of this volume,
and we will limit citation of the literature to those which are
directly related to understanding these findings.

For echolocation, the mustached bat produces biosonar signals
each consisting of a long constant-frequency (CF) component
followed by a short frequency-modulated (FM) component (Fig. 1).

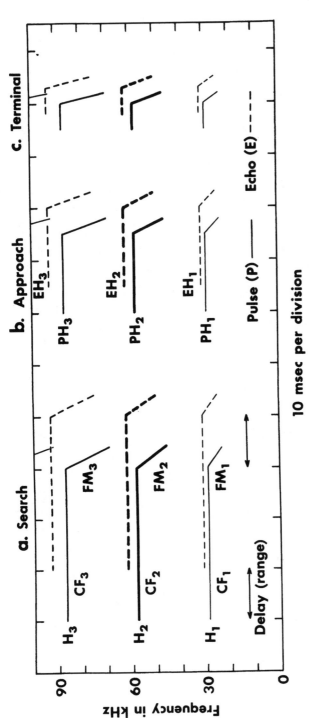

10 msec per division

Figure 1. Schematized sonagrams of the synthesized mustached bat biosonar pulses (P; solid lines) and echoes (E; dashed lines) mimicking three phases of target-oriented flight. The three harmonics of both the pulses (PH$_{1-3}$) and the echoes (EH$_{1-3}$) each contain a long CF component (CF$_{1-3}$) followed by a short component (FM$_{1-3}$). The fourth harmonic (H$_4$) is not shown in the figure. (a) search phase: CF and FM durations are 30 and 4 msec, respectively. (b) approach phase: CF and FM durations are 15 and 3 msec. (c) terminal phase: CF and FM durations are 5 and 2 msec. Repetition rates for (a), (b), and (c) were 10, 40, and 100 pairs per second, respectively. Thickness of the lines indicates the relative amplitudes of each harmonic in the pulses and echoes: H$_2$ is strongest, followed by H$_3$ (-6 dB) and H$_1$ (-12 dB). Echo delay is measured as the time interval between the onsets of corresponding components of the pulse and the echo in a stimulus pair.

Each component is composed of four harmonics (H_1-H_4), and therefore, there are eight components (CF_{1-4}, FM_{1-4}) in total. During the search phase, the frequency of the predominant second harmonic remains at about 61 kHz for 20 - 35 msec, then sweeps down to about 50 kHz within 2 - 4 msec. Since the reflected sound energy is highly concentrated at a single wavelength, the long CF tone is ideal for target detection and measurement of target velocity. The FM component, on the other hand, is more appropriate for ranging, localization, and characterization of a target, because of its short duration and wide bandwidth.

The mustached bat, like the horseshoe bat Rhinolophus ferrumequinum, performs a fascinating acoustic behavior called "Doppler-shift compensation" (Schnitzler, 1970). When there are no moving objects, the bat produces sounds with the second harmonic CF at about 61 kHz.* If the bat receives a Doppler-shifted echo, say at 63 kHz, from an approaching object, the bat reduces the frequency of the subsequent biosonar signals so as to stabilize the second harmonic echo CF at several hundred Hz above 61 kHz. Behaviorally the bat is extraordinarily sensitive to echoes from moving targets (Goldman & Henson, 1977).

To analyze the small shifts in the CF echo frequency produced by motion of the target, fine frequency resolution is required. Resolution at frequencies higher than 5 kHz is directly related to the sharpness of the frequency-tuning curves of auditory neurons. The cochlea of the Panamanian mustached bat is apparently specialized for fine frequency analysis of the CF components of the biosonar signal, because the cochlear microphonic response shows prominent ringing at frequencies at or near 61 kHz (Suga & Jen, 1977). As a consequence, single auditory nerve fibers most sensitive to 60 - 63 kHz, especially 61 kHz, show extremely sharp frequency-tuning. To a lesser extent, frequency-tuning curves of neurons sensitive to the 1st and 3rd harmonics are also sharper than others tuned to different frequencies (Suga, Simmons & Jen, 1976; Suga & Jen, 1977).

When the target is a flying insect, the returning echo CF may be modulated in amplitude and frequency by its wing beat. Sharply tuned peripheral neurons can code frequency modulation as low as $\pm$ 0.01%, i.e. $\pm$ 6.1 Hz for 61 kHz, by changing their discharge rate synchronously with the modulation (Suga & Jen,

---

* The resting frequency within a population of mustached bats may vary and 61 kHz is an approximate value for the Panamanian colony from which bats were taken. The average resting frequency of different populations may vary more widely (see Schnitzler, 1970).

1977). Since the Doppler-shift from the beating wing of a moth
can be as high as 800 Hz (Henson & Goldman, 1976), peripheral
auditory neurons code the wing beat of an insect by both synchronous
neural discharges of individual neurons and also synchronous
change in excitation from one group of neurons to another differing
in best frequency. The neurons sharply tuned at about 61 kHz are
potentially flying-insect detectors.

FUNCTIONAL ORGANIZATION OF THE CEREBRAL AUDITORY CORTEX OF
THE MUSTACHED BAT

     The auditory cortex reflects the remarkable specialization
of the peripheral auditory system (Fig. 2). There is a clear
tendency for high-frequency sensitive neurons to be located
anteriorly and low-frequency sensitive neurons to be located
posteriorly, as in the primary auditory cortex of other mammals
(Fig. 2B). However, in two respects this tonotopic organization
is quite unique: (1) neurons sensitive to 50 - 60 kHz, i.e.,
those processing the information carried by the dominant FM
component, FM_2, are not on the main tonotopic axis, but are
displaced anterodorsally (Fig. 2B,b). This FM processing area is
large and tonotopic representation in this area is vague and
complex; (2) 61-63 kHz sensitive neurons, i.e., those processing
the information carried by the main CF component, CF_2, occupy
about 30% of the primary auditory cortex, even though the range
of hearing is probably very broad, from a few kHz to 150 kHz.
Since this species' biosonar signals are quite distinct from its
communication sounds, the 61-63 kHz tuned area is undoubtedly
specialized for processing the main CF component of Doppler-
shifted echoes from moving targets. Accordingly, this area is
called the Doppler-shifted-CF (or DSCF) processing area (Fig. 2B,
c). This prominent disporportionate frequency representation of
the DSCF processing area was the first demonstration of a specific
functional organization in the auditory system and is comparable
to the over-representation of the fovea centralis of the eye in
the mammalian visual system and also of the hand areas in the
primate somatosensory and motor systems (Suga & Jen, 1976).

DSCF Processing Area

     In the DSCF processing area, each orthogonal microelectrode
penetration is characterized by neurons with nearly identical
best frequencies (BF's), excitatory frequency-tuning curves, and
impulse-count functions (Suga & Jen, 1976; Suga, 1977). The peak
of the impulse-count function in response to a tone burst at the
BF uniquely defines the best stimulus amplitude (BA) for maximum
excitation. Since all the neurons arranged orthogonally to the
cortical surface are characterized by nearly identical BF's and

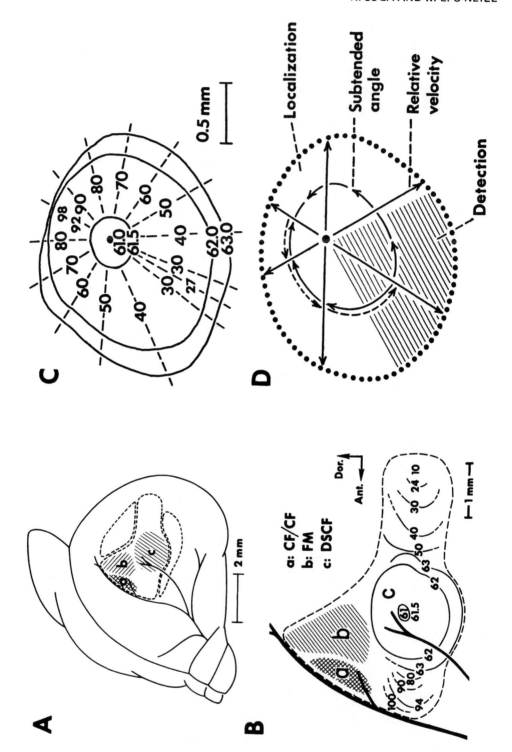

A

B

a: CF/CF
b: FM
c: DSCF

Dor.
Ant.

100
90
80
94
63
62

62
61.5
63
63
62
50
40
30
24
10

⌐1 mm⌐

⌐2 mm⌐

C

80
92 90
98
80
70
80
70
60
60
50
40
61.0
61.5
30 30
27
40
50
62.0
63.0

⌐ 0.5 mm ⌐

D

Localization

Subtended
angle

Relative
velocity

Detection

BA's, it is clear that for the above properties, the auditory cortex has a columnar organization.

In the DSCF processing area, the BF and BA of the columns systematically vary with the location on the cortical surface, so that there are systematic tonotopic and amplitopic representations: 61.0 kHz sensitive neurons are located at the center and 63.0 kHz sensitive neurons at the circumference, while neurons tuned to weak sounds of 30-40 dB SPL are located ventrally and those tuned to strong sounds of 80-90 dB SPL dorsally (Fig. 2C). Along the ventral margin of the DSCF processing area, neurons with BF's of 64-65 kHz are often found. Thus the amplitude spectrum of a signal is expressed in amplitude vs. frequency coordinates parallel to the cortical surface.* However, this amplitude spectrum representation is disproportionate, so that acoustic signals of 61.5 - 62.0 kHz and 30 - 50 dB SPL are projected to a much larger

-----

* The tonotopic representation in the DSCF processing area correlates with the "resting frequency" of the CF_2 emitted by the individual mustached bat. When a bat's resting frequency is found to be about 60.5 kHz, its DSCF processing area represents sounds mainly between 60.5 and 62.5 kHz. On the other hand, if the resting frequency of another bat is about 61.5 kHz, its DSCF processing area represents sounds mainly between 61.5 to 63.5 kHz. Thus, differences between individual bats in the frequency of the biosonar signal are reflected by the DSCF processing area.

Figure 2. (A) The dorso-lateral view of the left cerebral hemisphere of the mustached bat. The auditory cortex is the area surrounded by the dashed lines. (a) CF/CF, (b) FM, and (c) Doppler-shifted-CF (or DSCF) processing area. (B) The DSCF processing area makes up the major portion of the main tonotopic axis of A1, which is illustrated by the iso-best frequency contour lines representing frequencies from about 100 kHz to 10 kHz antero-posteriorly. (C) Amplitopic (---) and tonotopic (—) representations in the DSCF processing area. Two-digit and three-digit numbers show the best amplitude in dB SPL and best frequency in kHz, respectively. (D) The amplitude and frequency axes represent, respectively, the subtended angular and velocity information about a target. The echo amplitude is the result of both the target cross-sectional area and the inverse 4th power of the target distance. Poorly directional neurons specialized for echo detection (E-E neurons) are mainly found in the shaded area, while neurons directionally sensitive and specialized for target localization (I-E neurons) are mainly found in the unshaded area. (Based upon Suga, 1977 and Manabe, Suga, and Ostwald, 1978).

area.   The frequency and amplitude of echoes from prey may
predominantly be in this range from the moment of target detection
through the early part of the target-oriented flight. It should
be noted that the DSCF processing area is disproportionately
large in the auditory cortex and that furthermore it itself is
disproportionately organized to represent biologically important
echoes by the activity of many neurons for fine echo processing.
The functional significance of the tonotopic and amplitopic
representation demonstrated in Fig. 2C is that the frequency of
an echo, i.e., the target velocity information, is represented
along the radial axis, while the amplitude, which is related to
target range and size, is expressed along the circular axis (Fig.
2D).

In the area tuned to sound of 61.5 - 62.0 kHz and 30 - 50 dB
SPL, BF changes at a rate of 1-2 Hz per micron along the frequency
axis.   If the distance between neurons is assumed to be about 30
microns, BF would vary from neuron to neuron by only 30 - 60 Hz
steps along the frequency axis. The frequency representation,
therefore frequency resolution, is very fine.   Frequency
resolution is directly related to the sharpness of the frequency-
tuning curve.   Neural mechanisms exist in the mustached bat which
further sharpen the previously mentioned narrow frequency-tuning
curves of peripheral neurons whose BF's are near 61 kHz (Suga and
Manabe, in preparation).   The narrowest curve obtained so far in
the DSCF processing area has a level-tolerant bandwidth of 300
Hz, i.e., its bandwidth remains at 300 Hz over a broad amplitude
range (Suga, 1977).   Because of lateral inhibition, the slope of
the frequency-tuning curve is essentially infinite in a significant
number of neurons.   Neurons with such properties act as narrow-
band frequency detectors and as such would contribute to fine
processing of frequency information.

When flying toward a stationary object, the speed of the
mustached bat is expressed by the frequency difference between
the emitted sound and the returning Doppler-shifted echo, not by
the magnitude of the bat's Doppler-shift compensation.   The
frequency information of the emitted sound is available to the
bat in the form of vocal self-stimulation and perhaps efferent
copy, about which little is currently known.   The frequency
information in the Doppler-shifted echo is available regardless
of whether Doppler-shift compensation is performed or not, but
the measurement of its value becomes much more accurate with
compensation because the echo frequency is then subjected to
especially sharply tuned neural filtering near 61 kHz.

For velocity detection, a different mechanism is also conceivable in CF-FM bats.   The envelope modulation produced by the
overlap of the Doppler-shifted echo with the self-vocalized sound
may be coded by auditory neurons which can change their discharge

rate synchronously with envelope modulation as high as 3 kHz (Suga, Simmons and Jen, 1975). However, this mechanism might be unreliable for the measurement of relative velocity of a flying insect, because of the ambiguity due to local motion produced by the wing beat which evokes both amplitude and frequency modulation of the echo.

When a bat pursues a flying insect, the resulting Doppler-shifted echo consists of both D.C. and A.C. components: the former is related to the velocity difference between the bat and the insect, while the latter (periodic frequency modulation) results from the insect's beating wings. In many mechano-receptor systems, neurons have been found which are sensitive to either D.C. or A.C. components of mechanical stimuli. Therefore, the question may be raised whether cortical auditory neurons in the DSCF processing area are suited for detection of either the D.C. or the A.C. components in the echo from a flying insect. In the DSCF processing area, some neurons are insensitive to the A.C. component while still responding to the D.C. component. Others, however, are sensitive to both components, and change their discharge rate synchronously with frequency modulations as low as 0.05%, i.e., 31 Hz deviation from a 61 kHz carrier, easily evoked by the beating wing. The A.C. component, which also is always accompanied by amplitude modulation, is therefore represented by these synchronous discharges and by synchronous alternation in excitation from one group of neurons to another with different best frequencies and best amplitudes. All peripheral neurons tuned at about 61 kHz are sensitive to both the A.C. and D.C. components in the Doppler-shifted echoes from flying insects (Suga & Jen, 1977), while the auditory cortex contains some neurons sensitive to both components and others sensitive to the D.C. component only. There is, thus, some differentiation between those neurons for signal processing.

The analysis of the D.C. component is obviously confounded by the A.C. component. The D.C. component frequency is expressed by the average BF of a group of activated neurons with different best frequencies. If the positive and negative phases of the A.C. component are not symmetrical, the measurement of the average frequency may not accurately reflect the relative velocity of the insect.

Doppler-shift compensation behavior has been discussed extensively by Schnitzler (1968, 1970) and Simmons (1974). It has also been discussed in relation to the neurophysiological data (Grinnell, 1970; Johnson et al., 1974; Neuweiler, 1970; Schnitzler, Suga & Simmons, 1976; Suga, Simmons & Jen, 1975). Comprehensive physiological data on the peripheral auditory system in mustached bats are reported by Suga and Jen (1977).

They also discuss the role of sharply tuned neurons in relation
to Doppler-shift compensation. Undoubtedly, other contributors
to this symposium will discuss this problem in detail, so that we
will end our discussion by pointing out the following. When the
bat pursues a flying insect, it initially lengthens the duration
of the orientation sound from about 20 to 28-37 msec, and then
quickly but systematically shortens the duration during the
approach and terminal phases, down to 6-8 msec (Novick & Vaisnys,
1964). This increase in duration during the initial approach
phase would increase the information carried by the CF component.
When the bat flies toward a landing platform, however, the initial
lengthening does not occur (Schnitzler, 1970; O'Neill, Kuriloff,
Berry & Suga, unpubl.). If the increase in duration takes place
only during insect pursuit, then perhaps it is related to maximizing
the information carried by the A.C. component, not the D.C. component.
Doppler-shift compensation also probably starts at the initial
part of the approach phase. Subsequently, as the duration of the
orientation sounds gradually shortens, the bat receives less and
less information about the beating wing during the major part of
Doppler-shift compensation. Furthermore, Doppler-shift compensation
occurs to stationary objects when the mustached bat is flying
(Schnitzler, 1970) or is passively moved in a "bat-mobile" (O'Neill,
Kuriloff, Berry & Suga, unpubl.). That is, the bat compensates
for the Doppler-shift even when there is no A.C. component.
These acoustic behaviors indicate that Doppler-shift compensation
functions in (i) target detection at long range, (ii) simple target identification, and (iii) target velocity estimation.

Almost all neurons in the DSCF area are excited by acoustic
stimuli from the contralateral ear, and are either excited or
inhibited by stimuli to the ipsilateral ear. They are called E-E
and I-E neurons, respectively. Each cortical column is characterized not only by a BF and BA, but also by binaural interaction in
which (1) all neurons are E-E, (2) all are I-E, or (3) the interactions vary with depth. Thus in terms of aural representation,
this area consists of at least three different types of columns.
E-E neurons are poorly directional, but are tuned to weaker
echoes and integrate (or even multiply) signals from both ears
for effective target detection. On the other hand, I-E neurons
are directionally sensitive, and are tuned to stronger echoes.
E-E and I-E neurons form two functional subdivisions in the DSCF
processing area, one suited for target detection, the other for
target localization (Fig. 2D; Manabe, Suga & Ostwald, 1978).

FM Processing Area

The functional organization of the FM processing area is
quite different from that of the DSCF processing area (Suga,
O'Neill & Manabe, 1978). Tonotopic organization is vague and

amplitopic representation has not been found, although the neurons
are tuned poorly to particular frequencies and amplitudes. This
area is composed of functional clusters of neurons which respond
to combinations of elements contained in the emitted biosonar
signals and Doppler-shifted echoes. Although our mapping of this
area is not yet complete, we have found separate clusters containing
neurons which respond strongly to either FM_2, FM_3, or FM_4 only
when it is delivered following FM_1 with particular delays. For
example, neurons in the largest cluster show facilitation of
response to an FM sound of the 2nd harmonic (FM_2) when it is
preceded by the FM component of the 1st harmonic (FM_1) and/or a
CF component of the 1st harmonic (CF_1). Some neurons in the
clusters do not show noticeable excitatory responses when stimulated
with any single component delivered alone, but respond strongly
when two sounds are combined in particular frequency, amplitude,
and time relationships. We call these neurons FM_1-FM_n facilitation
neurons (n designates the specific FM harmonic(s) of the second
sound to be combined with FM_1 of the first sound for facilitation.
n = 2, 3, or 4, or their combinations).

Since many facilitation neurons in the FM processing area do
not respond to pure tones, FM sounds or noise bursts delivered
alone, the measurement of their frequency-tuning curves can only
be accomplished by delivering pairs of acoustic stimuli. While
the frequency and amplitude of one member of the stimulus pair is
held constant at values producing the strongest facilitation, the
frequency and amplitude of the other stimulus can be varied to
determine the frequency-tuning curve for facilitation. This
process is then reversed for the measurement of the tuning curve
of the other, initially fixed stimulus. We call the curves
measured in this way facilitation-tuning curves.

Fig. 3A is an example of the frequency-tuning curves of an
H_1-FM_2 neuron obtained by the above method. There were no excitatory responses in this neuron to pure tones or FM sounds delivered
alone at any available frequency and amplitude. However, a CF
tone (30 msec duration) with a frequency within the first harmonic
FM frequency sweep (FM_1: 6 kHz sweep), followed by a FM sound
mimicking the FM_2 of the biosonar signal (12 kHz sweep), elicited
a vigorous response. By holding the initial CF tone burst constant
at its BF and BA for facilitation, we could measure the facilitation-
tuning curve for FM_2 (curve b, Fig. 3A). The BF and BA for FM_2
was 55.74 kHz (center frequency) and 41 dB SPL, respectively.
Conversely, by presenting a FM sound sweeping from 61 to 49 kHz,
we measured the facilitation-tuning curve for the CF_1 component
(curve a, Fig. 3A). Its BF and BA were 27.80 kHz and 56 dB SPL,
respectively. Similar facilitation could be evoked by replacing
the 27.80 kHz CF_1 tone burst with H_1 or FM_1 with the FM center
frequency at 27.80 kHz. In other words, the combination of the

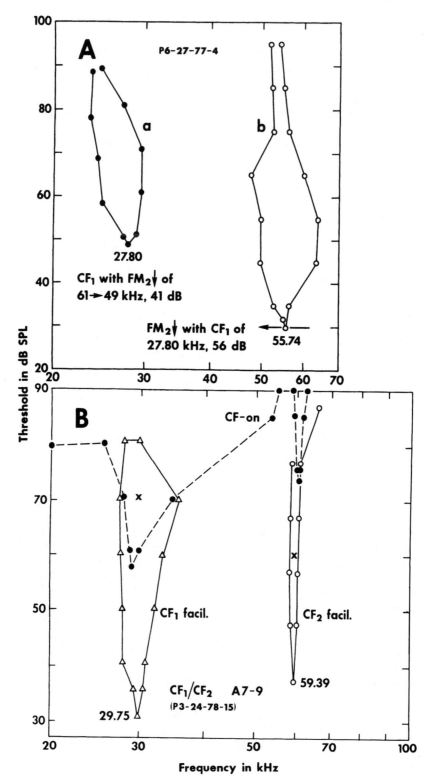

first harmonic H_1 (or its elements) and the second harmonic FM component was essential for the excitation of this neuron. Therefore, we call it a H_1-FM_2 facilitation neuron.

Since these neurons were maximally facilitated only when there was a delay between the FM_1 and FM_n, we hypothesized that during echolocation they would be "conditioned" by self-stimulation with the FM_1 component of the bat's emitted biosonar signal (hereafter, the pulse, P) to respond to the higher harmonic FM components in the time-delayed echo. In other words, these neurons process FM components in an echo in a temporal relation to the FM_1 of self-vocalized sound.

Since the primary cue for target range is the delay of an echo from an emitted sound, (e.g. Simmons, Howell & Suga, 1975), we tested the neurons in this area for delay sensitivity by delivering pulse-echo stimulus pairs which could contain all of the first three harmonics (H_1-H_3), or only the essential components which elicited facilitation (Suga, O'Neill & Manabe, 1978; O'Neill & Suga, 1979). These were delivered at three repetition rates and durations, corresponding to biosonar signals and their echoes in the search, mid-approach, and terminal phases (Fig. 1). By holding the frequency and intensity of the pulse and echo components constant at their BF's and BA's and then varying the delay of the echo component at each repetition rate, we found that the strength

Figure 3. (A) Frequency-tuning curves of an H_1-FM_2 facilitation neuron. This neuron did not respond at all to CF or FM sounds presented alone. The "facilitation" tuning curves shown were measured for the response when two sounds (CF_1 followed by FM_2) were presented in a pair. "a" is the tuning curve for CF_1 which caused facilitation in a combination with a 4 msec downward-sweeping FM sound (61-49 kHz) at an amplitude of 41 dB SPL. "b" is the tuning curve (center frequency) of an FM sound (12 kHz sweep) measured while the first sound in the pair (CF_1) was held at its BF (27.80 kHz) and 56 dB SPL (Suga, O'Neill, & Manabe, 1978; Copyright 1978 American Association for the Advancement of Science). (B) Excitatory (dashed lines) and facilitation (solid lines) tuning curves of a CF_1/CF_2 facilitation neuron. The excitatory areas were measured by delivering a 34 msec long CF tone. The facilitation areas were measured by delivering two CF tones simultaneously, holding one at a constant frequency and amplitude (x's) while measuring the other. The responses to CF tones presented alone were poor and inconsistent. Presenting two harmonically related CF tones at 29.75 and 59.39 kHz elicited a strong facilitation accompanied by a dramatic decrease in the threshold (Suga, O'Neill, & Manabe, 1979; Copyright 1978 American Association for the Advancement of Science).

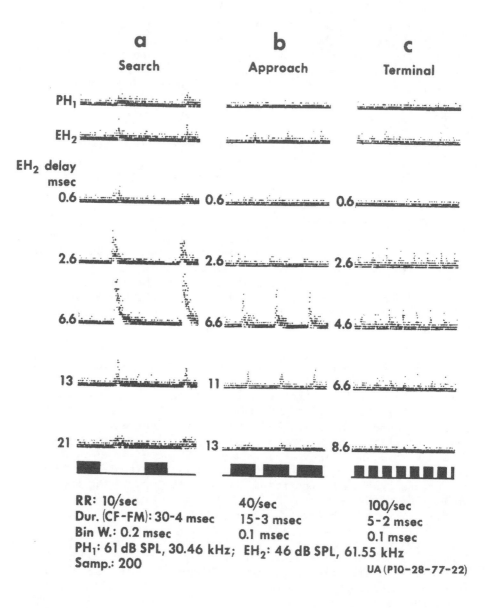

RR: 10/sec   40/sec   100/sec
Dur. (CF-FM): 30-4 msec   15-3 msec   5-2 msec
Bin W.: 0.2 msec   0.1 msec   0.1 msec
PH₁: 61 dB SPL, 30.46 kHz; EH₂: 46 dB SPL, 61.55 kHz
Samp.: 200
                                                        UA (P10-28-77-22)

of the facilitation, and therefore, the ability of the neuron to
follow at the repetition rate of the stimulus pairs, was a function
of the echo delay.

The PST histograms of Fig. 4 illustrate the responses of an
FM_1-FM_2 neuron to stimulus pairs repeated at 10/sec (CF, 30 msec;
FM, 4 msec duration), 40/sec (CF, 15 msec; FM 3 msec) and 100/sec
(CF, 5 msec; FM, 2 msec). Note that the delay resulting in the
best following to the stimuli (the best delay) at 40/second, for
instance, is 6.6 msec. The neuron shows some following over the
range from 2.6 to 13 msec delay. Also note that there is almost
no response to presentation of the pulse or echo components
alone.

In some neurons the best delay changed as a function of
repetition rate and duration of stimuli, while it remained relatively
constant in others. We call these two types tracking and range-
tuned neurons, respectively. Their response properties are best
illustrated by measuring their delay-tuning curves (Fig. 5).
These curves are obtained by holding the amplitude of the pulse
component (PH_1) constant at BA (71 $\pm$ 7 dB, N = 113) and measuring
the threshold of response to the echo as a function of echo
delay.

Range-tuned neurons have nearly the same delay-tuning curve
regardless of repetition rate. In the neuron shown in Fig. 5A,
the best delay (at the lowest threshold, 47 dB SPL) is 2.7 msec,
corresponding to a 46 cm target range. Since a target would be

---

Figure 4. Peri-stimulus-time (PST) histograms of the responses of
a tracking neuron (H_1-FM_2 specialized) to different echo delays
during simulated search, approach, and terminal phases of target-
oriented flight. The essential harmonics for facilitation of
this neuron were H_1 of the pulse (CF_1, 30.46 kHz) and H_2 of the
echo (CF_2, 61.55 kHz). PST histograms represent time on the
abscissa and number of impulses per bin on the ordinate. The
amplitudes of PH_1 and EH_2 for these responses were 61 and 46 dB
SPL, respectively. The upper two histograms in each column show
the poor response or lack of response of the neuron to PH_1 or
EH_2 delivered alone at each repetition rate. The five histograms
below them show the facilitation at various echo delays when PH_1
and EH_2 were delivered in a pair. Notice the decrease in the
range of delays over which the neuron followed the stimulus pair,
and also the shortening of the best delay as the repetition rate
increased from 10 to 100 pairs per second (corresponding to a
change in distance from 114 to 55 cm). Stimulus markers beneath
the histograms indicate the pulse of the stimulus pair only
(O'Neill & Suga, 1979; Copyright 1978 Am. Assoc. Adv. Sci.).

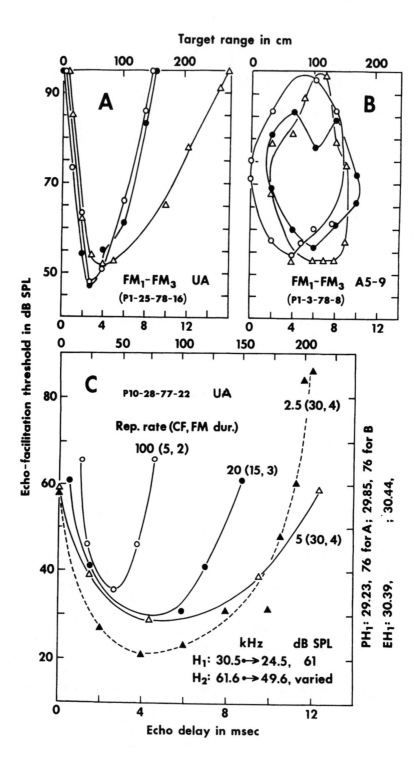

at this range only during the late approach phase, this neuron is apparently specialized to respond to the echo only around that time. Some range-tuned neurons have upper thresholds in response to an echo, and are then selective for both distance and target subtended angle (Fig. 5B). The slope of the leading edge of the delay tuning curve is steep, 24 dB/msec in Fig. 5A and infinite in Fig. 5B. The width of the delay-tuning curve at 10 dB above the minimum threshold is 2.8 msec for Fig. 5A. The tuning curve does not appear to be terribly sharp because our criterion for threshold was very low, i.e. about 0.1 response per stimulus (just-noticeable response). If the criterion for threshold was raised to 0.5 response per stimulus, the delay-tuning curve would become much narrower and its minimum threshold higher. Thus, for the neuron shown in Fig. 5B, the delay-tuning curve would become a tiny circle centered at 5 msec delay and 70 dB SPL.

The best delay differs from neuron to neuron. We now have data clearly indicating that target distance is expressed along an axis, like velocity and subtended angular information are expressed in the DSCF processing area (Suga, 1977), and also like azimuth and elevation of a sound source are systematically mapped in the midbrain auditory nucleus of the barn owl, Tyto alba (Knudsen &

Figure 5. Delay-tuning curves for three neurons in the FM processing area. (A) Range-tuned neuron (FM_1-FM_3 specialized). The delay-tuning curves for the search (10 pairs per second), mid-approach (40/second), and terminal (100/second) phases are very sharp and nearly identical for the approach and terminal phases. The best delay is also the same, 2.7 msec. This neuron is tuned to a target 46 cm away - that is, during the late approach to terminal phase. (B) Another range-tuned neuron (FM_1-FM_3 specialized) demonstrating upper thresholds for facilitation. The curves are similar, and the neuron is tuned to a target that returns an echo of about 65 dB SPL from a distance of about 85 cm. The frequencies and amplitudes of the CF_1 components of the pulses (PH_1) and the frequencies of the CF_1 components of the echoes (EH_1) used for the measurements are shown at the right. (O'Neill & Suga, 1979; Copyright 1978 Am. Assoc. Adv. Sci.). (C) A tracking neuron (FM_1-FM_2) shows broad delay-tuning with echo facilitation thresholds of 20-30 dB SPL and best delays of 3.8-5 msec (target range 70-85 cm) for search and early approach phases. (▲, 2.5/sec; △, 5/sec; ●, 20/sec: these data were taken at an early stage of the research before repetition rates were standardized). During the terminal phase (○, 100 pairs per second), the best delay shortens dramatically to about 2.5 msec (50 cm distance), the threshold increases to about 36 dB SPL, and the tuning curve becomes much narrower. (Suga, O'Neill, & Manabe, 1978; Copyright 1978 Am. Assoc. Adv. Sci.).

Konishi, 1978). Neurons tuned to shorter delays and which follow better to repetitive stimuli at higher repetition rates are located more anteriorly in the FM processing area.

In contrast to range-tuned neurons, a tracking neuron changes both its best delay and the shape of the delay-tuning curve as the repetition rate changes (Fig. 5C; Fig. 4). These neurons respond to targets over a broad range during the search and early approach phases. But during the late approach and terminal phases, the best delay becomes shorter and the tuning curve narrower. Thus these neurons can follow the echo during all phases of target pursuit. The most interesting feature of these neurons is the narrowing of the delay-tuning curve as the bat approaches the target, so that the neuron effectively "ignores" possible secondary echo sources from objects farther away.

Behavioral experiments on range discrimination clearly indicate that "CF-FM" bats use the FM component for ranging just as "FM" bats do (Simmons, 1971; Simmons, Howell & Suga, 1975). The essential components in the pulse and echo for the excitation of delay-sensitive neurons which we found are particular combinations of FM components, so that our neurophysiological data parallel the behavioral results. Further, delay-sensitive neurons, apparently of this type, have been found in the intercollicular nucleus of an FM bat, Eptesicus fuscus (Feng, Simmons & Kick, 1978), and in the auditory cortex of another FM bat, Myotis lucifugus (W.E. Sullivan, unpublished). The importance of FM signals in ranging in both CF-FM and FM bats is thus demonstrated neurophysiologically.

In order to produce the response properties of range-tuned or tracking neurons, lower order neurons with a broad spectrum of recovery cycles (Friend, Suga & Suthers, 1966; Suga & Schlegel, 1973) and facilitation to echoes (Grinnell, 1963b, 1970) are necessary. The neural network ranging model proposed by Suga & Schlegel (1973) consists of three neural levels: (1) tonic neurons with short recovery cycles, (2) phasic latency-constant neurons with a broad spectrum of recovery cycles, and (3) neurons with different gates (or time slots) for echo detection and ranging. Modifying this model in the light of our cortical data, the second level should contain neurons with a recovery cycle which includes a facilitation period preceded and/or followed by an inhibitory period, and the third level should consist of neurons showing facilitation to the pulse-echo pair without a response to the pulse or echo alone. At the third level, range-tuned neurons express the extent to which echo delays match their best delays by the magnitude of their excitation. It is not necessary for this type of neuron to be very phasic and latency-constant, like those at the second level. The model contains the minimum neural elements for ranging. It is, of course, subject to further elaboration in the future. Neurons which respond to both pulse and echo even at short echo delays represent the lowest level of neurophysiological activity in that they simply reflect the tonic responses of peripheral neurons. The auditory system undoubtedly processes information by more elegant means than this.

CF/CF Processing Area

In addition to the FM processing area, we have found a third functional division of the auditory cortex, the CF/CF processing area (Fig. 2B, a; Suga, O'Neill & Manabe, 1979). Neurons here are facilitated by two (or more) harmonically related CF tones, one of which is always CF_1. These harmonic-sensitive neurons are labeled CF_1/CF_n (where n is the particular upper harmonic, n = 2, 3, or 4 or their combinations). The slash means that the harmonic components must be delivered simultaneously to elicit the best facilitation. CF_1/CF_2 neurons form one cluster, CF_1/CF_3 another, and so on. Between clusters are neurons facilitated by either of two upper harmonics combined with the first, e.g., $CF_1/CF_{2,3}$ neurons.

The responses of a CF_1/CF_2 neuron are shown by PST histograms in Fig. 6. This neuron showed some response to CF_1 alone, but no response to CF_2 alone. When CF_1 and CF_2 were delivered simultaneously, however, the neuron responded strongly. The best facilitation occurred when CF_1 and CF_2 were combined in an appropriate amplitude relationship: 50-70 dB SPL for CF_1 and 56 dB SPL for CF_2. The best amplitude relationships were different for different neurons. For facilitation, a precise harmonic relationship between the two CF signals was required. For the CF_1/CF_2 neuron in Fig. 3B, for example, the best-facilitation frequency was 29.75 kHz for CF_1 and 59.39 kHz for CF_2. The CF_2 frequency-tuning curve was always especially sharp.

In contrast to FM_1-FM_n neurons, CF_1/CF_n neurons are facilitated by combinations of CF signals rather accurately harmonically related. Furthermore, their best facilitation occurs when there is no delay between the two components. Therefore, these neurons are not processing echo components in relation to the emitted sounds, nor are they sensitive to target range in the sense that FM_1-FM_n neurons are. Most CF_1/CF_n neurons do not follow well to paired components over about 40 per second, although some do so up to 100 pairs per second. The importance of harmonic-sensitive neurons to the bat is not immediately clear: one notion is that they are used to detect the presence of nearby echolocating conspecifics who may be competing for insect prey in the same area.

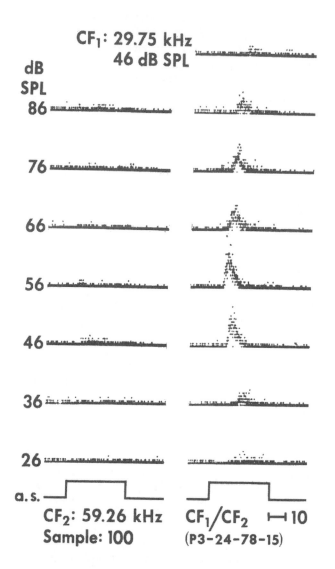

Figure 6. PST histograms of responses of a single CF_1/CF_2 facilitation neuron (tuning curves shown in Fig. 3B). The response to the CF_1 alone is shown at the top of the right column, and the responses to the CF_2 alone and CF_2 with CF_1 are shown in the left and right columns respectively. The acoustic stimuli (a.s.) are 34 msec in duration, 0.5 msec rise-decay time. The CF_1 is 29.75 kHz and 46 dB SPL, while the CF_2 is 59.26 kHz and is either 26, 36, 46, 56, 66, 76, or 86 dB SPL. Each PST histogram consists of neural activity for 100 presentations of the same sound. (Suga, O'Neill, & Manabe, 1979; Copyright 1978 Am. Assoc. Adv. Sci.).

The importance of the third and fourth harmonic CF and FM components to the response of significant numbers of FM_1-FM_n and CF_1/CF_n neurons in the cortex of the mustached bat is rather surprising, because these harmonics are thought to be comparatively weak in the bat's sounds. As has been recently reported for the long-eared bat, <u>Plecotus phyllotis</u> (Simmons & Farrell, 1977), perhaps the mustached bat enhances the first harmonic while flying in the open and emphasizes the higher harmonics in confined environments. FM_1-FM_3 and FM_1-FM_4 neurons are better suited for the ranging of smaller targets than FM_1-FM_2 neurons, and the bats may exploit this capacity by actively shunting energy into the harmonics which are appropriate for the required resolution.

J. Ostwald (Poster session abstracts, this volume) reports that, analogous to the mustached bat, there is also a disproportionate cortical representation of frequencies associated with the predominant CF component in the biosonar signals and echoes of the CF-FM horseshoe bat, <u>R. ferrumequinum</u>. He has also found a separate and large area devoted to processing the frequencies associated with the FM component. There appears to be, then, a remarkable convergence in the nervous systems of these two unrelated species, apparently related to the processing of their similarly-structured biosonar signals.

CONCLUSION

In the central auditory system, there are neurons with response properties both similar to and different from those at the periphery. Primary-like neurons of the central auditory system may function to maintain intact the coding of signals by primary auditory neurons, while non-primary-like neurons would be concerned with encoding the signals to extract meaningful information. As we have demonstrated, the acoustic information received by the mustached bat during echolocation is segregated by the central nervous system into at least three separate, functional divisions of the auditory cortex. Neurons in each division are thus specialized to process particular subsets of the information reaching the bat's ears which have obvious relationships to successful orientation and prey capture. That the auditory system encodes and segregates sonic information in this systematic way is a very recent discovery: the <u>amplitopic representation</u> found in the DSCF processing area of the mustached bat (Suga, 1977), the <u>neural map of target range</u> found in the FM processing area of the mustached bat, and the <u>neural map of auditory space</u> found in the barn owl (Knudsen & Konishi, 1978) were the first significant departures from the classical view of vertebrate auditory neurophysiology. Our uncovering of cortical areas devoted to processing complex sounds is particularly germane to the general problem of understanding the neural mechanisms for the processing of meaningful acoustic patterns, and suggests that similar discoveries may be made about auditory processing in other species by employing biologically meaningful stimuli to analyze the auditory system.

EPILOGUE

The finding of neurons with different response properties in different aggregates in the brain is one of the most important recent advances in auditory physiology. This is true to some extent even in the cochlear nucleus, the lowest center of integration. Before this was understood, microelectrodes had been usually inserted into auditory nuclei in a standard orientation and in more or less the same part of the nucleus, so that sampling of single neurons was not random. In such studies, therefore, it is really not significant that a certain type of neuron comprises a particular percentage of the population in a certain nucleus. The discovery of different types of neurons in the nucleus and the detailed description of their response properties are, however, very important.

As a next step, we physiologists should study how neurons with particular response properties are arranged in a particular nucleus, from where and which type of neurons they receive signals, and to where and which type of neurons they send signals. We should go beyond the more classical studies which rely on quantitative descriptions of neural responses only to pure tones. According to our own experience, neurons which are not "interesting" in terms of responses to pure-tone stimuli are often "very interesting" in terms of responses to complex acoustic stimuli relevant to animal behavior. We should therefore use acoustic stimuli which are biologically significant.

With a lack of knowledge of the functional organization of auditory nuclei, lack of adequate control experiments with standardization in stimulus delivery and data processing, and lack of statistical treatment of these data, researchers sometimes reach hasty and inappropriate conclusions and criticize other data as incorrect. These researchers should determine whether two sets of data are adequate for comparison and whether a statistically significant difference between them, if any, is only due to a single factor. Since our knowledge of the system is still limited, hasty judgements do not promote scientific understanding and is somewhat analogous to those reached by "blind men touching an elephant". In order to better understand the auditory system, we must incorporate the data obtained by different scientists in reasonable and constructive ways.

One of the most neglected studies on the echolocation system concerns its fine functional neuroanatomy. New neuroanatomical techniques, such as labelled deoxyglucose, promise to be extremely useful for mapping the neural pathways which are active or are activated by echoes during echolocation. Other labelling techniques such as radioactive amino acids and/or horseradish peroxidase should also be used for detailed analysis of neural connectivity. The combined approach of neuroanatomy with electrophysiology and ethology is essential to promote our understanding of the echolocation system.

This work supported by NSF grant BMS 75-17077 and BNS 78-12987 to N.S. and NICDS (PHS) training grant 1-T32-NS07057-01 to W.E.O.

Supplementary References

Knudsen, E.I. & Konishi, M. (1978). A Neural Map of Auditory Space in the Owl. Science 200, 795-797.

O'Neill, W.E. & Suga, N. (1979). Target Range-Sensitive Neurons in the Auditory Cortex of the Mustache Bat. Science 203, 69-73.

Rose, J.E., Greenwood, D.D., Goldberg, J.M., & Hind, J.E. (1963). Some Discharge Characteristics of Single Neurons in the Inferior Colliculus of the Cat. I. Tonotopic Organization, Relation of Spike-Counts to Tone Intensity, and Firing Patterns of Single Elements. J. Neurophysiol. 26, 294-320.

Suga, N., O'Neill, W.E., & Manabe, T. (1979). Harmonic-Sensitive Neurons in the Auditory Cortex of the Mustache Bat. Science 203, 270-274.

Suga, N. & O'Neill, W.E. (1979). Neural Axis Representing Target Range in the Auditory Cortex of the Mustached Bat. Science (submitted).

Suga, N. (1979). Functional Organization of the Auditory Cortex Beyond Tonotopic Representation. Fed. Proc. Fed. Am. Soc. Exp. Biol. (submitted).

Chapter V
Theories and Models of Echolocation

Chairman : B. Escudié

co-chairmen
R.W. Floyd   and   J.A. Simmons

- Underwater
   Models of cetacean signal processing.
      R.W. Floyd

   Models for echolocation.
      R.A. Altes

   Energy spectrum analysis in echolocation.
      R.A. Johnson

- Airborne
   The processing of sonar echoes by bats.
      J.A. Simmons

   Signal processing and design related to bat sonar
   systems.
      B. Escudié

   Models of spatial information processing in biosonar
   systems, and methods suggested to validate them.
      P. Greguss

   A new concept of echo evaluation in the auditory system
   of bats.
      K.J. Beuter

MODELS OF CETACEAN SIGNAL PROCESSING

Robert W. Floyd

Naval Ocean Systems Center

Kailua, Hawaii 96734

INTRODUCTION

There are several reasons for studying signal processing in dolphins in addition to the pure delight in understanding this mechanism. The most immediate benefit is in being able to predict and understand the results of echolocation tasks assigned to the animal. This is especially important in psychophysical experiments where one wishes to know if a threshold is the result of the behavior of the animal or is actually determined by the stimuli and the ability of the animals' receiver to detect it. Another reason for studying signal processing is to discover what principals could be used by man in such areas as aids to the blind, sonar, geophysical exploration, etc. Finally, on a metaphysical level, knowledge of how the dolphin processes echoes helps to understand how the animal perceives his environment.

Although most of the signal processing theories can be applied conceptually to both bats and dolphins, there are differences in the medium that make the practical applications of techniques different. The first major difference is that in air, there is a great difference in impedances between the source (bat), the medium (air), and the target, while in water the differences in acoustic impedance are much less. The result of this is that while it is possible for an animal in water to produce an impulse-like sound at relatively high power density levels, it is much more difficult to do so in air. Dolphins are able to transmit signals of RMS amplitudes of 10^6 ergs/cm^2 sec and total energies of ~ 80 ergs/cm^2 (Au 79). Bats, as reported by Simmons (Simmons, 1979) emit signals on the order of 10^{-3} ergs/cm^2 sec. Even assuming that bats use much higher signals in the field than in the lab, it is obvious that dolphins still have peak intensities several orders of magnitude greater than bats. Thus bats are forced to use much longer pulses to get a reasonable amount of energy into the medium.

This difference in time-bandwidth product (TBW) has important implications as to the optimality of a given processor. For dolphin signals, whose TBW approaches the theoretical minimum (Au, 1979), it has been demonstrated that an energy detector is as effective as a matched filter (Urkowitz, 1967). However, as the TBW product increases, the number of degrees of freedom increase also, and the effectiveness of an energy detector decreases. Even without a matched filter, detection can be improved if one has a prior knowledge of the time frequency distribution of the expected signal, as happens in FM sweeps of fixed duration. Attempts to demonstrate this in humans has failed, but such an effect has been demonstrated for Rhinolophus ferrumequinum (Vater, 1979).

The next important difference relates to the relative impedances between the fluid and the potential target. In air, the differences are great enough so that there is very little penetration. For targets in water, however, there is considerable penetration, and internal reflections give important contributions to the target echo. Finally, the velocity of sound in water is approximately five times as great as it is in air. Since Doppler stretch is related to the target velocity divided by sound velocity, Doppler resolvency or tolerance questions are much less important to a dolphin using a short pulse.

The classic method of detecting a known signal in noise is to whiten and match. In other words, the noise spectrum is measured, a pre-filter is constructed that will flatten (or whiten) the spectrum, and a correlator matched to the whitened version of the expected signal makes the detection. Aside from the lack of physiological evidence of a matched filter, there are several theoretical reasons why the matched filter model may be inappropriate. First, the interferring noise in most cases is not gaussian or stationary, but actually reverberation from the reflection of the animal's outgoing signal with the clutter in its environment. This makes the whitening part very difficult because the reverberation has an average spectrum similar to the outgoing pulse. Secondly, the target echo is almost never known exactly a priori, so that the receiver would usually not have a stored echo for correlation, unless the receiver had the capability of storing nearly infinite numbers of echoes for purposes of comparisons.

A better way of approaching the target recognition/detection problem would be to choose some feature set that could describe the target and base the signal processing scheme on these features. For humans, using a visual modality, the natural way to classify

targets is by size and shape. Seville Chapman (Chapman, 1968)
gives an interesting discussion on how this might be approached
with a sonar system. His discussion was related to holography,
where one of the objectives was to measure the scattering centers
of the target. True holography is unlikely in the case of the
dolphin, because it requires a signal with extremely good phase
coherence whose length is large compared to the dimensions of the
target, neither of which is true for the dolphin (Au 1979).
However, there are other ways for measuring highlights. Given a
broadband signal, one way to accomplish this is with a replica
correlator. This detector passes the received signal through a
filter matched to the outgoing signal. The output of this filter
is a series of spikes in the time domain related to the time of
occurrences of the scattering centers, and whose amplitude is
related to the strength of the scatterer and to how well it is
correlated with the original signal. This model is attractive
because it approaches a matched filter in detecting signal in
noise and gives an intuitively useful description of the target.
The work of Au and Hammer (Au, Hammer 1979) indicates the dolphin's
ability to discriminate metal cylinders in both qualitatively and
quantitatively related to the output of a replica correlator.
Unfortunately, the physiological and neurological evidence suggests
no method by which true replica correlation could be accomplished.

Another potentially useful feature set would be shape dependent
scattering functions. Although this technique had been applied to
radar signal analysis, R. A. Altes was the first to apply it to
animal sonar. The original model was the bionic coefficients
scheme, which represented the target transfer function as a power
series expansion in ω. This method was actually implemented on an
experimental sonar, the US Navy AN-SQQ27XN-2. The processor consisted of filters matched to a -5 to +5 expansion in ω of the
original waveform,

$$U(\omega) = \omega^{\nu} \exp \left\{ 2\pi jc \, \frac{\ln \omega}{\ln k} - \frac{(\ln(\omega))^2}{2 \ln (k)} \right\}$$

The arrival time of the echo was estimated and the filters sampled
at that instant. Since the filters overlapped, a matrix inversion
was performed to estimate the true value of each coefficient.
However, this system turned out to be impractical for several
reasons. First, the scattering model only applied to certain
idealized shapes and did not take into consideration the arrival
of components other than at one instant. Furthermore, the filters
were non-orthogonal. In other words, the energy in one filter
contributed to the output of other filters. The actual values
were separated by a matrix inversion, which left the system sensitive to noise. Some of these problems were alleviated by using
the geometric theory of diffraction (GTD) which used non-coherent

filters and a 1/2 power expansion in ω for the target transfer function. When the outputs of the filters combined with a timing parameter were used, this system performed reasonably well at moderate to high signal-to-noise ratios. However, it still suffers from not considering resonances and from having overlapping filters. In addition, it is difficult to find a known biological mechanism that could implement such a scheme.

Another group of theories are related to exploiting resonances in the target. This is at first glance appealing, because the modes of resonance are strictly related to the target structure. Furthermore, any receiver that can perceive pitch would be an effective detector. In actual fact, many resonances were stimulated by a broad band pulse and the echo spectrum depended on the aspect of the target. This method by itself did not work very well (Chestnut and Landsman, 1977). Another all pole model was the use of linear prediction theory. This model assumes that the output is the sum of the original signal plus delayed version multiplied by same coefficient added to it. This method has been used successfully in speech synthesis and recognition. The major theoretical objections to this method is that the linear prediction coefficients are not unique, i.e., the same output can be synthesized with many different combinations of coefficients. In spite of this, the method worked well at moderate signal-to-noise ratio (Chestnut and Landsman, 1977). These results are discussed further in R. A. Johnson's review.

All of the above theories have started from target models and then moved to signal processing methods based on exploiting these models. Another approach is to start with what we know about dolphin hearing and move from there. It is known, for instance, that the dolphins auditory system can be modeled as a bank of constant Q or proportional bandwidth filters (Johnson, 1968) which is functionally similar to the human auditory system. Each of these filters is followed by a rectifier or square law detector and then by an integrator whose time constant is approximately 1/BW (Jeffress, 1968). The output of such a bank of filters would represent a spectrogram if all the energy occurred during the integration period, or a time spectral history for multiple highlights following outside this interval. Although attractively simple, this method on the surface appears less than optimal. Its performance is inferior to that of a replica correlator because both signal and noise are summed together incoherently during the integration time of the filter. It also does not explain the demonstrably fine range resolution (Murchison, 1979) of echolocating animals. However, spectrogram models have proved the most successful of the models thus far at detecting and classifying echoes in introduced gaussian noise (Chestnut, Altes 1979).

In spite of the previously stated objections, there are ways in which the spectrogram models can result in correlation equivalent processing gain and fine range resolution. If a spectrogram is taken of two wide band signals separated by some delay τ, then ripples in the spectrum will appear at intervals $\Delta f=1/\tau$ (Johnson et. al., 1976). These ripples are perceived by the auditory system as a pitch whose frequency is approximately Δf. This phenomenon is known as time separation pitch (TSP) and has been demonstrated both in humans (McClellan, 1967) and in the bottlenosed dolphin, <u>Tursiops</u> <u>truncatus</u> (Caine, 1976). One virtue of this method is that it can be used to explain the results of many dolphin echolocation experiments. For instance, the results of the range resolution experiments (Murchison, 1979) were consistent with what one would predict using the TSP model. As a method of measuring range, the TSP model takes the spectrum, the outgoing signal and returning echo, and produces a tone whose pitch is inversely related to the distance. For very short distances, the range resolution would depend on the auto correlation function of the signal. As the range increased, resolution would be determined by the ability to resolve differences in the time difference

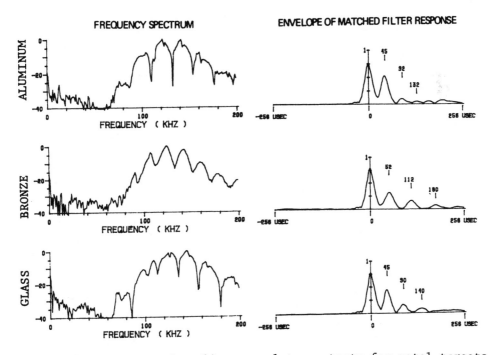

Figure 1. Spectra and replica correlator outputs for metal targets.

pitch. The results of Murchison fit this hypothesis if the porpoise's frequency resolution ($\Delta f/f$) equals .005. As mentioned
previously, the results of Hammer and Au indicate the classification performance closely follows a replica correlator. However,
the spectrogram model gives nearly equivalent results to the
replica correlator. Figure 1 shows the spectra and replica
correlator outputs of aluminum, bronze and glass cylinders. The
nulls in the spectrum occur at $1/\tau$ where τ is the time of arrival
of the secondary peak. The strength of this phenomena is
related to how well correlated the secondary peak is to the
primary. This implies that this model is correlation equivalent,
and may explain the success of animals in resisting jamming. One
problem this has as a complete model of echolocation is that it
cannot be used at longer ranges. First, as the integration
time increases between transmitted and received signal, the signal-
to-noise ratio becomes poorer. Secondly, the Δf decreases as
range increases until it is no longer discriminable.

SUGGESTIONS FOR FUTURE EXPERIMENTS

Briefly, there are four likely types of signal processing
models. The first, match filter or replica correlation, assumes
an ideal receiver and would give the best results for single or
multiple highlight targets in gaussian noise. Since the time
bandwidth (TBW) product of the dolphin is 1, another model that
would work as well for the single glint model is a simple energy
detector whose bandwidth equals that of the echo and whose integration time is 1/BW. However, this model would not perform as
well as a coherent receiver for multiple glint or greater TBW
product signals. A model based on the energy in a bank of constant
Q filters would require a 3 dB greater signal-to-noise ratio (S/N)
to have the same performance as the coherent models for either the
single or multiple glint echoes. The TSP model would predict also
a 3 dB poorer result for the single glint target, but for a
double highlight target the performance should approximate that
of a matched filter.

One problem in designing an experiment to test these hypotheses is that all of these are based on a single ping. If the
animal is allowed to emit multiple pings, one of three processes
could take place. First, the echoes could all be summed coherently, although this is likely only for the matched filter
hypothesis. In this case, the S/N would go up as log N, where N
is the number of echolocation clicks. A more likely case is that
the energy is summed noncoherently, in which case the S/N would
increase by $\log \sqrt{N}$. Finally, each echo could be treated as an
independent event, which would give approximately a $\log \sqrt[4]{N}$ for
an animal working near the minor diagonal on the Receiving
Operating Curve (Green and Swets, 1974). To avoid the problems

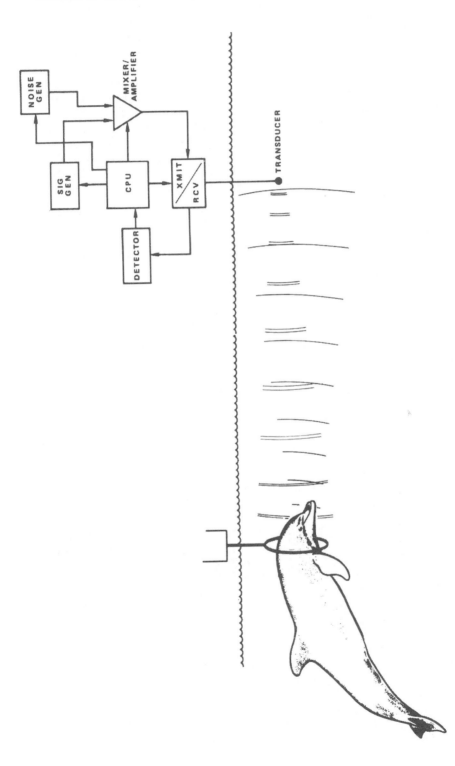

Figure 2. Experimental design for determining S/N at which dolphin can detect synthetic echoes.

of sorting out the multiple signal hypotheses, it would be preferable to have the animal make the decision on only one echolocation click. This could be accomplished by skillful training, or by using the following experimental set (Figure 2). The dolphin is trained to station in a hoop and to echolocate on a phantom target. This target is originally created by transmitting a synthetic echo each time the animal's echolocation pulse is detected. This is later changed so that a phantom echo is created only on the first pulse. This phantom echo always has the same energy, but may consist of either a single highlight or dual highlight. Gaussian noise is added to the stored echo, and the dolphin's performance is measured as a function of S/N. From this experiment, we would predict the following results:

If the matched filter hypothesis is true, then

	Single Glint	Double Glint
	$d^o = {}^{Eo}/No_2$	${}^{Eo}/No_2$
Energy detector, same Time-Band-width as transmitted signal	$d^1 = {}^{Eo}/No_2$	${}^{Eo}/\ 2\ No_2$
Bank of Constant Q filters	$d^1 = {}^{E}/_2 No2$	${}^{Eo}/2No_2$
Time Separation Pitch	$d = {}^{E}/2No_2$	${}^{Eo}/No_2$

REFERENCES

Au, W. W. L., 1979, Echolocation signals of the Atlantic bottlenosed dolphin (Tursiops truncatus) in open waters, this volume.

Au, W. W. L., and Hammer, C., 1979, Target recognition via echolocation by Tursiops truncatus, this volume.

Caine, N. G., 1976, Time separation pitch and the dolphin's sonar discrimination of distance, Ph. D. Diss., San Diego State University, San Diego, California.

Chapman, S., 1968, Dolphins and multifrequency, multiangular images, Science, 160:208.

Chestnut, P., and Landsman, H., 1977, Sonar target recognition experiment, ESL Inc. Report N° ESL-ER178, 495 Java Dr., Sunnyvale, California.

Green, D. M., and Swets, J. A., 1966, "Signal Detection Theory and Psychophysics", J. Wiley, New York.

Jeffress, L. A., 1968, Mathematical and electrical models of auditory detection, J. Acoust. Soc. Amer., 44:187.

Johnson, C. S., 1968, Relation between absolute threshold and dur-
      ation-of-tone pulses in the bottlenosed porpoise, J. Acoust.
      Soc. Amer., 43:757.
Johnson, C. S., 1968, Masked tonal thresholds in the bottlenosed
      porpoise, J. Acoust. Soc. Amer., 44:965.
Johnson, R. A., and Titlebaum, E. L., 1976, Energy spectrum analysis:
      a model of echolocation processing, J. Acoust. Soc. Amer.,
      60:484.
McClellan, M. E., and Small, A. M., 1967, Pitch perception of pulse
      pairs with random repetition rate, J. Acoust. Soc. Amer.,
      42:690.
Murchison, A. E., 1979, Detection range and range resolution of echo-
      locating bottlenosed porpoise (Tursiops truncatus), this vol-
      ume.
Simmons, J. A., 1979, Perception of echo phase information in bat
      sonar, in press.
Simmons, J. A., 1979, Processing of sonar echoes by bats, this volume.
Urkowitz, H., 1967, Energy detection of unknown deterministic signals,
      Proc. IEEE, 55:523.
Vater, M., 1979, Coding of sinusoidally frequency-modulated signals
      by single cochlear nucleus neurons of Rhinolophus ferrumequi-
      num, this volume.